Superlattices and Microstructures of Dielectric Materials

介电体超晶格 下

朱永元 王振林 陈延峰 陆延青 祝世宁

南京大学出版社

目 录

第五章 离子型声子晶体与超构材料 ·· 003
Chapter 5 Ionic-Phononic Crystals and Metamaterials ························ 007
   5.1 Optical Properties of an Ionic-Type Phononic Crystal ············· 012
   5.2 New Type of Polariton in a Piezoelectric Superlattice ············· 019
   5.3 Piezoelectric-Induced Polariton Coupling in a Superlattice ········ 026
   5.4 Phonon-polaritons in Quasiperiodic Piezoelectric Superlattices ··· 034
   5.5 Coupling of Electromagnetic Waves and Superlattice Vibrations in a Piezomagnetic Superlattice: Creation of a Polariton Through the Piezomagnetic Effect ··· 041
   5.6 Coupled Phonon Polaritons in a Piezoelectric-piezomagnetic Superlattice ··· 049
   5.7 Mimicing Surface Phonon Polaritons in Microwave Band Based on Ionic-type Phononic Crystal ··· 062
   5.8 Magnetic Plasmon Hybridization and Optical Activity at Optical Frequencies in Metallic Nanostructures ····································· 070
   5.9 Stereometamaterials ·· 078
   5.10 Magnetic Plasmon Propagation Along a Chain of Connected Subwavelength Resonators at Infrared Frequencies ························· 090
   5.11 Long-Wavelength Optical Properties of a Plasmonic Crystal ······ 098

第六章 准相位匹配量子光学与光子芯片 ·································· 109
Chapter 6 Review Article: Quasi-phase-matching Engineering of Entangled Photons ·· 113
   6.1 Transforming Spatial Entanglement Using a Domain-Engineering Technique ·· 128
   6.2 Compact Engineering of Path-Entangled Sources from a Monolithic Quadratic Nonlinear Photonic Crystal ····································· 136
   6.3 On-chip Steering of Entangled Photons in Nonlinear Photonic Crystals ······ 146
   6.4 Lensless Imaging by Entangled Photons from Quadratic Nonlinear Photonic Crystals ·· 156
   6.5 Observation of Quantum Talbot Effect from a Domain-engineered Nonlinear Photonic Crystal ··· 167

6.6 Mode-locked Biphoton Generation by Concurrent Quasi-phase-matching ·········· 175

6.7 Generation of N00N State with Orbital Angular Momentum in a Twisted Nonlinear Photonic Crystal ·········· 185

6.8 Tailoring Entanglement Through Domain Engineering in a Lithium Niobate Waveguide ·········· 197

6.9 On-Chip Generation and Manipulation of Entangled Photons Based on Reconfigurable Lithium-Niobate Waveguide Circuits ·········· 216

6.10 Generation of Three-mode Continuous-variable Entanglement by Cascaded Nonlinear Interactions in a Quasiperiodic Superlattice ·········· 226

第七章 介电体超晶格与畴工程学 ·········· 239
Chapter 7 Domain Engineering for Dielectric Superlattice ·········· 242

7.1 The Growth Striations and Ferroelectric Domain Structures in Czochralski-grown $LiNbO_3$ Single Crystals ·········· 246

7.2 Growth of Optical Superlattice $LiNbO_3$ with Different Modulating Periods and Its Applications in Second-harmonic Generation ·········· 256

7.3 Growth of Nd^{3+}-doped $LiNbO_3$ Optical Superlattice Crystals and Its Potential Applications in Self-frequency Doubling ·········· 261

7.4 Fabrication of Acoustic Superlattice $LiNbO_3$ by Pulsed Current Induction and Its Application for Crossed Field Ultrasonic Excitation ·········· 266

7.5 $LiTaO_3$ Crystal Periodically Poled by Applying an External Pulsed Field ·········· 273

7.6 Poling Quality Evaluation of Optical Superlattice Using 2D Fourier Transform Method ·········· 279

7.7 Frequency Self-doubling Optical Parametric Amplification: Noncollinear Red-green-blue Lightsource Generation based on a Hexagonally Poled Lithium Tantalate ·········· 289

7.8 Direct Observation of Ferroelectric Domains in $LiTaO_3$ Using Environmental Scanning Electron Microscopy ·········· 295

7.9 Nondestructive Imaging of Dielectric-Constant Profiles and Ferroelectric Domains with a Scanning-Tip Microwave Near-Field Microscope ·········· 301

第八章 光学超晶格的应用研究 ·········· 309
Chapter 8 Engineered Quasi-phase-matching for Laser Techniques ·········· 312

8.1 Efficient Continuous Wave Blue Light Generation in Optical Superlattice $LiNbO_3$ by Direct Frequency Doubling a 978 nm InGaAs Diode Laser ·········· 345

8.2 Femtosecond Violet Light Generation by Quasi-phase-matched Frequency Doubling in Optical Superlattice LiNbO$_3$ ⋯⋯ 349

8.3 Visible Dual-wavelength Light Generation in Optical Superlattice Er：LiNbO$_3$ through Upconversion and Quasi-phase-matched Frequency Doubling ⋯⋯ 354

8.4 Frequency Tuning of Optical Parametric Generator in Periodically Poled Optical Superlattice LiNbO$_3$ by Electro-optic Effect ⋯⋯ 360

8.5 Electro-optic Effect of Periodically Poled Optical Superlattice LiNbO$_3$ and Its Applications ⋯⋯ 366

8.6 High-power Red-green-blue Laser Light Source Based on Intermittent Oscillating Dual-wavelength Nd：YAG Laser with a Cascaded LiTaO$_3$ Superlattice ⋯⋯ 373

8.7 Diode-pumped 1988-nm Tm：YAP Laser Mode-locked by Intracavity Second-harmonic Generation in Periodically Poled LiNbO$_3$ ⋯⋯ 378

8.8 Efficiency-enhanced Optical Parametric Down Conversion for Mid-infrared Generation on a Tandem Periodically Poled MgO-doped Stoichiometric Lithium Tantalate Chip ⋯⋯ 387

8.9 Polarization-free Second-order Nonlinear Frequency Conversion Using the Optical Superlattice ⋯⋯ 395

8.10 Polarization Independent Quasi-phase-matched Sum Frequency Generation for Single Photon Detection ⋯⋯ 401

8.11 DFB Semiconductor Lasers based on Reconstruction-equivalent-chirp Technology ⋯⋯ 410

8.12 High Channel Count and High Precision Channel Spacing Multi-wavelength Laser Array for Future PlCs ⋯⋯ 418

第九章 总结与展望 ⋯⋯ 431
Chapter 9 Summary and Outlook ⋯⋯ 435

附录 最新的重要成果收录 ⋯⋯ 441

第五章 离子型声子晶体与超构材料
Chapter 5 Ionic-Phononic Crystals and Metamaterials

5.1 Optical Properties of an Ionic-Type Phononic Crystal

5.2 New Type of Polariton in a Piezoelectric Superlattice

5.3 Piezoelectric-Induced Polariton Coupling in a Superlattice

5.4 Phonon-polaritons in Quasiperiodic Piezoelectric Superlattices

5.5 Coupling of Electromagnetic Waves and Superlattice Vibrations in a Piezomagnetic Superlattice: Creation of a Polariton Through the Piezomagnetic Effect

5.6 Coupled Phonon Polaritons in a Piezoelectric-piezomagnetic Superlattice

5.7 Mimicing Surface Phonon Polaritons in Microwave Band Based on Ionic-type Phononic Crystal

5.8 Magnetic Plasmon Hybridization and Optical Activity at Optical Frequencies in Metallic Nanostructures

5.9 Stereometamaterials

5.10 Magnetic Plasmon Propagation Along a Chain of Connected Subwavelength Resonators at Infrared Frequencies

5.11 Long-Wavelength Optical Properties of a Plasmonic Crystal

第五章　离子型声子晶体与超构材料

陆延青

以准位相匹配理论的拓展作为物理基础，以具有有序调制畴结构的铁电晶体作为材料基础，以激光与非线性光学效应作为应用牵引，形成了在介电体超晶格光学方向研究探索的一条清晰思路。然而，从介电体超晶格这一名词中可以看出，我们是在介电体中建构一种与天然晶体结构物理对应的人工超结构，而不是仅仅寻求非线性光学效应的增强。正如描绘晶体的物性取决于其组元（即原子）和结构，介电体超晶格通过人工微结构的设计来调控物性，获得不同于其基质材料的崭新性能。从这一点看，介电体超晶格的理论和实验体系，打开了一扇通过微结构获得天然材料所不具备的超常特性的大门，这和现在十分热门的超构材料（或称之为超材料）殊途同归。

在介电体超晶格的理论体系中，离子型声子晶体概念的提出与验证、拓展，不但呈现出从介电体超晶格到超构材料的异曲同工的演进思路，也体现了我们在超构材料领域的研究特色。离子型声子晶体实际上是与离子晶体对应的一种人工电磁材料。当电磁波入射于离子晶体中，电磁场与晶格振动相互作用，形成与晶体横振动光学声子耦合的模量子，即声子极化激元。为纪念其奠基者黄昆先生，它又被称为"黄子"。这是最先得到研究的极化激元。随后激子极化激元、磁振子极化激元等也相继被提出。

1999 年，我们提出了"离子型声子晶体"的概念。详见本章第一篇代表性论文。在 5.1 中，我们将声子极化激元概念推广到介电体超晶格。与普通声子晶体不同，我们设计制备的离子型声子晶体是由自发极化方向首尾相接的具有周期畴结构的铌酸锂超晶格构成。由于相邻带电畴界可看作是异号的电荷中心，于是一维离子型声子晶体就可以完美对应一维异性离子链。不过，实际上，离子型声子晶体中超晶格振动与电磁场的耦合来源于压电效应。基于压电方程，我们得到了离子型声子晶体中超晶格和电磁波的耦合运动方程组及其极化激元的色散关系。其形式与离子晶体中反映晶格振动与电磁波耦合的黄昆方程完全一致。正是由于电磁波与超晶格的耦合导致的极化激元模，离子型声子晶体中出现了其基质材料不具备的诸多新颖物性，如微波波段的介电反常、电磁吸收，乃至类拉曼散射、类布里渊散射等。这就把若干原本存在于红外波段的经典物理概念推广到微波领域，进一步深化了人们对这些物理过程物理本质的认识。

以介电反常为例，正如 5.1 文中指出的，在微波波段，"出现了一个介电常数为负的带隙。其对应的折射率为虚数，从而相应波段的电磁波被反射。然而，它与光子晶体中的带隙不一样，却是源于声子和光子相互作用"。因此，基于局域的压电共振以及各单元之间的同相增强，可通过结构设计，获得所需波段的负介电常数。这与通过亚波长局域共振结构的设计，用有效介质方法，得到等效的负磁导率并结合金属的本征负介电常数实现负折射的超构材料在物理思路上出现了有趣的一致。只不过，我们是通过压电共振获得负介电常数，而常规超构材料是通过亚波长电磁共振获得负磁导率。如果我们不拘泥于负折射这一较狭义的超构材料定义，而是着眼于通过微结构设计获得与基质材料截然不同的奇异物性参数，我们就实现了从离子型声子晶体到超构材料的概念的自然过渡。基于离子型声子晶体带来的超晶格声子极化激元这一新思路，我们开展了系列相关的研究工作。在本章中，除 5.1 之外，我们还选取了另外 10 篇代表性论文，从中我们可以看出相关物理思想演进的自然脉络。

离子型声子晶体的概念着眼于压电体超晶格与离子晶体的物理对应。在此物理现象得到证实后，我们即围绕电磁场与超晶格的耦合过程，以微结构对极化激元的调控为主线，开始了进一步深入研究和拓展。5.2 是其中的一个重要工作。在离子晶体中，声子极化激元只能由横向光学声子与光子耦合形成，光子不可能与声学声子发生相互作用。我们发现，由于离子型声子晶体中的光、声耦合实际上源于压电效应，这就使得压电体超晶格中的声子极化激元具有不同的形成条件。除了色散关系曲线的一些典型不同外，在压电体超晶格中，不但横向超晶格声子能够与光子耦合形成声子极化激元，而且还存在纵向超晶格声子与光子耦合形成的声子极化激元，从而扩展了传统的极化激元的概念。其物理原因是，由于压电效应的各向异性，超晶格横振动和纵振动都有可能伴随着超晶格的横向极化，该极化与电磁波耦合，从而导致了新型声子极化激元的产生。

进一步的研究(5.3)表明，由于材料的压电效应，偏振方向相互垂直的电磁波在超晶格中传播时能够发生强烈耦合，这种耦合导致了压电体超晶格中两种新型极化激元的产生。耦合使其中一个模存在禁带，而另一个模则可以通过。这种奇特的现象在传统的晶格中并不存在，说明通过超晶格的设计不但可以在新波段拓展某些物理效应，而且通过合适的微结构和应用场景设计还可以实现更加新颖的物理过程。

当然，既然是超晶格，其结构的人工设计不但带来了特征尺度的可控，也提供了灵活设计超越普通周期性结构的可能。5.4 则是我们将离子型声子晶体和压电体超晶格的概念拓展到准周期结构的一个代表性工作。我们曾在准周期钽酸锂超晶格中成功实现了多波长二次谐波和高效的直接三倍频，将准周期超晶格结构引入压电体，同样出现了一个有趣的物理现象。我们从理论与实验两个方面，研究了具有推广的一维二组元准周期结构的压电体超晶格中的声子极化激元。结果显示，声子极化激元的频率位置由推广的一维二组元准周期结构的倒格矢的位置决定；而带隙大小由材料阻尼系数、机电耦合系数以及傅立叶展开系数的大小决定。小的阻尼系数，以及大的机电耦合系数以及傅立叶展开系数有利于带隙的展宽。结果还显示，具有推广的一维二组元准周期结构比一维周期，以及一维二组元的 Fibonacci 结构在应用上有着更大的灵活性，因此也拓宽了压电体超晶格的声子极化激元的潜在应用范围。我们结合具有推广的一维二组元准周期结构的压电体超晶格中声子极化激元的物理特征，阐述了其在光学、声学以及相关应用领域的可能应用，如制作偏光器、光通信中的多波复用器件，以及环境保护材料等。

同样电磁波与声子的耦合也可发生在压磁材料和压电/压磁复合材料中。在 5.5 一文中，我们与当时在 UCLA 的张翔教授合作，研究了压磁系数周期调制的超晶格中电磁波的传播特性。由于压磁效应，导致了电磁波与晶格振动的耦合，形成一种不存在普通磁性材料中的磁极化激元。令人兴奋的是，正如在离子型声子晶体中出现负介电常数一样，在这种压磁超晶格中，磁导率的异常色散也导致了在某些共振频率出现的负磁导率。于是，通过超晶格的设计，我们实现了负介电常数和负磁导率的人工调控，这何尝不是另一种实现"超构材料"的途径呢？

很自然地，5.6一文从理论和实验两方面对压电/压磁超晶格开展了讨论，该超晶格材料由压电和压磁薄膜交替生长而成，可以看作压电和压磁的复式超晶格。当电磁场与这种超晶格相互作用时，由于电磁场包括电场和磁场矢量，因此显示了新的特点。由于电场和磁场同时激发了超晶格振动，因而产生了两种宏观激化激元，这两种激化激元通过超晶格振动发生耦合，完全改变了原来各自的特征。实验表明，同一个超晶格可以实现两个谐振频率声波的激发，分别对应于电场激发和磁场激发的振动。这些有趣的结果昭示着压电/压磁超晶格不但是研究电、磁、声间复杂耦合作用的平台，也有着重要的应用前景。这方面的探索研究，特别是与其他超构材料的联系值得深入研究。

当然，光学材料的表面效应的研究吸引了众多关注，表面等离激元的研究可谓如火如荼，离子型声子晶体等材料的表面效应也是十分值得关注的问题。5.7一文是我们在此方向的一项基础性的工作。由于电磁波与压电超晶格的耦合能带来负介电常数，而正是负介电常数才使得金属表面等离激元（SPP）得以存在，那在离子型声子晶体中必然也存在着一种对应的表面模式。基于这一思想，我们预言压电超晶格表面声子极化激元的存在，并导出了其相应的色散关系。由于表面等离极化激元（SPP）在诸多超构材料、微纳光子器件中扮演了重要的角色，或许，我们预言的表面声子极化激元也可在微波、太赫兹（THz）波等重要频段的新型电磁材料和器件的设计上发挥出重要的作用。

在以上的这些代表性工作中，声子起到了十分关键的作用。组成超晶格的各单元可以看作一系列的振子，具有自身的谐振频率。它们之间又在物理上链接在一起，这样，光子通过压电或压磁效应与超晶格的集体振动能量量子即声子，发生相互作用，产生丰富的物性。然而，由于声学谐振系统的尺寸往往较大，对应的频率一般在微波波段，这就极大地限制了其应用范围。

为提高工作频率，要减小各微结构单元的尺寸和单元之间的间隔，这就必须考虑各共振单元之间的耦合。借助于超晶格研究中关注耦合效应的研究思路，研究中我们并不仅仅考虑磁共振单元共振性质的平均效应，而是更加关注单元之间的耦合效应所带来的新奇性质。5.8一文介绍了由两个完全相同的金属开口环组成的磁共振"超构分子"，两个磁共振"原子"之间会形成很强的耦合作用，导致原来单独磁原子的共振模劈裂为磁共振分子的杂化模：成键模与反键模，这种模式劈裂的大小决定于单元之间的距离。

进一步研究发现，两个开口环之间的耦合强度不但与两个单元之间距离有关，还与两个单元相互旋转形成的不同空间立体构型有关，从而可以形成立体超构分子。通过与德国斯图加特大学合作，利用高精密的电子束光刻对准技术，制备出了样品并进行了测量，实验与理论符合的较好。相关工作发表在 Nature Photonics 上，即论文5.9。

一维原子链是固体物理中分析原子之间耦合作用形成晶格振动波的经典理论模型。由"超构分子"过渡到类似于一维原子链的"超晶格"是十分自然的拓展。我们利用开口环作为结构单元组成一维超晶格。由于单元之间距离很小，我们发现单元之间磁共振耦合也会形成一种类似格波的传播模式，由于这种波是很多磁共振单元耦合形成的集体振荡，我们把它称为磁等离激元波。相关的工作参见论文5.10。

这样，在介电体超晶格和考虑单元耦合效应的驱动下，我们的超构材料研究由"超构分子"，向一维超构原子链、二维超构晶体演进。显然，向三维系统的过渡是十分自然的趋势。

由于金属纳米颗粒具有独特的光学响应,我们设想将金属纳米棒粒子镶嵌于介电材料中,形成三维立方晶格(见5.11)。由于电磁波能够激发纳米棒中自由电子的运动,且电子运动在纳米棒两端形成电荷聚集,故产生电偶极矩。后者能够辐射电磁波并与入射波进行干涉。因而在一定的条件下,入射电磁波与金属粒子的等离激元波(极化波)能够发生强烈的耦合。我们称这种独特的三维"超构晶体"为等离激元晶体。通过将等离激元晶体与离子晶体类比,我们发现适用于离子晶体的黄昆方程也能够被推广到等离激元晶体。原先建立于离子晶体的一系列长波光学性质,如极化激元激发、极化激元带隙、介电异常、红外吸收等,都能够在等离激元晶体中产生。

回顾1999年,我们在离子型声子晶体中利用黄昆方程来研究入射电磁波与超晶格的相互作用并导出极化激元、介电异常等新颖性质。等离激元晶体与离子型声子晶体呈现出惊人的一致。虽然材料体系、单元共振基质、工作频率发生了巨大的变化,但在这一步步变迁中蕴含的物理脉络却是如此的清晰。我们从离子型声子晶体到超构材料走出了一条特色的发展之路,并且还在不断的发展中。我们相信这类的物理体系中还蕴含着太多的新机制、新现象、新效应,值得我们不断研究探索。这证实了"介电体超晶格"这一材料体系中包含着丰富的物理思想和强大的生命力!

Chapter 5 Ionic-Phononic Crystals and Metamaterials

Yanqing Lu

Dielectric superlattice(DSL) has been a well-recognized research achievement. Although it covers a wide range from optics to acoustics, from classical phenomena to quantum world, the laser and nonlinear optical applications still dominate its progresses in the past years. However, DSL is not just a quasi-phase matched(QPM) nonlinear material. As can be seen from the unique term, "dielectric superlattices", we are actually building a correspondence between natural crystal structures and artificial superlattices, rather than merely seeking some enhanced nonlinear optical effects. The DSL actually supplies a platform to manipulate various physical properties through the intentional design of microstructures. It opened a door to achieve extraordinary properties that do not exist in natural materials, which is quite similar to a very hot topic, metamaterial.

In the entire theoretical matrix of DSL, the proposal, demonstration and expansion of "ionic type photonic crystal"(ITPC), not only reflect the concept evolution from DSL to metamaterial, but also represent our unique research style in the field of metamaterials.

The so-called ITPC, a counterpart of ionic crystal in the mesoscopic scale, is an artificial electromagnetic(EM) material. As long as the EM wave propagates in an ionic crystals, the EM field would interact with transverse lattice vibration then forms photon polaritons. This phenomenon and related mechanism were firstly revealed by a famous Chinese physicist, Prof. Huang Kun, in 1951. Afterwards, other types of polaritons such as exciton polaritons, magnon polaritons, etc. have also been proposed.

In 1999, we proposed the concept of ITPC that is composed of two piezoelectric media aligned periodically in a superlattice structure.(5.1) Just like a real ionic crystal, there exists the coupling between vibrations of the superlattice and the electromagnetic(EM) field when the EM wave propagates in the ITPC, which results in the long wavelength optical properties. After proposing a simple method for calculating the dispersion relation of a 1-dimentional(1D) phononic crystal, we studied the coupling effects in ITPC and obtained the fundamental equations that describe the coupling between the superlattice vibration and the EM wave. These equations have the same type as the Huang's equations that describe the coupling between the lattice vibration and EM wave in ionic crystals.

From the fundamental coupling equations in ITPC, some long-wavelength optical properties that originally exist in ionic crystal were predicted and experimentally demonstrated based on a LiNbO$_3$ superlattice. Among them, there are the LST relation, the EM wave absorption, the dielectric abnormality and the polariton mode. Even the Raman scattering and Brillouin scattering in the microwave band might also be expected in an ITPC. The only difference is different response frequencies.

Taking the dielectric abnormality as an example, as we pointed out in the paper, "There is a band gap in which the dielectric constant is negative. The corresponding refractive index becomes imaginary. The incident radiation with these frequencies will be reflected. However, this gap does not originate from the interference of EM waves due to the periodic structure but, rather, originates from the interaction of the photon and the transverse optics phonon." As a consequence, based on the local piezoelectric resonance and in-phase enhancement between different units, we may design suitable domain structure to obtain negative dielectric constant in corresponding wavelength bands. As we know, the key steps toward negative index are negative permeability and negative permittivity. Metallic materials normally have to be used, while our results give the possibility to achieve exotic negative parameter in a dielectric medium. In this case, if we do not stick to strict definition of negative index metamaterial, but rather focus on obtaining extraordinary material parameters through microstructure design, well, we thus realize a natural transition from ITPC to metamaterials.

Afterwards, starting from the induced polariton in ITPC, we further carried out a serial of related research work. In this chapter, we still picked other 10 representative papers in addition to paper 5.1. From these work, we may see a clear trend emerge from ITPC to material.

Paper 5.2 is an important work after the ITPC and its basic properties were demonstrated. In an ionic crystal, the phonon-polariton can only be accompanied with the transverse optical phonons, whereas in the piezoelectric superlattice(PSL), the coupling could occur between the photons and longitudinal phonons as well as transverse phonons. Physically, it is the polarization of dipoles induced by lattice vibrations that interferes with the EM waves in the ionic crystal, whereas it is the polarization of bound charges induced by superlattice vibrations that interferes with the EM waves in the PSL. Due to the anisotropic nature of the piezoelectricity, the polarization of bound charges can be induced by transverse superlattice vibrations, or longitudinal superlattice vibrations as well. In this case, even a transverse polarization can be induced by a longitudinal wave which couples strongly to the EM wave in some particular frequency regions, resulting in the creation of a new type of polariton that does not exist in ionic crystals. The forbidden band associated with the polariton is not due to the Bragg reflection, but rather to the coupling.

Further study(5.3) revealed that the superlattice vibration not only can be excited by and coupled strongly to the incident EM wave, but they can also induce coupling between two orthogonally polarized EM waves. This coupling results in two types of polariton modes, one is supported in the band gap and the other prohibited. We have noticed that in the literature two interesting cases occurred in the stop band. In one case, the coupled-thin-film structure can support guided-wave plasmon polaritons; and in the other, the propagation of a probe light beam is allowed within the polariton stop band via exciton-biexciton coupling. Our case presents another example, where the phonon polariton is involved. This unusual coupling effect is not present in real lattices.

On the other hand, as we know, the discovery of quasicrystals has fired up a new field of condensed-matter physics and given rise to many practical applications since 1984. For example, the multiwavelength second-harmonic generation and the direct third-harmonic generation have been realized by Zhu et al. in a Fibonacci superlattice. Furthermore, compared with a one-dimensional(1D) two-component Fibonacci superlattice, a 1D two component generalized quasiperiodic superlattice(GQPSL) possesses more freedom for applications. As a consequence, we further investigate the influence of structural variation of the piezoelectric superlattice upon the phonon polaritons on the basis of a 1D two-component GQPSL. (5.4) We measured the dielectric function of the GQPSL at the microwave region, thereby obtaining the physical information required to deduce the properties of polaritons. The possible applications are discussed, which include WDM devices in optical communications, and for EM wave(or sound wave)-proof materials in environmental protection.

Furthermore, our vision could go beyond piezoelectric materials. In paper 5.5, we collaborated with Prof. Xiang Zhang who was with UCLA at that time to study the propagation of an electromagnetic (EM) wave in a piezomagnetic superlattice with piezomagnetic coefficient being modulated. Because of the piezomagnetic effect, the coupling between the EM wave and vibration of superlattice also can be established, resulting in the creation of a new type of magnetic polariton that does not exist in ordinary magnetic material. At some resonance frequencies, the abnormality of dispersion of permeability introduces negative value and piezomagnetic superlattice can make a kind of negative permeability material. Therefore both negative permittivity and permeability could be realized through the design of suitable microstructures, which might lead to a possible way toward new electro-magnetic materials.

As a straightforward extension, in paper 5.6 we investigated the propagation of EM waves in a piezoelectric-piezomagnetic superlattice(PPS). In a PPS, the electric and magnetic vectors of EM waves could simultaneously couple with the identical superlattice vibration, respectively, due to piezoelectric effect and piezomagnetic effect, which results in magnetoelectric effect. Consequently, two orthogonally polarized EM waves could simultaneously couple with the identical vibration, which would give birth to coupled phonon polaritons. Attributed to this mechanism, in a PPS, the propagation of EM waves varies drastically. EM waves perpendicular to the PPS vector can propagate, while the propagation is inhibited along the PPS vector in the original band gap of either the piezoelectric superlattice or the piezomagnetic superlattice. The origin of the differences in propagation is analyzed and some potential applications are discussed. These phenomena and applications reveal that the dielectric superlattice has become a versatile platform to study the complex couplings among electronic, magnetic and mechanic processes.

However, just like the surface plasmon polariton mode resulted from the negative permittivity, we also predicted a surface phonon polariton(SPhP) based on ITPCs in paper 5.7. Due to the scale of the superlattice structure, this SPhP mode normally exists in megahertz-gigahertz band. This might be a quite fundamental work since many surface EM mode related phenomena thus could be artificially designed.

Among all these representative works above, the phonons play very critical roles. The superlattice units could be viewed as a serial of oscillators with corresponding resonance frequencies. Due to the physically tight link between adjacent units, the photons thus could interact with the superatttice's phonons though piezoelectric or piezomagnetic effects. Many interesting phenomena were observed. However, due to the large scale of mechanical/ultrasonic oscillation systems, the corresponding frequencies normally exist in microwave band, which limits their high frequency applications.

To raise the operation frequency then further induce coupling between different units, metallic split ring resonators are ordinarily used choices. Due to the background of optical superlattice, not only the magnetic plasmon oscillations of single units may be considered, but also a kind of magnetic dimer(MD) consists of two single split-ring resonators(SSRR) were reported in paper 5.8. The fundamental MD resonances are viewed as bonding and antibonding combinations of individual SSRRs eigenmodes. The so-called hybridization of the magnetic plasmon modes were demonstrated at near-infrared frequencies. A metamaterial made of a large number of coupled magnetic dimers could be utilized as a tunable optically active medium with possible applications in optical elements and devices.

Based on the ideas above, we propose a new concept in nanophotonics, namely stereometamaterials, which refer to metamaterials with the same constituents but different spatial arrangements. In collaboration with Universität Stuttgart in Germany, we theoretically and experimentally study a typical stereometamaterial, i.e., meta-dimers, which consist of a stack of two identical SSRRs with various twist angles.(5.9) We found that the twisting of stereometamaterials offers a way to engineer complex plasmonic nanostructures with a tailored electromagnetic response.

If more coupled units are taken into account, a linear SSRR chain is formed that is just similar to a 1D superlattice.(5.10) These SSRRs interact mainly through the exchange of conduction current, resulting in stronger coupling as compared to the corresponding magneto-inductive interaction. This configuration could be used to support propagation of long range MP polaritons with some new applications in subwavelength transmission lines for a wide range of integrated optical devices.

Eventually, our superlattice inspired metamaterial researches move to 3D cases as reported in representative paper 5.11. We studied the long-wavelength optical properties of a plasmonic crystal composed of 3D nanorod matrix. Just like the original ITPC, we emphasized the concept of the polariton, which is due to the coupling between the photons and the long transverse plasmon wave. The polariton stop band, associated with the coupling effect rather than the Bragg reflection, has been suggested. The results also show that the long-wavelength method developed for an ionic crystal can be applied to a plasmonic crystal and that the artificial and classic lattices may share a common physics. In fact, to study the effect, the Huang's equations were also used.

As we know, the Huang's equations were established originally in ionic crystals where exist a strong coupling between the photons and lattice vibrations. Look back to 1999, we adopted these equations to study the coupling between incident EM wave and the superlattice vibration in an ITPC. Since the similar equations are used, the long-wavelength optical properties suggested in the former, such as the dielectric abnormality, polariton excitation, etc., can also be found in the plasmonic lattice. Although the material systems, resonance units, and related frequencies are quite different, the evolution of the related concepts, mechanisms, and properties are so clear. We developed a unique way from ITPC to metamaterial under the framework of dielectric superlattice. We believe there are still many new physics and phenomena that deserve further and continuous investigation. These are just the powers and charms of dielectric superlattices.

Optical Properties of an Ionic-Type Phononic Crystal[*]

Yan-qing Lu, Yong-yuan Zhu, Yan-feng Chen, Shi-ning Zhu, Nai-ben Ming

National Laboratory of Solid State Microstructures

Yi-Jun Feng

Department of Electronic Science and Engineering, Nanjing University, Nanjing 210093,
People's Republic of China

An ionic-type phononic crystal composed of two ferroelectric media with opposite spontaneous polarization aligned periodically in a superlattice structure was studied theoretically and experimentally. The coupling between vibrations of the superlattice and the electromagnetic waves results in various long-wavelength optical properties, such as microwave absorption, dielectric abnormality, and polariton excitation, that exist originally in ionic crystals. The results show that this artificial crystal structure can be used to simulate the microscopic physical processes in real crystals.

Study of the periodic medium has long been a topic of interest. In a crystal, the periodic potential causes the energy structure of electrons to form a band structure with only those electrons in pass bands that are capable of moving freely. In artificial composites such as superlattices, the periodic modulation of the related physical parameters may also result in band structure and novel properties. Associated with the variation of dielectric constants is the photonic crystal[1], which is important for applications such as suppressing spontaneous emission, manipulating light in a specific path, and creating novel laser geometries[1,2]. The modulation of nonlinear optical coefficients results in a quasi-phase-matched frequency conversion that is more efficient than that with a birefringence phase-matching method[3,4]. Recently, interest in phononic crystal, a periodic elastic composite, has grown[5—7]. Attention has been given to phenomena such as Anderson localization[8] and possible applications such as acoustic filters and new transducers[5—7]. The structure modulation may be extended to quasi-periodic[9] or aperiodic structures, and the modulation parameters may be more complicated. For example, objects such as the ferroelectric domain or piezoelectric coefficient may be modulated. Even two or more parameters may be modulated together, which could result in some coupling effects.

In a real crystal, various couplings exist between the motions of electrons, photons, and phonons. For example, infrared absorption and polariton excitation results from the coupling between lattice vibrations and electromagnetic (EM) waves in an ionic crystal. If

---

[*] Science, 1999, 284(5421): 1822

the ferroelectric domain or piezoelectric coefficient is modulated in a phononic crystal, the interaction between the superlattice vibrations and EM wave may be established. Similar effects can be expected in such an artificial medium. This kind of phononic crystal is termed an ionic-type phononic crystal(ITPC).

To calculate the dispersion relation of phononic crystals, several effective theories have been proposed[6, 10, 11]. For a one-dimensional(1D) phononic crystal, a simple model can be suggested, which is similar to the 1D atom chain model in real crystals. As we know, there are infinite degrees of freedom(DOF) of vibration of a real object. However, the fundamental vibration may be characterized as that of an equivalent single or finite DOF system, which is the basis of the so-called normal mode method or lumped-parameter method for studying the vibration characteristics of a continuous system[12]. According to this idea, the fundamental thickness vibration of a free thin plate can be equivalent to the vibration of a spring with two identical mass dots at the terminals. The equivalent mass m^* and the equivalent force constant β^* are determined as $m^* = \rho A l/2$ and $\beta^* = (\pi^2 v^2/4l)\rho A$ for keeping the fundamental frequency unchanged, where ρ, A, l, and v are the density, cross-section area, thickness, and sound velocity of the plate, respectively. Thus, a periodic combination of two kinds of thin plates, A and B, can be viewed as a diatomic chain. The mass of each "atom" is $m = m_A^* + m_B^*$, and the force constants are β_A^* and β_B^*, respectively. The sites of the atoms are at the joints of adjacent plates, which means that the mass is viewed as being concentrated at the boundary of neighboring domains. From the wavelike behavior and traveling wave solution of the motion equation, the dispersion relation of a 1D phononic crystal is obtained with

$$\omega_\pm^2 = \frac{\beta_A^* + \beta_B^*}{m} \pm \frac{1}{m}\sqrt{\beta_A^{*2} + \beta_B^{*2} + 2\beta_A^* \beta_B^* \cos[k(l_A + l_B)]} \qquad (1)$$

where ω_+ and ω_- correspond to the optic branch and acoustic branch, respectively, and k is the wave vector. At the edge of the Brillouin zone $[k = \pi/(l_A + l_B)]$, a phononic band gap appears, which is the same as the result deduced by other methods.

Here, we consider a 1D phononic crystal combined with two ferroelectric media with their spontaneous polarization aligned in opposite directions[Fig. 1(a)]. The boundaries of neighboring domains are charged differently; thus, the domain boundary can be viewed not only as the concentration of mass but also as the charge center. This phononic crystal can be viewed as a 1D diatom chain with positive and negative "ions" connected periodically, similar to a real ionic crystal, therefore forming an ITPC.

In an ITPC, the two kinds of ions constitute dipoles whose relative motion can be influenced by an EM field, especially that of a microwave. Because the wave vector of a microwave is much smaller than the width of the Brillouin zone, the microwave interacts only with optic branch phonons at the center of the Brillouin zone($k \approx 0$), giving rise to some typical long wavelength optical properties. Similar to an ordinary ionic crystal, the fundamental equations of the coupled motion can be written as

$$\begin{cases} \ddot{W} = b_{11}W + b_{12}E & (2) \\ P = b_{21}W + b_{22}E & (3) \end{cases}$$

where $\ddot{W}$ is the second derivative of W with respect to time; W represents the relative motion of positive and negative ions; E is the electric field of the microwave; P is the polarization induced by the electric field and relative motion of different ions; and b_{11}, b_{12}, b_{21}, and b_{22} are undetermined parameters. In a real crystal, the counterpart equations are those first proposed[13] when the coupled motion of lattice and EM waves was studied. In an ITPC, the relative motion is caused by the piezoelectric effect, which implies that vibrations of ITPCs can interact with EM waves even if the constituents of ITPCs are piezoelectric and not ferroelectric.

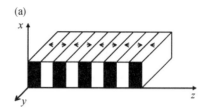

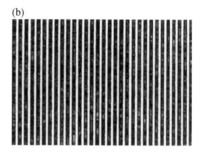

FIG.1.(a) Schematic of an ITPC consisting of two kinds of ferroelectric media. The arrows represent the orientation of the spontaneous polarization.(b) Optical microscope photograph showing the periodic ferroelectric domain structure of the etched y-cut face in an ITPC. The modulation period is 7.2 μm.

In piezoelectric materials, the fundamental equations are piezoelectric equations:

$$\begin{cases} \sigma = cS - eE & (4) \\ P = eS + \varepsilon_0(\varepsilon - 1)E & (5) \end{cases}$$

where S is strain, σ is stress, ε is the original dielectric constant, c is the elastic constant, and e is the piezoelectric coefficient. These equations reflect the coupling of the elastic wave and the electric field. Naturally, one would ask if there is any relation between the equations in[13] and the piezoelectric equations. The simplest case in which the ITPC is composed of two domains with the same thickness and elastic properties was chosen. The only difference between the neighboring domains is that their spontaneous polarizations are aligned oppositely. Because the piezoelectric coefficient is a third-order tensor, it changes its sign from the positive domains to the negative domains. The motion of domain boundaries with positive and negative charges is defined as U_+ and U_-, respectively. Under the condition of long-wavelength approximation, the motion of each primitive cell can be viewed as identical. Thus, using Newton's motion law and treating the ITPC as a 1D chain with discrete equivalent mass dots at domain boundaries, we obtain the equations of the relative motion of these mass dots:

$$\begin{cases} \ddot{W} = -\dfrac{\pi^2 v^2}{l^2}W + \dfrac{2e}{\sqrt{\rho}\, l}E & (6) \\ P = \dfrac{2e}{\sqrt{\rho}\, l}W + \varepsilon_0(\varepsilon - 1)E & (7) \end{cases}$$

where $W=\rho^{1/2}/2(U_+-U_-)$ and l is the domain thickness.

Comparing Eqs. 6 and 7 with Eqs. 2 and 3, one finds that they have the same format, which means that the equations in[13] and the piezoelectric equations are equivalent. Thus, parameters in Eqs. 2 and 3 are acquired as $b_{11}=-\pi^2 v^2/l^2$, $b_{12}=b_{21}=2e/\rho^{1/2}l$, and $b_{22}=\varepsilon_0(\varepsilon-1)$. According to the procedure for studying the long-wavelength optical properties of ionic crystal, the following results can be obtained from Eqs. 6 and 7 and Maxwell's equations.

The eigenfrequency of the transverse vibration without the coupling of external electric field is $\omega_{T0}=\omega_0=(-b_{11})^{1/2}$, whereas the eigenfrequency of the longitudinal wave is $\omega_{L0}=[-b_{11}+b_{12}^2/(\varepsilon_0+b_{22})]^{1/2}$; ω_{L0} is larger than ω_{T0}. Their ratio is described by the Lyddane-Sachs-Teller(LST) relation:

$$\frac{\omega_{L0}}{\omega_{T0}}=\left[\frac{\varepsilon(0)}{\varepsilon(\infty)}\right]^{1/2} \tag{8}$$

where $\varepsilon(0)$ is the dielectric constant at low frequency and $\varepsilon(\infty)$ represents the dielectric response that occurs at frequencies much higher than the eigenfrequency of ITPCs. They can be determined as $\varepsilon(\infty)=\varepsilon$ and $\varepsilon(0)=-b_{12}^2/b_{11}\varepsilon_0+\varepsilon$.

When an EM wave propagates in an ITPC, the electric field stimulates a long-wavelength optic branch vibration in the ITPC and then causes an intensive attenuation of the electric energy at a specific frequency. The absorption power is related to the imaginary part of the dielectric constant, which can be deduced from Eqs. 6 and 7. By including a damping term $-\gamma \dot{W}$ in the right side of Eq. 6, we get

$$\varepsilon''(\omega)=\frac{b_{12}^2\omega\gamma}{\varepsilon_0[(\omega^2+b_{11})^2+\omega^2\gamma^2]} \tag{9}$$

From Eq. 9, the absorption peak locates at $\omega=(-b_{11})^{1/2}=\omega_{T0}$ with the width of γ. This absorption in ITPCs is equivalent to the infrared absorption in ionic crystals, but the frequency determined by the period of ITPC is normally in the microwave band.

Near the eigenfrequency of ITPC, the EM wave interacts with the mechanical vibration strongly; thus, the transverse mode is neither a pure photon mode nor a pure optic branch phonon mode in this narrow range of k values. It is called a polariton mode, a coupling mode of photons and optic branch phonons. This was first predicted in real ionic crystals[13] and confirmed experimentally in 1964[14]. Our results show that the polariton also exists in ITPCs.

We have calculated the polariton dispersion curve of an ITPC based on periodically poled $LiNbO_3$(PPLN) with the period of 7.2 μm by taking $\gamma=0$, which makes ε and therefore k real(Fig. 2). The eigenfrequencies of the transverse wave and longitudinal wave are $f_{T0}=\omega_{T0}/2\pi=500$ MHz and $f_{L0}=\omega_{L0}/2\pi=532$ MHz, respectively[15]. There is a band gap in which the dielectric constant is negative. The corresponding refractive index becomes imaginary. The incident radiation with these frequencies will be reflected. However, this gap does not originate from the interference of EM waves due to the periodic structure but,

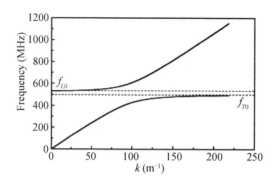

FIG.2. The calculated polariton dispersion curve of an ITPC with the period of 7.2 μm without consideration of damping. There is a frequency gap between f_{L0} and f_{T0} where no EM waves are permitted to propagate in the sample.

rather, originates from the interaction of the photon and the transverse optics phonon.

For verification of the predictions above, an ITPC based on PPLN with the period of 7.2 μm was fabricated by the growth striation method[4]. Its microscopic domain structure was revealed after hydrogen fluoride etching[Fig. 1(b)]. A y-cut 1.7-mm-thick sample with a pair of Ag electrodes(2.0 mm by 2.0 mm) deposited on each surface was selected for the experiments.

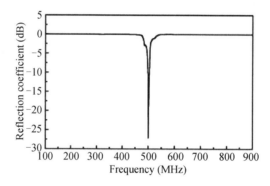

FIG. 3. The measured reflection coefficient of an ITPC in the microwave band. The minimum of the reflection coefficient indicates that there is a strong microwave absorption peak at 502 MHz.

With an HP8510C network analyzer, the reflection coefficient(S_{11}) of the ITPC was measured to simulate the situation in which a microwave beam propagates in the sample. Under this condition, the electric field was applied on the sample simultaneously($k=0$), which coincides with the long-wavelength approximation used above. As the microwave wavelength is much larger than the dimension of the electrodes, this assumption is reasonable. The reflection coefficient as a function of frequency(Fig. 3) shows that there is an absorption peak($S_{11}=-26$ dB) at 502 MHz, which is very close to the theoretical value $f=f_{T0}=500$ MHz.

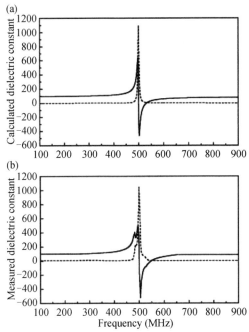

FIG. 4. The dielectric constant curves of an ITPC. The solid lines represent the real part, and the dashed lines represent the imaginary part. (a) The calculated result obtained by choosing proper damping ($\gamma=0.01\omega_{T0}$). (b) The measured dielectric constant.

The dielectric constant was measured by an impedance-analyzing method in order to demonstrate the polariton excitation in ITPC. The results (Fig. 4) are compared with the theoretical results by choosing the proper damping term ($\gamma=0.01\omega_{T0}$). The dielectric constant changed near ω_{T0}, caused by the superlattice in ITPC. The dielectric spectrum in this band has the same curve shape as the far-infrared dielectric constant of an ionic crystal caused by lattice vibration, showing that they have a similar origin. Much information on the ITPC can be obtained from the dielectric constant curve. The measured f_{T0}, f_{L0}, $\varepsilon(0)$, and $\varepsilon(\infty)$ values are 502 MHz, 547 MHz, 99.88, and 88.02, respectively. They all agree well to the theoretical predictions. The LST relation was also proved, which is $\omega_{L0}/\omega_{T0}=1.09$, only a slight deviation from the value of $[\varepsilon(0)/\varepsilon(\infty)]^{1/2}$ (1.07). The absorption peak in Fig. 3 is at the same position of the peak of the $\varepsilon''(\omega)$, just as predicted. There is a gap between ω_{T0} and ω_{L0} where $\varepsilon<0$ and incident EM waves will be strongly reflected. The phenomena above show that there is a polariton mode in ITPC.

From the similarity between the real ionic crystal and the ITPC, other long-wavelength optical properties (such as Raman and Brillouin scattering) might also be expected in an ITPC. The only difference is that they occur in different frequencies. For example, Raman scattering appears in the terahertz region for a real ionic crystal, whereas it might appear in the gigahertz region for an ITPC. Study on these effects is of fundamental interest in physics.

References and Notes

[1] E. Yablonovitch, *Phys. Rev. Lett.* **58**, 2059(1987).

[2] ___, T. J. Gmitter, K. M. Leung, *ibid.* **67**, 2295(1991); A. Mekis *et al.*, *ibid.* **77**, 3787(1996); Y. S. Chan *et al.*, *ibid.* **80**, 956(1998); M. M. Sigalas *et al.*, *Microwave Opt. Technol. Lett.* **15**, 153 (1997); J. S. Foresi *et al.*, *Nature* **390**, 143(1997). For a review, see J. J. Joannopoulos *et al.*, *Photonic Crystals* (Princeton Univ. Press, Princeton, NJ, 1995).

[3] M. M. Fejer *et al.*, *IEEE J. Quantum Electron.* **QE-28**, 2631(1992).

[4] D. Feng *et al.*, *Appl. Phys. Lett.* **37**, 607(1980); Y. L. Lu *et al.*, *ibid.* **59**, 516(1991); Y. L. Lu *et al.*, *ibid.* **68**, 1467(1996); Y. Q. Lu *et al.*, *ibid.* **69**, 3155(1996); Y. Lu *et al.*, *Science* **276**, 2004 (1997).

[5] L. Ye *et al.*, *Phys. Rev. Lett.* **69**, 3080(1992).

[6] M. S. Kushwaha *et al.*, *ibid.* **71**, 2022(1993).

[7] M. M. Sigalas *et al.*, *Phys. Rev. B* **50**, 3393(1994); M. S. Kushwaha and P. Halevi, *Appl. Phys. Lett.* **64**, 1085(1994); J. P. Dowling, *J. Acoust. Soc. Am.* **91**, 2539(1992); M. M. Sigalas, *ibid.* **101**, 1256(1997); F. R. Montero de Espinosa *et al.*, *Phys. Rev. Lett.* **80**, 1208(1998).

[8] P. Sheng, Ed., *Scattering and Localization of Classical Waves in Random Media* (World Scientific, Singapore, 1990).

[9] S. N. Zhu *et al.*, *Science* **278**, 843(1997); S. N. Zhu *et al.*, *Phys. Rev. Lett.* **78**, 2752(1997).

[10] K. M. Ho *et al.*, *Phys. Rev. Lett.* **65**, 3152(1990).

[11] J. B. Pendry and A. Mackinnon, *ibid.* **69**, 2772(1992); H. Dong and S. Xiong, *J. Phys. Condens. Matter* **5**, 8849(1993); M. M. Sigalas and C. M. Soukoulis, *Phys. Rev. B* **51**, 2780(1995).

[12] J. D. Turner and A. J. Pretlove, *Acoustics for Engineers* (Macmillan, Houndmills, UK, 1991), pp. 27-29; L. Meirovitch, *Elements of Vibration Analysis* (McGraw-Hill, New York, 1975), pp. 281-284.

[13] K. Huang, *Proc. R. Soc. London A* **208**, 352(1951); M. Born and K. Huang, *Dynamical Theory of Crystal Lattices* (Oxford Univ. Press, Oxford, 1954).

[14] C. H. Henry and J. J. Hopfield, *Phys. Rev. Lett.* **15**, 964(1965).

[15] $\varepsilon=84.1$, $\varepsilon_0=8.854\times10^{-12}$ F/m, $v=3600$ m/s, $e_{15}=3.8$ c/m^2, $\rho=4.64\times10^3$ kg/m^3 [Y. Nakagawa *et al.*, *J. Appl. Phys.* **44**, 3969(1973)].

[16] We are grateful to the State Key Program for Basic Research of China, the National Natural Science Foundation Project of China (contract 69708007), and the National Advanced Materials Committee of China for their support of this work.

New Type of Polariton in a Piezoelectric Superlattice[*]

Yong-yuan Zhu, Xue-jin Zhang, Yan-qing Lu, Yan-feng Chen, Shi-ning Zhu, and Nai-ben Ming

National Laboratory of Solid State Microstructures, Nanjing University, Nanjing 210093, China

We studied the propagation of an electromagnetic(EM) wave in a piezoelectric superlattice. Because of the piezoelectric effect, a transverse polarization can be induced by a longitudinal wave which couples strongly to the EM wave in some particular frequency regions, resulting in the creation of a new type of polariton that does not exist in ionic crystals. The forbidden band associated with the polariton is not due to the Bragg reflection, but rather to the coupling.

Study of the periodic medium has long been a topic of interest. In a crystal, the periodic potential causes the energy structure of electrons to form a band structure with only those electrons in passbands that are capable of moving freely. In artificial composites such as superlattices, the periodic modulation of the related physical parameters may also result in band structure. Associated with the variation of dielectric constants is the photonic crystal[1,2], which is important for applications such as suppressing spontaneous emission, manipulating light in a specific path, and creating novel laser geometries[1-3]. The modulation of nonlinear optical coefficients results in a quasi-phase-matched frequency conversion that is more efficient than that with a birefringence phase-matching method[4-6]. Recently, interest in phononic crystal, a periodic elastic composite, has grown[7-10]. The structure modulation may be extended to quasiperiodic or aperiodic or two-dimensional structures[11-14], and the modulation parameters may be more complicated. For example, objects such as the ferroelectric domain or piezoelectric coefficient may be modulated. Even two or more parameters may be modulated together, which could result in some coupling effects.

In a real crystal, various couplings exist between the motions of electrons, photons, and phonons. For example, infrared absorption and polariton excitation results from the coupling between lattice vibrations(transverse optical phonons) and electromagnetic(EM) waves(photons) in an ionic crystal[15]. If the ferroelectric domain or piezoelectric coefficient is modulated in a superlattice, the coupling between the superlattice vibrations and the EM wave may be established. Similar effects such as polariton excitation can be expected in such an artificial medium. The above idea has been verified for the coupling between the

---

[*] Phys.Rev.Lett.,2003,90(5):053903

transverse superlattice vibrations and the photons[16].

Then can an EM wave couple with a longitudinal lattice vibration? The problem is treated theoretically for a piezoelectric superlattice (PSL) in this Letter. A piezoelectric material is a material that becomes electrically polarized when it is strained or that becomes strained when placed in an electric field[17], which can be a result of propagation of an electromagnetic wave. That is, superlattice vibration can be excited by an EM wave. On the other hand, the superlattice vibrations will induce electrical polarization either longitudinally or transversely depending on the configuration of the PSL due to the piezoelectric effect. The lateral polarization in turn will emit EM waves that interfere with the original EM wave. In such a case, a longitudinal lattice vibration will couple strongly with the EM wave through the piezoelectric effect, resulting in polariton excitation, dielectric abnormality, etc., at microwave range.

In order to elucidate the above idea, let us consider the following case. Here we assume that the PSL is made of a periodically poled ferroelectric crystal[4−6] (taking a periodically poled $LiNbO_3$ as an example; a ferroelectric material is piezoelectric) arranged along the x axis and that the domain walls lie in the yz plane. The thicknesses of the positive and negative domains are the same(d). Figure 1 is a schematic diagram of the case. Here only three periods of the PSL have been shown. The direction of the spontaneous polarization is along the z axis, perpendicular to the sheet. Also we assume that the transverse dimensions are very large compared with an acoustic wavelength so that a one-dimensional model is applicable and that the EM wave propagates along the x direction. Under these conditions, a longitudinal acoustic wave (LAW) propagating along the x axis will be excited. The piezoelectric equations pertaining to this case are[17]

$$T_1 = C_{11}^E S_1 + e_{22}(x) E_2,$$
$$P_2 = -e_{22}(x) S_1 + \varepsilon_0 (\varepsilon_{11}^S - 1) E_2, \quad (1)$$

with

$$e_{22}(x) = \begin{cases} +e_{22}, \text{in positive domains} \left(0 \leqslant x < \frac{\Lambda}{2} = d\right), \\ -e_{22}, \text{in negative domains} \left(\frac{\Lambda}{2} \leqslant x < \Lambda = 2d\right), \end{cases}$$

where T_1, S_1, E_2, and P_2 are the stress, strain, electric field, and polarization, respectively. C_{11}^E, $e_{22}(x)$, and ε_{11}^S are the elastic, piezoelectric, and dielectric coefficients, respectively.

Previous study shows that under the action of an electric field, the positive and negative domains act differently[18,19]. That is, when the positive domains expand, the negative ones contract and vice versa, which results in the appearance of the charges of the same sign (positive or negative) on the same side of the two different domains [Figs. 1(b) and 1(c)]. That is, the PSL as a whole polarizes electrically synchronously. This fact can be seen in Eq. (1). It tells us that a transverse polarization P_2 can be induced by a longitudinal wave S_1 through the piezoelectric effect. And through Maxwell equation, this transverse polarization P_2 will in turn emit EM waves that interfere with the original EM wave. In other words, it is this transverse polarization P_2 that couples strongly with the EM wave. This resembles the lattice vibrations belonging to the transverse optical branches. In that case, the atoms carrying opposite charges vibrate against each other. This type of vibration can be coupled with an electric field of a light wave, resulting in a so-called polariton excitation[15].

With the use of Newton's law, the equation of motion for a vibrating medium can be obtained:

$$\rho \frac{\partial^2 S_1}{\partial t^2} - C_{11}^E \frac{\partial^2 S_1}{\partial x^2} = -\frac{\partial^2 [e_{22}(x) E_2]}{\partial x^2}, \quad (2)$$

where ρ is the mass density. Equation (2) indicates that the PSL is a forced oscillator. That is, a LAW propagating along the x axis will be excited by a transverse EM wave. The frequency of the acoustical wave will be that of the EM wave. By using the Fourier transformation

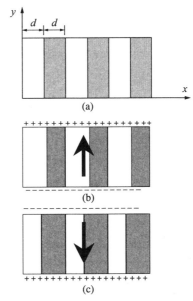

FIG.1.(a) A schematic diagram of a PSL. The white area represents the positive domain and the gray area the negative domain. (b),(c) At any instant of time under the action of an EM wave, the positive and negative domains act differently. That is, when the positive domains expand, the negative ones contract and vice versa, which results in the appearance of the of the two different domains. The arrow indicates the stress-induced electric polarization.

$$E_2(x,t) = \int E(k) e^{i(\omega t - kx)} dk,$$

$$S_1(x,t) = \int S(q) e^{i(\omega t - qx)} dq,$$

$$e_{22}(x) = \sum_{m \neq 0} \frac{i(1 - \cos m\pi) e_{22}}{m\pi} e^{-iG_m x} \quad \left(G_m = m \frac{2\pi}{\Lambda}\right). \quad (3)$$

Here different wave numbers for photons and phonons have been used. Equation (2) becomes

$$\int (\rho \omega^2 - C_{11}^E q^2) S(q) e^{-iqx} dq = \int - \sum \frac{i(1 - \cos m\pi) e_{22}}{m\pi} (G_m^2 + 2G_m k + k^2) E(k) e^{-i(k+G_m)x} dk, \quad (4)$$

where $e^{i\omega t}$ has been omitted. For photons with very long wavelength, that is $k \to 0$ or $k \ll G_m$, Eq.(4) becomes

$$\int (\rho \omega^2 - C_{11}^E q^2) S(q) e^{-iqx} dq = -\sum_m \frac{i(1 - \cos m\pi) e_{22}}{m\pi} G_m^2 E(k) e^{-i(k+G_m)x}. \quad (5)$$

In order for the two sides to be equal, q must take the form $q = k + G_m = k + m\frac{2\pi}{\Lambda}$. Then we obtain

$$S(q = k + G_m) = -\frac{i(1-\cos m\pi)e_{22}}{m\pi}\frac{G_m^2}{\rho\omega^2 - C_{11}^E G_m^2}E(k) \qquad (6)$$

and

$$S_1(x,t) = -\sum \frac{i(1-\cos m\pi)e_{22}}{m\pi}\frac{G_m^2}{\rho\omega^2 - C_{11}^E G_m^2}e^{iG_m x}E(x,t) = H(x)E(x,t). \qquad (7)$$

Substituting Eq.(7) into Eq.(1), we have

$$P_2 = \{-e_{22}(x)H(x) + \varepsilon_0(\varepsilon_{11}^S - 1)\}E_2(x,t) = \kappa(x)E_2(x,t), \qquad (8)$$

where $\kappa(x)$ is a function of the x coordinate. For EM waves with $k \to 0$ or their wavelength (λ) much larger than the length of the sample, the PSL can be taken to be homogeneous. The space average value of $\kappa(x)$ should be used:

$$\kappa = \overline{\kappa(x)} = \frac{1}{\Lambda}\int_0^\Lambda \kappa(x)dx = \varepsilon_0(\varepsilon_{11}^S - 1) + \frac{1}{\Lambda}\int_0^\Lambda e_{22}(x)H(x)dx$$

$$= \varepsilon_0(\varepsilon_{11}^S - 1) + \frac{1}{\Lambda}\sum \frac{(1-\cos m\pi)e_{22}^2}{m\pi}\frac{G_m}{\rho\omega^2 - C_{11}^E G_m^2}2(e^{-im\pi} - 1). \qquad (9)$$

The dielectric function obtained from Eq.(9) is

$$\varepsilon(\omega) = \varepsilon_{11}^S + \frac{4e_{22}^2/(d^2\rho\varepsilon_0)}{\omega_L^2 - \omega^2}, \qquad (10)$$

where $\omega_L = m\pi v_a/d$ ($m = 1,3,5,\ldots$), the resonance frequency of the LAW due to piezoelectric effect; $v_a = \sqrt{C_{11}^E/\rho}$ represents the velocity of the acoustic wave.

The polariton dispersion relation due to the coupling between the lateral polarization induced by a LAW and EM wave can be deduced from Eqs.(9) and (10) and Maxwell's equations:

$$\frac{c^2 k^2}{\omega^2} = \varepsilon_{11}^S + \frac{4e_{22}^2/(d^2\rho\varepsilon_0)}{\omega_L^2 - \omega^2}, \qquad (11)$$

where c is the EM wave velocity in vacuum.

This result resembles the dispersion relation in an ionic crystal where the EM wave couples strongly with the transverse optical phonons[15]. Figure 2(a) shows the coupled modes of photons and longitudinal phonons in the PSL described by Eq.(11). The solid line labeled $v_c = \frac{\omega}{k} = c/\sqrt{\varepsilon_{11}^S}$ corresponds to EM waves, but uncoupled to the lattice vibrations, and the dotted line represents the lattice vibration in the absence of coupling to the EM field due to the Brillouin zone folding of the PSL. The wave number of the acoustic wave $q = k + G_m$ is equivalent to $q = k$[15]. The region of the crossover [marked by A in Fig. 2(a)] of the solid line and the dotted line is the resonance region. By resonance, we mean that the frequency of the EM wave equals the acoustic resonance frequency of the PSL determined by the periodicity. At resonance the photon-phonon coupling entirely changes the character of the propagation. The heavy lines are the dispersion relations in the presence of coupling

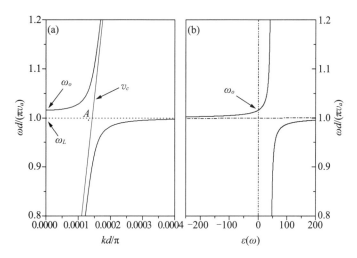

FIG.2. Calculated polariton dispersion(a) and dielectric abnormality(b). The dielectric constant is negative between ω_o and ω_L where no EM waves will be permitted to propagate in the sample, but will be reflected at the boundary. Thus, a band gap appears.

between the lateral polarization induced by a LAW and the EM wave. In the resonance region the propagation mode is neither a pure photon mode nor a pure longitudinal acoustic mode in a narrow range of k values(unlike the case in the ionic crystal, where the coupling occurs between the photons and the transverse optical phonons). The quantum of the coupled photon-phonon wave field is called a polariton. It is a new type of polariton. One effect of the coupling is to create a frequency gap between ω_o and ω_L. Here ω_o can be obtained by setting $\varepsilon(\omega_o)=0$ in Eq.(10). Thus, we have

$$\omega_o^2 = \omega_L^2 \left(1 + \frac{4}{(m\pi)^2}K^2\right) \tag{12}$$

and

$$\varepsilon(\omega) = \varepsilon_{11}^S \frac{\omega_o^2 - \omega^2}{\omega_L^2 - \omega^2}, \tag{13}$$

where K^2 is an electromechanical coupling coefficient[17],

$$K^2 = \frac{e_{22}^2}{C_{11}^E \varepsilon_0 \varepsilon_{11}^S} \tag{14}$$

For frequencies $\omega_L < \omega < \omega_o$, $\varepsilon(\omega)$ is negative and the corresponding refractive index becomes imaginary as can be seen in Fig. 2(b). The incident radiation with these frequencies will be reflected. However, this gap does not originate from the Bragg reflection due to the periodic structure, but rather originates from the coupling of the photon and the longitudinal phonon. Since K^2 is less than 1, Eq.(12) can be approximated to

$$\omega_o \approx \omega_L \left(1 + \frac{2}{(m\pi)^2}K^2\right). \tag{15}$$

From it the gap($\omega_o - \omega_L$) can be determined. The larger the K value is, the stronger the coupling and the wider the gap. Here for the coupling between the EM wave and the LAW in LiNbO$_3$, K is about 0.28[19]. Thus, the gap is about 1.6% for $m=1$. If the period of the PSL is 6.6 μm, then the polariton will be excited around the resonance frequency 1 GHz (determined by $\omega_L = \pi v_a/d$ with $m=1$) [19], which lies at microwave region. High-order polaritons ($m=3,5,\ldots$) can create in higher frequency regions, however, the corresponding gaps are much narrower that are inversely proportional to m^2 as indicated by Eq. (15). Please note that here m is an odd number. For even m, no polariton can be excited. The gap can be widened by use of some materials with larger electromechanical coupling coefficients. For Pb(ScNb)$_{0.5}$O$_3$:PbTiO$_3$, the K value for the coupling between the EM wave and the LAW can get to 0.48[20]. Theoretically, the corresponding gap is about 4.67%. An even larger gap can be realized by using the coupling between the EM wave and the transverse lattice vibration. For LiNbO$_3$, K can be as large as 0.76[19]. In that case, the gap is 11.7%. The above result might have potential applications for microwave reflectors based on polariton excitation.

For piezoelectric materials, there are still two coupled waves even without periodic modulation, each having both EM and acoustic fields[17]. In both cases the relationship between the EM and acoustic fields is given by the dispersion

$$(\rho\omega^2 - C_{11}^E k^2)\left(\frac{\varepsilon_{11}^S \omega^2}{c^2} - k^2\right) = \frac{e_{22}^2 \omega^2 k^2}{\varepsilon_0 c^2} \tag{16}$$

or

$$\left(\frac{\omega}{k}\right)^2 \approx \begin{cases} v_c^2\left(1 + \dfrac{v_a^2 K^2}{v_c^2}\right) & \text{quasielectromagnetic} \\ v_a^2\left(1 - \dfrac{v_a^2 K^2}{v_c^2}\right) & \text{quasiacoustic} \end{cases} \tag{17}$$

with $v_c^2 = c^2/\varepsilon_{11}^S$.

Compared with Eq. (11), the dispersion here is quite different. The dispersions represented by Eq. (17) are two straight lines passing through the origin. In the presence of piezoelectric coupling, the quasiacoustic phase velocity shifts to a lower value than the acoustic velocity and the quasielectromagnetic phase velocity shifts to a higher value than the electromagnetic velocity. However, the coupling is so weak that only very small shifts are produced, and no polariton is excited.

In summary, the propagation of an EM wave in a PSL was studied theoretically. A new type of polariton was proposed that comes from the coupling of the EM wave with the LAW. The polariton dispersion and dielectric abnormality were discussed. The origin of the band gap is not due to the Bragg reflection, but rather to the coupling.

References and Notes

[1] E. Yablonovitch, Phys. Rev. Lett. **58**, 2059(1987).
[2] S. John, Phys. Rev. Lett. **58**, 2486(1987).

[3] J. J. Joannopoulos, R. D. Meade, and J. N. Winn, *Photonic Crystals* (Princeton University, Princeton, NJ, 1995).
[4] D. Feng, N. B. Ming, J. F. Hong, Y. S. Yang, J. S. Zhu, Z. Yang, and Y. N. Wang, Appl. Phys. Lett. **37**, 607(1980).
[5] M. M. Fejer, G. A. Magel, D. H. Jundt, and R. L. Byer, IEEE J. Quantum Electron. **QE-28**, 2631 (1992).
[6] R. Byer, IEEE J. Sel. Top. Quantum Electron. **6**, 911(2000).
[7] J.O. Vasseur, P.A. Deymier, B. Chenni, B. Djafari-Rouhani, L. Dobrzynski, and D. Prevost, Phys. Rev. Lett. **86**, 3012(2001).
[8] M. Kafesaki, M. M. Sigalas, and N. Garcia, Phys. Rev. Lett. **85**, 4044(2000).
[9] J.Danglot, J.Carbonell, M.Fernandez, O.Vanbesien, and D. Lippens, Appl. Phys. Lett. **73**, 2712(1998).
[10] *Scattering and Localization of Classical Waves in Random Media*, edited by P. Sheng (World Scientific, Singapore, 1990).
[11] S. N. Zhu, Yong-yuan Zhu, and N. B. Ming, Science **278**, 843(1997).
[12] S.N.Zhu, Y.Y.Zhu, Y.Q.Qin, H.F.Wang, C.Z.Ge, and N. B. Ming, Phys. Rev. Lett. **78**, 2752(1997).
[13] V. Berger, Phys. Rev. Lett. **81**, 4136(1998).
[14] N.G. Broderick, G.W. Ross, H.L. Offerhaus, D.J. Richardson, and D.C. Hanna, Phys. Rev. Lett. **84**, 4345(2000).
[15] C. Kittel, *Introduction to Solid State Physics* (Wiley, New York, 1996), 6th ed.
[16] Y.Q.Lu, Y.Y.Zhu, Y.F.Chen, S.N.Zhu, N.B.Ming, and Y. J. Feng, Science **284**, 1822(1999).
[17] B. A. Auld, *Acoustic Fields and Waves in Solids* (Wiley, New York, 1973).
[18] Y.Y. Zhu and N. B. Ming, J. Appl. Phys. **72**, 904(1992).
[19] Y.Y.Zhu, S.N.Zhu, Y.Q.Qin, and N.B.Ming, J.Appl. Phys. **79**, 2221(1996).
[20] E.F. Alberta and A.S. Bhalla, Mater. Lett. **35**, 199(1998).
[21] This work was supported by a grant for the State Key Program for Basic Research of China and by the National Natural Science Foundation of China (No. 69938010 and No. 10021001).

Piezoelectric-Induced Polariton Coupling in a Superlattice*

Cheng-ping Huang[1,2] and Yong-yuan Zhu[1]

[1] *National Laboratory of Solid State Microstructures, Nanjing University, Nanjing 210093, China*
[2] *Department of Applied Physics, Nanjing University of Technology, Nanjing 210009, China*

Propagation of electromagnetic waves in a piezoelectric superlattice is studied. Because of the piezoelectric effect, a coupling between two orthogonally polarized electromagnetic waves is induced by the superlattice vibration. As a consequence of the strong coupling, two types of polariton modes are found: one is supported in the band gap while the other prohibited. This unusual coupling effect is not present in classical lattices.

The interaction of waves with matter, especially matter with a periodic structure, has been a fundamental topic in solid state physics. Study of this topic results in many important concepts, such as the Brillouin zone, band structure, etc. It is well known that this is due to the periodic variation of the physical parameters that causes the waves to Bragg reflect, forming band structures. In a real crystal, the periodic potential leads to a band gap, where the electrons are prohibited from moving freely[1]. In a photonic crystal with periodic modulation of dielectric constants, photons can be described in terms of a photonic band gap structure, which can be employed to control the propagation of light[2-4]. And similarly, the elastic coefficient modulation corresponds to the phononic crystal, with which potential applications such as noise-proof devices and sound filters can be developed[5-7].

The Bragg reflection may not be the sole mechanism for band structure formation. Some other effects such as the coupling of an electromagnetic(EM) wave with the internal degrees of freedom of the medium can also generate a band gap[8]. In an ionic crystal, the coupling between lattice vibration and the EM waves leads to the phonon polariton, where various optical properties in the infrared range are exhibited. One important effect accompanying the phonon polariton is that a stop band is produced where wave propagation is forbidden[9].

The phonon polariton exists in the superlattices as well[10]. When the periodicity of lattices is expanded from atomic scale to microns, e.g., piezoelectric superlattice(PSL), the counterpart phenomenon appears. The PSL is composed of periodically reversed ferroelectric domains, which can be fabricated by the crystal growth technique [11] or

---

* Phys. Rev. Lett., 2005, 94(11):117401

electric poling method[12]. In the PSL, the coupling between the transverse superlattice vibration and the photons can be established, resulting in the phonon polariton in the microwave region[13,14]. And similar long-wavelength optical properties such as microwave absorption, dielectric abnormality, etc., can be obtained. In addition, theoretical and experimental work has suggested that in the PSL an EM wave can also couple with longitudinal superlattice vibration, introducing a new type of polariton that does not exist in ionic crystals[15,16]. Therefore, there are not only similarities, but also differences between artificial superlattices and real lattices, implying rich physics of artificial microstructures.

In this Letter, we suggest another interesting effect in the PSL that has not been found in real crystals. That is, the superlattice vibration not only can be excited by and coupled strongly to the incident EM wave, but they can also induce coupling between two orthogonally polarized EM waves, thus bringing novel character to the polariton excitation.

To start this, we consider the following case. In the PSL(a periodically poled LiNbO$_3$ is taken as an example), the positive and negative domains have the same thickness, and they are aligned periodically along the x axis with the spontaneous polarization in the $\pm z$ direction. We assume that the transverse dimensions of the PSL are much larger than the acoustic wavelength, and then a one-dimensional model is applicable. When a y-polarized EM wave propagates along the x axis, due to piezoelectric effect, a longitudinal acoustic wave and a z-polarized EM wave propagating in the same direction will be excited. The propagation of acoustic and EM waves are governed by Newton's and Maxwell's equations, respectively[17], and the coupling between them is described with the piezoelectric equations:

$$T_1 = C_{11}^E S_1 + e_{22}(x)E_2 - e_{31}(x)E_3,$$
$$D_2 = -e_{22}(x)S_1 + \varepsilon_0 \varepsilon_{11}^S E_2, \qquad (1)$$
$$D_3 = e_{31}(x)S_1 + \varepsilon_0 \varepsilon_{33}^S E_3.$$

Here, T_1, S_1, $E_{2,3}$, and $D_{2,3}$ are the stress, strain, electric field, and displacement, respectively. $C_{11}^E = c_{11} - i\omega\eta_{11}$ is the elastic coefficient, where c_{11} represents the real part and η_{11} is the damping coefficient. ε_{ij}^S are the dielectric constants. $e_{ij}(x) = e_{ij}f(x)$ are the periodically modulated piezoelectric coefficients, where $f(x)$ is $+1$ or -1 in positive or negative domains, respectively.

Equations(1) imply that a longitudinal acoustic wave S_1 will be excited by and coupled to the y-polarized(original) EM wave. In addition, the acoustic wave can also induce a polarization in the z direction, and simultaneously this polarization will emit a z-polarized EM wave. The latter EM wave contributes to the superlattice vibration as well and further couples to the original one. This is the mechanism for polariton coupling discussed in this Letter.

By using Newton's equation, the equation of motion for a vibrating medium is obtained as

$$\rho \frac{\partial^2 S_1}{\partial t^2} - C_{11}^E \frac{\partial^2 S_1}{\partial x^2} = \frac{\partial^2}{\partial x^2}[(e_{22}E_2 - e_{31}E_3)f(x)]. \tag{2}$$

This is the fundamental wave equation that describes the excitation and propagation of superlattice vibration in the PSL. With the Fourier transformation, the modulation function is written as $f(x) = \sum_m f_m e^{iG_m x}$, where $f_m = i[\cos(m\pi) - 1]/m\pi$, $G_m = \pi m/d$, and d is the thickness of the positive or negative domains. In the long-wavelength approximation (the EM wavelength is much larger than the period of PSL), Eq. (2) becomes

$$C_{11}^E \frac{\partial^2 S_1}{\partial x_2} + \rho\omega^2 S_1 = \sum_m f_m G_m^2 (e_{22}E_2 - e_{31}E_3)e^{iG_m x}. \tag{3}$$

Equation (3) just resembles to that of a forced oscillation, with the solution:

$$S_1 = -\sum_m \frac{f_m G_m^2}{C_{11}^E G_m^2 - \rho\omega^2} (e_{22}E_2 - e_{31}E_3)e^{iG_m x}. \tag{4}$$

Thus a longitudinal superlattice vibration is excited; to which both y- and z-polarized EM waves make a contribution. Here, the summation includes two values of m, one is $+|m|$ and the other is $-|m|$. Physically, the former represents the forward wave and the latter the backward wave. Hence, S_1 is proportional to $\sin(G_m x)$. That means a stationary wave is produced in the PSL, with the nodes in the middle of each domain when the fundamental vibration is excited.

The coupling of superlattice vibration to EM waves can be embodied in the dielectric response. Substitution of Eq. (4) into (1) gives

$$\begin{aligned} D_2 &= \varepsilon_0 \varepsilon_{22}(x)E_2 + \varepsilon_0 \varepsilon_{23}(x)E_3, \\ D_3 &= \varepsilon_0 \varepsilon_{23}(x)E_2 + \varepsilon_0 \varepsilon_{33}(x)E_3, \end{aligned} \tag{5}$$

where

$$\begin{aligned}
\varepsilon_{22}(x) &= \varepsilon_{11}^S + \sum_m \frac{f_m G_m^2 e_{22}^2/\varepsilon_0}{C_{11}^E G_m^2 - \rho\omega^2} f(x)e^{iG_m x}, \\
\varepsilon_{23}(x) &= -\sum_m \frac{f_m G_m^2 e_{22}e_{31}/\varepsilon_0}{C_{11}^E G_m^2 - \rho\omega^2} f(x)e^{iG_m x}, \\
\varepsilon_{33}(x) &= \varepsilon_{33}^S + \sum_m \frac{f_m G_m^2 e_{31}^2/\varepsilon_0}{C_{11}^E G_m^2 - \rho\omega^2} f(x)e^{iG_m x}.
\end{aligned} \tag{6}$$

Here $\varepsilon_{ij}(x)$ are functions of the x coordinate. In the long-wavelength approximation, the PSL can be taken to be homogeneous in space. Thus the space average values are applicable, i.e., $\varepsilon_{ij}(\omega) = \langle \varepsilon_{ij}(x) \rangle = (1/\Lambda)\int_{n\Lambda}^{(n+1)\Lambda} \varepsilon_{ij}(x)dx$, where Λ is the period of the PSL. With this approximation, Eqs. (5) are changed to

$$\begin{pmatrix} D_2 \\ D_3 \end{pmatrix} = \varepsilon_0 \begin{pmatrix} \varepsilon_{22}(\omega) & \varepsilon_{23}(\omega) \\ \varepsilon_{23}(\omega) & \varepsilon_{33}(\omega) \end{pmatrix} \begin{pmatrix} E_2 \\ E_3 \end{pmatrix}. \tag{7}$$

Correspondingly, Eqs. (6) become

$$\varepsilon_{22}(\omega) = \varepsilon_{11}^S + \frac{8e_{22}^2/d^2\rho\varepsilon_0}{\omega_L^2 - \omega^2 - i\gamma\omega},$$

$$\varepsilon_{23}(\omega) = -\frac{8e_{22}e_{31}/d^2\rho\varepsilon_0}{\omega_L^2 - \omega^2 - i\gamma\omega}, \quad (8)$$

$$\varepsilon_{33}(\omega) = \varepsilon_{33}^S + \frac{8e_{31}^2/d^2\rho\varepsilon_0}{\omega_L^2 - \omega^2 - i\gamma\omega}.$$

Here, $\omega_L = m\pi v_a/d$ ($m = 1, 3, 5, \ldots$) is the resonance frequency of PSL, and $v_a = \sqrt{C_{11}/\rho}$ is the velocity of the longitudinal acoustic wave. The imaginary part in Eq. (8) implies the damping of the materials, where $\gamma = (\eta_{11}/C_{11})\omega_L^2$. One can see that there are resonances in the dielectric coefficients, indicating the strong response of superlattice vibration to EM waves.

Because of the nondiagonal term $\varepsilon_{23}(\omega)$ in Eq. (7), a coupling between E_2 and E_3 exists in the polariton excitation. The coupling brings the change of principal axes of the dielectric ellipsoid. Here, the principal axis x remains unchanged while the y and z axes rotate an angle $\theta(\omega) = \frac{1}{2}\mathrm{tg}^{-1}\{2\varepsilon_{23}(\omega)/[\varepsilon_{33}(\omega) - \varepsilon_{22}(\omega)]\}$ about the x axis to the y' and z' axes. Corresponding changes appear to the dielectric function and polariton dispersion. Using variable separation, the electric fields can be written in the form $E_j = E_j(x)e^{-i\omega t}$ ($j = 2, 3$). With Maxwell's equations and Eq. (7), we obtain

$$\frac{d^4 E_j(x)}{dx^4} + p\frac{d^2 E_j(x)}{dx^2} + qE_j(x) = 0, \quad (9)$$

where the two coefficients are $p = k_0^2[\varepsilon_{22}(\omega) + \varepsilon_{33}(\omega)]$, $q = k_0^4[\varepsilon_{22}(\omega)\varepsilon_{33}(\omega) - \varepsilon_{23}^2(\omega)]$, and k_0 is the wave vector in the free space. From Eq. (9), we obtain the polariton dispersion relation:

$$k_{\pm}^2 = \frac{\omega^2}{c^2}\varepsilon_{\pm}(\omega), \quad (10)$$

where

$$\varepsilon_{\pm}(\omega) = \frac{\varepsilon_{22}(\omega) + \varepsilon_{33}(\omega)}{2} \pm \sqrt{\left(\frac{\varepsilon_{22}(\omega) - \varepsilon_{33}(\omega)}{2}\right)^2 + \varepsilon_{23}^2(\omega)}. \quad (11)$$

The dielectric function and first-order polariton dispersion ($m = 1$) are plotted, respectively, in Figs. 1 and 2 as a function of normalized frequency (ω/ω_L), where both the real and imaginary parts are included[18]. Figure 1 shows two groups of curves of the dielectric functions, $\varepsilon_+(\omega)$ and $\varepsilon_-(\omega)$. For $\varepsilon_+(\omega)$, the real part $\varepsilon'_+(\omega)$ varies drastically at the resonance and exhibits negative values in the frequency gap (ω_L, ω_O), which is similar to the case where no coupling between E_2 and E_3 exists. Here $\omega_O \approx [1 + 4(K_2^2 + K_3^2)/\pi^2]\omega_L = 1.032\omega_L$, obtained by setting $\varepsilon_+(\omega_O) = 0$, K_2, and K_3 ($K_2 = e_{22}/\sqrt{C_{11}\varepsilon_0\varepsilon_{11}^S}$, $K_3 = e_{31}/\sqrt{C_{11}\varepsilon_0\varepsilon_{33}^S}$), are the electromechanical coupling coefficients. There is a maximum for the imaginary part $\varepsilon''_+(\omega)$, and indicating an absorption peak. While for $\varepsilon_-(\omega)$, $\varepsilon'_-(\omega)$ varies slenderly with the frequency and it is always positive in the considered frequency range; and $\varepsilon''_-(\omega)$ is much smaller and can be ignored. Correspondingly, there are two polariton modes for the coupled waves, one with the wave vector k_+ and the other k_-. The electric field and electric displacement associated with the first mode are both perpendicular to the xz' plane, and this mode can be called the ordinary polariton[8]. One can see from Fig. 2 that k''_+ owns large values near the frequency gap, where the related mode is not supported. The second mode, called the extraordinary polariton mode, has its electric field and electric displacement both in the z' direction. Its wave vector k_- has negligible imaginary parts, which means the propagation of this mode is allowed even in the gap. The results show that the coupling between E_2 and E_3 causes two types of polariton modes with different propagating character.

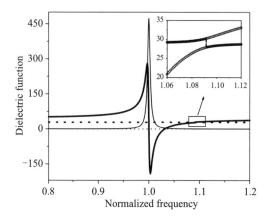

FIG.1. Calculated dielectric functions $[\varepsilon_\pm(\omega)=\varepsilon'_\pm(\omega)+i\varepsilon''_\pm(\omega)]$. The thicker and thinner solid lines represent $\varepsilon'_+(\omega)$ and $\varepsilon''_+(\omega)$; and the thicker and thinner dotted lines represent $\varepsilon'_-(\omega)$ and $\varepsilon''_-(\omega)$, respectively. The insert shows the real parts near the intersection, where the open circles stand for $\varepsilon'_+(\omega)$ and the filled circles $\varepsilon'_-(\omega)$. In the calculation, a proper damping constant $\gamma=0.006\omega_L$ is chosen.

We note that the dielectric function $\varepsilon'_\pm(\omega)$ exhibits a discontinuity near the intersection $\omega_A=1.091\omega_L$. To see this, we rewrite the second term on the right-hand side of Eq.(11) as $\sqrt{F'(\omega)+iF''(\omega)}$. When $F'(\omega)$ is positive and $F''(\omega)$ is changing its sign, a break in the dielectric functions appears. By setting $F''(\omega_A)=0$, we obtain $\omega_A \approx [1+8K_2^2\varepsilon_{11}^S/\pi^2(\varepsilon_{11}^S-\varepsilon_{33}^S)]^{1/2}\omega_L$. This can be understood physically that at the intersection the rotation angle $\theta(\omega)$ changes from $-\pi/4$ to $\pi/4$, resulting in the exchange of y' and z' axes. The discontinuity is also displayed in the polariton dispersion curves, correspondingly.

The EM fields consist of a linear combination of two polariton modes. With Maxwell's equations, we have

$$E_2(x)=a\mathrm{e}^{ik_+x}+b\mathrm{e}^{ik_-x},$$

$$E_3(x) = \frac{\varepsilon_+(\omega) - \varepsilon_{22}(\omega)}{\varepsilon_{23}(\omega)} a e^{ik_+ x} + \frac{\varepsilon_-(\omega) - \varepsilon_{22}(\omega)}{\varepsilon_{23}(\omega)} b e^{ik_- x},$$

$$H_2(x) = \frac{\varepsilon_{22}(\omega) - \varepsilon_+(\omega)}{\mu_0 c \varepsilon_{23}(\omega)} \sqrt{\varepsilon_+(\omega)} a e^{ik_+ x} + \frac{\varepsilon_{22}(\omega) - \varepsilon_-(\omega)}{\mu_0 c \varepsilon_{23}(\omega)} \sqrt{\varepsilon_-(\omega)} b e^{ik_- x},$$

$$H_3(x) = \frac{\sqrt{\varepsilon_+(\omega)}}{\mu_0 c} a e^{ik_+ x} + \frac{\sqrt{\varepsilon_-(\omega)}}{\mu_0 c} b e^{ik_- x}. \qquad (12)$$

Where, $k_\pm = (\omega/c)\sqrt{\varepsilon_\pm(\omega)}$. The coefficients a, b can be determined with the boundary conditions that $E_2(0) = E_{20}$, and $E_3(0) = 0$,

$$a = \frac{\varepsilon_{22}(\omega) - \varepsilon_-(\omega)}{\varepsilon_+(\omega) - \varepsilon_-(\omega)} E_{20}, \quad b = \frac{\varepsilon_+(\omega) - \varepsilon_{22}(\omega)}{\varepsilon_+(\omega) - \varepsilon_-(\omega)} E_{20}. \qquad (13)$$

We define the transmission of the electric fields as $t_{2,3}(L) = |E_{2,3}(L)/E_{20}|$. The dependence of $t_2(L)$ and $t_3(L)$ on the frequency is shown in Fig. 3(a), where the PSL has a length $L = 4000d$.

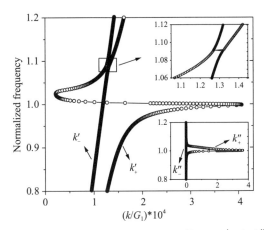

FIG.2. First-order polariton dispersion curves $[k_\pm = k'_\pm + i k''_\pm]$. The lower insert represents the imaginary parts of the polariton dispersion, and the upper one shows real parts near the intersection. The ordinary and extraordinary polariton are denoted by the open and filled circles, respectively, $(G_1 = \pi/d)$.

The total power flow density(PFD) of the coupled waves includes that of superlattice vibration $\overline{P}_{SL} = (-1/2) \times \text{Re}(v_1^* \cdot T_1)$ and EM waves $\overline{P}_{EM} = \overline{P}_2 + \overline{P}_3$[17]. Where $v_1 = \partial u_1/\partial t$ is the velocity of vibration of the medium, $\overline{P}_2 = (1/2)\text{Re}(E_2 \times H_3^*)$, and $\overline{P}_3 = (1/2)\text{Re}(E_3 \times H_2^*)$. A detailed calculation shows

$$v_1^* \cdot T_1 = -2i\rho\omega^3 G_m \left| \frac{(e_{22}E_2 - e_{31}E_3)f_m}{C_{11}^E G_m^2 - \rho\omega^2} \right|^2 \sin(2G_m x). \qquad (14)$$

The complex PFD of the superlattice vibration is a purely imaginary number. Therefore $\overline{P}_{SL}$ is zero, which agrees with the character of the stationary wave.

The PFD of EM waves can be calculated with Eqs.(12). When the frequency is far from the resonance(the dielectric coefficients and wave vectors thus can be treated as real numbers, approximately), $\overline{P}_2(x)$ and $\overline{P}_3(x)$ oscillate with a period of $2\pi/(k_+-k_-)$ with the length. And the total PFD is constant in the propagation direction, that is,

$$\overline{P}_{EM} = \frac{\varepsilon_{22}(\omega) + \sqrt{\varepsilon_+(\omega)\varepsilon_-(\omega)}}{2\mu_0 c[\sqrt{\varepsilon_+(\omega)} + \sqrt{\varepsilon_-(\omega)}]} |E_{20}|^2. \qquad (15)$$

When the frequency is close to the resonance, the attenuation of PFD becomes significant. The variation of PFD with the frequency is shown in Fig. 3(b). If we neglect the imaginary part of $\varepsilon_-(\omega)$ and the length of PSL is assumed to be infinite, the PFD of EM waves can be approximated as $\overline{P}_2(\infty) = (1/2)\varepsilon_0 c \sqrt{\varepsilon_-(\omega)} |b|^2$, and $\overline{P}_3(\infty) = |[\varepsilon_{22}(\omega) - \varepsilon_-(\omega)]/\varepsilon_{23}(\omega)|^2 \overline{P}_2(\infty)$. Hence a portion of energy can still travel through the PSL even in the gap.

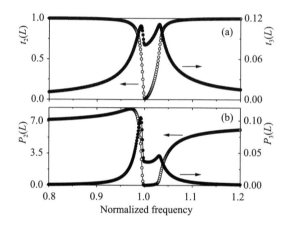

FIG. 3. Variation of the electric fields (a), and the PFD (normalized to $\frac{1}{2}\varepsilon_0 c |E_{20}|^2$) (b) with the frequency. The energy is transferred between E_2 and E_3 due to piezoelectric-induced coupling.

The interaction between wave and matter results in many elementary excitations. The polariton is one of them. In this Letter, the superlattice vibration due to piezoelectric effect bridges the coupling of two orthogonally polarized EM waves. This coupling results in two types of polariton modes, one is supported in the band gap and the other prohibited. We have noticed that in the literature two interesting cases occurred in the stop band. In one case, the coupled-thin-film structure can support guided-wave plasmon polaritons[19]; and in the other, the propagation of a probe light beam is allowed within the polariton stop band via exciton-biexciton coupling[20]. Our case presents another example, where the phonon polariton is involved. This unusual coupling effect is not present in real lattices.

References and Notes

[1] C. Kittel, *Introduction to Solid State Physics* (Wiley, New York, 1986), 6th ed.
[2] E. Yablonovitch, Phys. Rev. Lett. **58**, 2059(1987).
[3] S. John, Phys. Rev. Lett. **58**, 2486(1987).

[4] J. J. Joannopoulos, R. D. Meade, and J. N. Winn, *Photonic Crystals* (Princeton University, Princeton, NJ, 1995).

[5] M. Kafesaki, M. M. Sigalas, and N. Garcia, Phys. Rev. Lett. **85**, 4044(2000).

[6] Z. Liu, X. Zhang, Y. Mao, Y. Y. Zhu, Z. Yang, C. T. Chan, and P. Sheng, Science **289**, 1734(2000).

[7] M. Torres, F.R. Montero de Espinosa, and J.L. Aragón, Phys. Rev. Lett. **86**, 4282(2001).

[8] D. L. Mills and E. Burstein, Rep. Prog. Phys. **37**, 817(1974).

[9] M. Born and K. Huang, *Dynamical Theory of Crystal Lattice* (Clarendon, Oxford, 1954).

[10] R. Tsu and S. S. Jha, Appl. Phys. Lett. **20**, 16(1972).

[11] D. Feng, N. B. Ming, J. F. Hong, J. S. Zhu, Z. Yang, and Y. N. Wang, Appl. Phys. Lett. **37**, 607 (1980).

[12] M. Yamada, N. Nada, M. Saitoh, and K. Watanabe, Appl. Phys. Lett. **62**, 435(1993).

[13] Y. Q. Lu, Y. Y. Zhu, Y. F. Chen, S. N. Zhu, N. B. Ming, and Y. J. Feng, Science **284**, 1822(1999).

[14] J. Chen and J. B. Khurgin, Appl. Phys. Lett. **81**, 4742(2002).

[15] Y. Y. Zhu, X. J. Zhang, Y. Q. Lu, Y. F. Chen, S. N. Zhu, and N. B. Ming, Phys. Rev. Lett. **90**, 053903(2003).

[16] X. J. Zhang, R. Q. Zhu, J. Zhao, Y. F. Chen, and Y. Y. Zhu, Phys. Rev. B **69**, 085118(2004).

[17] B. A. Auld, *Acoustic Fields and Waves in Solids* (Wiley, New York, 1973).

[18] For the LiNbO$_3$ crystal, we use $C_{11}=2.03\times10^{11}$ N/m^2, $\varepsilon_{11}^S=44, \varepsilon_{33}^S=29, e_{22}=2.5$ C/m^2, $e_{31}=0.2$ C/m^2, $\rho=4.64\times10^3$ kg/m^3; (see Ref. [17]).

[19] M. A. Gilmore and B. L. Johnson, J. Appl. Phys. **93**, 4497(2003).

[20] S. Chesi, M. Artoni, G. C. La Rocca, F. Bassani, and A. Mysyrowicz, Phys. Rev. Lett. **91**, 057402 (2003).

[21] This work was supported by the State Key Program for Basic Research of China (Grant No. 2004CB619003), by the National Natural Science Foundation of China (Grant Nos. 60378017 and 10474042) and by the Natural Science Foundation of Jiangsu (Grant No. BK2004209).

Phonon-polaritons in Quasiperiodic Piezoelectric Superlattices*

Xue-jin Zhang, Yan-qing Lu, Yong-yuan Zhu, Yan-feng Chen, and Shi-ning Zhu

National Laboratory of Solid State Microstructures, Nanjing University, Nanjing 210093, P. R. China

> Phonon-polaritons are studied both theoretically and experimentally in a one-dimensional two-component generalized quasiperiodic piezoelectric superlattice. The experimental observation of phonon-polaritons through dielectric abnormality is carried out at the microwave region. Some potential applications are discussed.

As an elementary excitation in solid-state physics, the polariton is due to the coupling between the photon and the polar elementary excitation. Owing to the unusual properties, the polaritons are of great interest from both a fundamental and an applied perspective. Recently, a polariton laser based on exciton-polaritons has been demonstrated in a semiconductor microcavity.[1] As for the phonon-polariton,[2,3] the rapidly varying refractive index is made use of in constructing prisms for infrared spectroscopy.[4] Ensued from the study of artificial microstructure materials, much effort has been devoted to the research on the phonon-polariton. In the periodic superlattice, a periodic potential with a giant period, in contrast with the atomic period, results in the formation of the miniature Brillouin zone. By virtue of this, the far-infrared Raman laser and Reststrahlen filter made of AlAs/GaAs superlattices become realizable.[5] Another property of the phonon-polariton, significantly reduced group velocity, can be utilized in solid-state traveling wave devices.[6] Moreover, the photonic band gap (PBG), in which the propagation of electromagnetic(EM) waves is forbidden, will be affected by the presence of the phonon-polariton in photonic crystals composed of polar materials.[7-9] It is shown that the phonon-polariton coupling flattens the photonic bands and is favorable for opening up the PBG.

On the other hand, the discovery of quasicrystals has fired up a new field of condensed-matter physics and given rise to many practical applications since 1984.[10] For example, the multiwavelength second-harmonic generation and the direct third-harmonic generation have been realized in a Fibonacci superlattice.[11,12] In the field of photonic crystals, the complete PBG in 12-fold symmetric quasicrystals has been reported.[13] Furthermore, compared with a one-dimensional (1D) two-component Fibonacci superlattice, a 1D two-component generalized quasiperiodic superlattice (GQPSL)

---

* Appl.Phys. Lett., 2004, 85(16):3531

possesses more freedom for applications.[14]

In this Letter, we investigate the influence of structural variation of the piezoelectric superlattice upon the phonon-polaritons on the basis of a 1D two-component GQPSL. We measured the dielectric function of the GQPSL at the microwave region, thereby obtaining the physical information required to deduce the properties of polaritons.[15] The possible applications are discussed.

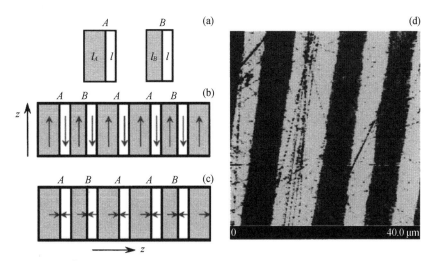

FIG.1. (Color online) Schematic of a 1D two-component GQPSL structure(a) composed of two building blocks A and B, (b) with side-by-side 180° reversed ferroelectric domains, or (c) with head-to-head 180° reversed ferroelectric domains. The arrows indicate the directions of spontaneous polarization. (d) Piezoresponse scanning force microscopy of a GQPSL based on the $LiTaO_3$ single crystal. It is the image of phase response.

In the absence of translational symmetry, the Bloch theory is no longer adequate for quasiperiodic structures. But according to the projection method,[16] the 1D quasiperiodic structure may be considered as the projection of a two-dimensional(2D) periodic structure. As an example, we consider a GQPSL with a $3m$ point group which consists of two building blocks, A and B, with each block made up of one positive and one negative ferroelectric domain(a ferroelectric material is piezoelectric). The widths of blocks A and B are l_A and l_B, respectively. We assume that the negative domain of blocks A and B has the same width l, shown in Fig. 1(a). Different from the Fibonacci sequence, the projection angle becomes an adjustable structure parameter in the GQPSL, and its tangent is not fixed as a golden ratio, i.e., $(1+\sqrt{5})/2$. Figures 1(b) and 1(c) show two schematic diagrams of the GQPSL structures: one is the so-called side-by-side configuration, the other the head-to-head configuration.[17] In both cases, the piezoelectric coefficient, as an odd-rank tensor, will change signs for domains with different spontaneous polarization directions, which arouses quasiperiodically modulated piezoelectric coefficients in the GQPSL.

With a GQPSL aligning along the x axis in Fig. 1(b), a vertically incident y-polarized EM wave propagates into it on the left-hand side surface. The properties of the phonon-polariton can be obtained from the following piezoelectric and motion equations:[18]

$$\begin{aligned} T_1(x,t) &= C_{11}^E S_1(x,t) + e_{22}(x) E_2(x,t), \\ P_2(x,t) &= -e_{22}(x) S_1(x,t) + \varepsilon_0(\varepsilon_{11}^S - 1) E_2(x,t), \\ \rho \frac{\partial^2 S_1(x,t)}{\partial t^2} &= \frac{\partial T_1(x,t)}{\partial x^2}, \end{aligned} \quad (1)$$

where T_1, S_1, E_2, and P_2 are the stress, strain, electric field, and polarization, respectively. C_{11}^E, $e_{22}(x)$, ε_{11}^S, and ρ are the elastic coefficient, piezoelectric coefficient, dielectric coefficient, and mass density, respectively. The damping of materials has been omitted here. The second equation of Eq. (1) implicates that a longitudinal wave S_1 introduces a transverse electric polarization P_2, which can interfere with the EM wave. For an infinite GQPSL structure, the quasiperiodically modulated piezoelectric coefficient $e_{22}(x)$ can be written, using Fourier transformation,[19] as $e_{22}(x) = e_{22} f(x) = e_{22} \sum_{m,n} f_{m,n} \exp(iG_{m,n}x)$ with

$$f_{m,n} = \frac{2(1+\tau)l}{D} \frac{\sin(G_{m,n}l/2)}{G_{m,n}l/2} \frac{\sin X_{m,n}}{X_{m,n}}, \quad (2)$$

where $G_{m,n} = 2\pi(m\tau + n)/D$ is the reciprocal vector (m, n are two integers), $D = \tau l_A + l_B$ is the average structure parameter of the GQPSL, $X_{m,n} = \pi(l+\tau)(nl_A + ml_B)/D$, $\tau = \tan\theta$, and θ is the adjustable projection angle. Using Eq. (1), the average dielectric function $\varepsilon_2(k,\omega)$ of the GQPSL is found to be

$$\varepsilon_2(k,\omega) = \varepsilon_{11}^S - \frac{e_{22}^2}{\varepsilon_0 L} \sum_{m,n,m',n'} f_{m,n} f_{m',n'} \times \frac{G_{m,n}^2 + 2G_{m,n}k + k^2}{\rho\omega^2 - C_{11}^E(G_{m,n} + k)^2} \int_0^L e^{i(G_{m,n}+G_{m',n'})x} dx, \quad (3)$$

where k and ω are the wave vector and angular frequency of EM waves. L is the whole length of the GQPSL, and m', n' are also two integers. From Eq.(3) we can see that the resonance, which engenders phonon-polaritons, will occur around $\omega^2 = G_{m,n}^2 v^2$ [$v = (C_{11}^E/\rho)^{1/2}$ is the phase velocity of the longitudinal superlattice vibration]. Hence, the positional distribution of resonance peaks reflects the quasiperiodicity of the GQPSL structure.

According to Maxwell's relation, the dispersion relation of the phonon-polariton in GQPSL can be gotten easily. That is,

$$c^2 k^2/\omega^2 = \varepsilon_2(k,\omega), \quad (4)$$

where c is the phase velocity of the EM wave in free space.

In the very long wavelength region, the wave vector is very close to zero, and to a good approximation the dielectric function can be considered to depend only upon the frequency of waves. Taking the damping of the material into account, Eq. (3) can be changed to the form

$$\varepsilon_2(\omega) = \varepsilon_{11}^S \left[1 - \frac{K^2}{L} \sum_{m,n,m',n'} \frac{f_{m,n} f_{m',n'} \omega_{m,n}^2}{\omega^2 - \omega_{m,n}^2 + i\omega\gamma} \times \int_0^L e^{i(G_{m,n}+G_{m',n'})x} dx \right], \quad (5)$$

where $K^2 = e_{22}^2/(C_{11}^E \varepsilon_0 \varepsilon_{11}^S)$ is an electromechanical coupling coefficient, $\omega_{m,n}^2 = G_{m,n}^2 v^2$, $\omega_{m,n}$ is resonance frequencies of longitudinal superlattice vibrations in the GQPSL, i.e., the frequency positions of phonon-polaritons, and γ represents a damping constant with the dimension of frequency.

The GQPSL based on the congruent LiTaO$_3$ single crystal is fabricated by the method of electric-field poling.[20] The GQPSL used in our experiment is 640 blocks with a total length of about 8.2 mm along the x axis. The other structure parameters are $\tau=0.5645$, $l=5.75$ μm, $l_A=14.41$ μm, and $l_B=10.04$ μm respectively. The widths of domains after poling are determined with piezoresponse scanning force microscopy, in which the inverse piezoelectric effect is used. The z surface of the GQPSL was polished, and the phase response was recorded, as shown in Fig. 1(d). In this Letter, the material constants of the congruent LiTaO$_3$ crystal are selected from Ref. [21]. The dispersion relation of the phonon-polariton in the GQPSL calculated from Eq.(4) is shown in Fig. 2(a). Figure 2(b) shows that for a phonon-polariton in the GQPSL, there are three branches forming two separate band gaps, while in the ionic crystal there are only two branches with one band gap. There are not only transverse phonon-polaritons, but also longitudinal phonon-polaritons in piezoelectric superlattices.[22,23] As shown above, the phonon-polaritons in this case are longitudinal phononpolaritions. When a y-polarized EM wave propagates along the x axis of the GQPSL, it will be strongly reflected as long as its frequency lies in the band gap of the phonon-polariton. By use of this property, the reflector and polarizer based on the phonon-polariton can be made.

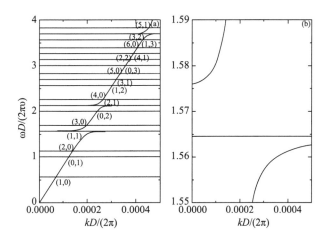

FIG.2. Normalized dispersion curves of (a) the longitudinal phonon-polaritons of the GQPSL and (b) the enlarged one labeled as (1,1).

The frequency distribution of the phonon-polaritons exhibits a quasiperiodic property of the GQPSL, in accordance with the reciprocal vector. In the periodic structure, the frequency positions of phonon-polaritons are equidistant.[23] Besides the piezoelectric coefficient, Eq.(3) also tells us that the Fourier coefficients are responsible for the size of band gaps, correlated with the coupling intensity. As we have seen, many adjustable parameters allow one to conveniently tune the frequency positions of phonon-polaritons and the magnitude of the Fourier coefficients according to specific applications. For example, the arbitrarily adjustable projection angle enables us to achieve tunable channels with a narrow wavelength interval in wavelength division multiplexing(WDM) devices. The major property of the phonon-polariton is prohibiting simultaneously incident EM waves or sound waves within some special frequency ranges from penetrating through materials. This makes it possible that the phonon-polariton material can be used as the barrier for harmful EM waves or noises. It is also beneficial to miniaturization of devices as the wavelength of EM waves is much larger than the size of building blocks, whereas they should be comparable with each other in photonic materials.

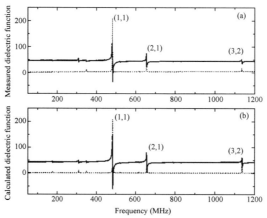

FIG.3. The dielectric function curves for the GQPSL. (a) The measured results and (b) the calculated results choosing damping constant $\gamma = 0.003\ \omega_{1,1}$. The solid lines represent the real part, and the dashed lines the imaginary part.

The experimental dielectric function of the GQPSL measured with an HP4291B RF impedance/material analyzer is shown in Fig. 3(a). There is a one-to-one correspondence between the frequency position and magnitude of abnormal dielectric function peaks in Fig. 3(a), and the frequency position and the band-gap size of phonon-polaritons in Fig. 2, especially between the three most intense peaks and the three largest band gaps, labeled as (1,1),(2,1), and(3,2). In order to simulate the experimental dielectric function, the damping constant γ of the GQPSL must be given beforehand. By virtue of the damped harmonic-oscillator model, the damping constant γ approximates to either the full width at half maximum of a peak in the imaginary part of the dielectric function, or the difference between the two frequencies corresponding to the maximum and minimum around a point of abnormality in the real part of the dielectric function. From the measured dielectric function peaks the damping constant y is approximately equal to $0.003\ \omega_{1,1}$. Figure 3(b) shows the calculated dielectric function of the GQPSL. The profiles of theoretical curves are quite similar to those measured.

The frequency region for phonon-polaritons of the GQPSL falls into the microwave band, which is determined by its average structure parameter. It is noteworthy that the epitaxial technique, such as magnetron sputtering, can bring the frequency of phonon-polaritons up to the far-infrared region by adjusting the average structure parameter of the GQPSL. Then, the frequeny of phonon-polaritons can span the operational range of most optoelectronic devices.

In conclusion, the phonon-polariton in the 1D two-component GQPSL was studied in theory and experiment. The quasiperiodic modulation gives rise to the quasiperiodic frequency distribution of phonon-polaritons. If the piezoelectric coefficient is modulated aperiodically, it will introduce aperiodical frequency distribution of phonon-polaritons. This property can lead to the practical applications, such as for WDM devices in optical communications, and for EM wave (or sound wave)-proof materials in environmental protection.

References and Notes

[1] G. Weihs, H. Deng, R. Huang, M. Sugita, F. Tassone, and Y. Yamamoto, Semicond. Sci. Technol. **18**, S386(2003).
[2] C. Kittle, *Introduction to Solid State Physics*, 6th ed.(Wiley, New York, 1986).
[3] M. Born and K. Huang, *Dynamical Theory of Crystal Lattice*(Clarendon, Oxford, 1954).
[4] J. R. Hook and H. E. Hall, *Solid State Physics*, 2nd ed.(Wiley, New York, 1991).
[5] R. Tsu and S. S. Jha, Appl. Phys. Lett. **20**, 16(1972).
[6] T. K. Ishii, *Practical Microwave Electron Devices*(Academic, San Diego, 1990).
[7] M. M. Sigalas, C. M. Soukoulis, C. T. Chan, and K. M. Ho, Phys. Rev. B **49**, 11080(1994).
[8] W. Zhang, A. Hu, X. Lei, N. Xu, and N. B. Ming, Phys. Rev. B **54**, 10280(1996).
[9] K. C. Huang, P. Bienstman, J. D. Joannopoulos, K. A. Nelson, and S. Fan, Phys. Rev. B **68**, 075209 (2003).
[10] D. Shechtman, I. Blench, D. Gratias, and J. W. Cahn, Phys. Rev. Lett. **53**, 1951(1984).
[11] S. N. Zhu, Y. Y. Zhu, Y. Q. Qin, H. F. Wang, C. Z. Ge, and N. B. Ming, Phys. Rev. Lett. **78**, 2752 (1997).
[12] S. N. Zhu, Y. Y. Zhu, and N. B. Ming, Science **27**, 8843(1997).
[13] M. E. Zoorob, M. D. B. Charlton, G. J. Parker, J. J. Baumberg, and M. C. Netti, Nature(London) **404**, 740(2000).
[14] C. Zhang, H. Wei, Y. Y. Zhu, H. T. Wang, S. N. Zhu, and N. B. Ming, Opt. Lett. **26**, 899(2001).
[15] D. L. Mills and E. Burstein, Rep. Prog. Phys. **37**, 817(1974).
[16] R. K. P. Zia and W. J. Dallas, J. Phys. A **18**, L341(1985).
[17] Y. Y. Zhu, S. N. Zhu, Y. Q. Qin, and N. B. Ming, J. Appl. Phys. **79**, 2221(1996).
[18] B. A. Auld, *Acoustic Fields and Waves in Solids*(Wiley, New York, 1973).
[19] Y. Y. Zhu and N. B. Ming, Opt. Quantum Electron. **31**, 1093(1999).
[20] M. Yamada, N. Nada, M. Saitoh, and K. Watanabe, Appl. Phys. Lett. **62**, 435(1993).
[21] A. W. Warner, M. Onoe, and G. A. Coquin, J. Acoust. Soc. Am. **42**, 1223(1967).
[22] Y. Y. Zhu, X. J. Zhang, Y. Q. Lu, Y. F. Chen, S. N. Zhu, and N. B. Ming, Phys. Rev. Lett. **90**, 053903(2003).

[23] X. J. Zhang, R. Q. Zhu, J. Zhao, Y. F. Chen, and Y. Y. Zhu, Phys. Rev. B **69**, 085118(2004).
[24] The authors gratefully acknowledge Professor Xiao-mei Lu and Dr. Peng Bao in the experiments. This work was supported by the National Natural Science Foundation of China(No. 60378017) and by the Natural Science Foundation of Jiangsu Province(No. BK2004209).

Coupling of Electromagnetic Waves and Superlattice Vibrations in a Piezomagnetic Superlattice: Creation of a Polariton Through the Piezomagnetic Effect[*]

H. Liu, S. N. Zhu, Z. G. Dong, Y. Y. Zhu, Y. F. Chen, and N. B. Ming

Department of Physics, National Laboratory of Solid State Microstructures,

Nanjing University, Nanjing 210093, People's Republic of China

X. Zhang

Mechanical and Aerospace Engineering, University of California,

Los Angeles, Los Angeles, California 90095, USA

We studied the propagation of an electromagnetic (EM) wave in a piezomagnetic superlattice with piezomagnetic coefficient being modulated. Because of the piezomagnetic effect, the coupling between the EM wave and vibration of superlattice can be established, resulting in the creation of a type of magnetic polariton that does not exist in ordinary magnetic material. At some resonance frequencies, the abnormality of dispersion of permeability introduces negative value and piezomagnetic superlattice can make a kind of negative permeability material.

The concept of negative permeability μ is of particular interest, not only because this is a regime not observed in ordinary materials, but also because such a medium can be combined with a negative permittivity ε to form a "left-handed" material (i.e., $\boldsymbol{E} \times \boldsymbol{H}$ lies along the direction of $-\boldsymbol{k}$ for propagating plane waves).[1-3] In recent work,[4] Pendry *et al.* have introduced the split ring resonator (SRR) medium whose dominant behavior can be interpreted as having an effective magnetic permeability. By making the constituent units resonant, the magnitude of μ is enhanced considerably, leading to a large positive effective μ near the low-frequency side of the resonance and, most strikingly, negative μ near the high-frequency side of the resonance. More recently, a magnetic response at terahertz frequencies has been achieved in a plannar structure composed of SRR elements.[5]

On the other hand, in artificial composites such as superlattices, the periodic modulation of related physical parameters may result in some coupling effects. For example, associated with the variation of dielectric constants is the photonic crystal[6-8] and the modulation of nonlinear optical coefficients results in a quasi-phase-matched frequency conversion.[9,10] Recently, a coupling between the superlattice vibrations and the electromagnetic (EM) wave was established in piezoelectric superlattices,[11-13] in which the

---

[*] Phys.Rev.B,2005,71(12):125106

piezoelectric coefficient is modulated. This coupling produced a new type of polariton, in which the resonance frequency was determined by the period of superlattice and the negative effective permittivity $\varepsilon_{eff}(\omega)$ can be got near the high-frequency side of the resonance.

In the present work, we will construct another kind of periodic structure, the piezomagnetic superlattice, with the piezomagnetic coefficient being modulated. In this system, the superlattice vibrations will induce spin waves due to the piezomagnetic effect. The lateral spin waves in turn will emit EM waves that interfere with the original EM wave. In such a case, the lattice vibration will couple strongly with the EM wave and result in polariton excitation. Near the piezomagnetic polariton resonance, a negative $\mu_{eff}(\omega)$ will be shown possible. Therefore, a kind of negative permeability material can be constiuted in piezomagnetic superlattice.

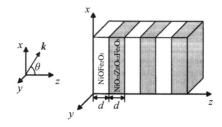

FIG.1. Schematic illustration of the piezomagnetic superlattice.

In order to elucidate the above idea, let us consider a one-dimensional(1D) periodic structure composed of alternating layers of $NiOFe_2O_3$ and $NiO_{0.8}ZnO_{0.2}Fe_2O_3$ along the z axis. Through changing the doping contents of NiO and ZnO, a period superlattice can be attained which is shown in Fig. 1. As illustrated in Table 1,[14] a different doping ratio introduces a different piezomagnetic coefficient. Therefore, the piezomagnetic coefficient is modulated periodically in the superlattice along the z axis. The piezomagnetic tensor matrix has the form

$$(q_{iJ}) = \begin{bmatrix} 0 & 0 & 0 & 0 & q_{15}(z) & 0 \\ 0 & 0 & 0 & q_{15}(z) & 0 & 0 \\ q_{31}(z) & q_{31}(z) & q_{33}(z) & 0 & 0 & 0 \end{bmatrix}. \quad (1)$$

Here, Voigt's notation is used in the representation of three-order tensor $q_{ijk} \rightarrow q_{iJ}(j, k \rightarrow J)$, which is specified in Table 2. In the layers $NiOFe_2O_3$ ($0 \leqslant z \leqslant d$), $q_{33}(z) = -212.4$ N/Am and $q_{31}(z) = -96.7$ N/Am, while in the layers $NiO_{0.8}ZnO_{0.2}Fe_2O_3$ ($d \leqslant z < \Lambda$), $q_{33(Z)} = -285.6$ N/Am and $q_{31}(z) = -125.7$ N/Am. The modification is along the z axis; the reciprocal vectors for periodic structure should be $\boldsymbol{G} = (0, 0, G_m)$ ($G_m = m\pi/d, m = 1, 3, 5, \ldots$) Also we assume that the transverse dimensions are very large compared with an acoustic wavelength so that a one-dimensional model is applicable. When an EM wave is radiated into this piezomagnetic superlattice, an acoustic wave will be excited through the piezomagnetic effect. The coupled interaction between the EM wave and acoustic wave can be described by the constitutive equations:

$$T_{ij} = c_{ijkl}\frac{\partial u_k}{\partial x_l} + q_{ijk}(z)H_k, B_i = \mu_{ij}^S H_j - q_{ijk}(z)\frac{\partial u_j}{\partial x_k}, \quad (2)$$

where T_{ij}, u, H, B, c_{ijkl}, and μ_{ij}^S are the stress tensor, displacement field, magnetic field, magnetic displacement, and the elastic and static permeability, respectively. With the use of Newton's law $\rho \partial^2 u_j / \partial t^2 = (\partial / \partial x_i) T_{ij}$, the equation of motion for a vibrating medium can be obtained:

$$\rho \frac{\partial^2 u_j}{\partial t^2} - c_{ijkl}\frac{\partial^2 u_k}{\partial x_i \partial x_l} = \frac{\partial (q_{ijk}(z)H_k)}{\partial x_i}, \quad (3)$$

where ρ is the mass density and c_{ijkl} is the stiffness tensor of superlattice. By using the Fourier transformation:

$$u_j = \int \widetilde{u}_j e^{i(\omega t - q \cdot r)} dq, H_k = \int \widetilde{H}_k e^{i(\omega t - k \cdot r)} dk,$$

$$q_{ijk}(z) = \sum_G \widetilde{q}_{ijk}(G) e^{-iG \cdot r}, \quad (4)$$

where q is the phonon wave vector, k is the electromagnetic wave vector, and G is the reciprocal vector of modulated structures, Eq. (2) can be expressed as

$$\int (-\rho \omega^2 \delta_{jk} + c_{ijkl} q_i q_l) \widetilde{u}_k e^{i(\omega t - q \cdot r)} = \sum_{\widetilde{G}} \int (-i)(k+G)_i \widetilde{q}_{ijk} \widetilde{H}_k e^{i[\omega t - (k+G) \cdot r]} dk. \quad (5)$$

TABLE 1. Piezomagnetic properties of the $NiO_x ZnO_{1-x} Fe_2 O_3$ superlattice.

$NiO, ZnO, Fe_2 O_3$	15:35:50	18:32:50	25:25:50	32:18:50	40:10:50	50:0:50
q_{33} (N/Am)	−113.1	−78.5	−128.2	−144.5	−212.4	−285.6
q_{31} (N/Am)	−44.1	−31.8	−61.9	−79.5	−96.7	−125.7

TABLE 2. Voigt's notation used in representation of $q_{ijk} \rightarrow q_{iJ}$.

(j,k)	(1,1)	(2,2)	(3,3)	(2,3) (3,2)	(1,3) (3,1)	(1,2) (2,1)
J	1	2	3	4	5	6

For photons with long wavelength, $|k| \ll G$, Eq.(5) becomes

$$\int (-\rho \omega^2 \delta_{jk} + c_{ijkl} q_i q_l) \widetilde{u}_k e^{i(\omega t - q \cdot r)} = \sum_G (-i) G_i \widetilde{q}_{ijk} \widetilde{H}_k e^{i[\omega t - (k+G) \cdot r]}. \quad (6)$$

In order for the two sides of Eq. (5) to be equal, q must satisfy $q = k + G \approx G$ for one of the reciprocal vectors of the periodic superlattice. Then we obtain, from Eq. (6),

$$(-\rho \omega^2 \delta_{jk} + c_{ijkl} G_i G_l) \widetilde{u}_k (q = k + G) = (-i) G_i \widetilde{q}_{ijk} \widetilde{H}_k \quad (7)$$

and we have

$$\widetilde{u}_k = i(\rho \omega^2 \delta_{jk} - c_{ijkl} G_i G_l)^{-1} G_i \widetilde{q}_{ijk} \widetilde{H}_k. \quad (8)$$

Then,

$$\frac{\partial u_k}{\partial x_j} = e^{-iG \cdot r} G_j (\rho \omega^2 \delta_{jk} - c_{ijkl} G_i G_l)^{-1} G_i \widetilde{q}_{ijk} H_k(r,t). \quad (9)$$

Substituting Eq. (9) into (2) and using the space average value, we get
$$B_i = \mu_{ik}(\omega) H_k, \tag{10}$$
where
$$\mu_{ik}(\omega) = \mu_{ik}^S + \tilde{q}_{ikj} G_j (-\rho\omega^2 \delta_{jk} + c_{ijkl} G_i G_l)^{-1} G_i \tilde{q}_{ijk}. \tag{11}$$
If we choose the first reciprocal vector $\boldsymbol{G} = (0, 0, G_1 = \pi/d)$, the permeability coefficients matrix can be attained as
$$\overleftrightarrow{\mu}(\omega) = \begin{bmatrix} \mu_\perp(\omega) & 0 & 0 \\ 0 & \mu_\perp(\omega) & 0 \\ 0 & 0 & \mu_\parallel(\omega) \end{bmatrix}. \tag{12}$$
Here,
$$\mu_\perp(\omega) = \mu_{11}^S \frac{\omega_{\perp,0}^2 - \omega^2}{\omega_\perp^2 - \omega^2}, \mu_\parallel(\omega) = \mu_{33}^S \frac{\omega_{\parallel,0}^2 - \omega^2}{\omega_\parallel^2 - \omega^2}, \tag{13}$$
where $\omega_\perp = G_1 \sqrt{c_{44}/\rho}$, $\omega_\parallel = G_1 \sqrt{c_{33}/\rho}$, $\omega_{\perp,0} = \sqrt{\omega_\perp^2 + \tilde{q}_{31}^2 G_1^2/(\rho\mu_{11}^S)}$, and $\omega_{\parallel,0} = \sqrt{\omega_\parallel^2 + \tilde{q}_{33}^2 G_1^2/(\rho\mu_{33}^S)}$. $\omega_\perp$ and $\omega_\parallel$ are the resonance frequency of the longitudinal vibration and transverse vibration due to piezomagnetic effect. Equation (13) exhibits the resonance structure produced in the piezomagnetic superlattice which profoundly affects the propagation of electromagnetic waves with frequency near $\omega_\perp$ and $\omega_\parallel$.

The effective dielectric tensor of piezomagnetic superlattice can be derived from the average-field method. Straightforward algebra gives the effective dielectric tensor in the well-known form[15]
$$\overleftrightarrow{\varepsilon} = \begin{bmatrix} \varepsilon_\perp & 0 & 0 \\ 0 & \varepsilon_\perp & 0 \\ 0 & 0 & \varepsilon_\parallel \end{bmatrix}. \tag{14}$$
In the discussion that follows, we shall assume for simplicity that the dielectric constant of the superlattice is independent of frequency near the magnetic resonance frequencies $\omega_\perp$ and $\omega_\parallel$.

The dispersion relation for polariton propagating in piezomagnetic superlattices follows from Maxwell's equations. After eliminating $\boldsymbol{E}$ in the curl equations, one obtains the following wave equation for $\boldsymbol{H}$:
$$\boldsymbol{k} \times [\overleftrightarrow{\varepsilon}^{-1} \cdot (\boldsymbol{k} \times \boldsymbol{H})] + \frac{\omega^2}{c^2} \overleftrightarrow{\mu}(\omega) \cdot \boldsymbol{H} = 0, \tag{15}$$
where c is the electromagnetic wave velocity in vacuum, ω the angular frequency, and $\overleftrightarrow{\varepsilon}^{-1}$ the inverse of Eq.(14). Since the susceptibility tensor is isotropic in the xy plane, there is no loss of generality if we choose the wave vector $\boldsymbol{k}$ in the xz plane, as shown in Fig. 1. The angle between the wave vector and the z axis will be denoted by θ, so that
$$k_x = k \sin\theta, k_z = k \cos\theta. \tag{16}$$
Then, if we proceed by the use of Eq.(15), we have
$$\left(\frac{\omega^2}{c^2 k^2} \mu_\perp \varepsilon_\perp - \cos^2\theta \right) H_x + \sin\theta \cos\theta H_z = 0, \tag{17a}$$

$$\left(\frac{\omega^2}{c^2 k^2}\mu_\perp \varepsilon_\perp - \sin^2\theta \frac{\varepsilon_\perp}{\varepsilon_\parallel} - \cos^2\theta\right) H_y = 0, \tag{17b}$$

$$\sin\theta\cos\theta H_x + \left(\frac{\omega^2}{c^2 k^2}\mu_\parallel \varepsilon_\perp - \sin^2\theta\right) H_z = 0. \tag{17c}$$

Equations (17) are a set of three homogeneous equations satisfied by **H** in the superlattice. The polariton dispersion relation due to the coupling between the spins and electromagnetic wave can be obtained by setting the determinant in Eqs. (17) equal to zero.

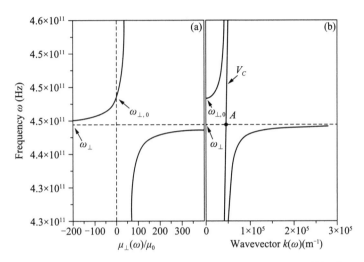

FIG.2. Calculated magnetic abnormality (a) and polariton disersion (b) of the coupled mode of the photon and pure transverse phonon.

From Eqs. (17), one sees that there are two propagating modes: one with the magnetic field vector normal to the xz plane and one with the magnetic field in the xz plane. Considering the first mode with the magnetic field normal to the xz plane, the magnetic field is in the y direction, parallel to the xy plane. We refer to this mode as the TM mode. This dispersion relation is obtained from Eqs. (17b) and (13):

$$\frac{c^2 k^2}{\omega^2} = \mu_{11}^S \varepsilon(\theta) \frac{\omega_{\perp,0}^2 - \omega^2}{\omega_\perp^2 - \omega^2}, \tag{18}$$

where $\varepsilon(\theta) = \sqrt{\sin^2\theta/\varepsilon_\parallel + \cos^2\theta/\varepsilon_\perp}$. The polariton with the magnetic field in the xz plane has a more complex dispersion relation for a general value of θ. We refer to this second mode as the TE polariton. When $\theta = 0$ (propagation along z axis), the dispersion relation is

$$\frac{c^2 k^2}{\omega^2} = \mu_{11}^S \varepsilon_\perp \frac{\omega_{\perp,0}^2 - \omega^2}{\omega_\perp^2 - \omega^2}. \tag{19}$$

It has same resonance frequency with that obtained for the TM polariton, which described the coupling between the EM mode, $V_c = c/\sqrt{\mu_{11}^S \varepsilon_\perp}$, and the transverse phonons. When $\theta = \pi/2$ (propagation along the x axis), the dispersion relation becomes

$$\frac{c^2 k^2}{\omega^2} = \mu_{33}^S \varepsilon_\perp \frac{\omega_{\parallel,0}^2 - \omega^2}{\omega_\parallel^2 - \omega^2}, \tag{20}$$

which shows coupling happens between the EM mode $V_c = c/\sqrt{\mu_{33}^S \varepsilon_\perp}$ and the longitudinal phonons. For any other direction ($\theta \neq 0$ and $\pi/2$), the resonance frequency is neither $\omega_\perp$ nor $\omega_\parallel$ and no pure longitudinal and transverse coupling occurs. EM wave is coupled to the admixture mode of longitudinal phonon and transverse phonon.

Figures 2(a) and 2(b) shows the coupled modes of photon and transverse phonons in the superlattice described by Eqs. (13) and (19). The solid line labeled $V_c = c/\sqrt{\mu_{11}^S \varepsilon_\perp}$

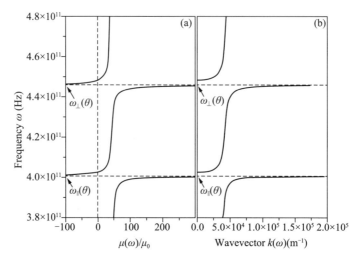

FIG. 3. Calculated magnetic abnormality (a) and polariton disersion (b) of the coupled mode of the photon and admixture of the longitudinal phonon and transverse phonon.

corresponds to EM waves, but uncoupled to the lattice vibrations, and the dotted line represents the lattice vibration in the absence of coupling to the EM field. The region of the crossover [marked by A in Fig. 3(b)] of the solid line and the dotted line is the resonance region. By resonance, we mean that the frequency of the EM wave equals the acoustic resonance frequency of the superlattice determined by the periodicity. At resonance the photon-phonon coupling entirely changes the character of the propagation. The heavy lines are the dispersion relations in the presence of coupling between the lateral spins induced by a lattice vibration and the EM wave. In the resonance region the propagation mode is neither a pure photon mode nor a pure acoustic mode in a narrow range of k values. The quantum of the coupled photon-phonon wave field is called a polariton. It is a type of polariton. One effect of the coupling is to create a frequency gap between $\omega_\perp$ and $\omega_{\perp,0}$. For frequencies $\omega_\perp < \omega < \omega_{\perp,0}$, $\mu(\omega)$ is negative as can be seen in Fig. 3(a). This negative gap originates from the coupling of the photon and the lattice vibration through the piezomagnetic effect. The resonance frequency $\omega_\perp = (\pi/d)\sqrt{C_{44}/\rho}$ is mainly determined by the periodic of the superlattice, $\Lambda = 2d$. As $\omega_\perp < \omega_{\perp,0}$, the band gap is $\omega_{\perp,0} - \omega_\perp$ wide, which is determined by $\Delta^2 = \tilde{q}_{31}^2 G_1^2/(\rho\mu_{11}^S)$. The larger the Δ value is, the stronger the coupling and the wider the gap. The negative band can be widened by use of some materials with larger piezomagnetic coupling coefficients. If the period is $\Lambda = 0.1$ μm, the first-order reciprocal vector of this periodic structure can be attained as $G_1 = \pi/d = 6.283 \times 10^7$ m^{-1}. The average mass density of superlattice is $\rho = 5.0 \times 10^3$ kg/m^3 and the piezomagnetic parameters used is given in Table 1. The transverse resonance frequency can be attained as $\omega_\perp = 4.44 \times 10^{11}$ Hz and the negative range is about 0.5×10^{10} Hz. By varying the period of the superlattice, a wide range of negative permeability can be achieved.

If the EM wave is radiated into the piezomagnetic superlattice with an oblique angle ($\theta \neq 0$ and $\pi/2$), the EM wave will not only couple with the longitudinal phonon but also couple with the transverse phonon. The dispersion relation of the admixture polariton can be calculated from Eqs. (17a) and (17c), which is shown in Figs. 3(a) and 3(b). It can be seen that there are two resonance regions in the curves. But the resonances are not equal to $\omega_\perp$ and $\omega_\parallel$, which are both functions of θ. In this general case, negative permeability can also be attained in the two resonance regions. The above results show that the piezomagnetic superlattice can be a kind of artificial material to produce negative permeability.

In this paper, we only consider one-dimensional piezomagnetic multilayer structure. The mathematical model introduced here can be extended to two-and three-dimensional structures, which can also be proved to produce negative permeability. On the other hand, the piezomagnetic superlattice can be combined with the piezoelectric superlattice to make left-handed material. Further study will be carried out to realize this object.

In summary, a kind of periodic structure, the piezomagnetic superlattice, is constructed in this paper. The propagation property of an EM wave in a piezomagnetic superlattice was studied theoretically. A type of magnetic polariton was proposed which originates from the coupling of the EM wave with the lattice vibration through the piezomagnetic effect. At some resonance frequency regions, negative permeability can be attained and the piezomagnetic superlattice can make a kind of negative permeability material.

References and Notes

[1] D. R. Smith, W. J. Padilla, D. C. Vier, S. C. Nemat-Nasser, and S. Schultz, Phys. Rev. Lett. **84**, 4184 (2000).
[2] R. A. Shelby, D. R. Smith, and S. Schultz, Science **292**, 77(2001).
[3] J. B. Pendry, Phys. Rev. Lett. **85**, 3966(2000).
[4] J. B. Pendry, A. J. Holden, D. J. Robbins, and W. J. Stewart, IEEE Trans. Microwave Theory Tech. **47**, 2075(1999).
[5] T. J. Yen, W. J. Padila, N. Fang, D. C. Vier, D. R. Smith, J. B. Pendry, D. N. Basov, and X. Zhang, Science **303**, 1494(2004).
[6] E. Yablonovitch, Phys. Rev. Lett. **58**, 2059(1987).

[7] S. John, Phys. Rev. Lett. **58**, 2486(1987).

[8] J. J. Joannopoulos, R. D. Meade, and J. N. Winn, *Photonic Crystals*(Princeton University, Princeton, NJ,(1995).

[9] S. N. Zhu, Y. Y. Zhu, and N. B. Ming, Science **278**, 843(2004).

[10] H. Liu, Y. Y. Zhu, S. N. Zhu, C. Zhang, and N. B. Ming, Appl. Phys. Lett. **79**, 728(2001).

[11] Y. Q. Lu, Y. Y. Zhu, Y. F. Chen, S. N. Zhu, N. B. Ming, and Y. J. Feng, Science **284**, 1822(1999).

[12] Y. Y. Zhu, X. J. Zhang, Y. Q. Lu, Y. F. Chen, S. N. Zhu, and N. B. Ming, Phys. Rev. Lett. **90**, 053903(2003).

[13] X. J. Zhang, R. Q. Zhu, J. Zhao, Y. F. Chen, and Y. Y. Zhu, Phys. Rev. B **69**, 085118(2004).

[14] Z. G. Zhou, *Ferrite Magnetic Materials*(Science Publishing House, Beijing, 1981).

[15] V. M. Agranovich and V. E. Kravstov, Solid State Commun. **55**, 85(1985).

[16] This work was supported by grants for the State Key Program for Basic Research of China, and by the National Natural Science Foundation of China under Contract No. 90201008 and of Jiangsu Province under Grant No. BK2002202, and by Jiangsu Planned Projects for Postdoctoral Research Funds.

Coupled Phonon Polaritons in a Piezoelectric-piezomagnetic Superlattice[*]

Jun Zhao, Ruo-Cheng Yin, Tian Fan, Ming-Hui Lu, Yan-Feng Chen,
Yong-Yuan Zhu, Shi-Ning Zhu, and Nai-Ben Ming

National Laboratory of Solid State Microstructures and Department of Materials Science and Engineering, Nanjing University, Nanjing 210093, China

Propagation of electromagnetic (EM) waves in a piezoelectric-piezomagnetic superlattice (PPS) has been investigated. In a PPS, the electric and magnetic vectors of EM waves could simultaneously couple with the identical superlattice vibration, respectively, due to piezoelectric effect and piezomagnetic effect, which results in magnetoelectric effect. Consequently, two orthogonally polarized EM waves could simultaneously couple with the identical vibration, which would give birth to coupled phonon polaritons. Attributed to this mechanism, in a PPS, the propagation of EM waves varies drastically. EM waves perpendicular to the PPS vector can propagate, while the propagation is inhibited along the PPS vector in the original band gap of either the piezoelectric superlattice or the piezomagnetic superlattice. The origin of the differences in propagation is analyzed and some potential applications are discussed.

1. Introduction

The interaction of waves with periodic condensed matter, especially artificial superlattices, has attracted much attention. It is well known that the periodic modulation of the given parameter may lead to band structures and make it possible to control the propagation of information carriers, for example, electrons and photons. Due to intense Bragg scatterings at the boundaries of the Brillouin zone, a photonic band gap can be introduced to control the propagation of light in photonic crystals.[1] In addition, an alternative to form band structures is the coupling effect of polaritons. For example, in an ionic crystal, the coupling between EM waves and lattice vibrations leads to phonon polaritons, whose band gaps appear in the infrared range.[2] It should be noticed that, different from the case in photonic crystals, the wave vector corresponding to the intense coupling in ionic crystals is close to the Γ point in the Brilloiun zone.

Polaritons can also exist in artificial superlattices,[3] such as piezoelectric (PE) superlattices[4—6] and piezomagnetic (PM) superlattices,[7] in which the periodicity of the

---

[*] Phys.Rev.B,2008,77(7):075126

lattice is artificially expanded from atomic scale to nanometers and microns. In PE and PM superlattices, both PE and PM phonon polaritons result from the coupling between EM waves and superlattice vibrations, and the propagation of EM waves is not allowed near resonant frequencies.

In this paper, we will demonstrate another kind of superlattice that alternates between PE and PM stacks, a complex structure made of a PE superlattice[4-6] and a PM superlattice.[7] In a PPS, owing to PE and PM effects, the electric and magnetic fields could simultaneously couple with the identical vibration, represented as PE phonons and PM phonons, which would result in piezoelectric (PE) polaritons, piezomagnetic (PM) polaritons, and magnetoelectric (ME) effect. Furthermore, due to the ME effect, the PE and PM polaritons would further couple with each other. The coupling between PE and PM polaritons would endow the propagation of electromagnetic (EM) waves with novel properties in the original band gaps of both PE polariton and PM polariton.

The organization of this paper is as the follows. In Sec. 2, we demonstrate how ME effect occurs naturally from PE and PM effects in a PPS through their mechanical strain by effective-medium expression. In Secs. 3 and 4, we combine the expression with Maxwell's equations to investigate the propagation of EM waves parallel and/or perpendicular to the stacking direction. In Sec. 5, we present the conclusion.

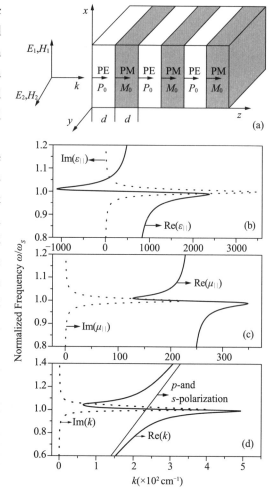

FIG. 1. (a) Schematics of the PPS and the two orthogonally polarized EM waves along the z axis. Due to ME effect, p polarization would be induced by the incident s polarization. (b), (c), and (d) are the calculated dispersion relations of permittivity, permeability, and wave vector k, described in the form of $X = \text{Re}(X) + i\text{Im}(X)$. Here, the thinner solid line in (d) represents the propagation of the degenerate p and s polarizations in the case without PE and PM effects.

2. Theoretical Treatment

The piezoelectric-piezomagnetic superlattice (PPS) presented in this paper [shown in Fig. 1(a)] is a one-dimensional periodic structure composed of alternate layers of PE and

PM materials along the z axis. To simplify the problem, we assume that PE and PM layers are polycrystalline, and the thickness for each layer is d. This PPS is electrically and magnetically poled along the z direction. Consequently, both PE and PM layers could be considered as the transversely isotropic systems with the same point group of ∞m symmetry.

Let us focus on the modulations of PE and PM constants. The interaction between EM waves and superlattice vibrations is described by the following equations:

$$T_I = C_{IJ} S_J(z) - e_{Ij} f(z) E_j - q_{Ij} g(z) H_j$$
$$D_i = e_{iJ} f(z) S_J(z) + \varepsilon_0 \varepsilon_{ij}^S(z) E_j \quad (i,j=1,2,3; I,J=1,2,\cdots,6). \tag{1}$$
$$B_i = q_{iJ} g(z) S_J(z) + \mu_0 \mu_{ij}^S(z) H_j$$

Here T, S, C, E, H, ε, μ, e, and q are the stress, strain, elastic stiffness coefficient, electric field, magnetic field, permittivity, permeability, and PE and PM constants, respectively. The modulation functions $f(z)$ and $g(z)$ owing to the PE and PM effects in Eq.(1) have the values of 1 and 0 in PE layers, respectively, while 0 and 1 in PM layers. The elastic tensor element is described as $C_{IJ} = c_{IJ} - i\omega\eta_{IJ}$, where c_{IJ} is the real part and η_{IJ} is the damping coefficient. It should be noticed that there is no ME effect involved in this stage. In our derivation, we have not supposed any a priori ME term that is used in previous approximations.[8,9] PE and PM effects can inherently interact with each other. ME effect results naturally from the stress that both adjacent PE and PM layers suffer, and it will be revealed in the following discussion.

We also assume that the transverse dimensions are much larger than the acoustic wavelength in the PPS, so a one-dimensional model can be applied properly. Then, taking the long-wavelength approximation into account and substituting T_5 into the equation of motion for superlattice vibrations, we can obtain

$$\rho \frac{\partial^2 S_5}{\partial t^2} - C_{44} \frac{\partial^2 S_6}{\partial z^2} = \frac{\partial^2}{\partial z^2} [-e_{15} f(z) E_1(z) - q_{15} g(z) H_1(z)], \tag{2}$$

where $E_1(z)$ and $H_1(z)$ are the x components of EM fields, the ρ and C_{44} are the effective mass density and the effective elastic tensor element, respectivley. With the Fourier transformation, the modulation functions are written as

$$f(z) = \sum_n f_n e^{iG_n z} = \frac{1}{2} + \frac{1}{2} \sum_{n \neq 0} \frac{i(1-\cos n\pi)}{n\pi} e^{iG_n z}$$
$$g(z) = \sum_n g_n e^{iG_n z} = \frac{1}{2} - \frac{1}{2} \sum_{n \neq 0} \frac{i(1-\cos n\pi)}{n\pi} e^{iG_n z} \quad \left(G_n = \frac{n\pi}{d}\right). \tag{3}$$

The PPS resembles a forced oscillator. The solution of Eq.(2) is described as

$$S_5(z) = \sum_n \frac{G_n^2}{\rho\omega^2 - C_{44} G_n^2} e^{iG_n z} [-e_{15} f_n E_1(z) - q_{15} g_n H_1(z)]. \tag{4}$$

Substitution of Eq.(4) into Eq.(1) gives

$$D_1(z) = \varepsilon_0 [\varepsilon_{11}^S(z) - ae_{15}^2 f(z) f_n / \varepsilon_0] E_1(z) - ae_{15} q_{15} f(z) g_n H_1(z)$$
$$= \varepsilon_0 \varepsilon_\parallel (z) E_1(z) + \alpha_{11}(z) H_1(z),$$
$$B_1(z) = \mu_0 [\mu_{11}^S(z) - aq_{15}^2 g(z) g_n / \mu_0] H_1(z) - ae_{15} q_{15} g(z) f_n E_1(z) \quad (5)$$
$$= \mu_0 \mu_\parallel (z) H_1(z) + \beta_{11}(z) E_1(z),$$

where

$$a = \sum_n \frac{G_n^2}{\rho \omega^2 - C_{44} G_n^2} e^{iG_n z}. \quad (6)$$

If the frequency of the incident EM wave is close to the fundamental resonant frequency of transverse vibration S_5, we can get $a = e^{iG_1 z} G_1^2 / (\rho \omega^2 - C_{44} G_1^2)$ by ignoring the other high-order reciprocal vectors. Owing to the boundary conditions, the transverse electric and magnetic fields are continuous at the interfaces of the layers. In the long-wavelength approximation (the EM wavelength is much larger than the period of the PPS), their variations inside each layer can be neglected. So, $E_1^{pe} = E_1^{pm} = \overline{E_1}$ and $H_1^{pe} = H_1^{pm} = \overline{H_1}$, where $\overline{E_1}$ and $\overline{H_1}$ represent their values averaged over the period; the superscripts pe and pm indicate the physical quantities of PE layers and PM layers, respectively. Thus the x components of electric displacement and magnetic field intensity averaged over the period are

$$\overline{D_1} = \frac{1}{2d} \left\{ \int_0^d [\varepsilon_0 \varepsilon_\parallel (z) E_1^{pe} + \alpha_{11}(z) H_1^{pe}] dz + \int_d^{2d} \varepsilon_0 \varepsilon_\parallel (z) E_1^{pm} dz \right\}$$
$$= \varepsilon_0 \varepsilon_\parallel E_1 + Ae_{15} q_{15} H_1,$$
$$\overline{B_1} = \frac{1}{2d} \left\{ \int_0^d \mu_0 \mu_\parallel (z) H_1^{pe} dz + \int_d^{2d} [\mu_0 \mu_\parallel (z) H_1^{pm} + \beta_{11}(z) E_1^{pm}] dz \right\} \quad (7)$$
$$= \mu_0 \mu_\parallel H_1 + Ae_{15} q_{15} E_1,$$

where

$$\varepsilon_\parallel = (\varepsilon_{11}^{S;pe} + \varepsilon_{11}^{S;pm})/2 - Ae_{15}^2 / \varepsilon_0,$$
$$\mu_\parallel = (\mu_{11}^{S;pe} + \mu_{11}^{S;pm})/2 - Aq_{15}^2 / \mu_0,$$
$$A = 2/[d^2 \rho (\omega^2 - \omega_S^2 + i\omega \gamma_S)]. \quad (8)$$

Here, $\omega_S = G_1 v_S$ and $v_S = \sqrt{C_{44}/\rho}$ are the fundamental resonant frequency and the effective velocity of the transverse acoustic wave in the PPS; $\gamma_S = (\eta_{44}/C_{44}) \omega_S^2$ is the relevant damping constant. After this, we can easily get the expressions of $\overline{D_2}$ and $\overline{B_2}$ for the transverse isotropy.

Similarly, the averaged values of the z components can be obtained. The dynamic equation for longitudinal vibration S_3 and its solution are

$$\rho \frac{\partial^2 S_3}{\partial t^2} - C_{33} \frac{\partial^2 S_3}{\partial z^2} = \frac{\partial^2}{\partial z^2} [-e_{33} f(z) E_3(z) - q_{33} g(z) H_3(z)],$$
$$S_3(z) = \sum_n \frac{G_n^2}{\rho \omega^2 - C_{33} G_n^2} e^{iG_n z} [-e_{33} f_n E_3(z) - q_{33} g_n H_3(z)]. \quad (9)$$

where $E_3(z)$ and $H_3(z)$ are the z components of EM fields, and ρ and C_{33} are the effective mass density and the effective elastic tensor element, respectively. Substitution of the expression of $S_3(z)$ into Eq.(1) gives

$$D_3(z) = \varepsilon_0[\varepsilon_{33}^S(z) - be_{33}^2 f(z)f_n/\varepsilon_0]E_3(z) - be_{33}q_{33}f(z)g_n H_3(z),$$
$$B_3(z) = \mu_0[\mu_{33}^S(z) - bq_{33}^2 g(z)g_n/\mu_0]H_3(z) - be_{33}q_{33}g(z)f_n E_3(z), \quad (10)$$

where

$$b = \sum_n \frac{G_n^2}{\rho\omega^2 - C_{33}G_n^2} e^{iG_n z}. \quad (11)$$

If the frequency is close to the fundamental resonant frequency of longitudinal vibration S_3, we can get $b = e^{iG_1 z}G_1^2/(\rho\omega^2 - C_{33}G_1^2)$ by ignoring the other high-order reciprocal vectors. In PE and PM layers, Eq.(10) can be expressed, respectively, as

$$D_3^{pe} = \frac{1}{d}\int_0^d [\varepsilon_0(\varepsilon_{33}^{pe} - be_{33}^2 f_1/\varepsilon_0)E_3^{pe} - be_{33}q_{33}g_1 H_3^{pe}]dz,$$

$$B_3^{pe} = \frac{1}{d}\int_0^d \mu_0\mu_{33}^{pe} H_3^{pe} dz,$$

$$D_3^{pm} = \frac{1}{d}\int_d^{2d} \varepsilon_0\varepsilon_{33}^{pm} E_3^{pm} dz,$$

$$B_3^{pm} = \frac{1}{d}\int_d^{2d}[\mu_0(\mu_{33}^{pm} - bq_{33}^2 g_1/\mu_0)H_3^{pm} - bq_{33}e_{33}f_1 E_3^{pm}]dz. \quad (12)$$

According to the boundary conditions, we have

$$D_3^{pe} = D_3^{pm} = \overline{D_3},$$
$$B_3^{pe} = B_3^{pm} = \overline{B_3},$$
$$\overline{E_3} = \frac{1}{2d}\left[\int_0^d E_3^{pe} dz + \int_d^{2d} E_3^{pm} dz\right],$$
$$\overline{H_3} = \frac{1}{2d}\left[\int_0^d H_3^{pe} dz + \int_d^{2d} H_3^{pm} dz\right]. \quad (13)$$

Solving the combined equations composed of Eqs.(12) and (13), we can get

$$\overline{D_3} = \varepsilon_0\varepsilon_\perp \overline{E_3} + \alpha_{33}\overline{H_3},$$
$$\overline{B_3} = \mu_0\mu_\perp \overline{H_3} + \beta_{33}\overline{E_3}, \quad (14)$$

where

$$\varepsilon_\perp = 2\varepsilon_{33}^{S,pm}(\varepsilon_0\varepsilon_{33}^{S,pe} - 2Be_{33}^2)(\mu_0\mu_{33}^{S,pe} + \mu_0\mu_{33}^{S,pm} - 2Bq_{33}^2)/C,$$
$$\alpha_{33} = 2\varepsilon_0\varepsilon_{33}^{S,pm}(\mu_0\mu_{33}^{S,pm} - 2Bq_{33}^2)Be_{33}q_{33}/C,$$
$$\mu_\perp = 2\mu_{33}^{S,pe}(\mu_0\mu_{33}^{S,pm} - 2Bq_{33}^2)(\varepsilon_0\varepsilon_{33}^{S,pe} + \varepsilon_0\varepsilon_{33}^{S,pm} - 2Be_{33}^2)/C,$$
$$\beta_{33} = 2\mu_0\mu_{33}^{S,pe}(\varepsilon_0\varepsilon_{33}^{S,pe} - 2Be_{33}^2)Be_{33}q_{33}/C,$$
$$B = 2/[d^2\rho(\omega^2 - \omega_L^2 + i\gamma_L\omega_L)],$$
$$C = (\varepsilon_0\varepsilon_{33}^{S,pe} + \varepsilon_0\varepsilon_{33}^{S,pm} - 2Be_{33}^2)(\mu_0\mu_{33}^{S,pe} + \mu_0\mu_{33}^{S,pm} - 2Bq_{33}^2) - 4B^2 e_{33}^2 q_{33}^2. \quad (15)$$

Here, $\omega_L = G_1 v_L$ and $v_L = \sqrt{C_{33}/\rho}$ are the fundamental resonant frequency and the effective velocity of the longitudinal acoustic wave in the PPS; $\gamma_L = (\eta_{33}/C_{33})\omega_L^2$ is the relevant longitudinal damping constant. If the top labels are omitted, the constitutive equation for the effective medium can be described as

$$\begin{bmatrix} D_1 \\ D_2 \\ D_3 \\ B_1 \\ B_2 \\ B_3 \end{bmatrix} = \begin{bmatrix} \varepsilon_0\varepsilon_\parallel & 0 & 0 & Ae_{15}q_{15} & 0 & 0 \\ 0 & \varepsilon_0\varepsilon_\parallel & 0 & 0 & Ae_{15}q_{15} & 0 \\ 0 & 0 & \varepsilon_0\varepsilon_\perp & 0 & 0 & \alpha_{33} \\ Ae_{15}q_{15} & 0 & 0 & \mu_0\mu_\parallel & 0 & 0 \\ 0 & Ae_{15}q_{15} & 0 & 0 & \mu_0\mu_\parallel & 0 \\ 0 & 0 & \beta_{33} & 0 & 0 & \mu_0\mu_\perp \end{bmatrix} \begin{bmatrix} E_1 \\ E_2 \\ E_3 \\ H_1 \\ H_2 \\ H_3 \end{bmatrix}. \quad (16)$$

From Eq.(16), we can find that a PPS is another kind of structure-induced bianisotropic medium,[10] in which both polarization and magnetization linearly depend on the magnetic field and the electric field, even though neither of the two layers has ME effect. The ME effect in the PPS has been obviously constructed by the fact that a PE layer is stressed in an electric field, which, in turn, induces the magnetization of the adjacent PM layer by PM effect. Similarly, the polarization in PE layers is also related to the magnetization in PM layers driven by the magnetic field. In other words, the mechanical strain and stress bridge the coupling between the electric field E and the magnetic field H in the PPS. The scenario of the ME effect in the PPS is electric-field-stress-magnetization and magnetic-field-stress-polarization.

3. Propagation Along The z Direction

In general, the propagation of EM waves in a bianisotropic medium is complicated. Here, we just discuss some special cases. First, we assume that the incident light is propagating along the z axis and polarized in the y direction [s polarization (E perpendicular to x-z plane)]. Due to ME effect, p polarization (E parallel to the x-z plane) will be excited by the s-polarized incidence, as shown in Fig. 1(a). From Eqs.(1) and (2) and the transverse isotropy, we can find that a transverse superlattice vibration S_4 oscillating in the y direction will be excited and synchronously induce both the transverse electric polarization P_2 due to PE effect and the transverse magnetic polarization M_2 due to PM effect. Through Maxwell's equations, the two transverse polarizations will, in turn, emit orthogonally polarized EM waves. Both of them interfere with the original EM waves. In fact, the same process will be induced by S_5, which is oscillating in the x direction. This is to say that PE and PM phonon polaritons could be excited, similar to the cases described, respectively, in Refs. [6] and [7]. In addition, the transverse vibrations would bridge the coupling between the two polaritons. Corresponding changes would appear in the polariton dispersion.

To get the polariton dispersion, we substitute Eq. (16) into Maxwell's equations. Thus, we can obtain

$$\frac{\partial E_1}{\partial z} - i\omega\mu_0\mu_\parallel H_2 - iA\omega e_{15}q_{15}E_2 = 0,$$

$$\frac{\partial E_2}{\partial z} + i\omega\mu_0\mu_\parallel H_1 + iA\omega e_{15}q_{15}E_1 = 0,$$

$$\frac{\partial H_1}{\partial z} + i\omega\varepsilon_0\varepsilon_\parallel E_2 + iA\omega e_{15}q_{15}H_2 = 0,$$

$$\frac{\partial H_2}{\partial z} - i\omega\varepsilon_0\varepsilon_\parallel E_1 - iA\omega e_{15}q_{15}H_1 = 0. \tag{17}$$

From Eq.(17), a dispersion relation can be obtained:

$$k^2/\omega^2 = \varepsilon_0\mu_0\varepsilon_\parallel\mu_\parallel - A^2 e_{15}^2 q_{15}^2. \tag{18}$$

The dispersion relation shown in Eq. (18) is a normal phonon polariton[2] with the modulation from the coupling between PE and PM polaritons (the second term of the right side).

A similar model of a PPS was proposed to realize double negative permittivity and permeability without considering elastic damping and ME effect.[11] Due to the neglect of ME effect, PE and PM polaritons would not couple with each other, and accordingly the modulation term would disappear in the case of Ref. [11]. In Ref. [11], the excited PE and PM polaritons will lead to individual band gaps near transverse resonant frequency ω_S. Double negativity can, in principle, be obtained in the overlapped band gap and result in a propagating state, which might be the situation of negative refraction as predicted firstly by Veselago[12] and has recently attracted intensive attention. However, for a real heterogeneous PPS, a proper damping constant γ_S and ME effect must be considered, which drastically modify the characters of the EM waves, as described in the following.

In order to have a better understanding of the above theoretical results, we perform a realistic numerical computation for a PPS with ∞m symmetry, such as a $BaTiO_3$-$CoFe_2O_4$ superlattice.[13] In the calculations of this paper, a proper damping constant $\gamma_S = 0.02\omega_S$ is chosen. Each layer in this PPS has a length of $d = 500$ nm. So, the transverse resonant frequency ω_S is calculated at 18.0 GHz according to Eq.(8). The abnormal dispersions of permittivity and permeability, as functions of normalized frequency (ω/ω_S), are shown in Figs. 1(b) and 1(c). The real part of permeability keeps positive, and the imaginary parts of permittivity and permeability are comparable to their real parts. Moreover, the coupling between the parallel electric and magnetic fields bridged by the transverse vibration gives birth to ME effect. Due to ME effect, PE and PM polaritons will further couple with each other, which leads to a new coupled polariton. The polariton dispersion relations are shown in Eq.(18) and Fig. 1(d). Near ω_S, the imaginary part of the wave vector is large and no propagating mode exists.

The ME coupling term shown in Eq.(18) dramatically changes the inherent characters of original PE and PM polaritons, even though their band gaps are still kept. Hence, it is hard to realize negative refraction via this approach, due to the huge loss and the intense ME coupling near ω_S.

4. Propagation Along The x Direction

In addition to the coupled polariton, as we described above, that is induced by EM waves propagating along the z direction of the PPS, the case for the propagation of EM wave along the x direction has also been investigated. We assume that the incident EM wave is polarized in the y direction and its frequency is near the transverse resonant frequency, far from the longitudinal one. Thus, the ME effect in the z direction can be negligible. Due to the ME effect in the y direction, p polarization (E parallel to the x-z plane) will be excited by the incident s polarization (E perpendicular to the x-z plane), as shown in Fig. 2(a). Similar to the analysis in Sec. 3, both two orthogonally polarized EM waves can contribute to the excitation of S_4. In other words, the orthogonally polarized EM waves will couple with the superlattice vibration S_4, which results in PE and PM phonon polaritons. The two polaritons then couple with each other due to ME effect via S_4, leading to the corresponding coupled polaritons.

Substituting our constitutive equation into Maxwell's equations, we can get

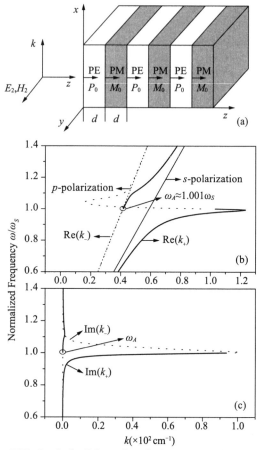

FIG. 2. (a) Schematic of the two orthogonally polarized EM waves along the x axis and the first-order polariton dispersion curves, and (b) the real part and (c) the imaginary part of the wave vector k. The coupling between EM waves and the vibration S_4 results in k_+ and k_- modes that are degenerating into s and p polarizations, respectively, when the frequency is away from ω_A.

$$-\frac{\partial E_3}{\partial x} + i\omega\mu_0\mu_\parallel H_2 + iA\omega e_{15}q_{15}E_2 = 0,$$

$$\frac{\partial E_2}{\partial x} + i\omega\mu_0\mu_\perp H_3 = 0,$$

$$\frac{\partial H_3}{\partial x} + i\omega\varepsilon_0\varepsilon_\parallel E_2 + iA\omega e_{15}q_{15}H_2 = 0,$$

$$\frac{\partial H_2}{\partial x} - i\omega\varepsilon_0\varepsilon_\perp E_3 = 0.$$

(19)

From Eq.(19), we obtain the following polariton dispersion relations:

$$2c_0^2 k_\pm^2 / \omega^2 = M_\pm = (\varepsilon_\parallel \mu_\perp + \varepsilon_\perp \mu_\parallel) \pm \sqrt{(\varepsilon_\parallel \mu_\perp - \varepsilon_\perp \mu_\parallel)^2 + 4A^2 e_{15}^2 q_{15}^2 \varepsilon_\perp \mu_\perp / \varepsilon_0 \mu_0}. \quad (20)$$

Here, c_0 is the phase velocity of EM waves in free space. The transverse permittivity and permeability have abnormal dispersion relations, shown in Figs. 1(b) and 1(c); while the longitudinal ones keep constant because the longitudinal ME effect is ignored. The imaginary and real parts of the first-order phonon polariton dispersions are plotted in Figs. 2(b) and 2(c), respectively. The propagations of p polarization and s polarization are illustrated in Fig. 2(b) too, which can be obtained according to our constitutive equations by setting all the elements of PE and PM tensors with the values of 0. It should be noticed that near ω_S, there is an exchange point (see Fig. 2) in the dispersion curves of the first-order polaritons, different from the conventional polaritons.[1] The second term of $M_\pm$ can also be expressed as $\pm\sqrt{D'(\omega) + iD''(\omega)}$. If $D'(\omega)$ is kept positive, the discontinuity of the wave vector would appear when $D''(\omega)$ changes its sign. The exchange point is set when $D''(\omega_A) = 0$. The first dispersion, named k_+ mode, has the large imaginary values but small values above ω_A. The second one named k_- mode, however, has correspondingly small and large imaginary values. When the frequency departs away from ω_A, k_+ and k_- modes degenerate into s polarization and p polarization, respectively. We can draw a conclusion that below ω_A, k_- mode is supported and k_+ mode is forbidden, while vice versa above ω_A. Such analysis of dispersion relations reveals that, different from the case in the pure PE or PM superlattice, EM waves can travel through the PPS along the x axis even in the original band gap, just by cleverly changing its propagating mode at the exchange point ω_A.

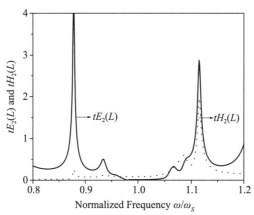

FIG.3. Variations of the parallel electric and magnetic fields in the y direction. A combination between the two fields occurs and the energy is transferred between electric and magnetic energies bridged by the vibration S_4.

To further investigate the characteristics of the PPS, a quantitative analysis of EM spectra has been calculated. We assume that the PPS has a length of $L = 5000d = 2.5$ mm in the x direction and is located in the free space. Using the dispersion relations and Maxell's equations, we find that the EM fields in the medium satisfy the following rules:

$$E_2(x) = ae^{-ik_+x} + ce^{ik_+x} + be^{-ik_-x} + de^{ik_-x},$$
$$H_3(x) = (ae^{-ik_+x} - ce^{ik_+x})/Z_+^S (be^{-ik_-x} - de^{ik_-x})/Z_-^S,$$
$$E_3(x) = P_+ (ae^{-ik_+x} - ce^{ik_+x}) + P_- (be^{-ik_-x} - de^{ik_-x}),$$
$$H_2(x) = P_+ (ae^{-ik_+x} + ce^{ik_+x})/Z_+^p + P_- (be^{-ik_-x} + de^{ik_-x})/Z_-^p, \quad (21)$$

where

$$P_\pm = \frac{\sqrt{M_\pm/2}(\varepsilon_\parallel \mu_\perp - M_\pm/2)}{Ae_{15}q_{15}\varepsilon_\perp \mu_\perp c_0},$$

$$Z_\pm^S = \sqrt{\frac{2\mu_0 \mu_\perp^2}{M_\pm \varepsilon_0}}, \quad Z_\pm^p = \mp\sqrt{\frac{M_\pm \mu_0}{2\varepsilon_0 \varepsilon_\perp^2}}. \quad (22)$$

In addition, we set the following boundary conditions:

$$E_2^I(0) = E_{20}, \quad E_3^I(0) = 0,$$
$$E_2^R(0) = e, \quad E_3^R(0) = f,$$
$$E_2^T(L) = g, \quad E_3^T(L) = h, \quad (23)$$

where the superscripts I, R, and T represent the physical quantities of the incidence, reflection, and transmission, respectively. In the expressions of EM fields, time harmonic factor $e^{i\omega t}$ is omitted. We can get eight equations including eight variations, according to the boundary conditions of the continuity of transverse electric and magnetic fields. The eight variations from a to h are first referred in Eqs. (21) and (23). By solving the eight equations, we can obtain eight expressions from a to h, respectively.

Now we can make a numerical analysis of the EM fields inside and outside PPS. We define the transmission of the EM fields as $tE_2(L) = |E_2(L)/E_{20}|$ and $tH_2(L) = |\sqrt{\mu_0} H_2(L)/(\sqrt{\varepsilon_0} E_{20})| = |E_3(L)/E_{20}|$. Their dependences on normalized frequency are shown in Fig. 3. One can find that at resonance, a very large conversion from the electric field to the magnetic field in the y direction occurs, though the efficiency is not quite high. Seen from Fig. 2(c), if there is a gap for EM waves, the imaginary value of the wave vector is about 100 cm^{-1} near the resonant frequency. For our PPS with the length $L = 2.5$ mm, the transmission index should be about e^{-25} that is much less than $tH_2(L)$ of 0.08 shown in Fig. 3. Hence, we can get a conclusion that EM waves can certainly travel through the PPS even in the original forbidden gap, just by wisely changing its propagating mode at the exchange point ω_A.

All above indicate that the coupling between polaritons, induced by PE and PM effects, makes it possible for EM waves to break the ban on the propagation near the resonant frequency. However, the results are quite different for the propagations along the x and z directions. Taking transverse isotropy into account and neglecting the higher-order reciprocal vectors, we can reduce Eq. (4) to $S_4 \propto (-e_{15} E_2 + q_{15} H_2)$. For the PE superlattice[4,6] with $q_{15}=0$, the magnitude of this vibration is proportional to the electric field. For the PM superlattice[7] with $e_{15}=0$, it is proportional to the magnetic field. For

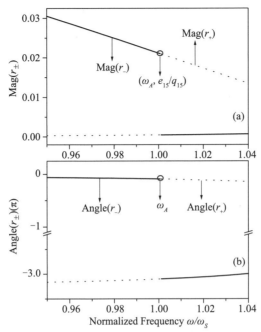

FIG. 4. The ratios of magnetic field to electric field in the y direction. (a) and (b) show the magnitudes and the phase angles of the ratios, respectively. Considering the dispersion curves, it is obvious that at resonance the magnitude and the phase angle of the supported mode have the values of about e_{15}/q_{15} and 0, respectively. With this given ratios of magnetic field to electric field in the y direction, the vibration is weak and the total dissipation is low.

the PPS, however, it is proportional to the linear combination of electric and magnetic fields. In the PE or the PM superlattice, intense reflective waves are excited to cut the vibration magnitude down, which results in band gaps. However, in the PPS, a route has been proposed above that EM waves can propagate near the resonant frequency by diminishing the magnitude of vibration S_4 with a given combination of electric and magnetic fields.

To see this clearly, we can define the ratios of the magnetic field to the electric field propagating in the y direction with different modes as $r_\pm(x) = P_\pm / Z_\pm^{TM}$, according to Eq. (21). We randomly choose a position far from the incident position along the x axis, such as $x = 0.5L$. The ratios at this position are calculated, and their magnitudes and phase angles are plotted in Figs. 4(a) and 4(b). It should be noticed that the identical exchange point ω_A appears again. Near this point, for the propagating modes, the magnitudes of the ratios are about e_{15}/q_{15}, and the phase angles are about 0; for the forbidden modes, they are about 10^{-4} and 3π, respectively. In this way, the vibration in the y direction S_4 is definitely weakened. This is to say that the EM waves propagating along the x axis can travel through the PPS with less dissipation even at resonant frequency. Unfortunately, when EM waves are propagating along the z axis, although one of the two degenerate transverse vibrations, S_4 and S_5, is weakened while the other is enhanced. So, the gap still exists, even though the coupling does occur. Generally, in a PPS, intense coupling between EM waves and superlattice vibrations does not always mean intense vibrations.

5. Conclusion

To summarize, we have demonstrated clearly how ME effect occurs naturally from PE and PM effects through their mechanical strain in a PPS and deduced the constitutive equation for this effective medium. With the constitutive equation and Maxwell's equations, we have studied the propagation of EM waves in a PPS. It turns out that the superlattice vibrations can couple synchronously with the electric field and the magnetic field of EM waves, which results in coupled phonon polaritons. Because of the coupling between phonon polaritons, we find that the propagation of EM waves is remodeled, especially near the resonant frequency: The propagation along the x axis is permitted, while it is forbidden along the z axis due to the degenerate transverse phonons. The coupling in a PPS presents an approach for EM waves to break the ban on the propagation near the resonant frequency, exhibiting rich physics in artificial microstructures. The properties can give birth to some practical applications, such as for wave division multiplexing[16] devices in optical communications and for microwave absorption.

References and Notes

[1] S. G. Johason and J. D. Joannopoulos, *Photonic Crystals: The Road from Theory to Practice* (Kluwer Academic, Boston, 2002).
[2] M. Born and K. Huang, *Dynamical Theory of Crystal Lattice* (Clarendon, Oxford, 1954).
[3] R. Tsu and S. S. Jha, Appl. Phys. Lett. **20**, 16(1972).
[4] Y. Y. Zhu, N. B. Ming, W. H. Jiang, and Y. A. Shui, Appl. Phys. Lett. **53**, 1381(1988).
[5] Y. Y. Zhu and N. B. Ming, J. Appl. Phys. **72**, 904(1992).
[6] Y. Q. Lu, Y. Y. Zhu, Y. F. Chen, S. N. Zhu, N. B. Ming, and Y. J. Feng, Science **284**, 1822(1999).
[7] H. Liu, S. N. Zhu, G. Dong, Y. Y. Zhu, Y. F. Chen, and N. B. Ming, Phys. Rev. B **71**, 125106 (2005).
[8] C. W. Nan, Phys. Rev. B **50**, 6082(1994).
[9] M. I. Bichurin, V. M. Petrov, and G. Srinivasan, J. Appl. Phys. **92**, 7681(2002).
[10] J. A. Kong, *Theory of Electromagnetic Waves* (Wiley, New York, 1975).
[11] H. Liu, S. N. Zhu, Y. Y. Zhu, Y. F. Chen, and N. B. Ming, Appl. Phys. Lett. **86**, 102904(2005).
[12] V. G. Veselago, Sov. Phys. Usp. **10**, 509(1968).
[13] For the $BaTiO_3$ ceramic, we use $\rho = 5700$ kg m^{-3}, $c_{33} = 162$ GPa, $c_{44} = 43$ GPa, $\varepsilon_{11}^S = 1264$, $\varepsilon_{33}^S = 1423$, $\mu_{11}^S = 4$, $\mu_{33}^S = 8$, and $e_{15} = 11.6$ C m^{-2}; for the $CoFe_2O_4$ ceramic, we use $\rho = 5000$ kg m^{-3}, $c_{33} = 269.5$ GPa, $c_{44} = 45.3$ GPa, $\varepsilon_{11}^S = 9$, $\varepsilon_{33}^S = 10$, $\mu_{11}^S = 470$, $\mu_{33}^S = 125$, and $q_{15} = 550$ N A^{-1} m^{-1}; for the effective medium, we use $\rho = 5350$ kg m^{-3}, $c_{33} = 204$ GPa, and $c_{44} = 44$ GPa(see Refs. 14 and 15).
[14] J. H. Huang and W. S. Kuo, J. Appl. Phys. **81**, 1378(1997).
[15] E. Pan, ZAMP **53**, 815(2002).

[16] B. Mukherjee, IEEE J. Sel. Areas Commun. **18**, 1810(2000).

[17] This work was jointly supported by the State Key Program for Basic Research of China (Grant No. 2007CB613202) and National Natural Science Foundation of China (Grants No. 50632030 and No. 50602022). We acknowledge the support by the Program for Changjiang Scholars and Innovative Research Team in University(PCSIRT).

Mimicing Surface Phonon Polaritons in Microwave Band Based on Ionic-type Phononic Crystal[*]

Xi-kui Hu, Yang Ming, Xue-jin Zhang, Yan-qing Lu, and Yong-yuan Zhu

National Laboratory of Solid State Microstructures, Nanjing University,
Nanjing 210093, People's Republic of China

  We propose an approach to scale the frequency of surface phonon polariton to megahertz-gigahertz region via an artificial microstructure, ionic-type phononic crystal(ITPC). The period of ITPC can be intentionally controlled on all relevant length scales, which allows the creation of surface phonon polariton with almost arbitrary dispersion in frequency and space. A field of surface phonon polariton optics in microwave band is expected with similar optical properties to those of ionic crystals in infrared.

  Surface polariton (SP) is collective excitation when electromagnetic (EM) surface waves propagate along the surface of materials that enable confinement and control of EM energy at subwavelength scales.[1—6] So far, researches of SP have even extent to a single atomic layer.[7] Surface phonon polariton (SPhP)[8—10] reveals the coupling between EM field and lattice vibration around the surface of ionic crystals. Although it attracts less attention than surface plasmon polariton (SPP), there are still many useful prospects in surface enhanced infrared (IR) absorption and transmission,[11] high-density IR data storage,[12] coherent thermal emission,[13] subwavelength scale phononic photonics,[14] and negative index metamaterials.[15] However, due to the material's natural characteristics, SPhP is normally excited in a narrow and fixed frequency range in infrared. Subwavelength microwave phenomena are hard to be observed. It would be much desired to effectively adjust and control the excitation frequency of SPhPs just like the mimicking low frequency SPPs.[16,17]

  Generally, ionic crystals are constituted of dipoles that determine the dielectric response.[18] When EM waves propagate along the surface of an ionic crystal, the dipoles are polarized to form the charge distribution. Therefore the SPhP is excited in negative permittivity region to satisfy the EM boundary condition. As the dipoles are usually composited of positive and negative ions, they are in the lattice scales. Therefore, the excited SPhPs are typically in infrared. If the dipole size is enlarged, the corresponding wavelength would be larger with lower frequency. On the other hand, ionic-type phononic crystal(ITPC), scilicet piezoelectric superlattice, has been proposed to be a counterpart of

---

[*] Appl. Phys. Lett., 2012, 101(15): 151109

ionic crystal in microwave band.[19] It has been realized for arbitrary polarized and structured piezoelectric superlattice,[20] and its phonon polariton dispersion has been deeply studied.[21]

ITPC is a kind of artificial media with ferroelectric domain or piezoelectric coefficient being modulated, in which the interaction between the superlattice vibration and EM wave may be established. And it exactly provides size increased "dipoles". In comparison with ionic crystal, ITPC has similar but tunable dielectric characteristics. The corresponding frequency with abnormal negative permittivity moves to megahertz-gigahertz band. Therefore, in this letter, we propose to realize ITPC based SPhP mode in microwave band. Similar optical properties to those of ionic crystals in infrared can be expected.

Typically, we can explain the dielectric properties of ionic crystals by one-dimensional (1D) diatomic chain structure with positive and negative ions. In the simplest case, when an EM field is supplied, positive and negative ions have relative displacement $u_+ - u_-$ due to lattice vibrations, and then the dielectric response of EM field can be described by Huang's equations:[18]

$$\begin{cases} \ddot{W} = b_{11}W + b_{12}E \\ P = b_{21}W + b_{22}E \end{cases}, \quad (1)$$

where $W = \rho^{1/2}(u_+ - u_-)$ indicates crystal vibrations, ρ is equivalent density, E is external electric field intensity, and P is crystal polarization. In an ITPC,[19—20] as a scale enlarged ionic crystal, the similar properties have been proven. As shown in Fig. 1, the typical structure is composited of two piezoelectric domains with a periodic superlattice structure. Compared with lattice vibrations of ionic crystals, the relative motion is caused by the piezoelectric effect.

In piezoelectric materials, set the strain S and electric field intensity E as the independent variables, the fundamental equations are piezoelectric equations:[22]

$$\begin{cases} T = cS - e^T E \\ D = eS + \varepsilon E \end{cases}, \quad (2)$$

where T is stress, D is electric displacement vector, ε is the dielectric constant, c is the elastic constant, and e is the piezoelectric coefficient. As shown in Fig. 1, the simplest case in which the ITPC is composed of two domains with the same thickness and elastic properties was chosen. The two kinds of "ions" constitute "dipoles," scilicet two piezoelectric domains, whose relative motion can be influenced by an EM field, especially that of a microwave. This process is equivalent to the polarization in ionic crystals. These two kinds of "ions" constitute dipoles, parallel to positive and negative ions in ionic crystals. The motion of domain boundaries with positive and negative charges is defined as S_+ and S_-, respectively. Under the condition of long-wavelength approximation, the motion of each primitive cell can be viewed as identical. Thus, using Newton's motion law and treating the ITPC as a 1D chain with discrete equivalent mass dots at domain boundaries, in the lossless limit, we obtain the equations of the relative motion of these mass dots

$$\begin{cases} \ddot{W} = b_{11}W + b_{12}E \\ P = b_{21}W + \varepsilon_0(\varepsilon - 1)E \end{cases}, \tag{3}$$

where $W = \rho^{1/2} l (S_+ - S_-)/2$, herein l is the domain thickness.

FIG. 1. Schematic of an ITPC consisting of two kinds of ferroelectric media. The arrows represent the orientation of the spontaneous polarization. The blue arc represents that there is SPhP mode at the interface between ITPC and vacuum.

Comparing Eq.(1) with Eq.(3), we find that they have the same format, which means that the Huang's equations and piezoelectric equations are equivalent. Therefore, parameters in Eq.(3) are acquired as $b_{11} = -\pi^2 v^2/l^2$, $b_{12} = b_{21} = 2e/\rho^{1/2}l$, and $b_{22} = \varepsilon_0(\varepsilon - 1)$. The spoof angular eigenfrequency of the transverse vibration without the coupling of external EM field is $\omega_{TO} = (-b_{11})^{1/2}$, and the spoof angular eigenfrequency of the longitudinal wave is $\omega_{LO} = [-b_{11} + b_{12}^2/(\varepsilon_0 + b_{22})]^{1/2}$. Generally, ω_{LO} is larger than ω_{TO}, because the electrostatic-like field formed by longitudinal vibration increase the resilience of the oscillator. Then, the dielectric response will be got from Eq.(3):

$$\varepsilon(\omega) = \varepsilon(\infty) \frac{\omega_{LO}^2 - \omega^2}{\omega_{TO}^2 - \omega^2}. \tag{4}$$

According to the Eq.(4) above, the dielectric response of an ITPC agrees perfectly with the Lyddane-Sachs-Teller relationship in ionic crystal. That is to say the ITPC's dielectric response is analogous to those of ionic crystal. Therefore, ITPC and ionic crystal have similar optical properties. But for an ITPC, both ω_{TO} and ω_{LO} can be changed by varying the domain thickness or choosing different piezoelectric media. Thus, the $\varepsilon(\omega)$ is controllable, indicative of tunable optical properties in ITPC.

In the actual crystals, due to scattering, absorption, and other processes, the loss is inappropriate to be ignored.[23] Without loss of generality, by employing a damping term $-i\gamma \dot{W}$ in the right side of Eq.(3), we get

$$\begin{cases} \ddot{W} = -i\gamma \dot{W} + b_{11}W + b_{12}E \\ P = b_{21}W + \varepsilon_0(\varepsilon - 1)E \end{cases}. \tag{5}$$

The dielectric response will be modified to be

$$\varepsilon(\omega) = \varepsilon(\infty) \frac{\omega_{LO}^2 - i\gamma\omega - \omega^2}{\omega_{TO}^2 - i\gamma\omega - \omega^2}. \tag{6}$$

Figure 2 shows the dielectric dispersion curve of an ITPC based on a Z-directional modulated periodically poled LiNbO$_3$ (PPLN) with the period of 7.2 μm by choosing a proper damping term ($\gamma = 0.01\omega_{TO}$).[19] The dielectric spectrum in this band (400—600 MHz) shows the same curve shape as the far-infrared dielectric constant of an ionic crystal caused by lattice vibration, meaning that they have a similar origin. It is possible that the ITPC generates spoof phonons in microwave band. The dielectric response $\varepsilon(\omega)$ are insensitive in the low or high frequencies, while they evidently change with frequencies around the transverse eigenfrequency $f_{TO} = \omega_{TO}/2\pi = 500$ MHz, owing to the photon-phonon interaction. The real part of $\varepsilon(\omega)$ have two zero points, one is in transverse eigenfrequency f_{TO}, and another one is in longitudinal eigenfrequency $f_{LO} = \omega_{LO}/2\pi = 532$ MHz. From the figure, $\varepsilon(\omega)$ is negative between f_{TO} and f_{LO}, giving the possibility to generate spoof SPhP on the surface of ITPC in this range.

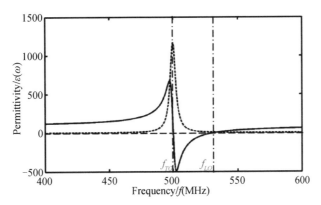

FIG.2. The calculated dielectric response curves of an ITPC. The solid line represents the real part of $\varepsilon(\omega)$, and the dashed line represents the imaginary part.

We also calculate the dispersion curves of both bulk and surface polaritons as shown in Fig. 3. The black solid lines represent the real parts and black dashed lines represent the imaginary parts. However, the ITPCs bulk effect has a band gap with negative dielectric response between the frequencies f_{TO} and f_{LO}, analogy to ionic crystal in IR. Although light cannot propagate in the ITPC, SPhP-like mode can be generated on the surface of ITPC to fill in the band gap. That is because surface mode could exist with certain negative values of ε.[1] Just considering a simple case, this surface mode is clipped at the interface of vacuum and ITPC; here, both vacuum and ITPC are semi-infinite. SPhP can be generated only with TM mode, whose electrical field distribution can be written as

$$\begin{cases} E_{z1} = E_{z1}^0 \exp(-\alpha_1 x)\exp(i(k_{SPhP}z - \omega t)) \\ E_{z2} = E_{z2}^0 \exp(-\alpha_2 x)\exp(i(k_{SPhP}z - \omega t)) \end{cases}. \quad (7)$$

The symbols E_{z1} and E_{z2} represent vacuum and ITPC, respectively, E_z represents a z-directional polarized EM field, E^0 represent the initial EM field at the interface, ω is the angular frequency, and k_0 is the corresponding wave number in vacuum. And the three important parameters for SPhP: $k_{SPhP} = (\varepsilon_1\varepsilon_2/(\varepsilon_1+\varepsilon_2))^{1/2}k_0$ is the wave number scilicet

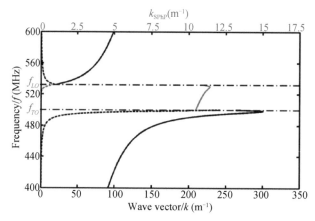

FIG.3. Dispersion relation(f-k curves) of an ITPC. Black lines represent the bulk polariton dispersion of the ITPC and red lines represent the SPhP. Solid lines represent the real parts, and dashed red lines represent the imaginary parts.

the propagation constant β along z direction and $\alpha_1 = (k_{SPhP}^2 - \varepsilon_1 k_0^2)^{1/2}$ and $\alpha_2 = (k_{SPhP}^2 - \varepsilon_2 k_0^2)^{1/2}$ are the corresponding lateral attenuation factors in vacuum and ITPC, respectively. As shown in Fig. 3, it is clearly that the spoof SPhP lies in the forbidden gap of the bulk phonon polariton. This is caused by the different mechanisms of the bulk and surface polariton effects in an ITPC. The real part of k_{SPhP} increases until a maximum value in the frequency f_s, represented by red solid line. The imaginary of k_{SPhP} also increases until a maximum value, as shown by red dashed line.

In our calculation, we take frequency $f_s = 529$ MHz, with wavelength $\lambda = 0.5667$ m. The corresponding permittivity of ITPC can be calculated from Eq.(4) $\varepsilon_2 = -8.6500 + 8.5871i$.[19] In this case, the wave number $k_{SPhP} = 11.4085 + 0.3498i$ m^{-1}, wherein the

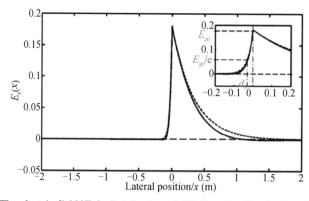

FIG.4. The electric field(E_z) distribution of SPhP mode, herein $f_s = 529$ MHz with $\alpha_1 = 2.9831 + 1.3377i$ m^{-1} and $\alpha_2 = 37.2906 - 14.0459i$ m^{-1}. The interface of ITPC and vacuum is set at $x = 0$, left part represents ITPC and right part represents vacuum. The solid line represents real part of E_z, the dashed line represents absolute value of E_z. The illustration in the upper right corner depicts the E_z distribution near $x = 0$ more clearly.

imaginary part represents the absorption. And related lateral attenuation factors are $\alpha_1 = 2.9831+1.3377i$ m^{-1} and $\alpha_2 = 37.2906-14.0459i$ m^{-1}. Figure 4 shows that the EM field is confined at the interface of vacuum and ITPC. The EM field is strongest at the interface and attenuates to both sides in ITPC and vacuum simultaneously. In particular, the EM field attenuates very quickly in ITPC. d_0 is defined to depict the EM field distribution normal to the interface where the EM field decay to $1/e$. Therefore, in the ITPC $d_0 = 1/\text{Re}(\alpha_2) = 0.0268$ m. It is seen that d_0 is far less than λ, and $d_0 \approx 0.05\lambda$. EM field is confined in a region much smaller than wavelength. That is to say, subwavelength optical phenomenon in microwave band is achieved by employing ITPC. As a consequence, we extend the frequency range of SPhP to microwave band which is impossible in natural ionic crystals.

At sufficiently low frequencies, dissipation must take charge. Since the imaginary part of k_{SPhP}, the EM energy will be attenuated when the spoof SPhP propagates forward. Figure 5 describes the change of EM field(E_z) at the interface($x=0$) when the spoof SPhP propagates forward along z direction. Generally, the EM energy is proportional to the square modulus of E_z. Thus, it is clear that the EM field is keeping attenuation with propagation forward. That is to say, there is an effective propagation length(l_0) for this spoof SPhP. l_0 is defined where the EM field decay to $1/e$. In our simulation, $k_{\text{SPhP}} = 11.4085 + 0.3498i$ m^{-1}, wherein the imaginary part is $\kappa = \text{Im}(k_{\text{SPhP}}) = 0.3498$ m^{-1} which determines the decay ratio of EM field. According to the definition of l_0, it is straightforward that $l_0 = 1/\kappa = 2.8588$ m, l_0 is larger than λ, $l_0 \approx 5.0\lambda$. That is to say, the spoof SPhP can propagate a certain distance at the interface of ITPC and vacuum although l_0 is not long enough. This is significant in practical applications.

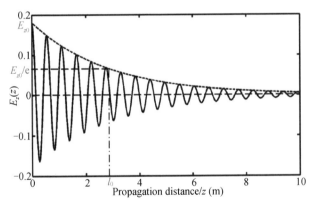

FIG.5. The propagation of SPhP, herein $f_s = 529$ MHz with $k_{\text{SPhP}} = 11.4085 + 0.3498i$ m^{-1}. The solid line represents real part of E_z, the dashed line represents absolute value of E_z.

In order to extend the propagation length, we should know that the vacuum is lossless and dissipation is resulted by imaginary part of ITPC dielectric permittivity. According to Eq.(6), we get the imaginary part:

$$\text{Im}(\varepsilon) \approx \gamma\omega\varepsilon(\infty) \frac{\omega_{LO}^2 - \omega_{TO}^2}{(\omega^2 - \omega_{TO}^2)^2}. \tag{8}$$

That is to say, if the materials chose in ITPC has a smaller damping factor γ, the imaginary part of dielectric permittivity ε of ITPC is decreased. Usually, $\omega_s \sim \omega_{LO}$, Eq.(8) can be further modified:

$$\text{Im}(\varepsilon(\omega_s)) \approx \frac{\gamma\omega_s\varepsilon(\infty)}{\omega_{LO}^2 - \omega_{TO}^2}. \tag{9}$$

With larger difference between ω_{LO} and ω_{LO}, the imaginary part of dielectric permittivity $\varepsilon(\omega_s)$ of ITPC is decreased. Therefore, the imaginary part of k_{SPhP}, namely κ, caused by the material dissipation will reduce via smaller damping factor γ or larger angular eigenfrequency squared difference $\omega_{LO}^2 - \omega_{TO}^2$. The propagation distance l_0 is extended.

In an actual case, the PPLN has the typical thickness around 0.5 - 1 mm which is smaller than d_0. In this case, the sample has a sandwich structure. A thin layer of ITPC is covered by both sides of normal dielectric cladding with low loss, e.g., air or glass. The EM field will penetrate more into the cladding. Assume that there is no loss in the cladding media, the propagation distance l_0 of spoof SPhP would increase remarkably, similarly to the long range surface plasmon polariton(LRSPP) waveguide.[24]

In this work, we only considered a simple ITPC with 1D periodicity. However, the modulation of ITPC could have different types and structures, which even may be extended to quasiperiodic, aperiodic, or two-dimensional(2D) superlattice.[20] In addition, since the SPhP gives rise to many subwavelength photonic devices, various subwavelength microwave device are also expected based on spoof SPhP effects in ITPCs. Borrowing the mature 1D/2D domain engineering in $LiNbO_3$, we believe the PPLN based SPhP device will play an important role in future acoustic and microwave applications.

In conclusion, the ITPC that is comprised of two kinds or two domains of piezoelectric media is equivalent to a scale enlarged ionic crystal. Just like normal ionic crystals, polaritions could be excited due to the coupling between EM field and lattice or superlattice vibration. In comparison with the previously reported bulk polariton, our current findings in this work predict a spoof SPhP mode. Both fundamental mechanism, such as the dispersion relation, and technical applications exhibit some differences. For example, the frequency of the generated spoof SPhP is right at the forbidden band gap of bulk polariton with negative permittivity, as shown in Figs. 2 and 3. Since the structure and period of an ITPC could be designed and fabricated freely, the corresponding polariton frequency region thus is adjustable. Interesting EM effects such as spoof SPhP in the megahertz region is predicted, which gives opportunities to control radiation at surfaces over a wide spectral range. In addition, the EM field enhancement associate with the spoof SPhP further results in many promising applications in microwave frequencies, such as subwavelength scale waveguide, EM sensing and detection, signal processing, and even some nonlinear devices.

References and Notes

[1] G. Borstel and H. J. Falge, Appl. Phys. **16**, 211(1978).
[2] D. K. Gramotnev and S. I. Bozhevolnyi, Nat. Photonics **4**, 83(2010).
[3] E. Ozbay, Science **311**, 189(2006).
[4] W. L. Barnes, A. Dereux, and T. W. Ebbesen, Nature(London) **424**, 824(2003).
[5] S. H. Kwon, J. H. Kang, S. K. Kim, and H. G. Park, IEEE J. Quantum Electron. **47**, 1346(2011).
[6] T. Holmgaard and S. I. Bozhevolnyi, Phys. Rev. B **75**, 245405(2007).
[7] Z. Fei, A. S. Rodin, G. O. Andreev, W. Bao, A. S. McLeod, M. Wagner, L. M. Zhang, Z. Zhao, M. Thiemens, G. Dominguez, M. M. Fogler, A. H. Castro Neto, C. N. Lau, F. Keilmann, and D. N. Basov, Nature(London) **487**, 82(2012).
[8] A. Huber, N. Ocelic, D. Kazantsev, and R. Hillenbrand, Appl. Phys. Lett. **87**, 081103(2005).
[9] K. Torii, T. Koga, T. Sota, T. Azuhata, S. F. Chichibu, and S. Nakamura, J. Phys.: Condens. Matter **12**, 7041(2000).
[10] A. Mohamed, G. E. Aitzol, S. Martin, H. Rainer, and A. Javier, Chin. Sci. Bull. **55**, 2625(2010).
[11] S. Shen, A. Narayanaswamy, and G. Chen, Nano Lett. **9**, 2909(2009).
[12] N. Ocelic and R. Hillenbrand, Nat. Mater. **3**, 606(2004).
[13] J. J. Greffet, R. Carminati, K. Joulain, J. P. Mulet, S. Mainguy, and Y. Chen, Nature(London) **416**, 61(2002).
[14] T. Taubner, F. Keilmann, and R. Hillenbrand, Nano Lett. **4**, 1669(2004).
[15] G. Shvets, Phys. Rev. B **67**, 035109(2003).
[16] J. B. Pendry, A. J. Holden, W. J. Stewart, and I. Youngs, Phys. Rev. Lett. **76**, 4773(1996).
[17] J. B. Pendry, L. Martin-Moreno, and F. J. Garcia-Vidal, Science **305**, 847(2004).
[18] K. Huang, Proc. R. Soc. London, A **208**, 352(1951).
[19] Y. Q. Lu, Y. Y. Zhu, Y. F. Chen, S. N. Zhu, N. B. Ming, and Y. J. Feng, Science **284**, 1822(1999).
[20] Y. Y. Zhu, X. J. Zhang, Y. Q. Lu, Y. F. Chen, S. N. Zhu, and N. B. Ming, Phys. Rev. Lett. **90**, 053903(2003).
[21] X. J. Zhang, R. Q. Zhu, J. Zhao, Y. F. Chen, and Y. Y. Zhu, Phys. Rev. B **69**, 085118(2004).
[22] H. F. Tiersten, J. Acoust. Soc. Am. **70**, 1567(1981).
[23] M. Hase and M. Kitajima, J. Phys.: Condens. Matter **22**, 073201(2010).
[24] A. Boltasseva, T. Nikolajsen, K. Leosson, K. Kjaer, M. S. Larsen, and S. I. Bozhevolnyi, J. Lightwave Technol. **23**, 413(2005).
[25] This work is supported by 973 Programs under Contract No. 2011CBA00205 and 2012CB921803, and the PAPD, Fundamental Research Funds for the Central Universities. Yan-qing Lu appreciates the support from National Science Fund for Distinguished Young Scholars.

Magnetic Plasmon Hybridization and Optical Activity at Optical Frequencies in Metallic Nanostructures[*]

H. Liu,[1] D. A. Genov,[1] D. M. Wu,[1] Y. M. Liu,[1] Z. W. Liu,[1] C. Sun,[1] S. N. Zhu,[2] and X. Zhang[1]

[1] *5130 Etcheverry Hall, Nanoscale Science and Engineering Center, University of California, Berkeley, California 94720 - 1740, USA*

[2] *Department of Physics, National Laboratory of Solid State Microstructures, Nanjing University, Nanjing 210093, People's Republic of China*

The excitation of optical magnetic plasmons in chiral metallic nanostructures based on a magnetic dimer is studied theoretically. Hybridization of the magnetic plasmon modes and a type of optical activity is demonstrated at near-infrared frequencies. A linearly polarized electromagnetic wave is shown to change its polarization after passing through an array of magnetic dimers. The polarization of the transmitted wave rotates counterclockwise at incident light frequencies corresponding to the low energy and clockwise at the high energy magnetic plasmon state. A metamaterial made of a large number of coupled magnetic dimers could be utilized as a tunable optically active medium with possible applications in optical elements and devices.

In 1999, Pendry *et al.* reported that a nonmagnetic metallic element, referred to as double split ring resonator (DSRR), with size smaller than the wavelength of radiation, exhibits a strong resonant response to the magnetic component of an incident electromagnetic field.[1] In such systems, there are no free magnetic poles, however, the excitation of displacement currents in the DSRR results in induction of a magnetic dipole moment that is comparable to a bar magnet. Similarly to the electric plasmons (EPs) excited in nanosize particles, an effective media made of DSRRs could support magnetic plasmon (MP) oscillations at GHz frequency, which can be used as frequency selective devices at microwave.[2] Combined with an electric response that has negative permittivity, such a system could ultimately lead to development of metamaterials with effective negative indexes of refraction.[3,4] It was also suggested that a combination of magnetic response and chirality could be used as an alternative route to negative refraction.[6] Various electromagnetic chiral structures have been reported in the microwave spectral range, such as helical wire spring,[5] swiss-role metal structure,[6] and rotating rosette shape.[7,8] Optical activity was also found in fractal aggregates of nanoparticles, where

[*] Phys.Rev.B,2007,76(7):073101

broad spectral response could be achieved due to the self-similarity of the underlying structure.[9] Recently, metallic elements have been demonstrated, with magnetic response in the near-infrared and visible spectral region.[10—17] This provides new possibilities to engineer magnetically coupled systems and realize artificial chiral effect at optical frequency.

On the other hand, it was shown that the calculation of the resonances modes of a complex metallic nanosize system is equivalent to estimating the electromagnetic interactions in nanostructures of simpler geometry.[18—24] The resonances of intricately shaped nanoparticles constitute hybridization states similar to the energy states of molecular systems. Thus, the hybridization principle could provide a simple conceptual approach for the rational design of nanostructures with desired plasmon resonances. This method has already been successfully used to describe the plasmon resonance in nanoshell,[18,19] nanoparticle dimer,[20] nanoshell dimer,[21,22] and nanoparticle and/or metallic surfaces.[23,24]

In this paper we introduce a hybridization model to investigate the magnetic response of a subwavelength nanostructure, referred to as a magnetic dimer(MD). The magnetic dimer consists of two single split-ring resonators (SSRR) coupled through magnetic induction. The fundamental MD resonances are viewed as bonding and antibonding combinations of individual SSRRs eigenmodes. A type of optical activity is observed in the coupled system which is not an inherent property of the individual magnetic resonators. For instance, a linearly polarized light flips into an elliptically polarized state as it passes through an array of magnetic dimers. Overall, all types of wave polarization are accessible with the proposed magnetic plasmon (MP) based metamaterial.

In Fig. 1(a), we present the general configuration of a MD, made of two SSRRs separated by a finite distance D. The slits in the rings are positioned perpendicular to each other, thus providing the platform for a unique set of physical phenomena to take place as discussed below. All characteristic system sizes are provided in the figure with D varying from 100 nm up to 600 nm. A second nanostructure, comprised of magnetic dimers positioned in a planar square array, is depicted in Fig. 1(b). The lattice period is set at 800 nm, and the incident electromagnetic radiation is assumed normal to the x-y plane.

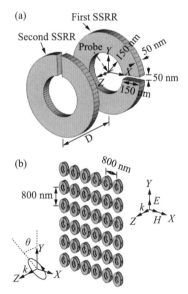

FIG.1. (color online)(a) Structure of a single magnetic dimer and (b) chiral metamaterial made of identical dimer elements. All sizes characterizing the system and the direction and polarization of the incident light are shown.

To study the magnetic response of the above described systems, we introduce a MP hybridization model similar to that proposed in the case of EP.[18—24] In our approach we use the Lagrangian formalism, first calculating the magnetic energy of a single SSRR and later expanding the theory for a system of resonators. For simplicity, in the analysis we view the SSRR as an ideal LC circuit, composed of a magnetic loop(the metal ring) with inductance L and a capacitor with capacitance C(corresponding to the gap). The resonance frequency of the structure is well known $\omega_0 = 1/\sqrt{LC}$, and the magnetic moment of the SSRR originates from the oscillatory behavior of the currents induced in the resonator. If we define the total charge Q accumulated in the slit as a generalized coordinate, the Lagrangian corresponding to a single SSRR is written as $\Im = L\dot{Q}^2/2 - Q^2/2C$, where $\dot{Q}$ is the induced current, $L\dot{Q}^2/2$ relates to the kinetic energy of the oscillations, and $L\omega_0^2 Q^2/2$ is the electrostatic energy stored in the SSRR's gap. Similarly, the Lagrangian that describes the MD is a sum of the individual SSRR contributions with added interaction term:

$$\Im = (L/2)(\dot{Q}_1^2 - \omega_0^2 Q_1^2) + (L/2)(\dot{Q}_2^2 - \omega_0^2 Q_2^2) + M\dot{Q}_1\dot{Q}_2, \quad (1)$$

where Q_i ($i = 1, 2$) are the oscillatory charges and M is the mutual inductance. By substituting $\Im$ in the Euler Lagrange equations:

$$(d/dt)(\partial\Im/\partial\dot{Q}_i) - \partial\Im/\partial Q_i = 0 \, (i=1,2), \quad (2)$$

it is straightforward to obtain the magnetic plasmon eigenfrequencies $\omega_{+/-} = \omega_0/\sqrt{1\mp\kappa}$, where $\kappa = M/L$ is a coupling coefficient. The high energy or antibonding mode $|\omega_+\rangle$ is characterized by antisymmetric charge distribution ($Q_1 = -Q_2$), and the opposite is true for the bonding or low energy $|\omega_-\rangle$ magnetic resonance ($Q_1 = Q_2$). Naturally, the frequency split $\Delta\omega = \omega_+ - \omega_- \approx \kappa\omega_0$ is proportional to the coupling strength.

The Lagrangian (hybridization) formalism provides a phenomenological picture of the electromagnetic response of the system. To study quantitatively the resonance behavior, and check the model, we rely on a set of finite-difference time-domain(FDTD) numerical simulations using a commercial software package, CST Microwave Studio (Computer Simulation Technology GmbH, Darmstadt, Germany). In the calculations, the metal permittivity is given by the Drude model: $\varepsilon(\omega) = 1 - \omega_p^2/(\omega^2 + i\omega_\tau\omega)$, where ω_p is the bulk plasma frequency and ω_τ is the relaxation rate. For gold, the characteristic frequencies, fitted to experimental data, are $\omega_p = 1.37\times 10^4$ THz and $\omega_\tau = 40.84$ THz.[25]

For excitation of the magnetic dimer(MD) we use a plane wave, with **E** field polarized in y direction and **H** field in x direction, as shown in Fig. 1(a). For a normal incidence, the magnetic field vector is in the plane of the SSRRs and direct magnetic response is unattainable. However, the electric component of the incident field excites an electric response in the slit and thus a magnetic field could be indirectly induced.[11] To study the local magnetic field we position probes at the center of the first SSRR, and vary the incident frequency. The recorded magnetic response is shown in Fig. 2 where the distance between the resonators is set at $D=250$ nm. As expected, two distinctive resonances with eigenfrequencies $\omega_-=61.6$ THz and $\omega_+=73.3$ THz are recorded. The magnetic response of the constituent SSRR is also depicted, showing a fundamental resonance at $\omega_0=66.7$ THz.

The hybridization of the magnetic response in the case of a dimer is mainly due to inductive coupling between the SSRRs. If each SSRR is regarded as a quasiatom, then the MD can be viewed as a hydrogenlike quasimolecule with energy levels ω_- and ω_+ originating from the hybridization of the original (decoupled) state ω_0. The strength of the inductive coupling depends strongly on the distance between the quasiatoms and for the considered geometry is estimated as $\kappa \approx 0.17$. The specific nature of the MP eigenmodes is studied in Fig. 2 where the local magnetic field distributions in y-cut planes are depicted for the low energy ω_- and high energy ω_+ states, respectively. In accordance to the prediction based on the Lagrangian approach the SSRRs oscillates in phase for the bonding mode $|\omega_-\rangle$, and out of phase for the antibonding mode $|\omega_+\rangle$. Additional information on the eigenmodes magnetic field configuration is given in the supplementary material.[28]

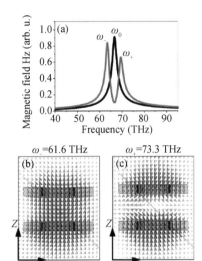

FIG. 2. (color online) (a) Local magnetic field as detected at the center of a single SSRR (black line), and a magnetic dimer with $D=250$ nm [gray (red) line]. A split in the magnetic plasmon resonance is observed due to inductive coupling. The local magnetic field profiles for the (b) bonding and (c) antibonding MP modes are depicted with the symmetric and antisymmetric field distributions are clearly seen.

Since the mutual inductance M decreases dramatically with distance, one should expect a strong change in the resonance frequencies $\omega_\pm$. This phenomenon is demonstrated in Fig. 3, where MP eigenfrequencies $\omega_\pm$ and the frequency change $\Delta\omega=\omega_+-\omega_-$ are calculated. With decreasing separation between the SSRR an increase in the frequency gap $\Delta\omega$ is observed. The opposite effect takes place at large distances where the magnetic response is decoupled. The specific profile of the frequency gap could be explained by estimating the self- and mutual inductance of the SSRRs: $\Delta\omega \approx \omega_0\kappa=\omega_0 M/L \propto \int_0^\infty dk\, e^{-kD} J_1^2(kR)$ where R is the SSRR's radius.[27] For $D> 2R$, we can expand the integral in series and write $\Delta\omega \propto (R/D)^3 - 3(R/D)^5 + 9.38(R/D)^7$. As evident from Fig. 3(b), this approximated relationship, based on the hybridization method, fits the numerical data quite well.

It is important to mention that the SSRR's size and shape, considered here, have been chosen in order to optimize the magnetic response and allow for a successful nanofabrication. There are several ways to tune the MP properties of the structure, with geometrical modifications having the most significant effect on the MP resonances. For instance, scaling down the size of the resonator R increases the resonance frequency $\omega_0 = 1/\sqrt{LC} \sim 1/R$ due to a linear increase of the inductance $L \sim R$ and capacitance $C \sim R$. However, the kinetic energy of the electrons in the metal is no longer negligible in the infrared range and results in saturation of the magnetic resonance frequency at about 400 THz for noble metals.[14] Another way to tune the resonance is to use different materials, both metals and insulators. For instance, encapsulating the MDs with high index materials cause redshift of the magnetic resonances (for $D=250$ nm and $\varepsilon=2$, $\omega_-=43.3$ THz and $\omega_+=51.5$ THz). However, our studies show that optimal magnetic dipole moment is achieved for silver-air or gold-air systems (due to the relatively low intrinsic loss) and a circular shaped SSRR.

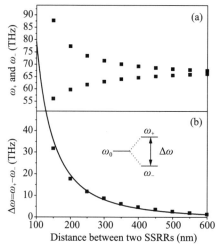

FIG. 3. The dependence of (a) the resonance frequencies $\omega_\pm$ and (b) the frequency gap $\Delta\omega$ on the distance D between the constituent SSRRs. The frequency gap predicted by the hybridization model (solid line) follows very well the FDTD result (dots). A diagram showing the energy states of the equivalent quasimolecule is included as an inset.

Having determined the fundamental response of the proposed MD, next we study the propagation of electromagnetic (EM) wave in a metamaterial composed of periodically arranged elements [see Fig. 1(b)]. A plain EM wave is incident on the system and the amplitude and phase change of the transmission wave is detected in far field as shown in Figs. 4(a) and 4(b). The host media is assumed to be air but other dielectric materials such as glass may also be applied. Although the incident light is linearly polarized ($\boldsymbol{E}=E_y\,\hat{y}$), the transmission wave is found to acquire both x and y electric field components in the resonance frequency range and some phase difference between the two orthogonal components. This change in polarization and phase delay origins from the specific three dimensional (3D) chiral arrangement of two SSRRs [see Fig. 1(a)]: one SSRR is shifted a

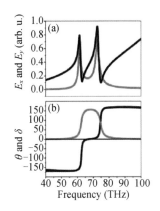 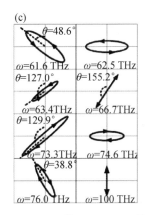

FIG. 4. (color online) (a) Electric field amplitudes $|E_x|$ [gray (red) line] and $|E_y|$ (black line) for a plain wave passing through a metamaterial made of magnetic dimers. (b) The phase difference δ between the E_y and E_x components of the transmitted wave, and the angle θ between the major polarization axes of the incident and transmitted radiation are also shown with black and gray (red) solid curves, respectively. (c) Polarization states acquired by a plain wave of frequency ω passing through a metamaterial made of periodically arranged MDs ($D = 250$ nm).

distance from the other and rotates 90°. The electric field in the slit of the first SSRR is aligned along the y axis, and thus a y polarized incident wave is electrically coupled into the system. At resonance, strong magnetic interaction between the SSRRs helps to transfer the energy from the front resonator to the back SSRR. Since the electric field in the slit gap of the second SSRR is along $\hat{x}$, the electric dipole radiation carry the same polarization. Thus the transmitted wave, detected in far field, is a superposition of x and y polarized light. Unlike the transmission minimum in a waveguide, the E_y appears as a Lorentz line shape around the resonances, which is a typical resonant scattering in free space. On the other hand E_x reaches maximum at the resonance range and becomes zero for off-resonance frequencies due the diminishing coupling between the SSRRs. According to the classic model developed by Born and Kuhn,[26] two spatially separated coupled oscillators, with a chiral symmetry, will induce optical activity for an impinging EM wave. In our system, the hybridization of the MP states constitutes a mechanism for achieving optical activity in the near-infrared range, thus opening opportunities for designing chiral metamaterials. Actually, Pendry's DSRR structure was also reported to be able to excite two orthogonal polarization states simultaneously,[2] but it is a planar achiral structure and the necessary retardation effect for optical activity[26] is absent, thus it cannot be considered as optically active elements.

The polarization states accessible with the proposed chiral media are presented in Fig. 4(b) where the phase difference δ between the components of the electric field is calculated for the transmitted wave. Remarkably δ undergoes a steplike type of behavior, flipping from $-165°$ to $0°$ at ω_- and again from $0°$ to $165°$ at ω_+. The change in polarization is easily understood by observing the time evolution of the end point of the electric field vector as it travels through space. For an observer facing the approaching wave, the track of the end point is described by the well known relationship:[27]

$$\left(\frac{E_x}{|E_x|}\right)^2 + \left(\frac{E_y}{|E_y|}\right)^2 - \frac{2E_xE_y}{|E_x||E_y|}\cos\delta = \sin^2\delta. \tag{3}$$

The polarization state of the EM wave is thus determined by δ, with the end point of the electric field revolving in a clockwise direction for $\sin\delta > 0$ and in a counterclockwise direction for $\sin\delta < 0$. Concurrently for $\omega < \omega_0$, the transmitted wave has right-hand polarization and the dimer constitutes an isomer of d type, while for higher frequencies $\omega > \omega_0$, the wave acquires a left-handed polarization and the isomer is of the l type. From Fig. 4(b) it is also evident that the transmitted wave passes through a linearly polarized state($\delta = 0$) at the fundamental frequency $\omega_0 = 66.7$ THz. Despite the energy dissipation in the metal, over 40% of the incident energy is transmitted around the resonance frequency.

In Fig. 4, we have calculated the amplitude and phase change of the transmitted wave for the configuration in Fig. 1(a). Another configuration is also considered in our simulations, in which the first and second SSRRs exchange their places. For y-polarized incident wave, the first SSRR is turned with its slit in the "wrong" position for the incident E field and the second "correctly" for that incident field.[28] In this case $|E_y|$ still appears to have a Lorentz line shape. However, as the two SRRs exchange their places, the phase difference $\delta = \delta_y - \delta_x$ alters its sign. This effect is due to the absence of inversion-invariance in our 3D chiral structure.

In general, the principal axes of the ellipse described by Eq.(3) are not aligned with the x and y directions. The angle θ between the major polarization axis and $\hat{y}$ [see Fig. 1 (b)] can be calculated from the equation $\tan 2\theta = 2|E_x||E_y|\cos\delta/(|E_x|^2 - |E_y|^2)$,[27] with the numerically obtained result shown in Fig. 4(b). A distinctive, bell-type shape is observed with θ reaching a maximum value 156.7° at $\omega_m = 68.4$ THz. Although $\cos\delta$ reaches a maximum at ω_0, the angle θ optimizes at $\omega_m \neq \omega_0$ due to the frequency dependence of the electric field amplitudes. At frequencies 62.5 THz and 74.6 THz, one has $\theta = 90°$ and the major axes of the wave polarization coincides with the x or y spatial directions, respectively. A detail schematics showing all polarization states that can be acquired by the transmitted wave is depicted in Fig. 4(c). Additionally, an animation of the change in polarization is accessible online.[28]

In conclusion, we propose a magnetic plasmon dimer(MD) made of single split ring resonators showing a magnetic response at infrared frequencies. The resonance properties of the MD are successfully described by a quasimolecular hybridization model. A split in the system resonance is observed due to inductive interaction, which strongly depends on the separation distance between the resonators. Under conditions of strong magnetic coupling, optical activity is demonstrated at near-infrared frequencies. A linearly polarized light flips into an elliptically polarized state when it passes through a periodic array of MDs. Thus, an artificial chiral metamaterial made of a large number of resonators could be utilized to tune the polarization of an incident light over a wide range of angles. Optical elements, such as tunable polarizers and switches, are possible applications of the proposed MD based complex media.

References and Notes

[1] J. B. Pendry et al., IEEE Trans. Microwave Theory Tech. **47**, 2075(1999).
[2] R. Marques et al., J. Opt. A, Pure Appl. Opt. **7**, S38(2005).
[3] D. R. Smith et al., Phys. Rev. Lett. **84**, 4184(2000).
[4] R. A. Shelby, D. R. Smith, and S. Schultz, Science **292**, 77(2001).
[5] I. Tinoco and M. P. Freeman, J. Phys. Chem. **61**, 1196(1957).
[6] J. B. Pendry, Science **306**, 1353(2004).
[7] A. Papakostas et al., Phys. Rev. Lett. **90**, 107404(2003).
[8] A. V. Rogacheva et al., Phys. Rev. Lett. **97**, 177401(2006).
[9] V. P. Drachev et al., J. Opt. Soc. Am. B **18**, 1896(2001).
[10] T. J. Yen et al., Science **303**, 1494(2004).
[11] S. Linden et al., Science **306**, 1351(2004).
[12] N. Katsarakis et al., Opt. Lett. **30**, 1348(2005).
[13] C. Enkrich et al., Phys. Rev. Lett. **95**, 203901(2005).
[14] J. Zhou et al., Phys. Rev. Lett. **95**, 223902(2005).
[15] A. Ishikawa, T. Tanaka, and S. Kawata, Phys. Rev. Lett. **95**, 237401(2005).
[16] T. Tanaka, A. Ishikawa, and S. Kawata, Phys. Rev. B **73**, 125423(2006).
[17] H. Liu et al., Phys. Rev. Lett. **97**, 243902(2006).
[18] E. Prodan et al., Science **302**, 419(2003).
[19] E. Prodan and P. Nordlander, J. Chem. Phys. **120**, 5444(2004).
[20] P. Nordlander et al., Nano Lett. **4**, 899(2004).
[21] D. W. Brandl, C. Ouber, and P. Nordlander, J. Chem. Phys. **123**, 024701(2004).
[22] C. Ouber and P. Nordlander, J. Phys. Chem. B **109**, 10042(2005).
[23] P. Nordlander and E. Prodan, Nano Lett. **4**, 2209(2004).
[24] F. Le et al., Nano Lett. **5**, 2009(2005).
[25] M. A. Ordal et al., Appl. Opt. **22**, 1099(1983).
[26] E. U. Condon, Rev. Mod. Phys. **9**, 432(1937).
[27] J. D. Jackson, *Classical Electrodynamics* (Wiley, NY, 1999).
[28] EPAPS Document No. E-PRBMDO-76-041731 for supplementary material. For more information on EPAPS, see http://www.aip.org/pubsvrs/epaps.html.
[29] This work was supported by AFOSR MURI (Grant No. FA9550-04-1-0434), SINAM, and NSEC under Contract No. DMI-0327077.

Stereometamaterials[*]

Na Liu[1], Hui Liu[2], Shining Zhu[2] and Harald Giessen[1]

[1] *Physikalisches Institut, Universität Stuttgart, D-70569 Stuttgart, Germany*
[2] *Department of Physics, Nanjing University, Nanjing 210093, People's Republic of China*

The subdiscipline of chemistry that studies molecular structures in three dimensions is called stereochemistry. One important aspect of stereochemistry is stereoisomers: materials with the same chemical formula but different spatial arrangements of atoms within molecules. The relative positions of atoms have great influence on the properties of chemical substances. Here, in analogy to stereoisomers in chemistry, we propose a new concept in nanophotonics, namely stereometamaterials, which refer to metamaterials with the same constituents but different spatial arrangements. As a model system of stereometamaterials, we theoretically and experimentally study meta-dimers, which consist of a stack of two identical split-ring resonators in each unit cell with various twist angles. The interplay of electric and magnetic interactions plays a crucial role for the optical properties. Specifically, the influence of higher-order electric multipoles becomes clearly evident. The twisting of stereometamaterials offers a way to engineer complex plasmonic nanostructures with a tailored electromagnetic response.

The word 'stereo' in Greek means 'relating to space' or 'three-dimensional'. In stereochemistry, the characteristics of organic compounds depend not only on the nature of the atoms comprising the molecules (constitution) but also on the three-dimensional arrangement of these atoms in space (configuration)[1]. For example, the spatial structure of a protein molecule determines its biological activities. In photonics, metamaterials are structured media, consisting of artificial 'atoms' with unit cells much smaller than the wavelength of light[2—4]. Such metamaterial atoms can be designed to yield electric as well as magnetic dipole moments, leading to effective negative permittivity and negative permeability. A medium with simultaneous negative permittivity and negative permeability can exhibit negative refraction and unique reversed electromagnetic properties[5,6]. As well as their applicability in constructing effective media, metamaterials have also been used as prototypes for studying coupling effects between artificial atoms in three-dimensional structures[7—9]. However, the mechanism of interactions arising from different spatial arrangements of such atoms has thus far not been examined. Inspired by the concept of stereochemistry, we now investigate the coupling effects of artificial atoms in three-dimensional metamaterials from a novel perspective. We study a set of

[*] Nat.Photon.,2009,3:157

stereometamaterials, each having unitcells consisting of two stacked split-ring resonators (SRRs) with identical geometry (same constitution); however, the two SRR atoms are arranged in space using different twist angles to achieve various structures (different configurations). We term these structures stereo-SRR dimers. We theoretically and experimentally demonstrate that the optical properties of these stereo-SRR dimers can be substantially modified by altering the twist angles between the two SRR atoms, arising from the variation of electric and magnetic interactions between them. Specifically, we investigate how the electric and magnetic interactions depend on the spatial arrangement of these SRR constituents. The nontrivial magnetic interaction makes metamaterials more versatile in nanophotonics than stereoisomers in chemistry, where generally only electric interactions are taken into account. Furthermore, we show that the inclusion of the higher-order electric multipolar interactions is essential to understanding the physical implications of the twisting dispersion. A theoretical model based on a Langrangian formalism is used to interpret the evolution of the coupling effects as a function of twist angle.

Stereometamaterial Design

Figure 1(a) illustrates the geometry of the stereo-SRR dimer metamaterials together with their design parameters. Each unit cell consists of two spatially separated SRRs, which are twisted at an angle φ with respect to one another. The SRRs are embedded in a homogeneous dielectric with $\varepsilon = 1$ (that is, air). For excitation of these SRR dimer metamaterials, we use normally incident light with its polarization along the x-direction as shown in Fig. 1(a). In order to gain insight into the resonant behaviour as well as coupling mechanisms for various stereo-SRR dimer metamaterials, we first study three specific dimer systems with twist angles $\varphi = 0°$, $90°$ and $180°$. The insets of Fig. 1(b)-(d) present the schematics of the three structures, in which the vertical distance between two SRRs is set at $s = 100$ nm. Numerical simulations were performed based on a commercial finite-integration timedomain algorithm, and the simulated transmittance spectra are shown in Fig. 1(b)-(d). For each system there are apparently two observable resonances (ω^- and ω^+). To understand these spectral characteristics, current and magnetic field distributions at the relevant resonances are calculated. For the 0° twisted SRR dimer metamaterial, the electric component of the incident light can excite circulating currents along the two SRRs, giving rise to induced magnetic dipole moments in the individual SRRs. As shown in Fig. 2(a), the electric dipoles excited in the two SRRs oscillate anti-phase and in-phase at resonances ω_0^- and ω_0^+, respectively. The resulting magnetic dipoles are aligned antiparallel at resonance ω_0^-, whereas they are parallel at resonances ω_0^+. The above phenomenon can be interpreted as the plasmon hybridization[10-12] between the two SRRs due to their close proximity. In the hybridization scheme, each SRR can be regarded as an artificial atom. The two SRR atoms are bonded into an SRR dimer or SRR 'molecule' due to the strong interaction between them. Such interaction leads to the formation of new plasmonic modes,

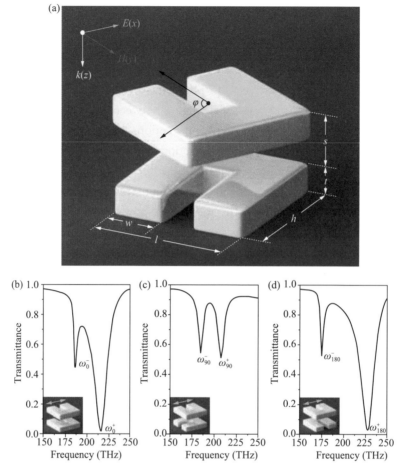

FIG. 1. Structural geometry and numerical simulation. (a), Schematic of the stereo-SRR dimer metamaterials with definitions of the geometrical parameters: $l=230$ nm, $h=230$ nm, $w=90$ nm, $t=50$ nm, and $s=100$ nm. The periods in both x and y directions are 700 nm. (b)-(d), Simulated transmittance spectra for the 0°(b), 90°(c) and 180°(d) twisted SRR dimer metamaterials. All the structures are embedded in air.

arising from the hybridization of the original state of an individual SRR. For the configuration of the 0° twisted SRR dimer system, the two excited electric dipoles are transversely coupled, while the two magnetic dipoles are longitudinally coupled. In the case of a transverse dipole-dipole interaction, the antisymmetric and symmetric modes are at the lower and higher resonance frequencies, respectively. In contrast, in the case of longitudinal dipole-dipole interaction, the two magnetic dipoles should align parallel at the lower resonance frequency and antiparallel at the higher resonance frequency[12]. It is evident that for the 0° twisted SRR dimer system[see Fig. 2(a)], the resonance levels are determined according to the picture of transverse electric dipole-dipole interaction, with the antisymmetric(symmetric) mode having the lower (higher) resonance frequency. In essence, the two coupling mechanisms, that is, the electric and magnetic dipolar interactions, counteract one another and the electric interaction dominates in this system.

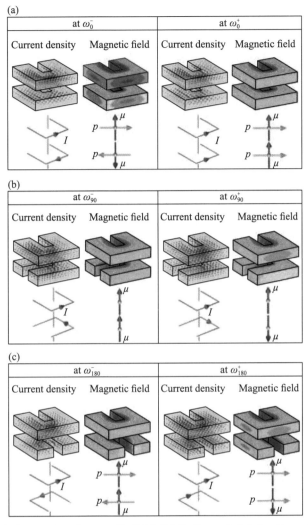

FIG. 2. Numerical current and magnetic field distributions. (a)-(c), Current and magnetic field distributions at respective resonances for the 0°(a), 90°(b) and 180°(c) twisted SRR dimer metamaterials. Lower left: schematics of currents (I) in two SRRs. Lower right: schematics of the alignments of the magnetic(μ) and electric(p) dipoles. At 0°, transverse electric and longitudinal magnetic interactions work against one another, whereas at 180° they add together. At 90°, only longitudinal magnetic interaction is present.

For the 90° twisted SRR dimer metamaterial, circular currents in the underlying SRR cannot be directly excited by the incident light due to its orientation with respect to the external electric field. In a sense, the underlying SRR can be regarded as a 'dark atom' at the resonant frequency of the ring[13]. Nevertheless, for the coupled dimer system, on resonance, excitation from the upper SRR can be transferred to the underlying one by the interaction between the two SRRs, which can also lead to the formation of new plasmonic modes(ω_{90}^- and ω_{90}^+). Interestingly, because the electric fields in the slit gaps of the two SRRs are perpendicular to one another, the electric dipole-dipole interaction equals zero. In addition, as the higher-order multipolar interaction is negligible in a first approximation, the electric coupling in the 90° twisted SRR dimer system can thus be ignored. As a consequence, the resonance levels are determined in line with the picture of longitudinal magnetic dipole-dipole coupling. As shown in Fig. 2(b), at resonances ω_{90}^- and ω_{90}^+, the resulting magnetic dipoles in the two SRRs are aligned parallel and antiparallel, respectively.

For the 180° twisted SRR dimer metamaterial, the interaction between the two SRRs results in new plasmonic modes, ω_{180}^- and ω_{180}^+. Notably, from the current and magnetic field distributions as shown in Fig. 2(c), resonances ω_{180}^- and ω_{180}^+ are associated with the excitation of the electric dipoles in the two SRRs oscillating anti-phase and in-phase, respectively. The two resulting magnetic dipoles are aligned parallel and antiparallel, accordingly. In essence, the transverse electric and longitudinal magnetic interactions contribute positively in the 180° twisted SRR dimer system. This leads to the largest spectral splitting, which is a direct indication of the coupling strength. Based on the above discussions, we infer that the optical properties of stereometamaterials depend dramatically on the spatial arrangement of metamaterial constituents. Specifically, the possibility of tuning the resonant behaviour by simply varying the relative twist angles makes stereometamaterials particularly interesting as model systems for exploring and comprehending different coupling mechanisms in complex three-dimensional plasmonic structures.

Twist Angle

To provide deeper insight, the dependence of the optical properties of the stereo-SRR dimer metamaterials on twist angle is investigated. Figure 3 presents the simulated twisting dispersion curves (black squares) of these stereometamaterials, in which the resonance positions are extracted from the transmittance spectra of different structures. It is apparent that by increasing the twist angle φ, the two resonance branches first tend to converge, with the ω^+ branch shifting to lower frequencies, while the ω^- branch shifts to higher frequencies. An avoided crossing is observed at φ_t, which is around 60°. Subsequently, the two branches shift away from one another. In order to clarify the underlying physics of the twisting dispersion curves, we introduce a Lagrangian formalism[14]. We start from the analysis of a single SRR and then expand it to coupled stereo-SRR dimer systems. One SRR can be modelled by an equivalent LC circuit with a resonance frequency $\omega_f = 1/(LC)^{1/2}$. It consists of a magnetic coil (the metal ring) with inductance L and a capacitor (the slit of the ring) of capacitance C. If we define the total charge Q accumulated in the slit as a generalized coordinate, the Lagrangian of an SRR can be written as $\varGamma = (L\dot{Q}^2/2) - (Q^2/2C)$. Here, $L\dot{Q}^2/2$ refers to the kinetic energy of the oscillations, and $Q^2/2C$ is the electrostatic energy stored in the slit. Consequently, the Lagrangian of the coupled SRR dimer systems is a combination of two individual SRRs with the additional electric and magnetic interaction terms:

$$\Gamma = \frac{L}{2}(\dot{Q}_1^2 - \omega_f^2 Q_1^2) + \frac{L}{2}(\dot{Q}_2^2 - \omega_f^2 Q_2^2) + M_H \dot{Q}_1 \dot{Q}_2 - M_E \omega_f^2 Q_1 Q_2 \cdot$$
$$(\cos\varphi - \alpha \cdot (\cos\varphi)^2 + \beta \cdot (\cos\varphi)^3). \qquad (1)$$

Here, Q_1 and Q_2 are oscillating charges in the respective SRRs, and M_H and M_E are the mutual inductances for the magnetic and electric interactions, respectively. Apart from the electric dipole-dipole interaction, the contributions from the higher-order electric multipolar[15] interactions are also included. α and β are the coefficients of the quadrupolar and octupolar plasmon interactions[16], respectively. They serve as correction terms to the electric dipolar interaction. It is straightforward to derive from equation(1) that the major interaction items for 0° and 180° cases are $M_H \dot{Q}_1 \dot{Q}_2 - M_E \omega_f^2 Q_1 Q_2$ and $M_H \dot{Q}_1 \dot{Q}_2 + M_E \omega_f^2 Q_1 Q_2$, respectively. It is in accord with the above simulation results that the magnetic and electric interactions contribute oppositely and positively for 0° and 180° twisted SRR dimer metamaterials, respectively. For the 90° twisted SRR dimer metamaterial, only the magnetic interaction plays a key role, as represented by the interaction term $M_H \dot{Q}_1 \dot{Q}_2$. Subsequently, by solving the Euler-Lagrange equations:

$$\frac{d}{dt}\left(\frac{\partial \Gamma}{\partial \dot{Q}_i}\right) - \frac{\partial \Gamma}{\partial Q_i} = 0, i = 1, 2 \qquad (2)$$

the eigenfrequencies of these stereo-SRR dimer systems can be obtained as

$$\omega_\pm = \omega_0 \cdot \sqrt{\frac{1 \mp \kappa_E \cdot (\cos\varphi - \alpha \cdot (\cos\varphi)^2 + \beta \cdot (\cos\varphi)^3)}{1 \mp \kappa_H}}, \qquad (3)$$

where $\kappa_E = M_E/L$ and $\kappa_H = M_H/L$ are the coefficients of the overall electric and magnetic interactions, respectively. By fitting the twisting dispersion curves, the corresponding coefficients are estimated to be $\kappa_E = 0.14$, $\kappa_H = 0.09$, $\alpha = 0.8$ and $\beta = -0.4$. Notably from Fig.3, the fitting curves(in red lines) reproduce the numerical data quite well and the avoided crossing is clearly observable around 60°. This shows that the Lagrangian model can quantitatively corroborate the results from the numerical simulations. It is of crucial importance that the higher-order electric multipolar interactions account for the existence of the avoided crossing. Owing to the finite length of the SRR ring, discrete electric plasmon modes characterized by different spatial symmetries can be excited by the incident light. The surface charges in the SRR ring are a superposition of such fundamental plasmon modes of the ring[16]. To reveal the significant role of the higher-order electric multipolar interactions, the grey lines in Fig. 3 display the twisting dispersion curves, in which only the dipolar coupling effect is taken into account; that is, $\alpha = 0$ and $\beta = 0$. The best fit is achieved with $\kappa_E = 0.2$ and $\kappa_H = 0.09$. Obviously, despite the fact that the grey curves can fit most parts of the numerical data, no avoided crossing is predicted. Instead, the ω^+ and

ω^- branches converge at φ_t. Therefore, it has to be emphasized that although the electric and magnetic dipolar interactions are the essential mechanisms, the higher-order electric multipolar interactions should also be carefully considered for fully understanding the origin of the spectral characteristics of the stereometamaterial systems.

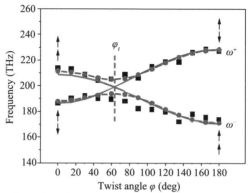

FIG.3. Twisting dispersion of the stereo-SRR dimer metamaterials. Black squares represent the numerical data. Red lines represent the fitting curves calculated from the Lagrangian model, in which the avoided crossing is clearly visible at φ_t. The black arrows represent the alignment of the magnetic dipoles at lower and higher resonance frequencies at twist angles $\varphi = 0°$ and $180°$. The grey lines represent the fitting curves calculated from the Lagrangian model without considering the higher-order electric multipolar interactions. No avoided crossing is observable in this case.

The angle where the avoided crossing occurs in the twisting dispersion spectrum is correlated with the geometry of the SRRs as well as the vertical distance between the two SRRs. For the specific stereo-SRR dimer metamaterials we investigated here, the avoided crossing appears at ~60°. Based on detailed simulated field distribution studies, we found that this angle is also a transition angle where the higher and lower frequency modes exchange their magnetic dipole alignments from parallel to antiparallel. At angles smaller than angle φ_t, the two magnetic dipoles are aligned parallel(antiparallel) at resonance ω^+ (ω^-). The electric coupling effect dominates in this regime. With continuous increase of the twist angle, due to the displacement of the two SRRs, the electric coupling contributes less effectively. Consequently, the splitting of the two resonance branches starts to decrease. This situation remains until the transition angle is reached, where the electric and magnetic dipole-dipole interactions cancel one another. The higher-order electric multipolar interactions account for the avoided crossing of the two resonance branches. After angle φ_t, the electric coupling continues to decrease. As a result, the resonance levels are determined according to the scheme of magnetic dipole-dipole coupling, that is, the parallel and antiparallel alignments of the magnetic dipoles in the two SRRs correspond to the lower and higher frequency resonances, respectively. When the twist angle reaches 90°, the electric coupling quenches and is negligible. This represents a purely magnetic dipole-dipole coupling situation. Subsequently, with further increase of the twist angle, the displacement of the two SRRs is reduced and the electric coupling comes into play again. Because of the orientation of the two SRRs, the electric and magnetic coupling can contribute positively, giving rise to a larger splitting of the two resonance branches with increasing twist angle. The splitting finally reaches its maximum at $\varphi=180°$.

The structures of stereometamaterials are compatible with nanofabrication stacking techniques[7,9]. We fabricated three stereo-SRR dimer metamaterials with specific twist angles $\varphi=0°$, $90°$ and $180°$, as illustrated in the insets of Fig. 1(b)-(d). In the experiment, the structures were fabricated on a glass substrate. Gold SRRs were embedded in a photopolymer(PC403), which served as the dielectric spacer. A spacer of $s=120$ nm was applied in order to achieve surface planarization for stacking the second SRR layer. The electron micrographs of the fabricated SRR dimer metamaterials were obtained by field-emission scanning electron microscopy. Figure 4(a)-(c) presents oblique views of the $0°$, $90°$ and $180°$ twisted SRR dimer metamaterials, in which the underlying SRRs are clearly visible. The insets of Fig. 4(a)-(c) show the normal views, demonstrating the good accuracy of lateral alignment for the different SRR layers. To experimentally investigate the optical properties of these SRR dimer metamaterials, the near-infrared transmittance spectra of the samples at normal incidence were measured by a Fourier-transform infrared spectrometer with electric field polarization as illustrated in Fig. 1. The measured transmittance spectra are presented by black curves in Fig. 4(d)-(f) and the simulated spectra as red dashed curves. The resonance positions are redshifted compared to those of the corresponding resonances in Fig. 1(b)-(d) due to the presence of glass substrate and dielectric spacers. For a reasonable comparison with the experiment, in the simulations in Fig. 4(d)-(f), gold with a three times higher damping constant as that used in Fig. 1(b)-(d) was used to account for the surface scattering and grain boundary effects in the thin film of the real systems[17]. The overall qualitative agreement between experimental and simulated results is quite good. The discrepancies are most likely due to tolerances in fabrication and assembly, as well as significant broadening in the experiment. For $0°$ and $180°$ twisted SRR dimer structures, the lower resonances ω_0^- and ω_{180}^- are less distinctly visible than the higher resonances ω_0^+ and ω_{180}^+, respectively[see spectra in Fig. 4(d),(f)]. This is due to the fact that for both dimer structures, the electric coupling plays a key role. At the lower resonance frequencies (ω_0^- and ω_{180}^-), the electric dipoles in the two SRRs oscillate anti-phase. Such resonances are not easily excited by light. On the other hand, at the higher resonance frequencies (ω_0^+ and ω_{180}^+), the electric dipoles in the two SRRs oscillate in-phase. Such resonances can strongly couple to light. For the $90°$ twisted SRR dimer structure, the splitting of the resonances is clearly observable when an analyser is applied behind the sample, which is rotated by $75°$ with respect to the polarization of the incident light. This is due to the polarization rotation effect arising from the chirality[18,19] of the $90°$ twisted structure.

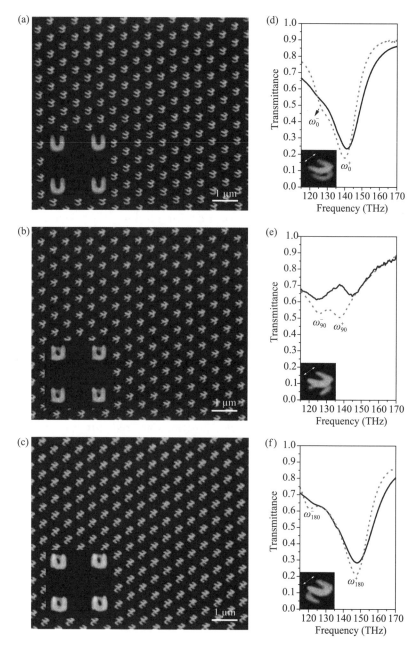

FIG.4. field-emission electron microscopy images and experimental measurement. (a)-(c), Oblique views of the 0°(a), 90°(b) and 180°(c) twisted gold SRR dimer metamaterials. Insets: normal views. The structures were fabricated on a glass substrate. The SRRs were embedded in a photopolymer (PC403), which served as the dielectric spacer. (d)-(f), Experimental transmittance spectra for the 0°(d), 90°(e), and 180°(f) twisted SRR dimer metamaterials. The black and red curves represent the experimental and simulated results, respectively. For the 90° twisted SRR dimer structure, an analyser is applied behind the sample, which is rotated by 75° with respect to the polarization of the incident light.

The new concept of stereometamaterials adds a significant degree of freedom through the interplay of electric and magnetic interactions, and tremendously enhances the versatility of nanophotonic structures. Stereometamaterials allow us to use higher-order electric multipolar as well as magnetic interactions, which can be nearly as large as the electric dipolar interaction. This is completely different from molecules, where electric dipolar interaction is the essential contribution determining optical properties. It will also be interesting to study the geometry and distance dependence of the different coupling effects. Our concept can be extended to more complex artificial molecules, such as stereotrimers, stereoquadrumers and so on. The tuneability of the resonant behaviour of these new artificial materials by altering the spatial arrangement of their constituents offers great flexibility for exploring useful metamaterial applications, such as chiral structures with negative refraction[20], invisibility cloaks[21] and magneto-optically active materials[22]. Stereometamaterials open up the potential for optical polarization control, which so far has been dominated by stereoisomers[1] and liquid crystals[23]. (See Supplementary Information for more details on optical stereoisomers as well as left- and right-handed enantiomers.) Stereometamaterials might also serve as artificial nanosystems for emulating the optical properties of complex biomolecules, such as double helix DNA chiral proteins and drug enzymes, which have profound application potentials in biophotonics, pharmacology, as well as diagnostics.

References and Notes

[1] Robinson, M. J. T. *Organic Stereochemistry* (Oxford Univ. Press, 2000).

[2] Smith, D. R., Pendry, J. B. & Wiltshire, M. C. K. Metamaterials and negative refractive index. *Science* **305**, 788–792(2004).

[3] Soukoulis, C. M., Linden, S. & Wegener, M. Negative refractive index at optical wavelengths. *Science* **315**, 47–49(2007).

[4] Shalaev, V. M. Optical negative-index metamaterials. *Nature Photon.* **1**, 41–48(2007).

[5] Veselago, V. G. The electrodynamics of substances with simultaneously negative values of ε and μ. *Sov. Phys. Usp.* **10**, 509–514(1968).

[6] Pendry, J. B. Negative refraction makes a perfect lens. *Phys. Rev. Lett.* **85**, 3966–3969(2000).

[7] Liu, N. *et al*. Three-dimensional photonic metamaterials at optical frequencies. *Nature Mater.* **7**, 31–37(2008).

[8] Liu, N. *et al*. Plasmon hybridization in stacked cut-wire metamaterials. *Adv. Mater.* **19**, 3628–3632 (2007).

[9] Liu, N., Fu, L. W., Kaiser, S., Schweizer, H. & Giessen, H. Plasmonic building blocks for magnetic molecules in three-dimensional optical metamaterials. *Adv. Mater.* **20**, 3859–3865(2008).

[10] Prodan, E., Radloff, C., Halas, N. J. & Nordlander, P. A hybridization model for the plasmon response of complex nanostructures. *Science* **302**, 419–422(2003).

[11] Wang, H., Brandl, D. W., Le, F., Nordlander, P. & Halas, N. J. Nanorice: a hybrid plasmonic nanostructure. *Nano Lett.* **6**, 827–832(2006).

[12] Nordlander, P., Oubre, C., Prodan, E., Li, K. & Stockman, M. I. Plasmon hybridization in

nanoparticle dimers. *Nano Lett.* **4**, 899-903(2004).

[13] Liu, N., Kaiser, S. & Giessen, H. Magnetoinductive and electroinductive coupling in plasmonic metamaterial molecules. *Adv. Mater.* **20**, 4521-4525(2008).

[14] Liu, H. et al. Magnetic plasmon hybridization and optical activity at optical frequencies in metallic nanostructures. *Phys. Rev.* B **76**, 073101(2007).

[15] Rockstuhl, C. et al. On the reinterpretation of resonances in split-ring-resonators at normal incidence. *Opt. Express* **14**, 8827-8836(2006).

[16] Hao, F. et al. Shedding light on dark plasmons in gold nanorings. *Chem. Rev. Lett.* **458**, 262-266 (2008).

[17] Zhang, S. et al. Demonstration of metal-dielectric negative-index metamaterials with improved performance at optical frequencies. *J. Opt. Soc. Am.* B **23**, 434-438(2006).

[18] Rogacheva, A. V., Fedotov, V. A., Schwanecke, A. S. & Zheludev, N. I. Giant gyrotropy due to electromagnetic-field coupling in a bilayered chiral structure. *Phys. Rev. Lett.* **97**, 177401(2006).

[19] Decker, M., Klein, M. W., Wegener, M. & Linden, S. Circular dichroism of planar chiral magnetic metamaterials. *Opt. Lett.* **32**, 856-858(2007).

[20] Pendry, J. B. A chiral route to negative refraction. *Science* **306**, 1353-1355(2004).

[21] Schurig, D. et al. Metamaterial electromagnetic cloak at microwave frequencies. *Science* **314**, 977-980 (2006).

[22] Svirko, Y. P. & Zheludev, N. I. *Polarization of Light in Nonlinear Optics*(Wiley, 1998).

[23] Scharf, T. *Polarized Light in Liquid Crystals and Polymers* (Wiley, 2007).

[24] The authors would like to thank M. Stockman, T. Pfau, F. Giesselmann and M. Dressel for useful discussions and comments. We thank S. Linden for stimulating us to study the twisted SRRs with different angles. We acknowledge S. Hein for his metamaterial visualizations. We gratefully thank M. Hirscher and U. Eigenthaler at the Max-Planck-Institut für Metallforschung for their electron microscopy support. We acknowledge S. Kaiser, H. Graebeldinger and M. Ubl for technical assistance. This work was financially supported by Deutsche Forschungsgemeinschaft(SPP1113 and FOR557), Landesstiftung BW and BMBF(13N9155 and 13N10146). The research of H.L. and S.Z. was financially supported by the National Natural Science Foundation of China(no. 10604029, no. 10704036 and no. 10874081) and the National Key Projects for Basic Researches of China (no. 2009CB930501, no. 2006CB921804 and no. 2004CB619003).

[25] Supplementary Information accompanies this paper at www.nature.com/naturephotonics.

[26] Methods: Structure fabrication. Three(or more) gold alignment marks(size 4×100 μm) with a gold thickness of 250 nm were first fabricated using a lift-off process on a quartz substrate. The substrate was then covered with a 50-nm gold film using electron-beam evaporation. Next, SRR structures were defined in negative resist (AR-N, ALLRESIST GmbH) by electron-beam lithography. Ion beam etching(Ar^+ ions) of the gold layer was then performed to generate the gold SRR structures. Subsequently, a 120-nm-thick spacer layer was applied on the first layer by spin-coating. A solidifiable photopolymer, PC403(JCR), was used as the planarized spacer layer. A pre-baking process in which the baking temperature was continuously increased from 90℃ to 130℃ was first performed to remove the solvent from the polymer. A sufficiently long bake at a higher temperature(30 min in a 180℃ oven) further hardened the layer. A 50-nm gold film and a spin-coated AR-N resist layer were subsequently deposited on the sample. Next, the stacking alignment using the gold alignment marks was applied to ensure accurate stacking of the second SRR layer. The procedures of in-plane fabrication were repeated

with the final layer being PC403. All structures had a total area of 200×200 μm.

Optical and structure characterization. Transmittance spectra were measured with a Fourier-transform infrared spectrometer(Bruker IFS 66v/S, tungsten lamp) combined with an infrared microscope($\times 15$ Cassegrain objective, NA = 0.4, liquid N_2-cooled MCT 77 K detector, infrared polarizer). The measured spectra were normalized with respect to a bare glass substrate.

The simulated transmittance spectra and field distributions were performed using the software package CST Microwave Studio. For the spectra in Fig. 1(b)-(d), the permittivity of bulk gold in the infrared spectral regime was described by the Drude model with plasma frequency $\omega_{pl} = 1.37 \times 10^{16}$ s^{-1} and the damping constant $\omega_c = 4.08 \times 10^{13}$ s^{-1}. For the spectra in Fig. 4(d)-(f), owing to the surface scattering and grain boundary effects in the thin film of the real systems, the simulation results were obtained using a damping constant that was three times larger than the bulk value. The optical parameters were the refractive index of PC403 $n_{PC403} = 1.55$ and the quartz substrate refractive index $n_{glass} = 1.5$.

The electron micrographs of the fabricated structures were taken with an FEI-Nova Nanolab 600 scanning electron microscope.

Magnetic Plasmon Propagation Along a Chain of Connected Subwavelength Resonators at Infrared Frequencies[*]

H. Liu,[1] D. A. Genov,[1] D. M. Wu,[1] Y. M. Liu,[1] J. M. Steele,[1] C. Sun,[1] S. N. Zhu,[2] and X. Zhang[1]

[1] 5130 Etcheverry Hall, Nanoscale Science and Engineering Center, University of California, Berkeley, California 94720-1740, USA

[2] Department of Physics, Nanjing University, Nanjing 210093, People's Republic of China

  A one-dimensional magnetic plasmon propagating in a linear chain of single split ring resonators is proposed. The subwavelength size resonators interact mainly through exchange of conduction current, resulting in stronger coupling as compared to the corresponding magneto-inductive interaction. Finite-difference time-domain simulations in conjunction with a developed analytical theory show that efficient energy transfer with signal attenuation of less then 0.57 dB/μm and group velocity higher than $1/4c$ can be achieved. The proposed novel mechanism of energy transport in the nanoscale has potential applications in subwavelength transmission lines for a wide range of integrated optical devices.

  A fundamental problem of integrated optics is how to transport electromagnetic(EM) energy in structures with transverse dimensions that are considerably smaller than the corresponding wavelength of illumination. The main reason to study light guiding in the nanoscale has to do with the size of transmission lines being a limiting factor for substantial miniaturization of integrated optical devices. Planar waveguides and photonic crystals are currently key technologies enabling a revolution in integrated optical components[1,2]. However, the overall size and density of the optical devises based on these technologies is limited by the diffraction of light, which sets the spatial extend of the lowest guided electromagnetic mode at about half wavelength.

  Recently, a new method of electromagnetic energy transport has been proposed that allows size reduction of the optical devices to below the diffraction limit [3—5]. The EM energy is coherently guided via an array of closely spaced metal nanoparticles due to a near-field coupling. Metal particles are well known to support collective electronic excitation, surface plasmon(SP) with resonance frequencies depending on the particle size and shape. Owing to the SP resonances, metal nanoparticles exhibit strong light absorption with absorption cross section far exceeding their geometrical sizes. Thus, metal nanostructures

---

[*] Phys.Rev.Lett.,2006,97(24):243902

could efficiently convert EM energy into oscillatory electron motion, which is a necessary condition for strong coupling of light into a waveguiding structures.

In 1999, [6] Pendry reported that nonmagnetic metallic element, double split ring resonator(DSRR), with size below the diffraction limit, exhibits strong magnetic response and behaves like an effective negative permeability material. While in such systems, there are no free magnetic poles; the excitation of displacement currents in the DSRR results in induction of a magnetic dipole moment that is somehow similar to a bar magnet. In analogy to the SP resonances in metal nanoparticles, an effective media made of DSRRs could support resonant magnetic plasmon (MP) oscillations at GHz [6—8] and THz frequencies[9—12]. Combined with an electric response, characterized by negative permittivity, such systems could lead to development of metamaterials with negative indexes of refraction[7,8].

In this letter, we propose a subwavelength size metal structure, referred to as a single split ring resonator(SSRR) which(a) demonstrates magnetic resonance in the THz range, and(b) could be used to support propagation of long range MP polaritons. It is well known that radiation loss of a magnetic dipole is substantially lower as compared to the radiation of an electric dipole of similar size[13]. Thus, application of MP for guiding EM energy for long distances has great potential for direct application in novel subdiffraction size transmission lines. Indeed, magnetic plasmons have been already shown to play an important role in the excitation of magneto-inductive[14] and electro-inductive waves [15] in the microwave range. In this letter, we show that at high frequencies, a coupling mechanism based on exchange of conduction current between specially designed resonators may be utilized to efficiently transfer energy along a subwavelength sized metal nanostructures. The interaction due to the conduction current is found to be much stronger than the corresponding magneto-inductive coupling, which leads to significant improvement in the properties of the guided MP wave.

Figure 1(a) presents a novel design of a SSRR characterized by two half-space metal loops with tails adjacent to their ends. Pendry's double split ring structure has no tail; nevertheless, the space between the rings acts as a capacitor allowing the flow of a displacement current[6]. In the present design, it is the gap in the tails that plays the role of a capacitor. Excitation of magnetic response in a system of SSRRs fabricated on a planar substrate results in induction of magnetic dipoles moments that are perpendicular to the substrate plane(see Fig. 1). Parallel dipoles are characterized with small spatial field overlap, and consequently the magneto-inductive interactions between them are expected to be rather weak. To substantially increase coupling between the dipoles, we physically connect the SSRRs, as shown in Fig. 1(b). The contact between the rings serves as a "bond" for conduction current to flow from one SSRR to its neighbor. Thus, in addition to the magneto-inductive coupling, our system interacts directly by exchange of conduction current. This type of coupling is somewhat similar to the electron exchange interaction

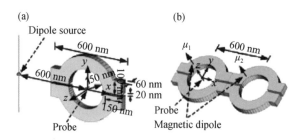

FIG.1. (color online) Schematic illustration of a single split ring resonator(a) and a connected unit for implementation in subwavelength sized MP transmission lines (b). The geometrical characteristics and position of the dipole source excitation are included.

between two magnetic atoms in a ferromagnetic material[16]. As shown below, the direct physical link between the resonators leads to stronger interaction between the SSRRs and improves the EM energy transport along a chain of SSRRs.

To study the EM response of the proposed SSRR, we perform a set of finite-difference time-domain (FDTD) calculations using a commercial software package CST Microwave Studio (Computer Simulation Technology GmbH, Darmstadt, Germany). In the calculations, we rely on the Drude model to characterize the bulk metal properties. Namely, the metal permittivity in the infrared spectral range is given by $\varepsilon(\omega)=1-\omega_p^2/(\omega^2+i\omega\omega_\tau)$, where ω_p is the bulk plasma frequency and $\omega\tau$ is the relaxation rate. For gold, the characteristic frequencies fitted to experimental data are $\hbar\omega_p = 9.02$ eV and $\hbar\omega_\tau = 0.027$ eV[17]. In the numerical calculations, a dipole source excites the SSRR, and a probe is employed to detect the local field at the center of SSRR(see Fig. 1). In the case of a single SSRR, we obseve a well pronounced resonance at $\hbar\omega_0 = 0.36$ eV with resonance bandwidth $\hbar\Gamma = 10$ meV (see Fig. 2). Introduction of a second SSRR, shown in Fig. 1(b), results in splitting of the MP resonance due to interaction. The two nondegenerate MP modes have eigenfrequencies $\hbar\omega_1 = 0.24$ eV and $\hbar\omega_2 = 0.38$ eV, respectively.

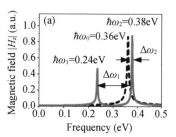

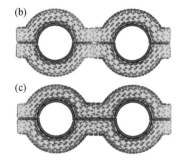

FIG.2. (color online) (a) The magnetic field amplitude $|H_z|$ detected at the center of the resonators is plotted vs excitation frequency. Frequency split $\Delta\omega_{1,2}$ is observed for the case of two physically connected SSRRs (solid line). The local current distributions are calculated for the (b) antisymmetric and (c) symmetric MP modes.

To better understand the interactions involved in the splitting of the MP resonance, we develop a comprehensive semianalytic theory based on the attenuated Lagrangian formalism. If q_m is the total oscillation charge in the m-th SSRR, L is the induction of the ring, and C is the capacitance of the gap, then we can write the Largangian of the coupled system as

$$\Im = \frac{1}{2}L(\dot{q}_1^2 + \dot{q}_2^2) - \frac{1}{2C}(q_1^2 + q_2^2) + M\dot{q}_1\dot{q}_2 - \frac{1}{4C}(q_1 - q_2)^2, \tag{1}$$

where the first two terms correspond to the energy stored in the inductors and end capacitors, the interaction term $M\dot{q}_1\dot{q}_2$ is due to magneto-inductive coupling, and the static energy stored in the shared middle capacitor is also included. By introducing Ohmic dissipation $\Re = \frac{1}{2}\gamma(\dot{q}_1^2 + \dot{q}_2^2)$ and substituting Eq.(1) in the Euler-Lagrangian equations:

$$\frac{d}{dt}\left(\frac{\partial \Im}{\partial \dot{q}_m}\right) - \frac{\partial \Im}{\partial q_m} = -\frac{\partial \Re}{\partial \dot{q}_m}, m = 1, 2 \tag{2}$$

we obtain a coupled equations for the magnetic moments $\mu_m = A\dot{q}_m$ (A is a constant related to the area of SSRR and it geometry):

$$\ddot{\mu}_1 + \omega_0^2\mu_1 + \Gamma\dot{\mu}_1 = \frac{1}{2}\kappa_1\omega_0^2(\mu_1 + \mu_2) - \kappa_2\ddot{\mu}_2, \tag{3a}$$

$$\ddot{\mu}_2 + \omega_0^2\mu_2 + \Gamma\dot{\mu}_2 = \frac{1}{2}\kappa_1\omega_0^2(\mu_1 + \mu_2) - \kappa_2\ddot{\mu}_1, \tag{3b}$$

where $\omega_0^2 = 2/(LC)$ and $\Gamma = \gamma/L$ are the degenerated MP mode eigenfrequency and bandwidth, respectively. Clearly, the electromagnetic coupling between the resonators is governed by two separate mechanisms. The first term to the right side of Eqs.(3), corresponds to interaction due to exchange of conduction current while the second term is the magneto-inductive contribution. The coupling coefficients are related to the equivalent circuit characteristics of the SSRR. For instance, $\kappa_2 = M/L$ depends on the SSRR's mutual and self inductance and for an ideal circuit $\kappa_1 = 1/2$[see Eq.(1)].

Eqs.(3) yield solutions in the form of damped harmonic oscillations $\mu = \mu_{i0}\exp\left(-\frac{1}{2}\Gamma_i t + i\omega_i t\right)$, where the index $i = 1, 2$ specifies the MP mode. Using that $\Gamma/2\omega_0 \ll 1$, it is straightforward to estimate from Eqs.(3) the system eigenfrequencies as $\omega_1 = \omega_0\sqrt{(1-\kappa_1)/(1+\kappa_2)}$ and $\omega_2 = \omega_0/\sqrt{1-\kappa_2}$, where the coupling coefficients $\kappa_1 \approx 0.5$ and $\kappa_2 \approx 0.1$ are retrieved numerically. For the high frequency (antisymmetric) mode ω_2, one has $\mu_1 = -\mu_2$, and the exchange current interaction term in Eq.(1) is negligible. Consequently, the observed frequency shift $\Delta\omega_2$ is predominantly due to the magneto-inductive coupling between the SSRRs. This phenomenon is depicted in Fig. 2(b), where we have plotted the local current density inside the resonators. Two distinctive current loops, each closed through a displacement current at the resonator tails, are formed. No conduction current is shared between the SSRRs. The opposite is true for the low-frequency (symmetric) MP mode ω_1, where $\mu_1 = \mu_2$, and both exchange of conduction current and magneto-inductive interactions contribute to the frequency shift $\Delta\omega_1$. The unimpeded flow of current between the SSRRs is clearly seen in Fig. 2(c). Comparison between the frequency shifts $\Delta\omega_1 \gg \Delta\omega_2$ and the absolute value of the coupling constants $\kappa_1 \gg \kappa_2$ undoubtedly shows that the exchange of conduction current is the dominant coupling mechanism for the proposed SSRRs system.

The magnetic dipole model described above can also be applied to investigate a finite or infinite chain of connected SSRRs. Indeed, if a magnetic dipole μ_m is assigned to each resonator and only nearest neighbor interactions are considered then the Lagrangian and the dissipation function of the system can be written as

$$\Im = \sum_m \left(\frac{1}{2} L \dot{q}_m^2 - \frac{1}{4C}(q_m - q_{m+1})^2 + M \dot{q}_m \dot{q}_{m+1} \right),$$

$$\Re = \sum_m \frac{1}{2} \gamma \dot{q}_m^2. \tag{4}$$

Substitution of Eq. (4) in the Euler-Lagrangian equations yields the equations of motion for the magnetic dipoles:

$$\ddot{\mu}_m + \omega_0^2 \mu_m + \Gamma \dot{\mu}_m = \frac{1}{2} \kappa_1 \omega_0^2 (\mu_{m-1} + 2\mu_m + \mu_{m+1}) - \kappa_2 (\ddot{\mu}_{m-1} + \ddot{\mu}_{m+1}). \tag{5}$$

The general solution of Eq. (5) corresponds to an attenuated MP wave: $\mu_m = \mu_0 \exp(-m\alpha d) \exp(i\omega t - imkd)$, where ω and k are the angular frequency and wave vector, respectively, α is the attenuation per unit length, and d is the SSRR's size. By substituting $\mu_m(t)$ into Eq. (5) and working in a small damping approximation ($\alpha d \ll 1$), simplified relationships for the MP dispersion and attenuation are obtained:

$$\omega^2(k) = \omega_0^2 \frac{1 - \kappa_1 [1 + \cos(kd)]}{1 + 2\kappa_2 \cos(kd)}, \tag{6a}$$

$$\alpha(\omega) = \frac{\omega \Gamma}{\kappa_1 \omega_0^2 + 2\kappa_2 \omega^2} \frac{1}{\sin(kd) \cdot d}. \tag{6b}$$

The range of applicability and overall accuracy of the predicted relationships are compared in Fig. 3 to FDTD results for a finite chain of SSRRs. The chain size is restricted to 50 resonators lengths, which assure reliable estimates of the system properties without imposing overwhelming computational constraints. The MP polariton is excited by a dipole source placed at a distance of 600 nm from the center of the leading SSRR element, and the H field along the chain is analyzed to determine the wave vector k of the propagating mode. Numerically and analytically estimated MP dispersion and attenuation curves are depicted in Figs. 3(b) and 3(c), respectively.

In contrast to the EP polariton in a linear chain of nanosized metal particles,[3-6] where both transverse and longitudinal modes could exist, the magnetic plasmon is exclusively a transversal wave. It is manifested by a single dispersion curve[black solid line in Fig. 3(b)] which covers a broad frequency range $\omega \in (0, \omega_c)$, with a cutoff frequency $\hbar\omega_c \approx 0.4$ eV. The precise contribution of each coupling mechanism in the MP dispersion can be studied readily using Eq. (6a). Exclusion of the magneto-inductive term results in

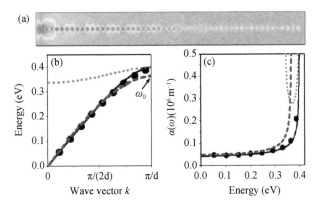

FIG.3. (color online) (a) FDTD simulation of a MP propagation along a connected chain of 50-SSRRs at $\hbar\omega = 0.3$ eV; (b) dispersion $\omega(k)$, and (c) attenuation coefficient $\alpha(\omega)$. The analytical result [Eq. (6)] including conduction current and magneto-inductive interactions, black solid curve, matches well with the FDTD numerical data (circle dots). The predicted MP characteristics based singularly on exchange current interactions ($\kappa_2 = 0$) or magneto-inductive interactions ($\kappa_1 = 0$) are presented with dashed and dotted curved lines, respectively.

slight decrease in the cutoff frequency $\omega_c \to \omega_0$ [dashed curved line in Fig. 3(b)]. On the other hand, if the SSRRs interact only through the magneto-inductive force, a dramatic change in the MP dispersion is observed [dotted curved line in Fig. 3(b)]. Namely, the propagating band shrinks to a very narrow range of frequencies $\Delta\omega \cong 2\omega_0 \kappa_2$ centered around ω_0. Such relatively short bandwidths are characteristic for the EP[5] and follows from the rapid fall of the magneto-inductive force with the distance.

Strong wave dissipation has been one of the major obstacles for utilization of surface plasmons in optical devices. The subdiffraction-sized MP transmission line, proposed in this letter, promises a considerable improvement in the wave transmission as shown in Fig. 3(c). For most of the propagation band, $\alpha(\omega)$ stays constant and has relatively low value. For instance, at an incident frequency $\hbar\omega = 0.3$ eV, the MP attenuation coefficient is $\alpha = 0.65 \times 10^5 \, \text{m}^{-1}$ (signal attenuation 0.57dB/μm), which gives a field decay length of 15.4 μm (3.7 free space wavelengths) or 25.7 unit cells. For comparison, the gold nanoparticle system presented in reference[5] manifests at $\omega = 2.4$ eV, a field decay length of about 410 nm(signal attenuation 21.4 dB/μm) which also corresponds to 5.4 unit cells or 0.8 free space wavelengths. Thus, the proposed MP transmission line performs better compared to the EP not only in terms of the absolute value of the propagation length, but also in its relation to the operation free space wavelength and the size of each individual resonator. The reason behind this improvement in MP transmission is easily understood by looking at the expected attenuation when one of the coupling mechanisms is artificially impeded. Clearly, MPs excited entirely by inductive coupling, similarly to the EPs, exhibit strong attenuation, while introduction of direct physical link between the resonators improves transmission[dashed and dotted curved lines in Fig. 3(c)]. This effect is also manifested in the MP group velocity $v_g = \partial\omega/\partial k$, which reaches values up to $0.25c$ at the center of the propagation band, and is a factor of 4 faster than the result reported for EP[5]. Thus, compared to the EP, a MP pulse could travel at higher speeds and propagates greater distances.

Finally, it is important to mention that the MP properties can be tuned by changing the material used and the size and shape of the individual SSRRs. For instance, at $\omega = 0.3$ eV, utilization of silver instead of gold results in longer MP's field decay length of about 16.6 μm. However, silver is easily oxidized, which make MP transmission lines based on this metal less versatile and difficult to integrate with the current CMOS technology. From our simulations, not presented here, it is also clear that system size manipulation such as down scaling or change of the capacitor gap width has strong effect on the magnetic response. Generally, the MP resonance frequency increases linearly with the decrease of the overall SSRR size. Unfortunately, for high frequencies ($\hbar\omega > 1.2$ eV), this scaling tends to saturate as shown in Ref. [18]. An alternative way is to employ more complicated in shapes magnetic resonators or to change the dielectric constant of the surrounding media. All those prospective solutions and their effect on the MP propagation require further studies.

In conclusion, we have proposed and studied a one-dimensional magnetic plasmon propagating in a linear chain of novel single split ring resonators. We show that at infrared frequencies, a coupling mechanism based on exchange of conduction current could be used to improve energy transmission. A comprehensive analytical model is developed for calculation of MP dispersion, attenuation coefficient, and group velocity. The theory is consistent with the performed FDTD simulations, representing a direct evidence of effective energy transfer below the diffraction limit. Excitation of MP could be a promising candidate for the development of a wide range of optical devices, including in-plane, CMOS compatible subwavelength optical waveguides, fast optoelectronic switches, and transducers.

References and Notes

[1] B. E. A. Saleh and M. C. Teich, *Fundamentals of Photonics* (Wiley, New York, 1991).

[2] A. Mekis *et al.*, Phys. Rev. Lett. **77**, 3787(1996).

[3] M. Quinten *et al.*, Opt. Lett. **23**, 1331(1998).

[4] M. L. Brongersma, J. W. Hartman, and H. A. Atwater, Phys. Rev. B **62**, R16356(2000).

[5] S. A. Maier, P. G. Kik, and H. A. Atwater, Phys. Rev. B **67**, 205402(2003).

[6] J. B. Pendry *et al.*, IEEE Trans Microwave Theory Tech. **47**, 2075(1999).

[7] D. R. Smith, W. J. Padilla, D. C. Vier, S. C. Nemat-Nasser, and S. Schultz, Phys. Rev. Lett. **84**, 4184 (2000).

[8] R. A. Shelby, D. R. Smith, and S. Schultz, Science **292**, 77(2001).

[9] T. J. Yen *et al.*, Science **303**, 1494(2004).

[10] S. Linden *et al.*, Science **306**, 1351(2004).

[11] C. Enkrich et al., Phys. Rev. Lett. **95**, 203901(2005).
[12] V. M. Shalaev et al., Opt. Lett. **30**, 3356(2005).
[13] J. D. Jackson, *Classical Electrodynamics* (John Wiley & Sons, Inc., New York, 1999).
[14] O. Sydoruk et al., Appl. Phys. Lett. **87**, 072501(2005).
[15] M. Beruete et al., Appl. Phys. Lett. **88**, 083503(2006).
[16] N. Majlis, *The Quantum Theory of Magnetism* (World Scientific, Singapore, 2000).
[17] M. A. Ordal et al., Appl. Opt. **22**, 1099(1983).
[18] J. Zhou et al., Phys. Rev. Lett. **95**, 223902(2005).
[19] This work was supported by AFOSR MURI(Grant No. FA9550-04-1-0434), SINAM and NSEC under Grant No. DMI-0327077.

Long-Wavelength Optical Properties of a Plasmonic Crystal[*]

Cheng-ping Huang,[1,2] Xiao-gang Yin,[1] Qian-jin Wang,[1] Huang Huang,[1] and Yong-yuan Zhu[1]

[1] *National Laboratory of Solid State Microstructures, Nanjing University, Nanjing 210093, People's Republic of China*

[2] *Department of Applied Physics, Nanjing University of Technology, Nanjing 210009, People's Republic of China*

The optical properties of a plasmonic crystal composed of gold nanorod particles have been studied. Because of the strong coupling between the incident light and vibrations of free electrons, the long-wavelength optical properties such as the dielectric abnormality and polariton excitation etc., which were suggested originally in ionic crystals, can also be present in the plasmonic crystal. The results show that the plasmonic and ionic lattices may share a common physics.

Recently, plasmonic materials composed of nanostructured metals have been intensively explored in realizing enhanced optical transmission, optical magnetism, as well as negative refractive index[1—3]. Employing plasmonic photonic crystals to produce a photonic stop band also opens up a new way for manipulating the motion of photons. In a metal surface, the surface-plasmon polariton(SPP) mode, originating from the interaction of light with the surface charges, can be supported. Stimulated by the concept of dielectric photonic crystal, a two-dimensional SPP crystal, a metal surface textured with periodic particles or nanoholes(the lattice period is comparable with the SPP wavelength), has been constructed[4,5]. Near the Brillouin zone boundary, a stop band for the surface mode can be created due to the Bragg reflection. Such a structure has been used to make subwavelength plasmonic waveguides, highly dispersive photonic elements, and other devices [6,7]. In addition, because of the larger dielectric contrast between the metal and dielectric medium, metallic nanostructures are also favorable for designing a three-dimensional plasmonic photonic crystal that operates in the optical frequencies[8—11].

The Bragg reflection may not be the sole mechanism for stop-band formation[12]. In this letter, we suggest that a strong coupling effect, between the incident light and vibrations of free electrons, may exist in a plasmonic crystal, i.e., a three-dimensional array of nanorod particles. Because of the coupling effect(rather than the Bragg reflection), a(polaritonic) stop band can be supported. Compared with a common photonic crystal, here the lattice period is of deep subwavelength. To study the effect, we propose that the

---

[*] Phys.Rev.Lett.,2010,104(1):016402

Huang-Kun equation, which was established originally in ionic crystals where a strong coupling between the photons and lattice vibrations is present, can be extended to the plasmonic crystal. Accordingly, the long-wavelength optical properties suggested in the former [13,14], such as the dielectric abnormality, polariton excitation, etc., can also be found in the plasmonic lattice. The proposed effect has been demonstrated by the numerical simulations. Thus our results also bridge the connection between the artificial(plasmonic) and real(ionic) crystals.

The plasmonic crystal under our study is composed of gold nanorod particles, which are well separated and arranged in a simple-cubic lattice in a host medium. Figure 1(a) shows the schematic view of the structure. Here, the lattice constant is d, the permittivity of the host medium is ε_d, and the nanorod has a length of l and radius of r_0. Supposing that the radius of the nanorod is smaller than the skin depth($r_0 < \delta \sim 22$ nm), the fields inside the nanorod can be taken to be homogeneous. We also assume that the sizes of nanorod are much larger than the Fermi wavelength($r_0 \gg \lambda_F \sim 0.5$ nm) and that the mean level spacing or Kubo gap($\Delta E = 4E_F/3N$, where E_F is the Fermi energy and N is the number of electrons) is very small compared with the thermal energy kT (For a very small metal particle with a larger Kubo gap, the quantum size effect will be dominant[15]). Consequently, the quantum effect in the nanorods can be ignored and a classic description of the effect is applicable.

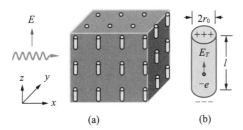

FIG.1. (color online) Schematic view of the structure under study: (a) in the plasmonic crystal, gold nanorod particles are arranged in a simple-cubic lattice. The incident light propagates in the x direction with the electric field along the rod axis; (b) the free electrons in the nanorod can be excited, leading to the accumulation of positive and negative charges on the opposite sides and the formation of a dipole moment.

When the incident light propagates with the electric field along the rod axis, the motion of free electrons can be excited and positive and negative charges will accumulate on the opposite sides of the nanorods, thus forming the electric dipole moment. This dipole moment stemming from electronic motion is equivalent to that caused by the relative motion of ions in an ionic crystal. On the other hand, the electric dipoles will emit electromagnetic waves, which further interfere with the incident light. The above effect can be enhanced in certain conditions. As we know, due to the near-field coupling, a metallic particle will interact with its neighbors such that a particle plasmon wave(or polarization wave) of transverse or longitudinal nature can propagate along the array of particles[16]. The dispersion relation for the transverse particle plasmon wave, which is similar to that of the optical branch phonon mode in an ionic crystal, will intersect the light line. Consequently, near the crossing of dispersion curves, the incident light couples strongly to the transverse plasmon or polarization wave of the rod lattice. In that case, the propagation mode is neither a pure photon mode nor a pure plasmon mode. It is called a polariton mode, a mixture of the photons and polarization waves[14].

The plasmon resonance in a gold nanorod(along the rod axis) can occur at a much longer wavelength[17]. Here, we are interested in the case that the lattice constant is very small compared with the light wavelength and thus, the photons interact strongly with the transverse plasmon wave near the center of the Brillouin zone. Similar to an ionic crystal[13], the fundamental equations governing the coupling effect in a plasmonic crystal may be given, in the long-wavelength approximation, as

$$\ddot{W} = b_{11}W + b_{12}E,$$
$$P = b_{21}W + b_{22}E. \tag{1}$$

Here, W represents the motion of free electrons in the nanorod, E is the electric field of the light, and P is the dielectric polarization induced by the electronic motion and the electric field. And, b_{11}, b_{12}, b_{21}, and b_{22} are four unknown coefficients.

To illustrate the above idea, we first consider the motion equation of free electrons in a nanorod [see Fig. 1(b)]. Note that, in the long-wavelength limit, the electronic motions of all nanorods can be taken to be almost identical. This enables us to use one parameter to characterize the long transverse plasmon waves. Moreover, different from an array of nanorod pairs which has a magnetic response[2,18], the effect of light magnetic field can be neglected in our structure, as there is only one nanorod in a unit cell. Under the action of an electric field of light, the motion of free electrons obeys Newton's equation $md^2z/dt^2 = -E_T e - \gamma m dz/dt$. Here, m, e, and γ are the mass, charge, and collision frequency of the free electrons respectively, z is the electronic displacement relative to the equilibrium positions, and E_T is the total electric field in the nanorod. Previously, we have shown that a gold nanorod can be modeled as an LC circuit having a self-inductance $L = (\mu_0 l/2\pi)\ln(l/2r_0)$ and a capacitance $C = 5\pi\varepsilon_0\varepsilon_d r_0/2$, where the two end faces of the nanorod become a circular capacitor[19]. Thus, the total electric field can be expressed as $E_T = E_{\text{eff}}^{(1)} + E_L + E_C$, where $E_{\text{eff}}^{(1)}$ represents the effective electric field imposed on the nanorod(caused by light and nanorod polarization), $E_L = -(L/l)d^2q/dt^2$ is the induced electric field associated with the self-inductance(q is the charge carried by the capacitor), and $E_C = -q/Cl$ is the electric field resulting from the circular capacitor. Noticing that $q = -nesz$ (n is the density of free electrons in the gold, $s = \pi r_0^2$ is the nanorod cross-sectional area), one obtains

$$(m + \alpha L)\ddot{z} = -(\alpha/C)z - \gamma m \dot{z} - E_{\text{eff}}^{(1)} e, \tag{2}$$

where $\alpha = ne^2 s/l$ is a coefficient dependent on the nanorod size and the electron density (for gold, a realistic value of $n = 5.90 \times 10^{28}$ m^{-3} is used in the calculation).

It can be seen from Eq.(2) that the free electrons confined in the nanorod will behave as the forced harmonic oscillators, which are characterized by an effective restoring force $F = -k_{eff} z$ ($k_{eff} = \alpha/C$) and an increased effective mass m_{eff} ($m_{eff} = m + \alpha L$). The restoring force is related to the nanorod capacitance, where the accumulated charges in the capacitor will prohibit the directional motion of electrons. The effective mass of electrons is increased due to the self-inductance of the nanorod, an electromagnetic inertia of the system (the increased electron mass has also been suggested for the periodic metallic wires[20]). For a nanorod with the length 150 nm and diameter 30 nm, for example, the effective electron mass attains $m_{eff} = 1.38\, m$. According to Eq.(2), the resonance frequency of free electrons in a single nanorod is $\omega_o = (k_{eff}/m_{eff})^{1/2}$. Equation(2) also suggests that the optical response of the free electrons in a nanorod is similar to that of the bounded electrons in a classic atom. Thus the gold nanorods can also be taken as the plasmonic "atoms".

The dielectric polarization of the plasmonic crystal comes from both electronic displacement in the nanorods and polarization in the host medium. The motion of free electrons along the rod axis gives rise to a dipole moment $p = ql = -nelsz$. In the long-wavelength approximation, the macroscopic polarization due to the free electrons is $P_{rod} = -nelsz/\Omega$, where $\Omega = d^3$ is the volume of unit cell. On the other hand, the macroscopic polarization of the host medium (related to the bounded electrons) can be written as $P_{host} = \varepsilon_0 \chi E_{eff}^{(2)}$, where χ is the molecule polarizability per unit volume and $E_{eff}^{(2)}$ is the effective electric field acting on the medium molecules. Hence, the total dielectric polarization of the crystal reads

$$P = -(nels/\Omega)z + \varepsilon_0 \chi E_{eff}^{(2)}. \tag{3}$$

To establish the relationship between Eqs.(1)—(3), the effective electric field imposed on the nanorods and medium molecules should be determined. According to the Lorentz effective-field model, we have $E_{eff}^{(1)} = E + P_{rod}/3\varepsilon_0 \varepsilon_d$ for the nanorod and $E_{eff}^{(2)} = E + P_{host}/3\varepsilon_0$ for the medium molecule. It is not difficult to find that, with the use of the effective-field formula, Eqs.(2) and (3) can be transformed to the following from

$$\ddot{W} = -\left(\omega_o^2 - \frac{Q^2}{3\varepsilon_0 \varepsilon_d M\Omega}\right)W + \frac{Q}{(M\Omega)^{1/2}} E,$$
$$P = \frac{Q}{(M\Omega)^{1/2}} W + \varepsilon_0 (\varepsilon_d - 1) E. \tag{4}$$

Here, $W = (M/\Omega)^{1/2} z$ is the motion parameter proportional to the electronic displacement, $M = nlsm_{eff}$ and $Q = -nlse$ are, respectively, the total effective mass and total charge of free electrons in a nanorod. In deriving Eq.(4), the Clausius-Mossotti formula $\chi = 3(\varepsilon_d - 1)/(\varepsilon_d + 2)$ has been used and the loss from the electronic collision in the nanorod has been neglected. One can see from Eqs.(1) and (4) that they have the same format, thus demonstrating the validity of the original hypothesis. The unknown coefficients in Eq.(1)

are then determined, respectively, to be $b_{11} = -(\omega_o^2 - Q^2/3\varepsilon_0\varepsilon_d M\Omega)$, $b_{12} = b_{21} = Q/(M\Omega)^{1/2}$, and $b_{22} = \varepsilon_0(\varepsilon_d - 1)$.

Equation(4) becomes the basic equations that describe the coupling effect between the photons and the long transverse plasmon wave. With the use of Eq.(4), the dielectric function can be obtained as

$$\varepsilon(\omega) = \varepsilon_d + \frac{fm/m_{\text{eff}}}{\omega_s^2 - \omega^2}\omega_p^2. \tag{5}$$

Here, $\omega_s = (-b_{11})^{1/2} = (\omega_o^2 - fm\omega_p^2/3m_{\text{eff}}\varepsilon_d)^{1/2}$ is the eigenfrequency of the system, which is smaller than that of a single gold nanorod (ω_o) due to the collective effect; $\omega_p = (ne^2/m\varepsilon_0)^{1/2}$ is the bulk plasma frequency of the gold; and $f = ls/\Omega$ is the filling ratio of the nanorods. Moreover, the dispersion relation of the polariton mode can be deduced with Maxwell equations and Eq.(5), yielding

$$\frac{c^2k^2}{\omega^2} = \varepsilon_d + \frac{fm/m_{\text{eff}}}{\omega_s^2 - \omega^2}\omega_p^2, \tag{6}$$

where c is the light velocity in vacuum. The result resembles that obtained in an ionic crystal where the photons couple strongly with the transverse optical phonons and phonon-polariton dispersion is induced.

The polariton dispersion and dielectric abnormality are plotted, respectively, in Figs. 2(a) and 2(b) as a function of normalized frequency (ω/ω_s). Without loss of generality, here the lattice constant is 80 nm, the permittivity of the host medium is 2.25, and the length and diameter of the nanorod is 40 nm and 10 nm, respectively. [Note that, here, the Kubo gap ($\Delta E \sim 40\ \mu eV$) is 3 orders of magnitude smaller than the thermal energy at room temperature ($kT \sim 26$ meV), thus justifying the classical approach]. This gives a resonance wavelength of $\lambda_s = 2\pi c/\omega_s = 960$ nm, significantly larger than the lattice constant

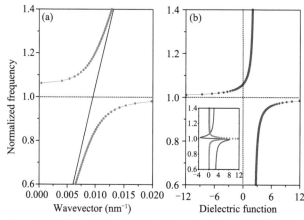

FIG.2. (color online) Calculated polariton dispersion(a) and dielectric abnormality (b). Inset shows the complex dielectric function where the loss of gold is considered; the solid line and circles represent, respectively, the real and imaginary part of the function (here $\gamma = 5 \times 10^{13}$ rad/s has been used).

(the resonance wavelength of a single nanorod is $\lambda_o=940$ nm). In Fig. 2(a), the inclined solid line and the flat dotted line correspond, respectively, to the light wave and the long transverse plasmon wave without mode coupling. The region of crossover of the solid and dotted lines is the resonance region, where the photons couple strongly to the long transverse plasmon wave(the solid circles represent the coupled mode). Near the resonance the propagation mode is not a pure photon mode or a pure plasmon mode but a coupled wave field consisting of both components. The quantum of this coupled mode is called a polariton.

One important effect of the coupling is that a polaritonic stop band will be created in the frequency range$[\omega_s,\omega_t]$, where the dielectric function is negative [see Fig. 2(b)] and the wave vector becomes imaginary(thus the light propagation will be forbidden). Here, the upper cutoff frequency is determined, by setting $\varepsilon(\omega)=0$, to be $\omega_t=(\omega_s^2+fm\omega_p^2/m_{\text{eff}}\varepsilon_d)^{1/2}$. With the used parameters, the corresponding cutoff wavelength is $\lambda_t=905$ nm (here the effective electron mass is $m_{\text{eff}}=1.04\ m$). The relative band width, which is defined as the absolute band width divided by the eigenfrequency, is approximately $\eta=fm\omega_p^2/2m_{\text{eff}}\varepsilon_d\omega_s^2$. From it, the relative band width is found to be $\eta=6.3\%$, which is 1 order of magnitude larger than the particle filling ratio($f=0.6\%$). In addition, when the loss from the free electrons is accounted [see Eq.(2)], a damping term $-i(m/m_{\text{eff}})\gamma\omega$ will appear in the denominator of dielectric function. This leads to a maximal imaginary part of dielectric function [see inset of Fig. 2(b)] and a peak of absorption locating at the eigenfrequency. The absorption is equivalent to the infrared absorption in an ionic crystal.

To verify the polaritonic stop-band effect, we have calculated the transmission spectrum of a plasmonic crystal film analytically and compared it with the numerical results(the lattice parameters mentioned above are used). For a free-standing film with the thickness h, the transmission efficiency of a normally incident light is

$$t=\left|\frac{4k_0k\exp(ikh)}{(k_0+k)^2-(k_0-k)^2\exp(2ikh)}\right|^2. \tag{7}$$

Here, k_0 is the wave vector in free space and k is the wave vector of the polariton mode. The transmission spectrum calculated by Eqs.(6) and(7) is presented as the solid circles in Fig. 3(a), showing a polaritonic stop band between 905 nm and 960 nm(the film thickness is 1600 nm, twenty unit cells thick). In the pass band, amplitude oscillation is observed due to the film Fabry-Perot resonance. As a comparison, Fig. 3(a) also presents the spectrum(the open circles) of the same structure simulated with the commercial software package FDTD Solutions 6.5 (Lumerical Solutions, Inc., Canada). In the numerical simulation, the gold is modeled with a lossless Drude model $\varepsilon_m=\varepsilon_\infty-\omega_p^2/\omega^2$, where $\varepsilon_\infty=7$ and $\omega_p=1.37\times10^{16}$ rad/s are used according to the experimental data of gold[21]. One can see that the numerical simulation agrees well with the analytical calculations, concerning the opening of stop band and amplitude oscillation in the pass band.

Additional insight can be provided by studying the transmission spectrum with a

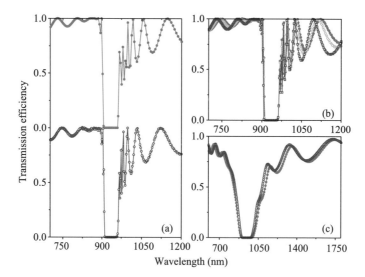

FIG. 3. (color online) (a) Analytically calculated (the solid circles) and numerically simulated (the open circles) transmission spectra (normal incidence and loss-free); (b) Simulated spectra (loss-free) with the incident angle being 0 (the solid circles), 15 (the open circles), and 25 (the open squares) degrees; (c) Spectra (normal incidence) accounting for the loss of gold: the solid and open circles represent the analytical and numerical results, respectively. Here, a free-standing plasmonic crystal film with the thickness 1600 nm was used.

varying incident angle. Figure 3(b) presents the numerical results for TE polarization, where the electric field is fixed along the rod axis to maximize the coupling effect. The results show that the stop-band formation is not dependent on the incident angle or following the Bragg diffraction. This can be understood, as the wavelength is much larger than the lattice constant and the light only "feels" an average response. In addition, we also calculated the transmission spectra, as shown in Fig. 3(c), of the structure considering the loss of gold (normal incidence). Here, the solid circles represent the analytical results of Eq. (7) and the open circles the numerical results using a lossy Drude dispersion (in both cases, $\gamma = 5 \times 10^{13}$ rad/s was used). A nearly perfect agreement between them is found. The results show that the stop band still survives but is enlarged due to the absorption.

In summary, the long-wavelength optical properties of a plasmonic crystal composed of nanorod particles have been studied. We emphasized the concept of the polariton, which is due to the coupling between the photons and the long transverse plasmon wave. The polaritonic stop band, associated with the coupling effect rather than the Bragg reflection, has been suggested. The results also show that the long-wavelength method developed for an ionic crystal can be applied to a plasmonic crystal and that the artificial and classic lattices may share a common physics.

References and Notes

[1] C. Genet and T. W. Ebbesen, Nature (London) **445**, 39 (2007).

[2] V. M. Shalaev, Nat. Photon. **1**, 41(2007).
[3] W. A. Murray and W. L. Barnes, Adv. Mater. **19**, 3771(2007).
[4] S. C. Kitson, W. L. Barnes, and J. R. Sambles, Phys. Rev. Lett. **77**, 2670(1996).
[5] L. Feng *et al.*, Appl. Phys. Lett. **93**, 231105(2008).
[6] S. I. Bozhevolnyi *et al.*, Phys. Rev. Lett. **86**, 3008(2001).
[7] V. Mikhailov *et al.*, Phys. Rev. Lett. **99**, 083901(2007).
[8] A. Moroz, Phys. Rev. Lett. **83**, 5274(1999).
[9] W.Y. Zhang *et al.*, Phys. Rev. Lett. **84**, 2853(2000).
[10] J. G. Fleming *et al.*, Nature(London) **417**, 52(2002).
[11] A.S.P. Chang *et al.*, Opt. Express **15**, 8428(2007).
[12] Y. Q. Lu *et al.*, Science **284**, 1822(1999).
[13] M. Born and K. Huang, *Dynamical Theory of Crystal Lattices*(Oxford Univ. Press, Oxford, 1954).
[14] D. L. Mills and E. Burstein, Rep. Prog. Phys. **37**, 817(1974).
[15] W.P. Halperin, Rev. Mod. Phys. **58**, 533(1986).
[16] S. A. Maier, *Plasmonics: Fundamentaland Applications*(Springer, New York, 2007).
[17] S.W. Prescott and P. Mulvaney,J. Appl. Phys. **99**, 123504(2006).
[18] V. M. Shalaev *et al.*, Opt. Lett. **30**, 3356(2005).
[19] C.P. Huang *et al.*, Opt. Express **17**, 6407(2009).
[20] J. B. Pendry *et al.*, Phys. Rev. Lett. **76**, 4773(1996).
[21] E. D. Palik, *Handbook of Optical Constants in Solids*(Academic, Boston, 1991).
[22] This work was supported by the National Natural Science Foundation of China(Grants No. 10874079, No. 10804051, and No. 10523001), by the State Key Program for Basic Research of China(Grants No. 2010CB630703 and No. 2006CB921804).

第六章 准相位匹配量子光学与光子芯片
Chapter 6 Review Article: Quasi-phase-matching Engineering of Entangled Photons

6.1 Transforming Spatial Entanglement Using a Domain-Engineering Technique

6.2 Compact Engineering of Path-Entangled Sources from a Monolithic Quadratic Nonlinear Photonic Crystal

6.3 On-chip Steering of Entangled Photons in Nonlinear Photonic Crystals

6.4 Lensless Imaging by Entangled Photons from Quadratic Nonlinear Photonic Crystals

6.5 Observation of Quantum Talbot Effect from a Domain-engineered Nonlinear Photonic Crystal

6.6 Mode-locked Biphoton Generation by Concurrent Quasi-phase-matching

6.7 Generation of NOON State with Orbital Angular Momentum in a Twisted Nonlinear Photonic Crystal

6.8 Tailoring Entanglement Through Domain Engineering in a Lithium Niobate Waveguide

6.9 On-Chip Generation and Manipulation of Entangled Photons Based on Reconfigurable Lithium-Niobate Waveguide Circuits

6.10 Generation of Three-mode Continuous-variable Entanglement by Cascaded Nonlinear Interactions in a Quasiperiodic Superlattice

第六章 准相位匹配量子光学与光子芯片

祝世宁

基于对准位相匹配的研究积累,本世纪初我们开始将介电体超晶格的研究从非线性光学拓展到量子光学。这一方面是研究工作本身的自然延伸,另一方面也是对后摩尔时代发展新型量子信息技术需求的一种呼应。量子纠缠是 1935 年由奥地利物理学家 Schrodinger 提出来的一种奇特的量子现象,它反映了微观粒子的基本属性:相干性、或然性和非定域性,被称之为"量子力学的精髓",也是量子信息中最重要的资源。人们从理论和实验上都证明通过非线性光学过程-自发参量下转换(SPDC)产生的光子对就是一个标准的双光子纠缠态(EPR 态)。EPR 双光子态函数可写为

$$|\Psi\rangle = \psi_0 \sum_{k_s, k_i} \delta(\omega_s + \omega_i - \omega_p) \delta(\bm{k}_s + \bm{k}_i - \bm{k}_p) \hat{a}_{k_s}^+ \hat{a}_{k_i}^+ |0\rangle \tag{1}$$

式中下标 p 表示入射的高频光子,s 和 i 分别表示从高频光子转换成的两个低频光子(分别称为信号和闲频光子),两个 δ 函数分别对应 SPDC 过程中的能量守恒与动量守恒条件,其中动量守恒即相位匹配可以利用非线性晶体的双折射来实现。而在介电体超晶格中同样能发生 SPDC 过程,只是三波相位失配可以不用晶体的双折射而用超晶格的倒格矢来补偿,也就是准相位匹配 SPDC 过程。这时对应的双光子波函数应为

$$|\Psi\rangle = \psi_0 \sum_{k_s, k_i} \delta(\omega_s + \omega_i - \omega_p) \delta(\bm{k}_s + \bm{k}_i + \bm{G} - \bm{k}_p) \hat{a}_{k_s}^+ \hat{a}_{k_i}^+ |0\rangle \tag{2}$$

其中 G 为超晶格的倒格矢,它是超晶格的结构参数。对于周期为 Λ 的一维结构,$G = 2\pi/\Lambda$。(2)式表明:介电体超晶格中通过准相位匹配自发参量下转换产生的光子的纠缠特性也与超晶格结构有关。因而我们可以通过超晶格的结构设计,即构建不同的准相位匹配条件来制备双光子态,调节其波前与位相,实现我们所需的频率、动量、位置、偏振等不同纠缠特性,这对量子光学和量子信息研究十分有用。

由于介电体超晶格大都使用铁电晶体作为基质材料,其中的超晶格结构是用铁电畴构建成的超结构,运用畴工程技术,介电体超晶格在双光子的产生、纠缠特性调控方面具有极大的优势和灵活性。最先将准相位匹配引入量子光学领域的是瑞士 N. Gisin 小组,他们在 2001 年报道了使用一种介电体超晶格——周期极化铌酸锂(PPLN)波导制备出了高亮度、通信波段的纠缠双光子。当时我们研究关注的是另一种介电体超晶格——周期极化钽酸锂(PPLT),我们在该体系中也证实了高通量的纠缠双光子产生及其关联特性、高的纠缠光子产率及对纠缠光子的调控能力。基于铌酸锂(LN)、钽酸锂(LT)和磷酸氧钛钾(KTP)等铁电晶体的介电体超晶格纠缠光子源的研究开始引起关注。虽然我们不是最早涉足这一领域的研究组,但由于在准相位匹配理论、实验和畴工程技术方面的长期积累,我们很快在介电体超晶格新型量子光源及其集成化功能芯片方面取得了重要进展,形成了自己的研究特色

和研究体系。先后制备出了高通量、频率可控、波前或路径可调的不同类型纠缠光子,在双光子、多光子高维纠缠和超纠缠,高精度时间、频率、位置,动量纠缠等方面取得重要进展,预示了介电体超晶格在超越经典极限的信息处理、测量和成像方面都会有重要应用。进一步的研究还发现结合畴工程技术和光波导工艺能将光子的汇聚、分束、干涉等功能在单片超晶格芯片上进行集成,这样的芯片能集纠缠光子的产生与调控于一体,直接完成量子模拟与量子计算功能,这正是量子信息技术实用化所追求的目标。

本章共选择了11篇代表性论文,其中第1篇是应《AIP ADVANCE》之邀撰写的综述文章(Review Article):"Quasi-phase-matching engineering of entangled photons"。该文介绍了准相位匹配双光子纠缠态产生的基础理论,以及我们在利用畴工程技术调控纠缠光子的特性,如相位、频率和空间纠缠等方面的进展,讨论了基于介电体超晶格的量子光源在光量子信息处理和量子成像方面可能的应用。该文是我们在这一领域工作的阶段性总结,故以此文作为本章的英文序言。

6.1是我们发表的第一篇有关介电体超晶格在量子光学研究方面的实验论文。该文首次报道使用多通道周期极化钽酸锂超晶格产生高维路径纠缠双光子对的研究结果。该实验观察到双光子高维路径之间的干涉效应,首次验证了晶体结构可以对纠缠特性进行相干调控。尽管在这之前我们已有多篇准相位匹配SPDC纠缠光子理论工作发表,但就其重要性及对实验室发展来讲该文是标志性的。

除了用多通道产生路径纠缠外,还可以用非共线的准相位匹配直接产生路径纠缠。6.2一文就介绍了另外一种产生路径纠缠光子的方案,所用晶体是一片二维六角极化的钽酸锂晶片。由于六角结构能同时提供多组有效的倒格矢,利用单个泵浦激光入射可以同时实现多个共线和非共线的准相位匹配SPDC过程。实验上我们利用此样品直接得到了单光子路径纠缠态、双光子N00N态和单光子高维路径纠缠态等。由于该结构利用非共线相位匹配方法,光子在发生非线性转换的同时自动分束,这相当于将分束器和量子干涉器都集成到该晶片中,展示了畴结构具有多功能集成的特征。该研究预言了二维超晶格自动分束功能可用于研制高效可预知的单光子源以及束斑质量良好的多光子N00N态,这在实用化的量子信息中非常有用。

6.3是上一篇有关二维结构工作的拓展,在二维对称的畴结构的横向引入一定的对称性破缺,将畴的横向设计成抛物线分布,这使得产生的参量光光子的波前的相位得到一定的调制,因而能汇聚到主光轴上的某一点,如同经过了一块透镜聚焦。进一步的理论和实验研究还发现,在超晶格主光轴两侧,还对称分布着一些次光轴,有一定的平移周期性。当入射光束位于两个主光轴之间的中间位置时,在泵浦光两侧会出现聚焦的双光子束。在这种条件下,一块这样的超晶格芯片相当于集成了纠缠光源、透镜和分束器等多种功能。不过上述所及效应仅对纠缠双光子有效,对基波光无效。

"鬼成像"是一种非经典效应,一种新原理成像技术,有重要的应用价值。"鬼成像"一般需要采用由纠缠光源、透镜和分束镜等光学元件组成的复杂光学系统。6.4一文采用6.3文中报导的畴横向呈抛物线分布的超晶格晶体,首次实验演示了无透镜的"鬼成像",还研究了无透镜一物多像的成像原理。一物多像能大大简化成像系统,这一工作将"鬼成像"的研究又向前推进一步。

"泰堡(Talbot)效应"是一种经典光学效应,是指如光栅这样的周期结构在平行光的照射下在近场区域产生的自成像效应,该效应起源于 Fresnel 近场衍射。有报道介绍采用非经典光如通过 SPDC 过程产生的简并参量光代替经典光,也能观察到对应的"量子泰堡效应"。采用符合测量,其效果等效于入射光的波长减半时经典光的效果。本章 6.5 的论文介绍了将经典泵浦激光耦合进一片具有等间隔多通道周期畴结构的光学超晶格晶体,没有观察到入射光的经典泰堡效应,但观察到了 SPDC 过程产生的简并参量光所造成的量子泰堡效应。其原因是显然的,因为超晶格调制的是二阶非线性系数,并没有对应的折射率调制,入射光不能感受到这些周期结构的存在。简并参量光的量子泰堡效应,本质上则是各超晶格通道产生的纠缠光子对在空间相互干涉的结果。6.5 的论文对"量子泰堡效应"的机理做了详细分析,预言了如"双色量子泰堡效应"、"泰堡拍频效应"等在超晶格晶体体系中所特有的一些新效应,并对"量子泰堡效应"一些可能的应用进行了讨论。

超晶格结构的最大优点是能实现功能集成,在产生纠缠光子的同时完成一些处理功能,制备出原先可能需要通过复杂的分立光路技术才能制备出的特殊纠缠光子。6.6 给出了一例,介绍了如何设计一种非周期超晶格来制配锁模双光子态。其基本原理其实很简单,即利用非周期结构提供的多个倒格矢,在一块超晶格中同时实现多个准相位匹配的 SPDC 过程,利用不同下转换模式的干涉产生锁模双光子态。这种设计对畴的分布有一定要求,即既要保证满足双光子在频谱上是等间隔的梳状分布,同时也要求不同模式之间的相位是相干的。为了得到极窄的时间关联,所设计的双光子频谱要求覆盖得尽量宽。当梳状频谱覆盖 442 nm(中心波长在 800 nm)时,设计的关联时间可以达到 4.2 fs。锁模双光子态在时钟同步、量子精密测量、量子相干层析以及量子计算等方面有着重要的应用。

对具有规则结构的介电体超晶格中的 SPDC 过程通常可采用准相位匹配原理来处理,这其中超晶格的倒格矢可参与参量过程的相位匹配。而对于非规则结构的超晶格,没有确定的倒格矢,我们也发展了一种方法,即所谓的局域准相位匹配或非线性 Huygens-Fresnel 原理,利用这一原理可以有效地控制参量光的波前与相位,详见第二章。这种方法不仅能用于频率上转换,也可用于 SPDC 过程,用于产生具有特殊纠缠特性的纠缠光子对。6.7 一文介绍了一种具有"叉形光栅"结构的介电体超晶格,这种结构没法给出确定的倒格矢,只能用非线性 Huygens-Fresnel 原理来处理。当泵浦光垂直入射到通过叉形光栅畴结构的铌酸锂晶体时,通过 SPDC 过程产生的纠缠光子的波前被这种横向叉型结构所调制,光子携带上了轨道角动量,形成所谓的 Laguerre-Gaussian(LG)模,这种模的模数(维度)可用量子数 L 来标度。这种叉型结构的一个重要应用是制备高维角动量的光子来构建双光子 N00N 态,这种高维双光子 N00N 态在许多场合可代替很难实现的多光子,用于量子信息处理,因此有重要价值。

具有非线性光学效应的介电体超晶格也同时具有电光效应,这一效应往往也具有重要应用。6.8 一文从功能集成的角度出发,提出了在介电体超晶格波导中利用电光效应调控光子偏振态的方案,使得产生的纠缠光子的偏振态可通过外加电信号加以调控,实现正交偏振和平形偏振纠缠态的实时切换,纠缠度高,带宽可调。该文提出的这种方案为下一代集成量子光源的设计和制备提供新思路。

基于上述工作,我们在 2014 年研制出基于铌酸锂晶体的集成量子光学芯片,6.9 报道了使用超晶格结构和质子交换光波导将纠缠光子源、光子分束器、电光调制器、光子干涉仪等功能单元成功集成到同一块铌酸锂光子芯片上,通过电光效应实现了芯片上纠缠光子聚束态和分离态的高效产生与快速切换。该芯片可在室温稳定工作,工作电压仅为 0 - 3.55 V,调控速率可达 40 GHz,纠缠光子对产率为 1.1×10^7 Hz nm^{-1} mW^{-1},其关键技术指标与国际上同期研制出的硅基光子芯片相比具有明显优势。该芯片输入输出端口可与光纤直接连接并固化,稳定性高,便于携带,接近实用化。这一工作是全固态量子芯片研究方面的重要进展,大大推动了铌酸锂光量子逻辑器件、光量子模拟芯片的研究。

以上各文讨论了分离变量的量子纠缠,涉及的是单光子的非经典特性和光子之间的纠缠。近些年来光场的连续变量纠缠也受到广泛关注,这是因为连续变量纠缠考虑的是两束或两束以上光束之间的多组分纠缠。这种光场纠缠具有亮度高、易于探测等独特的优势。基于压缩态光场的连续变量纠缠一般是通过光参量放大产生的,这使得基于准相位匹配的介电体超晶格在连续变量纠缠研究方面也有用武之地。我们基于自行发展的多重准相位匹配理论和准周期、非周期和二维超晶格,重点将二组分连续变量纠缠拓展至三组分、多组分直至频率梳之间的纠缠。6.10 是我们第一篇关于三组分连续变量纠缠的理论文章,该文具体阐述了如何利用准周期结构完成参量放大与和频的级联过程,以及这一级联过程所产生的三束光场之间的关联特性及其纠缠判据。该文奠定了后续介电体超晶格中多组分连续变量纠缠研究的理论基础,也为多组分连续变量纠缠的实验研究奠定了基础。

Chapter 6 Review Article: Quasi-phase-matching Engineering of Entangled Photons[*]

—In Lieu of a Preface

P. Xu and S. N. Zhu

National Laboratory of Solid State Microstructures and School of Physics, Nanjing University, Nanjing 210093, China

Quasi-phase-matching(QPM) technique has been successfully applied in nonlinear optics, such as optical frequency conversion. Recently, remarkable advances have been made in the QPM generation and manipulation of photon entanglement. In this paper, we review the current progresses in the QPM engineering of entangled photons, which are finished mainly by our group. By the design of concurrent QPM processes insides a single nonlinear optical crystal, the spectrum of entangled photons can be extended or shaped on demand, also the spatial entanglement can be transformed by transverse inhomogeneity of domain modulation, resulting in new applications in path-entanglement, quantum Talbot effects, quantum imaging etc. Combined with waveguide structures and the electro-optic effect, the entangled photons can be generated, then guided and phase-controlled within a single QPM crystal chip. QPM devices can act as a key ingredient in integrated quantum information processing.

1. Introduction

The nonlinear crystals with modulated quadratic nonlinear coefficients $\chi^{(2)}$ are called quasi-phase-matching(QPM) materials. The concept was first referred to by Armstrong *et al.* in 1960s[1,2] and first experimentally verified by Feng and Ming *et al.*[3,4] in LiNbO$_3$ crystals with periodic ferroelectric domains by the Czochralski method in 1980s. Berger later extended the two-dimensional(2D) QPM concept from one dimension(1D) to 2D, and proposed the concept of $\chi^{(2)}$ nonlinear photonic crystal(NPC)[5] in order to contrast and compare it with a regular photonic crystal having a periodic modulation on the linear susceptibility. Since then, this artificial micro-structured material has been widely applied to the fields of nonlinear optics and laser, in particular, recently in quantum optics. The rapid development of QPM materials actually result from the breakthrough of the fabrication technique-the electrical poling technique at room temperature in 1990s.[6-8] The sign of $\chi^{(2)}$ in such a crystal is modulated by reversing the orientation of ferroelectric domain according to some sequence. The motivation for such a modulation is to achieve either a significant enhancement of nonlinear frequency conversion efficiency by QPM,[9-11]

---

[*] AIP Advances, 2012, 2(4):041401

or a required wavefront of the parametric wave by nonlinear Huggens-Fresnel principle,[12—14] or for both. In history, the study for the domain modulation in a ferroelectric crystal was extended from 1D to 2D,[5, 15] from periodic to quasi-periodic,[10, 11, 16, 17] aperiodic,[18, 19] even more complicated structures.[12—14] Many novel nonlinear phenomena, such as third harmonic generation,[10] nonlinear light scattering,[20, 21] nonlinear Čerenkov radiation,[22] nonlinear Talbot effect[23, 24] etc. were discovered from such artificial materials. Nowadays, the domain-engineered crystal has been utilized in the field of quantum optics which indicates that the study of QPM technique has entered a new regime. The bright entangled photon pairs have been generated from 1D optical superlattice by spontaneously parametric down-conversion (SPDC).[25] Moreover, the generated entangled photons can been controlled with full freedom offered by designed domain structures in crystals, demonstrating two-photon focusing,[26,27] beam-splitting and other novel effects,[28—39] which can hardly be realized in a uniform nonlinear crystal. This will bring revolutionary impacts on quantum optics and quantum information science in future.

In principle, the QPM technique is endowed with several remarkable advantages in the engineering of entangled photon source. First, different from birefringence phase matching (BPM), the arbitrary polarization configuration is possible in QPM materials as long as the poling period is feasible, thereby entangled photons can be generated efficiently at any designed wavelength by using the largest nonlinearity. Secondly, spectral and spatial entanglement may be transformed by the structure design of NPC.[26—39] High-dimensional entanglement or hyper-entanglement will be generated under multiple concurrent QPM SPDC processes. In recent years, great attentions have been paid to the domain-engineered NPC for its special functionality in the integrated spectral and spatial control of entangled photons, which is inherent to the SPDC process. Finally, combined with waveguide structures and the electro-optic effect,[40] the entangled photons can be generated, then guided and phase-controlled within a single lithium niobate(LN) or lithium tantalate(LT) chip, therefore QPM materials can act as a key ingredient in integrated quantum optics. The integrated engineering of photon entanglement at the source is of fundamental importance for improving the quality of photon source and enabling the generation of new types of entangled photon source which may play a key role in the science of quantum optics and photonic quantum technologies.[41]

2. The General Two-Photon State From QPM Materials

In general, the second-order nonlinearity of a two-dimensional NPC can be expressed by

$$\chi^{(2)}(x,z) = d_{\text{eff}} \sum_n F(g_n) U(x, g_n) e^{ig_n z}. \tag{1}$$

The effective nonlinearity d_{eff} is determined by the polarization configuration between the pump, the signal and idler. For LN or LT, the maximum nonlinearity d_{33} can be used when the interacted waves are all $\hat{e}$-polarized. The pump propagates along $\hat{z}$ direction. The periodic modulation of $\chi^{(2)}$ along this direction ensures the generation of entangled photon pairs by the QPM condition $k_p - k_s - k_i + g_n = 0$, in which $g_n = n \frac{2\pi}{\Lambda} (n = \pm 1, \pm 2, \cdots)$ is the nth-order reciprocal vector with corresponding Fourier coefficient $F(g_n)$, and k_p, k_s, and k_i are wavevectors of the pump, the signal and the idler, respectively. $U(x, g_n)$ represents the transverse structure of the NPC which will determine the transverse amplitude and phase profile of entangled photon pair. For some simple two-dimensional structures, $U(x, g_n)$ is independent with the longitudinal structure and $\chi^{(2)}(x, z)$ follows a two-dimensional lattice with j-fold ($j = 3, 4, 6$) symmetry.

According to the interaction Hamiltonian $H_I = \varepsilon_0 \int_V d\mathbf{r} \chi^{(2)}(x, z) \{E_p^{(+)} E_s^{(-)} E_i^{(-)}\} + h.c.$, we obtain a general two-photon state based on the first-order perturbation theory:[42]

$$|\psi\rangle \propto \psi_0 \iint d\omega_s d\omega_i \iint d\mathbf{q}_s d\mathbf{q}_i \sum_n \phi_n(\omega_s, \omega_i; \mathbf{q}_s, \mathbf{q}_i) \hat{a}_{\omega_s, \mathbf{q}_s}^{(+)} \hat{a}_{\omega_i, \mathbf{q}_i}^{(+)} |0\rangle. \quad (2)$$

All the slowly varying terms and constants are absorbed into ψ_0. Here we assume more than one reciprocal vectors are participating into the QPM-SPDC processes. $\phi_n(\omega_s, \omega_i; \mathbf{q}_s, \mathbf{q}_i) = E_p(\omega_s + \omega_i) F(g_n) \int dz e^{i(k_p - k_s - k_i - g_n)z} \int d\boldsymbol{\rho} U(x, g_n) e^{i(\mathbf{q}_s + \mathbf{q}_i) \cdot \boldsymbol{\rho}}$ is the two-photon mode function which can be factorized into the spectral mode function $\phi_n(\omega_s, \omega_i) = E_p(\omega_s + \omega_i) F(g_n) \int dz e^{i(k_p - k_s - k_i - g_n)z}$ and spatial mode function $\phi_n(\mathbf{q}_s, \mathbf{q}_i) = \int d\boldsymbol{\rho} E(\boldsymbol{\rho}) U(x, g_n) e^{-i(\mathbf{q}_s + \mathbf{q}_i) \cdot \boldsymbol{\rho}}$ under the condition that the magnitude of transverse wave vectors $\mathbf{q}_s (\mathbf{q}_i)$ is much smaller than $k_s(k_i)$, i.e. satisfying $\mathbf{k}_j \approx k_j \hat{e}_z + \mathbf{q}_j$.

In the above calculations, the spectral function of pump takes the form of $E_p(\omega_p)$ and the spatial distribution follows $E_p(\boldsymbol{\rho})$. When the transverse size of pump is taken to be infinite, the spatial mode function is mainly determined by the NPC's transverse structure function $U(x, g_n)$. From Eq. (2), it is obvious that both the spectral and spatial mode functions of entangled two-photon state can be tailored by the design of NPC's structure, therefore it offers a new strategy to manipulate the entanglement in most of the degrees of freedom, like the polarization, spatial mode, frequency etc. New types of entangled state for various applications in quantum technologies can be generated. In the paper, we will review recent progress in this field and mainly focus on the work done by our group.

3. Integrated Spectral Engineering of Entangled Photons by QPM Technique

The multiple QPMs device can dramatically expand the bandwidth of SPDC source.[35—39] Usually, the bandwidth of entangled photons which is mainly determined by the phase-matching condition is on the order of several THz or hundreds of GHz. To decrease or increase the bandwidth are both of great importance. Narrowband entangled photons of roughly MHz are designed to match the bandwidth of atomic ensemble-based quantum memories.[43—45] This can be achieved by an external Fabry-Perot cavity. The QPM material itself shows little advantages in narrowing the bandwidth of entangled photons. However, QPM technique is superior in engineering broadband entangled photons with relatively high efficiency, which is not possible for BPM crystals. By designing chirped QPM nonlinear crystals,[35—38] ultra broadband two-photon source can be achieved. It corresponds to the ultrashort temporal correlation toward single-cycle limit which is extraordinarily useful in clock synchronization,[46] quantum metrology,[47] and quantum optical coherence tomography.[35] An alternative and equivalent broadband source is the two-photon frequency comb.[39] One can design multiple QPM SPDC processes happening inside a single nonlinear crystal, then multiple equal spacing two-photon frequency modes can be achieved, which supplies as a natural mode-locked biphoton source since all the two-photon frequency modes come from the same original pump photon and inherit the pump's phase. The crystal can be aperiodically poled following the structure function:

$$\chi^{(2)}(z) = \chi^{(2)} \text{sgn}\left\{\sum_n a_n \cos(g_n z + \phi_n)\right\}, \tag{3}$$

in which g_n is the nth reciprocal vector to ensure the nth two-photon frequency mode generation. a_n and ϕ_n are adjustable parameters representing amplitude and phase for each g_n, respectively. For mode-locked two-photon source, we set reciprocal vectors to be in phase. The two-photon temporal correlation is calculated to be[37]

$$G^{(2)}(\tau) \propto \text{rect}\left(\frac{2\tau}{DL}\right)\left\{\sin\left(\frac{N\Delta\omega\tau}{2}\right)/\sin\left(\frac{\Delta\omega\tau}{2}\right)\right\}^2. \tag{4}$$

For the photon pair with a single frequency pair, the time correlation function is a rectangle function in the form of $\text{rect}\left(\frac{2\tau}{DL}\right)$, whose width is determined by the group velocity dispersion $D = \frac{1}{u_s} - \frac{1}{u_i}$ and the crystal length L. u_s and u_i are the group velocities of signal and idler photons. For LT, the entangled photon pair from a 10 mm length crystal is synchronized by picoseconds precision, while for frequency-comb entangled photons from an aperiodically poled LT (APLT), the time correlation presents ultrashort pulse train in the form of $\left\{\sin\left(\frac{N\Delta\omega\tau}{2}\right)/\sin\left(\frac{\Delta\omega\tau}{2}\right)\right\}^2$. Figure 1[39] is a schematic second-order temporal correlation of mode-locked biphotons as well as the nondegenerate HOM interference pattern. The width of two-photon correlation peaks is determined by the mode spacing $\Delta\omega$ and the number of frequency pairs N. If the frequency comb covers as broadly as the central frequency, the width of two-photon correlation peak will approach the level of femtosecond, i.e., a single-cycle. Here, the group velocity dispersion for each frequency pair is taken to be constant, which is valid for some small dispersion materials, or the frequency comb covering not too broadly. So engineering the spectrum of entangled photons by QPM technique supplies as a unique way to shape the two-photon temporal waveform, which can be applied for exploiting new types of quantum light source, studying nonlinear optical process at single photon level and playing an important role in quantum computing.[48—50]

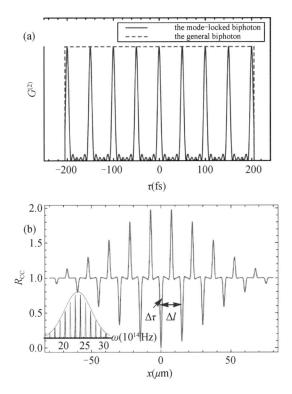

FIG.1. (a) Comparison between time correlation of mode-locked biphotons from the APLT and that of the general biphotons from a single plain crystal. In the numerical calculation, we take $\lambda_p = 400$ nm, $\Delta\omega = 1.26 \times 10^{14}$ Hz, $L = 10$ mm and the number of frequency comb $N = 6$. The overall frequency comb covers 442 nm. The two-photon then will be mode-locked to 4.2 fs. Suppose each mode has equal spectrum width $\sigma = 6.3 \times 10^{12}$ Hz. (b) HOM interference pattern of mode-locked two-photon state when the number of frequency pairs is 6. The inset is the corresponding frequency spectrum of optical frequency comb with Gaussian Envelope. Fig. 1 is selected from literature[39] (Reproduced with permission from Y. F. Bai et al., Phys. Rev. A 85, 053807(2012). Copyright 2012 American Physical Society).

For a monochromatic pump, the photon pair is anti-correlated in frequency. However, this situation will be changed when extended phase matching condition is applied.[51,52] The photon pair can be correlated in frequency or noncorrelated with each other. Suppose the pump follows a spectral function $E_p(\omega_p)$ and the group velocities between the pump, signal and idler satisfies a certain relationship such as $\frac{2}{u_p}=\frac{1}{u_s}+\frac{1}{u_i}$, the two-photon state will take the form of

$$|\psi\rangle \propto \psi_0 \iint d\omega_s d\omega_i E_p(\omega_s+\omega_i) H(\omega_s-\omega_i, L) \hat{a}_{\omega_s}^{(+)} \hat{a}_{\omega_i}^{(+)} |0\rangle \tag{5}$$

in which $H(\omega_s-\omega_i, L) = \mathrm{sinc}\left\{\frac{(\omega_s-\omega_i)L}{2(u_p-u_s)}\right\}$ is the phase matching function ($\sin c(x) = \sin(x)/x$). Here the zero-order expansion of wavevetors has been assumed to be matched. It is straightforward that how the frequencies of photon pair are correlated depends on the comparison between the bandwidths of pump and the phase-matching functions. For a monochromic pump, the bandwidth of pump is narrower than that of phase-matching function, thus the pair is always anticorrelated in frequency. However, frequency correlation is expected when the phase-matching condition is dominant and noncorrealation exists when the two bandwidths are comparative. Identical frequency of photon pair can improve the performance of HOM interference under a pulsed pump dispensing with any spectral filters[51,52] while the noncorrelated entangled photons will certainly benefit the applications of single photon source with better spectral purit.[53,54] However, to transform the frequency correlation requires the group velocity and phase velocity to be matched simultaneously. This may be difficult for conventional BPM crystals. But the QPM material is competent since the reciprocal vector can be designed independently with the group velocity matching condition. The two conditions never conflict in QPM materials.

4. Integrated Spatial Engineering of Entangled Photons by QPM Technique

Now we turn to the discussion of spatial entanglement of photon pairs from QPM materials. When studying on the spectral properties, we may assume only one spatial mode is considered, which is true for a long crystal under the plane wave pump. For a focused pump, the single spatial mode may be justified by proper spatial filters. Here, when discussing the spatial entanglement, we always assume the single frequency mode is concerned, which can be approached by narrowband spectral filters. Then we have the two-photon state in the following form:

$$|\psi\rangle \propto \psi_0 \sum_{q_s, q_i} H_{tr}(q_s, q_i) \hat{a}_{q_s}^{(+)} \hat{a}_{q_i}^{(+)} |0\rangle. \tag{6}$$

As discussed earlier, the spatial mode function $H_{tr}(q_s, q_i) = \int d\rho E(\rho) U(x) e^{-i(q_s+q_i)\cdot\rho}$ is mainly determined by the transverse function of domain modulation and the pump profile. Several interesting two-photon effects are observed from a two-dimensional NPC including the two-photon focusing,[27] lensless ghost imaging[28] quantum Talbot effect,[29] the sub-wavelength diffraction in the far filed etc.[30]

For a simple multi-stripe PPLT (MPPLT), on one hand it works as an efficient platform for entangled photon generation under QPM condition, on the other hand its transverse periodicity engenders quantum Talbot effect directly.[29] So the quantum Talbot effect emerges dispensing with a real grating. This compact and stable self-image can be further employed for a lenless and contactless diagnosis of ferroelectric domains. The transverse structure function of modulation of MPPLT follows a grating function $U(x) = \sum_{n=-\infty}^{\infty} \text{rect}[(x-nd)/a]$ along x-axis with period d and stripe width a. Here $\text{rect}(x)$ is 1 for $|x| \leqslant 1/2$ and 0 for other values. The Fourier expansion of $U(x)$ is $\sum_{n=-\infty}^{\infty} c_n e^{i2\pi n x/d}$, where $c_n = \sin(\pi n a/d)/(\pi n)$ is the Fourier coefficient of the n-th harmonic. Suppose signal and idler photons are captured by detectors D_1 and D_2, respectively. When calculating the two-photon spatial correlation in the Fresnel zone, we found at certain distance from the output surface of MPPLT, coincidence counting rate between two detectors will retrieve the transverse structure of MPPLT. As long as $\frac{1}{\lambda_s z_s} + \frac{1}{\lambda_i z_i} = \frac{1}{2md^2}$, where $\lambda_{s,i}$ is the wavelength of signal or idler and m is an integer indicating the m-th Talbot plane, the two-photon coincidence counting rate

$$R_{c.c.}(x_s, z_s; x_i, z_i) = |\langle 0 | E_1^{(+)} E_2^{(+)} | \psi \rangle|^2 \propto \left| \int dx_0 e^{i\pi\eta(x_0-\xi)^2} U(x_0) \right|^2$$

$$\propto \left| \sum_{n=-\infty}^{\infty} c_n e^{i2\pi n \xi/d} e^{-i\pi n^2/(d^2\eta)} \right|^2 \tag{7}$$

will turn into the grating function $U(\xi)$. A reproduction of grating function, thus the quantum Talbot effect is directly observable after the MPPLT crystal. $\xi = \left(\frac{x_s}{\lambda_s z_s} + \frac{x_i}{\lambda_i z_i} \right) / \left(\frac{1}{\lambda_s z_s} + \frac{1}{\lambda_i z_i} \right)$ is an associate coordinate, which means the Talbot self-image of the domain structure is magnified and the exhibited period depends on the way of detection and the wavelengths of the photon pair. In the above calculation, $E_j^{(+)}(\mathbf{r}_j, t_j)$ is the electric field evaluated at the two detectors' spatial coordinate $\mathbf{r}_j(\boldsymbol{\rho}_j, z_j)(j=s,i)$. z_s and z_i stand for the distances from the crystal to the detection planes of D_1 and D_2, respectively. The propagation of the two free-space electric fields is $E_j^{(+)}(\mathbf{r}_j, t_j) = \sum_{k_j} E_j e^{-i\omega_j t_j} g(\mathbf{k}_j, \omega_j; \boldsymbol{\rho}_j, z_j) \hat{a}_{k_j}$, in which the Green function[42] takes the form of $g(\mathbf{k}_j, \omega_j; \boldsymbol{\rho}_j, z_j) = \frac{-i\omega_j}{2\pi c} \frac{e^{i(\omega_j/c)z_j}}{z_j} \int d\boldsymbol{\rho}_0 e^{i\frac{\omega_j}{2cz_j}|\boldsymbol{\rho}_j-\boldsymbol{\rho}_0|^2} e^{i\mathbf{k}_j \cdot \boldsymbol{\rho}_0}$, where $\boldsymbol{\rho}_0$ is the transverse coordinate at the output face of the crystal.

Figure 2(a)[29] is the quantum Talbot carpet observed after the MPPLT when pumped by 532 nm at 170 ℃. The two detectors scanned in step for capturing a pair of 1064 nm photons. The transverse stripe interval is $\Lambda_{tr} = 160$ μm with stripe width $b = 20$ μm and stripe length $L = 10$ mm. The longitudinal periodicity is $\Lambda = 7.548$ μm for all stripes Talbot length is calculated as $z_T = 96.2$ mm. The experimental two-photon Talbot carpet consists well with the theoretical simulations.

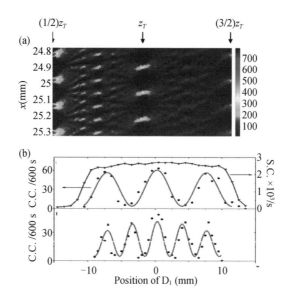

FIG. 2. The near field and far field two-photon interference pattern. The longitudinal periodicity is $\Lambda = 7.548$ μm for the 1064 nm degenerate entangled photons generation. (a) Experimental two-photon Talbot carpet from $(1/2)z_T$ (48 mm) to $(3/2)z_T$ (144 mm). The transverse stripe interval is $\Lambda_{tr} = 160$ μm with stripe width $b = 20$ μm and stripe length $L = 10$ mm. (b) Experimental results of two-photon far-field interference in the detection plane $z = 1.4$ m. The transverse stripe interval is $\Lambda_{tr} = 200$ μm with stripe width $b = 30$ μm and stripe length $L = 6$ mm. The upper one corresponds the case that one detector is fixed and the other scans, while the lower one corresponds to the case that two detectors scan in-step. Figs. 2(a) and 2(b) are selected from literature[29] (Reprinted with permission from H. Jin et al., Appl. Phys. Lett. 101, 211115(2012). Copyright 2012 American Institute of Physics) and literature[30] (Reproduced with permission from X. Q. Yu et al., Phys. Rev. Lett. 101, 233601 (2008). Copyright 2008 American Physical Society), respectively.

When exploring the two-photon far field diffraction pattern,[30] we found the coincidence counting rate takes the form of

$$R_{c.c.}(x_s, z_s; x_i, z_i) \propto \left| H\left(\frac{\omega_s x_s}{c z_s} + \frac{\omega_i x_i}{c z_i}\right) \right|^2, \tag{8}$$

in which $H(q) = \dfrac{\sin(Nq\Lambda_{tr}/2)}{\sin(q\Lambda_{tr}/2)} \mathrm{sinc}(qb/2)$ is the far field interference-diffraction pattern for a grating, which mathematically is the Fourier transform of $U(x)$. In Eq.(8), $\xi = \dfrac{\omega_s x_s}{c z_s} + \dfrac{\omega_i x_i}{c z_i}$ is an associate coordinate which indicates that the observed two-photon interference pattern depends on the detection scheme. When the two detectors scanned in step, the subwavelength diffraction will be observed[30] [Fig. 2(b)]. From the near field and far field two-photon diffraction pattern, it is obvious that the domain structure can shape the spatial waveform of entangled photons. The two-photon spatial entanglement can be transformed for the purpose of certain quantum technologies.

Spatial entanglement will dramatically change when the distorted transverse structure is introduced into the NPC. By designing a transversely parabolic domain structure,[26, 27] $U(x) = e^{-ig_n\alpha x^2}$, the photon pair will be self-focused after a certain distance from the crystal. In this case, the engineered crystal is equivalent to a homogeneous nonlinear crystal and a focusing lens. The spatial correlation at the focal plane is

$$R_{c.c.}(x_s, x_i)\bigg|_{z=f_{eff}} \propto \delta^2\left(\frac{\omega_s x_s}{cz_s} + \frac{\omega_i x_i}{cz_i}\right), \tag{9}$$

under the condition of

$$\frac{1}{\lambda_s z_s} + \frac{1}{\lambda_i z_i} = \frac{1}{\lambda_p f_{eff}}. \tag{10}$$

The pump and crystal size is assumed to be infinite. The equivalent focal length f_{eff} equals $\frac{\pi}{g_n \alpha \lambda_p}$. f_{eff} is relevant to the pump wavelength λ_p, the curvature of the parabolic NPC α and the reciprocal vector g_n for the longitudinal QPM condition. If we use a two-photon detector to capture the two-photon probability after the crystal, two-photon self-focusing will be observed as shown in Figure 3.[27] If we use two independent single photon detectors to examine the spatial correlation between the signal and idler photons, we will obtain a well-defined point to point correspondence when Eq. (10) is satisfied. This suggests an important application in lensless ghost imaging following the Gaussian thin lens equation of Eq. (10). When we put an object in one of the path, the image will be recovered in the idler path with the magnification of $-\frac{\lambda_i z_i}{\lambda_s z_s}$. When $z_s = z_i = f_{eff}$, an equal size image will be produced as shown by Fig. 4.[28]

For a multi-stripe parabolic NPC,[27,28] new characters will be brought to the two-photon lens, such as the transverse periodicity. When the pump incident at certain transverse position, a dual-focusing phenomena was observed in which the two-photon is focused onto either of two symmetric directions. In this case, the engineered crystal serves as the entangled photon source, lens and beam splitter simultaneously.[27] This multifunctional integration is free from any bulk optical elements and, therefore, may be exploited for on-chip integrated quantum optics. With these inherent linear optical elements, a lensless twin-image is observed after the same crystal.[28]

5. Integrated Engineering of Polarization-, Path- and Hyper-Entanglement

As we know, polarization-entanglement is widely used in testing the foundations of quantum mechanics[55, 56] and for developing quantum technologies.[41] A typical method to generate entangled photons relies on the type-II BPM in a nonlinear crystal[57] or two type-

I crystals.[58] However, this source is less efficient since the photons pair emitted conically and only a small fraction of the cone can be collected for use. Regarding of this, beam-like entangled photons are more attractive, which can be achieved by coherently combining two SPDC sources at a polarizing beam splitter,[59—65] by manipulating polarization ququarts,[66] by overlapping two cascaded PP crystals.[67,68] However, a compact, postselection-free and bright polarization-entangled photons source is more valuable for practical applications. The cascaded[69,70] or concurrent[31,71,72] SPDC processes in a single QPM crystal can meet all the demands as shown by recent several works.

For a periodically poled LN or LT, usually only a single reciprocal vector is used to fulfill the phase-matching condition $k_p - k_s - k_i - g_1 = 0$. Degenerate or nondegenerate photon pair is generated under a certain polarization configuration. But for a dual-periodically poled crystal or other structured crystals, two concurrent QPM conditions can be satisfied simultaneously, therefore, the entangled photon pair can be generated under two possible polarization configurations, achieving $|\hat{e}_{\omega_1}\rangle|\hat{o}_{w_2}\rangle + |\hat{e}_{\omega_2}\rangle|\hat{o}_{\omega_1}\rangle$. Cascaded by a dichroic mirror, the nondegenerate polarization entanglement is thus produced.[59,60] Furthermore, a scheme for narrowband counterpropagating polarization entangled photon pairs is proposed for realizing the natural spatial separation of degenerate photon pair.[31] So multiple QPM processes provide a new solutions for the realization of compact, beam-like and high-brightness source of polarization entangled photon pairs. In addition, by multi-stripe arrangement in a single QPM crystal, high-dimensional frequency entanglement together with polarization entanglement can be simultaneously achieved, thus a frequency-polarization hyper-entangled state may possibly be engineered,[34] which may find potential applications in quantum communication with better security and higher capacity. Furthermore, when two concurrent beam-like QPM SPDC processes exist, the cross-polarized photon pair will contribute to a bright and integrate polarization entangled state when a polarizaiton-beam-splitting is cascaded.[33] The concurrent multiple QPMs provided by the nonlinear photonic crystal open up a way to integrated quantum light sources.

The simple structured two-dimensional NPC, like a two-dimensional periodical poled LN or LT with j-fold($j=3,4,6$) symmetry is provided with the inherent advantages in the engineering of path-entanglement since multiple parametric down-conversions can happen simultaneously with a collinear or noncollinear geometry, which is crucial for path-entangled state generation.[32,33] The two-dimensional reciprocal space is expanded by

$$\chi^{(2)}(x,z) = \sum_{m,n} F(\boldsymbol{G}_{m,n}) e^{-i\boldsymbol{G}_{m,n} \cdot \boldsymbol{\rho}} \tag{11}$$

in which $\boldsymbol{G}_{m,n} = m\,\boldsymbol{e}_1 + n\,\boldsymbol{e}_2$ is reciprocal vectors. Suppose N reciprocal vectors can simultaneously participate into the SPDC processes, in general this will result in $2N$ spatial modes for the signal and idler photons. However, when some of the spatial modes are identically overlapped, then a two-photon multimode-entangled source will be generated. For three QPM SPDC processes happening in a rectangle NPC like Fig. 5,[32] a two-photon

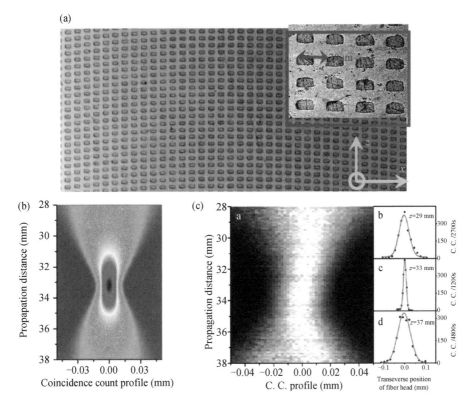

FIG.3. (a) The micrograph of parabolic NPC. The longitudinal periodicity of NPC is $\Lambda = 13.917$ μm. The third-order reciprocal vector ensures the degenerate 914 nm photon pair generation. The focal length of two-photon lens is designed to be $f_{eff} = 33.3$. The stripe interval Λ_{tr} is 20 μm, stripe width b is 10 μm and stripe length L is 6 mm. (b) The theoretical simulation of two-photon focusing dynamics under a full pump width of 0.82 mm. (c) The experimental results of two-photon focusing. The two-photon focusing spot is 28 μm, which is consistent with the theoretical value. Fig. 3(a) is selected from literature[28] (Reproduced with permission from P. Xu et al., Phys. Rev. A. 86, 013805(2012). Copyright 2012 American Physical Society). Figs. 3(b) and 3 (c) are from literature[27] (Reproduced with permission from H. Y. Leng et al., Nature Commun. 2, 429 (2011). Copyright 2011, Rights Managed by Nature Publishing Group).

path-entangled state will be achieved:[32]

$$|\psi\rangle = (|2,0\rangle + |0,2\rangle + \gamma |1,1\rangle)/\sqrt{2+\gamma^2}. \tag{12}$$

When seeded by a two-mode coherent state, this crystal can produce two-, three-, four-, or five-photon path-entangled states in a postselection-free way.[32] In particular, up to five-photon NOON state can be generated, which enables phase supersensitive measurements at the Heisenberg limit. When a different combination of QPM geometries is adopted, a heralded single-photon multipartite entanglement can be achieved. For the multi-photon entanglement under different QPM conditions will be more interesting, especially when the path-entanglement and polarization-entanglement are simultaneously engineered.

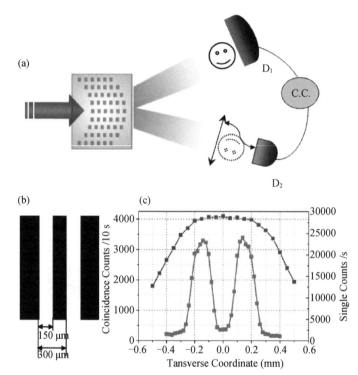

FIG. 4. (a) The schematic setup for lensless ghost imaging using entangled photons from parabolic NPC. (b) The double slit object. The slit interval is 300 μm and the slit width is 150 μm. (c) The recovered imaging by coincidence counting. Figs. 4(a) and 4(c) are selected from literature[28] (Reproduced with permission from P. Xu et al., Phys. Rev. A. 86, 013805(2012). Copyright 2012 American Physical Society).

6. Conclusion

In a summary, the concurrent QPM technique enables that the flexible engineering of spatial and spectral entanglement towards the full control of photons over most degrees of freedom, enabling new types of entanglement in polarization, spatial mode, and frequency and the hyper-entanglement over them. The new type of entangled photon source based on QPM technique may find important applications in testing quantum fundamentals, quantum communication, quantum imaging and quantum computing. It is worth noting that although the spectral and spatial properties of photon pairs in our discussion are independent, they actually affect each other.[73] For some peculiar situations, the spectral and spatial mode function can not be factorized, which deserves further consideration. Besides the natural integration function benefiting from domain engineering, LN or LT will exhibits more advantages for the integrated realization of quantum circuits after the new components like waveguides and the electro-optic effect are introduced. The electro-optics effect can offer a fast phase control of photons,[40] with a standard level of 40 GHz. The

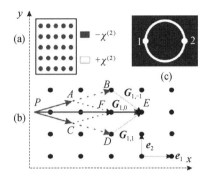

FIG.5. (a) Schematic of a 2D NPC with a rectangular inverted domain structure. (b) the reciprocal lattice and the concurrent three QPM SPDC processes. Reciprocal lattice of the crystal. (c) Transverse pattern of the parametric light in the Fourier plane. Fig. 5 is selected from literature[32] (Reproduced with permission from Y. X. Gong *et al.*, Phys. Rev. A. **86**, 023835(2012). Copyright 2012 American Physical Society).

entangled source can be generated, then guided and rapidly phase-controlled within a single LN crystal. The QPM LN or LT chip will certainly be applied for integrated generation and manipulation of quantum bits soon. This will bring revolutionary impacts on quantum optics and quantum information science in future.

References and Notes

[1] J. Armstrong, N. Bloembergen, J. Ducuing, and P. S. Pershan, Phys. Rev. **127**, 1918(1962).

[2] P. A. Franken and J. F. Ward, Rev. Mod. Phys. **35**, 23(1963).

[3] N. B. Ming, *Physical Fundamentals of Crystal Growth* (Shanghai Scienti®c & Technical Publishers, Shanghai, 1982)(in Chinese).

[4] D. Feng, N. B. Ming, J. F. Hong, J. S. Zhu, Z. Yang, and Y. N. Wang, Appl. Phys. Lett. **37**, 607 (1980).

[5] V. Berger, Phys. Rev. Lett. **81**, 4136(1998).

[6] M. Yamada, N. Nada, M. Saitoh, and K. Watanabe, Appl. Phys. Lett. **62**, 435(1993).

[7] J. Webjörn, V. Pruneri, P. St. J. Russell, J. R. M. Barr, and D. C. Hanna, Electron. Lett. **30**, 894 (1994).

[8] S. N. Zhu, Y. Y. Zhu, Z. Y. Zhang, H. Shu, H. F. Wang, J. F. Hong, C. Z. Ge, and N. B. Ming, J. Appl. Phys. **77**, 5481(1995).

[9] M. M. Fejer, G. A. Magel, D. H. Jundt, and R. L. Byer, IEEE J. Quantum Electron **28**, 2631(1992).

[10] S. N. Zhu, Y. Y. Zhu, and N. B. Ming, Science **278**, 843(1997).

[11] S. N. Zhu, Y. Y. Zhu, Y. Q. Qin, H. F. Wang, C. Z. Ge, and N. B. Ming, Phys. Rev. Lett. **78**, 2752 (1997).

[12] J. R. Kurz, A. M. Schober, D. S. Hum, A. J. Saltzman, and M. M. Fejer, IEEE J. Sel. Top. Quantum Electron. **8**, 660(2002).

[13] Y. Q. Qin, C. Zhang, Y. Y. Zhu, X. P. Hu, and G. Zhao, Phys. Rev. Lett. **100**, 063902(2008).

[14] T. Ellenbogen, N. Voloch-Bloch, A. Ganany-Padowicz, and A. Arie, Nature Photon. **3**, 395(2009).

[15] N. G. R. Broderick, G. W. Ross, H. L. Offerhaus, D. J. Richardson, and D. C. Hanna, Phys. Rev. Lett. **84**, 4345(2000).

[16] Y. Y. Zhu, and N. B. Ming, Phys. Rev. B **42**, 3676(1990).

[17] Y. Y. Zhu and N. B. Ming, Opt. Quantum Electron. **31**, 1093(1999).

[18] A. Norton and C. de Sterke, Opt. Express **12**, 841(2004).

[19] J. Liao, J. L. He, H. Liu, J. Du, F. Xu, H. T. Wang, S. N. Zhu, Y. Y. Zhu, and N. B. Ming, Appl. Phys. B **78**, 265(2004).

[20] P. Xu, S. H. Ji, S. N. Zhu, X. Q. Yu, J. Sun, H. T. Wang, J. L. He, Y. Y. Zhu, and N. B. Ming, Phys. Rev. Lett. **93**, 133904(2004).

[21] P. Xu, S. N. Zhu, X. Q. Yu, S. H. Ji, Z. D. Gao, G. Zhao, Y. Y. Zhu, and N. B. Ming, Phys. Rev. B **72**, 064307(2005).

[22] Y. Zhang, Z. D. Gao, Z. Qi, S. N. Zhu, and N. B. Ming, Phys. Rev. Lett. **100**, 163904(2008).

[23] Y. Zhang, J. Wen, S. N. Zhu, and M. Xiao, Phys. Rev. Lett. **104**, 183901(2010).

[24] Z. Chen, D. Liu, Y. Zhang, J. Wen, S. N. Zhu, and M. Xiao, Opt. Lett. **37**, 689(2012).

[25] S. Tanzilli, H. De Riedmatten, W. Tittel, H. Zbinden, P. Baldi, M. De Micheli, D. B. Ostrowsky, and N. Gisin, Electron. Lett. **37**, 26(2001).

[26] J. P. Torres, A. Alexandrescu, S. Carrasco, and L. Torner, Opt. Lett. **29**, 376(2004).

[27] H. Y. Leng, X. Q. Yu, Y. X. Gong, P. Xu, Z. D. Xie, H. Jin, C. Zhang, and S. N. Zhu, Nature Commun. **2**, 429(2011).

[28] P. Xu, H. Y. Leng, Z. H. Zhu, Y. F. Bai, H. Jin, Y. X. Gong, X. Q. Yu, Z. D. Xie, S. Y. Mu, and S. N. Zhu, Phys. Rev. A **86**, 013805(2012).

[29] H. Jin, P. Xu, J. S. Zhao, H. Y. Leng, M. L. Zhong, and S. N. Zhu, Appl. Phys. Lett. **101**, 211115(2012).

[30] X. Q. Yu, P. Xu, Z. D. Xie, J. F. Wang, H. Y. Leng, J. S. Zhao, S. N. Zhu, and N. B. Ming, Phys. Rev. Lett. **101**, 233601(2008).

[31] Y. X. Gong, Z. D. Xie, P. Xu, X. Q. Yu, P. Xue, and S. N. Zhu, Phys. Rev. A **84**, 053825(2011).

[32] Y. X. Gong, P. Xu, Y. F. Bai, J. Yang, H. Y. Leng, Z. D. Xie, and S. N. Zhu, Phys. Rev. A **86**, 023835(2012).

[33] Y. X. Gong, P. Xu, J. Shi, L. Chen, X. Q. Yu, P. Xue, and S. N. Zhu, Opt. Lett. **37**, 4374(2012).

[34] J. Shi, S. J. Yun, Y. F. Bai, P. Xu, and S. N. Zhu, Opt. Commun. **285**, 5549(2012).

[35] S. Carrasco, J. P. Torres, L. Torner, A. Sergienko, B. E. A. Saleh, and M. C. Teich, Opt. Lett. **29**, 2429(2004).

[36] S. E. Harris, Phys. Rev. Lett. **98**, 063602(2007).

[37] M. B. Nasr, S. Carrasco, B. E. A. Saleh, A. V. Sergienko, M. C. Teich, J. P. Torres, L. Torner, D. S. Hum, and M. M. Fejer, Phys. Rev. Lett. **100**, 183601(2008).

[38] S. Sensarn, G. Y. Yin, and S. E. Harris, Phys. Rev. Lett. **104**, 253602(2010).

[39] Y. F. Bai, P. Xu, Z. D. Xie, Y. X. Gong, and S. N. Zhu, Phys. Rev. A **85**, 053807(2012).

[40] D. Bonneau, M. Lobino, P. Jiang, C. M. Natarajan, M. G. Tanner, R. H. Hadfield, S. N. Dorenbos, V. Zwiller, M. G. Thompson, and J. L. O'Brien, Phys. Rev. Lett. **108**, 053601(2012).

[41] J. L. O'Brien, A. Furusawa, and J. Vučković, Nature Photon. **3**, 687(2009).

[42] M. H. Rubin, Phys. Rev. A **54**, 5349(1996).

[43] Z. Y. Ou and Y. J. Lu, Phys. Rev. Lett. **83**, 2556(1999).
[44] C. E. Kuklewicz, F. N. C. Wong, and J. H. Shapiro, Phys. Rev. Lett. **97**, 223601(2006).
[45] X. H. Bao, Y. Qian, J. Yang, H. Zhang, Z. B. Chen, T. Yang, and J. W. Pan, Phys. Rev. Lett. **101**, 190501(2008).
[46] A. Valencia, G. Scarcelli, and Y. Shih, Appl. Phys. Lett. **85**, 2655(2004).
[47] V. Giovannetti, S. Lloyd, and L. Maccone, Science **306**, 1330(2004).
[48] N. C. Menicucci, S. T. Flammia, and O. Pfister, Phys. Rev. Lett. **101**, 130501(2008).
[49] M. Pysher, Y. Miwa, R. Shahrokhshahi, R. Bloomer, and O. Pfister, Phys. Rev. Lett. **107**, 030505 (2011).
[50] M. Pysher, A. Bahabad, P. Peng, A. Arie, and O. Pfister, Opt. Lett. **35**, 565(2010).
[51] O. Kuzucu, M. Fiorentino, M. A. Albota, F. N. C. Wong, and F. X. Kärtner, Phys. Rev. Lett. **94**, 083601(2005).
[52] V. Giovannetti, L. Maccone, J. H. Shapiro, and F. N. C. Wong, Phys. Rev. Lett. **88**, 183602(2002); Phys. Rev. A **66**, 043813(2002).
[53] Zhang *et al.*, Nature Photon. **5**, 628(2011).
[54] A. M. Brańczyk, A. Fedrizzi, T. M. Stace, T. C.Ralph,and A. G. White, Opt.Express **19**, 55(2011).
[55] M. Genovese, Phys. Rep. **413**, 319(2005).
[56] Y. Shih, Rep. Prog. Phys. **66**, 1009(2003).
[57] P. G. Kwiat, K. Mattle, H. Weinfurter, A. Zeilinger, A. V. Sergienko, and Y. Shih, Phys. Rev. Lett. **75**, 4337(1995).
[58] P. G. Kwiat, E. Waks, A. G. White, I. Appelbaum, and P. H. Eberhard, Phys. Rev. A. **60**, R773 (1999).
[59] A. Yoshizawa and H. Tsuchida, Appl. Phys. Lett. **85**, 2457(2004).
[60] F. König, E. J. Mason, F. N. C. Wong, and M. A. Albota, Phys. Rev. A **71**, 033805(2005).
[61] Y.-K. Jiang and A. Tomita, J. Phys. B **40**, 437(2007).
[62] H. C. Lim, A. Yoshizawa, H. Tsuchida, and K. Kikuchi, Opt. Express **16**, 12460(2008); **16**, 16052 (2008).
[63] S. Sauge, M. Swillo, M. Tengner, and A. Karlsson, Opt. Express **16**, 9701(2008).
[64] M. Hentschel, H. Hübel, A. Poppe, and A. Zeilinger, Opt. Express **17**, 23153(2009).
[65] M. Fiorentino and R. G. Beausoleil, Opt. Express **16**, 20149(2008).
[66] E. V. Moreva, G. A. Maslennikov, S. S. Straupe, and S. P. Kulik, Phys. Rev. Lett. **97**, 023602 (2006).
[67] M. Pelton, P. Marsden, D. Ljunggren, M. Tengner, A. Karlsson, A. Fragemann, C. Canalias, and F. Laurell, Opt. Express **12**, 3573(2004).
[68] D. Ljunggren, M. Tengner, P. Marsden, and M. Pelton, Phys. Rev. A **73**, 032326(2006).
[69] T. Suhara, G. Nakaya, J. Kawashima, and M. Fujimura, IEEE Photon. Technol. Lett. **21**, 1096 (2009).
[70] W. Ueno, F. Kaneda, H. Suzuki, S. Nagano, A. Syouji, R. Shimizu, K. Suizu, and K. Edamatsu, Opt. Express **20**, 5508(2012).
[71] K. Thyagarajan, J. Lugani, S. Ghosh, K. Sinha, A. Martin, D. B. Ostrowsky, O. Alibart, and S. Tanzilli, Phys. Rev. A **80**, 052321(2009).
[72] Z. H. Levine, J. Fan, J. Chen, and A. L. Migdall, Opt. Express **19**, 6724(2011)
[73] C. I. Osorio, A. Valencia, and J. P. Torres, New J. Phys. **10**, 113012(2008).

Transforming Spatial Entanglement Using a Domain-Engineering Technique[*]

X. Q. Yu,[1,2] P. Xu,[1] Z. D. Xie,[1] J. F. Wang,[1] H. Y. Leng,[1] J. S. Zhao,[1] S. N. Zhu,[1] and N. B. Ming[1]

[1] *National Laboratory of Solid State Microstructures and Department of Physics,*
Nanjing University, Nanjing, 210093, China
[2] *Physics Department, Southeast University, Nanjing, 211189, China*

We study the spatial correlation of a two-photon entangled state produced in a multistripe periodically poled LiTaO₃ crystal by spontaneous parametric down-conversion. The far-field diffraction-interference experiments reveal that the transverse modulation of domain patterns transforms the spatial mode function of the two-photon state. This result offers an approach to prepare a novel type of two-photon state with a unique spatial entanglement by using a domain-engineering technique.

The two-photon state, which is produced by spontaneous parametric down-conversion (SPDC), exhibits spatial entanglement in a finite-or infinite-dimensional Hilbert space[1—5]. Mathematically, such a spatial entanglement is embedded in the mode function of a two-photon state and plays an important role in the study of foundational quantum mechanics[6] and quantum communication[7,8]. Many nonclassical behaviors of the two-photon pair, such as quantum interference and quantum imaging, also rely heavily on the spatial structure of the mode function. Therefore an important issue in quantum optics is how to control the mode function of a two-photon pair or, in other words, how to prepare two-photon pairs with required spatial entanglement. As reported previously, in some cases, such a goal has been achieved by manipulating the beam profile of the pump laser[9,10] or by placing the spatial light modulators in the arms of the two-photon pair[11]. Recently, another scheme was put forward by Torres *et al.* in theory[12]. This scheme allows the two-photon spatial mode to be tailored and manipulated by transversely patterned quasi-phase-matched(QPM) gratings. But the issue is still open due to a lack of related studies, in particular, on the experiment aspect.

It is known that, in a QPM device, the modulated nonlinearity would affect the time-frequency and space-momentum properties of the generated two-photon pairs[13—19]. Especially, transverse modulation of nonlinearity can greatly influence the spatial properties of the downconverted beam. Fejer *et al.* and Qin *et al.* extended Huygens-Fresnel principle(HFP) from linear optics to parametric processes induced by quadratic

---

[*] Phys.Rev.Lett.,2008,101(23):233601

nonlinearity [20,21]. It can be described that each point on the primary wave front acts as a source of secondary wavelets of the pump, as well as a source of the signal and idler waves in a parametric process. In this case the HFP can be used to engineer domain patterns in crystal for the aims of parametric beam focusing, shaping, or multifunction integration. In this letter, we study the entangled state generated from a multistripe periodically poled LiTaO$_3$(MPPLT) crystal, in which the longitudinal modulation of nonlinearity works for QPM-SPDC, whereas the transverse modulation is used to manipulate the two-photon spatial mode. In our case, the pump, signal, and idler photons are all e-polarized, so the maximum quadratic nonlinearity component d_{33} is used. The work actually links up the HFP with a particular domain pattern to create and at the same time transform the spatial entanglement of photon pairs. The experimental results prove that the structure information of domain patterns is transferred into the spatial mode of an entangled state. By using this technique, it should also be convenient to control the effective finite Hilbert space related to the freedom of the transverse degree, such as the orbital angular momentum, and increase the efficiency of the quantum communications[22].

The MPPLT sample is sketched in Fig. 1(a), in which N stripes are periodically poled for collinear SPDC and arranged in parallel. When the pump beam waist is large enough, at a certain time, the two-photon pair can be produced from any one of the illuminated stripes, which can be regarded as a coherent SPDC subsource. The generated two-photon state is a superposition of all of the possible states: $\Psi = \sum_{j=1}^{N} c_j e^{i\phi_j} |j,j\rangle$, where $|j,j\rangle$ corresponds to a photon pair created from the jth stripe and c_j is the relative amplitude and

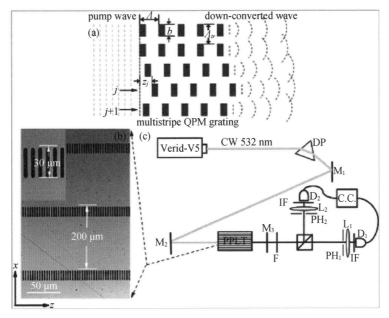

FIG.1. (color online)(a) Schematic diagram of the nonlinear HFP principle and the multistripe QPM structure.(b) Micrograph of the MPPLT.(c) Experimental setup.

is proportional to the jth stripe volume V_j and the intensity of the pump beam. $\varphi_j = 2\pi z_j/\Lambda$ is the relative phase, where z_j is the initial coordinate of the jth stripe along the z axis and Λ the longitudinal modulation period of domain. By adjusting the volume and initial position of each stripe, in principle, we could get any spatial mode of down-converted photon pairs according to the HFP as shown in Fig. 1(a). In the following parts, we will derive the mode function of this case.

In the interaction picture, the effective Hamiltonian for SPDC is given by

$$H_1 = \varepsilon_0 \int_V d^3 r \chi^{(2)} E_p^{(+)} E_s^{(-)} E_i^{(-)} + \text{H.c.}, \qquad (1)$$

where H.c. stands for the Hermitian conjugate and V is the interaction volume. $\chi^{(2)}$ is the quadratic nonlinear susceptibility of the crystal. The positive frequency part of the quantized field is $E_j^{(+)} = \sum_{k_j} E_j e^{i(k_j \cdot r - \omega_j t)} a_j(k_j)$, where ω_j, k_j ($j = s, i, p$) are the angular frequencies and the wave vectors of signal, idler, and pump, respectively, $a_{s(i)}$ is the annihilation operator of signal(idler) photon, $E_j = i\sqrt{\hbar \omega_j/(2\varepsilon_0 n_j^2 V_Q)}$, and V_Q is the quantization volume. For simplicity, we assume the pump field to be nondepleted, monochromatic, and classical.

The longitudinal and transverse modulation of $\chi^{(2)}$ in the MPPLT is separable under our experimental condition. We can show that

$$\chi^{(2)}(r) = U(\boldsymbol{\rho}) \sum_m f(G_m) e^{-iG_m z}, \qquad (2)$$

where the subscript m refers to the mth component of the Fourier series with coefficient $f(G_m)$ and the reciprocal vector $G_m = 2m\pi/\Lambda$ and $U(\boldsymbol{\rho})$ is the transverse modulation function of the multistripe grating. Generally, the first-order reciprocal vector G_1 is designed to satisfy the longitudinal phase-matching condition. Under the first-order perturbation approximation, the two-photon state is

$$|\Psi\rangle = g \sum_{k_s, k_i} \Phi(\Delta k_z L) H_{tr}(k_s + k_i) \delta(\omega_p - \omega_s - \omega_i) \hat{a}_s^+ \hat{a}_i^+ |0\rangle, \qquad (3)$$

where

$$H_{tr}(k_s + k_i) = (1/A) \int_A d^2 \rho E_p(\boldsymbol{\rho}) U(\boldsymbol{\rho}) e^{-i(k_s + k_i) \boldsymbol{\rho}} = \mathcal{F}\{E_p(\boldsymbol{\rho}) U(\boldsymbol{\rho})\}. \qquad (4)$$

We have assumed that the beam size A is large enough. The longitudinal mode function $\Phi(\Delta k_z L) = \sin(\Delta k_z L)/(\Delta k_z L)$, where $\Delta k_z = k_p - k_{sz} - k_{iz} - G_1$ is the phase mismatching in longitudinal z direction, L the stripe length, and $\kappa_{s(i)}$ the transverse wave vector of the signal(idler) photon. It is noteworthy that the transverse mode function H_{tr} is the Fourier transform of the function $E_p \cdot U$, so the transverse information of the MPPLT is transferred to the spatial mode of the generated two-photon state. In our case, there is no modulation in the y axis, so we just consider the modulation along the x axis, and then H_{tr} is expressed by

$$H_x(k_{s,x} + k_{i,x}) \propto \frac{\sin[N(k_{s,x} + k_{i,x})\Lambda_{tr}/2]}{\sin[(k_{s,x} + k_{i,x})\Lambda_{tr}/2]} \text{sinc}[(k_{s,x} + k_{i,x})b/2]. \qquad (5)$$

In Eq. (5), Λ_{tr} and b represent the stripe interval and the strip width, respectively. This is a standard far-field diffraction-interference pattern of a multislit. Thus the coincidence counting rate in the far-field region can be derived:

$$R_c(x_1, x_2) = \langle \Psi | E_1^{(-)} E_2^{(-)} E_2^{(+)} E_1^{(+)} | \Psi \rangle \propto \left| H_x \left(\frac{\omega_s x_1}{c z_1} + \frac{\omega_i x_2}{c z_2} \right) \right|^2, \quad (6)$$

where $E_1(E_2)$ is the field operator at detector $D_1(D_2)$, c the velocity of light in vacuum, $x_1(x_2)$ the horizontal position of detector $D_1(D_2)$ in the detecting plane, and $z_1(z_2)$ the distance between the output face of the crystal and the detecting plane.

The theoretical result above was proved by the coincidence measurement on the two-photon pairs in the far-field region. Figure 1(b) is the micrograph of the domain pattern of the etched MPPLT sample, showing the stripe interval $\Lambda_{tr} = 200$ μm and stripe width $b = 30$ μm. The whole stripe length $L = 6$ mm, and the effective stripe number $N = 4$ for the pumping beam waist of 1 mm. All stripes had the same longitudinal modulation period $\Lambda = 7.548$ μm. The sample was operated at 174.6 ℃, which was the phase-matching temperature for the SPDC process. The degenerate 1064 nm photon pairs were generated collinearly as the sample was illuminated by a cw-532 nm laser. Figure 1(c) shows the schematic setup of the experiment. The generated photon pairs were first separated from the pump beam by mirror M_3, which was coated with high reflectivity for 532 nm and high transmissivity for 1064 nm. A cutoff glass filter F was then used to suppress the pump photon further. The photon pairs were separated by a 50/50 beam splitter and then were, respectively, collected into two single-photon detectors (Perkin Elmer SPCM-AQR-14), which were preceded by a pinhole (PH_1/PH_2) with 1 mm diameter, a collection lens (L_1/L_2) with $f = 50$ mm, and a 20 nm bandwidth interference filter (IF) centered at 1064 nm as shown in Fig. 1(c). The components above can be regarded as an assembly detector D_1 (D_2), whose detection position is defined by the position of pinhole PH_1 (PH_2). The distance z from the output surface of MPPLT to each detection plane was 1.4 m. The output electric pulses from D_1 and D_2 were then sent to a time correlation circuit with a coincidence window of 2.4 ns.

The two-photon far-field diffraction-interference pattern was measured by two schemes. In the first one, detector D_2 was fixed and D_1 scanned in the horizontal direction by the step size of 1 mm. Figure 2(a) shows the D_1 single counts and $D_1 - D_2$ coincidence counts versus the position of detector D_1. In this figure, the D_1 single counts have an even distribution in a wide range, whereas the $D_1 - D_2$ coincidence counts exhibits the interference pattern in the range. The peak interval is about 7.41 mm, and the visibility of the fringe is 0.82 ± 0.03. The resultant two-photon diffraction-interference pattern strongly resembles the first-order diffraction-interference pattern of a multislit grating (with the identical Λ_{tr} and b as the tested sample) by 1064 nm light. For comparison, moreover, we performed the same single and coincidence counting measurement on a periodically poled $LiTaO_3$ crystal without transverse modulation as shown in the inset of Fig. 3. As a result, the interference pattern was erased, and the coincidence counts demonstrated a single peak as shown in Fig. 3. The Gaussian-like coincidence peak results from the Fourier transformation of the pump proflie in this case.

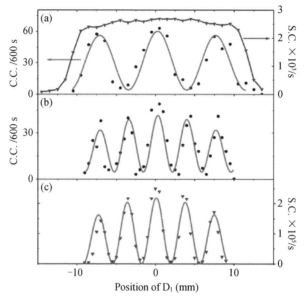

FIG.2. (color online)(a),(b) Measurement of single and coincidence counts versus the position of detector D_1. (a) D_2 was fixed. (b) D_1 and D_2 scanned in step. (c) Measurement of diffraction-interference pattern for the sample mask illuminated by the classical 532 nm light in the same experimental setup. The solid red curves in(a),(b), and(c) are theoretical fittings.

In the second scheme, both detectors D_1 and D_2 scanned along the same x direction by the step size of 0.5 mm synchronously. As expected, the $D_1 - D_2$ coincidence counts presented a second-order interference pattern as shown in Fig. 2(b). The peak interval is about 3.72 mm, exactly half of that in Fig. 2(a). This is a so-called two-photon subwavelength interference effect[23]. The two-photon behaves as an effective single photon but with the wavelength of a half. In comparison, a classical interference experiment of a multislit grating was performed using a 532 nm laser source. The multislit grating had the identical transverse modulation with the measured MPPLT. The result is shown in Fig. 2 (c). Although the wavelength of the two-photon case is 1064 nm, which is twice that of classical 532 nm light, the diffraction-interference patterns are almost identical for these two cases.

The above results consist well with Eq.(6). The two-photon pairs can be generated from any one of the illuminated stripes in the MPPLT. This is similar in that two-photon pairs pass through a multislit grating and are diffracted by these slits at the same time. When one detector was fixed and the other moved, the optical path difference of the two-photon pairs from different strips depended only on the position of the moving detector. While the two detectors scanned in step, the optical path difference increased twice, so the peak interval of the interference pattern in the later situation should be reduced twice. In previous studies, the two-photon interference was observed by manipulating the pump beam profile[10,24], using two SPDC crystals[2] or placing a double slit at the output surface of the crystal[23]. In our experiment, the N-stripe grating was embedded intrinsically inside the nonlinear crystal and transformed the generated two-photon spatial mode. Hence no accessional multislit grating or pump beam reforming was necessary. It suggests a new method to control and tailor the spatial entanglement of two-photon pairs by the technique of domain engineering.

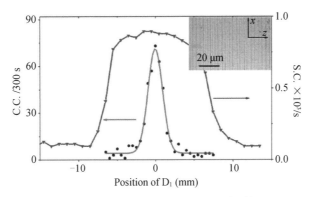

FIG.3. (color online) Measurement of single and coincidence counts for the PPLT grating with no transverse modulation. Detector D_1 scanned, while D_2 was fixed. The inset shows the PPLT grating structure.

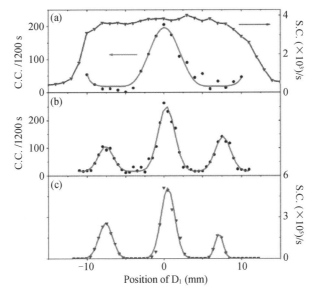

FIG.4. (color online) (a), (b) Measurement of single and coincidence counts versus the position of detector D_1 for the MPPLT with $\Lambda_{tr} = 100$ μm, $b = 20$ μm, and $L = 4$ mm. (a) D_2 was fixed. (b) D_1 and D_2 scanned in step. (c) Measurement of diffraction-interference pattern for the sample mask illuminated by the classical light 532 nm.

To confirm the above conclusion, we took a similar measurement on another MPPLT with $\Lambda_{tr}=100$ μm, $b=20$ μm, and crystal length $L=4$ mm. The result is shown in Fig. 4. The peak interval of the diffraction-interference pattern is about 14.62 mm when only one detector scans. The interval is reduced twice when D_1 and D_2 scanned in step. The visibility of the coincidence fringe was 0.83 ± 0.02. The experimental results also agree well with theoretical values according to Eq.(6).

In conclusion, we have experimentally studied the spatial mode of a two-photon entangled state produced from the transversely engineered QPM gratings. The spatial information of the MPPLT was transferred to the two-photon spatial mode, which was proved by the far-field diffraction-interference experiment. The QPM technique allows large variation of the spatial mode function over a small spatial scale in a single nonlinear crystal, so it is very favorable for integration and application. Moreover, by combining the longitudinal and transverse engineering of a QPM grating, we could construct the various space-momentum, time-energy, and spatial shape entanglement and even prepare the hyperentangled states[25]. The QPM technique would be widely used in the design of quantum optics devices.

References and Notes

[1] M. H. Rubin, Phys. Rev. A **54**, 5349(1996).

[2] P. H. S. Ribeiro, M. F. Santos, P. Milman, and A. Z. Khoury, J. Opt. B **4**, S437(2002).

[3] A. Mair, A. Vaziri, G. Weihs, and A. Zeilinger, Nature(London) **412**, 313(2001).

[4] C. K. Law, I. A. Walmsley, and J. H. Eberly, Phys. Rev. Lett. **84**, 5304(2000).

[5] S. P. Walborn and C. H. Monken, Phys. Rev. A **76**, 062305(2007).

[6] D. Kaszlikowski, P. Gnaciński, M. Żukowski, W. Miklaszewski, and A. Zeilinger, Phys. Rev. Lett. **85**, 4418(2000); D. Collins, N. Gisin, N. Linden, S. Massar, and S. Popescu, Phys. Rev. Lett. **88**, 040404(2002).

[7] N. J. Cerf, M. Bourennane, A. Karlsson, and N. Gisin, Phys. Rev. Lett. **88**, 127902(2002).

[8] G. Molina-Terriza, A. Vaziri, R. Ursin, and A. Zeilinger, Phys. Rev. Lett. **94**, 040501(2005).

[9] T. B. Pittman et al., Phys. Rev. A **53**, 2804(1996).

[10] C. H. Monken, P. H. Souto Ribeiro, and S. Pádua, Phys. Rev. A **57**, 3123(1998).

[11] L. Neves et al., Phys. Rev. Lett. **94**, 100501(2005); A. Vaziri, G. Weihs, and A. Zeilinger, J. Opt. B **4**, S47(2002).

[12] J. P. Torres, A. Alexandrescu, S. Carrasco, and L. Torner, Opt. Lett. **29**, 376(2004).

[13] S. Tanzilli et al., Electron. Lett. **37**, 26(2001).

[14] M. C. Booth et al., Phys. Rev. A **66**, 023815(2002).

[15] S. Tanzilli et al., Nature(London) **437**, 116(2005).

[16] S. E. Harris, Phys. Rev. Lett. **98**, 063602(2007).

[17] M. B. Nasr et al., Phys. Rev. Lett. **100**, 183601(2008).

[18] S. Carrasco et al., Opt. Lett. **29**, 2429(2004).

[19] A. V. Burlakov et al., Phys. Rev. A **56**, 3214(1997).

[20] J. R. Kurz, A. M. Schober, D. S. Hum, A. J. Saltzman, and M. M. Fejer, IEEE J. Sel. Top.

Quantum Electron. **8**, 660(2002).

[21] Y. Q. Qin, C. Zhang, Y. Y. Zhu, X. P. Hu, and G. Zhao, Phys. Rev. Lett. **100**, 063902(2008).

[22] G. Molina-Terriza, J. P. Torres, and L. Torner, Phys. Rev. Lett. **88**, 013601(2001); S. Gröblacher, T. Jennewein, A. Vaziri, G. Weihs, and A. Zeilinger, New J. Phys. **8**, 75(2006).

[23] M. D'Angelo, M. V. Chekhova, and Y. H. Shih, Phys. Rev. Lett. **87**, 013602(2001).

[24] E. J. S. Fonseca, C. H. Monken, and S. Pádua, Phys. Rev. Lett. **82**, 2868(1999).

[25] J. T. Barreiro, N. K. Langford, N. A. Peters, and P. G. Kwiat, Phys. Rev. Lett. **95**, 260501(2005).

[26] The authors thank Y. H. Shih for helpful discussions and technical assistance. This work is supported by the National Natural Science Foundation of China(No. 10534020) and by the National Key Projects for Basic Researches of China(No. 2006CB921804 and No. 2004CB619003).

Compact Engineering of Path-Entangled Sources from a Monolithic Quadratic Nonlinear Photonic Crystal[*]

H. Jin,[1] P. Xu,[1] X.W. Luo,[1] H.Y. Leng,[1] Y.X. Gong,[2] W.J. Yu,[1] M.L. Zhong,[1] G. Zhao,[3] and S.N. Zhu[1]

[1] *National Laboratory of Solid State Microstructures, College of Physics, and National Center of Microstructures and Quantum Manipulation, Nanjing University, Nanjing 210093, China*

[2] *Department of Physics, Southeast University, Nanjing 211189, China*

[3] *College of Engineering and Applied Sciences, Nanjing University, Nanjing 210093, China*

An integrated realization of photonic entangled states becomes an inevitable tendency toward integrated quantum optics. Here we report the compact engineering of steerable photonic path-entangled states from a monolithic quadratic nonlinear photonic crystal. The crystal acts as a coherent beam splitter to distribute photons into designed spatial modes, producing the heralded single-photon and appealing beamlike two-photon path entanglement. We characterize the path entanglement by implementing quantum spatial beating experiments. Such a multifunctional entangled source can be further extended to the high-dimensional fashion and multiphoton level, which paves a desirable way to engineering miniaturized quantum light sources.

Nowadays a consequent tendency toward practical quantum information processing is to manipulate photons including entangled photons on a monoplatform like an integrated waveguide chip[1-11]. However, most of the optical chips require external quantum light sources, which are usually bulky and involved with a lot of optical elements; thereby, as a further step toward integrated quantum optics, it is necessary to miniaturize the external light sources for improving their performance like multifunction, stability, and portability. A solid strategy for achieving integrated multifunctional quantum light sources turns to the traditional nonlinear optical crystals especially the domain-engineered quadratic nonlinear photonic crystals (NPC)[12,13]. By the domain-engineering technique, the spatial and temporal properties of entangled photons can be controlled inherently during the quasi-phase-matching(QPM) spontaneous parametric down-conversion(SPDC) processes[14-23], resulting in the generation of new types of photonic entanglement.

In this work, we mainly concentrate on the integrated engineering of single-photon and two-photon path-entangled states. The path entanglement is a valuable resource which applies the spatial mode to encode information, and has been harnessed for a variety of applications, such as quantum precise phase measurement[24,25] and super-resolution quantum lithography[26,27]. Even for the simple single-photon path entanglement[28-31],

---

[*] Phys.Rev.Lett.,2013,111(2):023603

people also find interesting applications in quantum teleportation[32] and quantum networks[33]. However, for achieving path entanglement, extra endeavors should be paid after the entangled photons are generated from the nonlinear crystal. For example, a two-photon NOON state, i.e. a two-photon maximally path-entangled state $(|2,0\rangle+|0,2\rangle)/\sqrt{2}$, which can be usually generated by Hong-Ou-Mandel interference[1,34,35], requires a beam splitter to cascade after the crystal and hold on at a balanced position; therefore, inte-grated realization of such states is of essential importance.

Here we experimentally demonstrate the direct generation of single-and two-photon path-entangled sources from a monolithic domain-engineered NPC. Resulting from the concurrent multiple QPM SPDC processes, versatile spatial forms of down-converted beams are achieved and different types of photonic path-entangled states can be generated and transformed inside the same crystal wafer. Specifically, the heralded single-photon path entanglement can be steered into the beamlike two-photon path entanglement. We further demonstrate the valuable extension of such path-entangled states to the high-dimensional entanglement. These results present unique advantages of such monolithic domain-engineered crystals in the engineering of novel integrated quantum light sources, which are geared to integrated quantum optics.

Experimentally, a hexagonally poled lithium tantalate(HPLT) crystal is designed for the engineering of integrated path-entangled photon source. Figure 1(a) shows the micrograph of the etched congruent HPLT with the poling period $a=7.507$ μm, the crystal length $L=18$ mm and a reversal factor of $r/a \sim 28\%$ (r is the radius of round domain-inverted area), which is qualified for the efficient generation of degenerate 1064 nm photon pairs when the pump laser of 532 nm propagates along the y axis under a polarization configuration of $e \rightarrow e + e$. The quadratic nonlinear coefficient of our crystal can be expressed as Fourier series of $\chi^{(2)}(\boldsymbol{r}) = d_{33} \sum_{m,n} f_{m,n} e^{i\boldsymbol{G}_{m,n} \cdot \boldsymbol{r}}$ corresponding to a reciprocal lattice with sixfold symmetry as sketched in Fig. 1(b). Multiple reciprocal vectors $\boldsymbol{G}_{m,n}$ (m, n are integers) can ensure multiple QPM geometries for the entangled photons generation. In this work, we mainly focus on $\boldsymbol{G}_{1,0}$ and $\boldsymbol{G}_{0,1}$, which have equal Fourier coefficients $f_{1,0} = f_{0,1} = 0.32$[12]. The involved pair of QPM geometries follow $\boldsymbol{k}_p - \boldsymbol{G}_{1,0} - \boldsymbol{k}_s - \boldsymbol{k}_i = 0$ and $\boldsymbol{k}_p - \boldsymbol{G}_{0,1} - \boldsymbol{k}_s - \boldsymbol{k}_i = 0$, wherein $\boldsymbol{k}_p$, $\boldsymbol{k}_s$, and $\boldsymbol{k}_i$ are the wave vectors of the pump, signal, and idler photons, respectively. Generally, the photon pair will emit as either one of conical beams with principle axis along $\boldsymbol{k}_p - \boldsymbol{G}_{1,0}$ or $\boldsymbol{k}_p - \boldsymbol{G}_{0,1}$. Here only two types of concurrent processes are of our interest, which are depicted in Figs. 1(c) and 1(d), indicating the direct generation of single-photon and two-photon path-entangled states, respectively. In Fig. 1(c), two SPDC processes share the same idler mode which propagates collinearly with the pump, thereby, the corresponding signal photon belongs to either mode 1(s_1) or mode 2(s_2). Figure 1(d) shows the other case. Notably, in this case the photon pair emit together into beamlike modes which are desirable for the high-efficiency collection and high degree of path entanglement since $|1,1\rangle$ term in this sample is prohibited by the QPM rule.

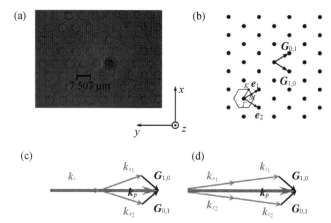

FIG. 1. (color online) (a) Micrograph of the etched congruent HPLT with dimensions of 18 mm×6 mm×0.5 mm. (b) The reciprocal lattice of the crystal. (c) and (d) are QPM conditions for the generation of single-photon and two-photon path entanglement, respectively.

Under the first-order perturbation approximation[36], the two-photon state from concurrent two SPDC processes can be written as

$$|\psi\rangle = \Psi_0 \int d\omega_s \int d\omega_i \phi(\omega_s,\omega_i)[\hat{a}_{s_1}^+(\boldsymbol{k}_{s_1})\hat{a}_{i_1}^+(\boldsymbol{k}_{i_1}) + \hat{a}_{s_2}^+(\boldsymbol{k}_{s_2})\hat{a}_{i_2}^+(\boldsymbol{k}_{i_2})]|0\rangle, \quad (1)$$

in which Ψ_0 is a normalization constant, and the subscripts represent two different QPM processes. The two-photon mode functions of two processes take the same form $\phi(\omega_s,\omega_i) = \mathrm{sinc}(\Delta k_y L/2)\delta(\omega_p - \omega_s - \omega_i)$, wherein L is the length of the crystal, Δk_y is the phase mismatching in the longitudinal direction. Here, for a focused pump with 60 μm beam size, the longitudinal phase-matching condition takes the dominant role in the two-photon mode function, so we omit the transverse phase-matching part. For the single-photon path entanglement indicated in Fig. 1(c), we can take a partial trace of $\hat{\rho} = |\psi\rangle\langle\psi|$[37], which is the density matrix operator of two-photon state, and write the density matrix of the signal as

$$\hat{\rho}_s = \Psi_0^2 \int dv\, |\phi(v)|^2 \left[a_{s_1}^+\left(\frac{\omega_p}{2}+v\right) + a_{s_2}^+\left(\frac{\omega_p}{2}+v\right)\right]|0\rangle\langle 0|\left[a_{s_1}\left(\frac{\omega_p}{2}+v\right) + a_{s_2}\left(\frac{\omega_p}{2}+v\right)\right], \quad (2)$$

in which we introduced a detuning frequency v. The longitudinal phase mismatching in $\phi(v)$ can be written as

$$\Delta k_y = \frac{v(1-\cos\theta_1)}{u_s} - \frac{v^2(1+\cos\theta_1)}{2}\frac{d}{d\omega}\left(\frac{1}{u}\right)\Big|_{\omega=\omega_p/2},$$

wherein $u_s = \frac{d\omega}{dk}\big|_{\omega=\omega_p/2}$ is group velocity of signal and $\theta_1 = 2.192°$ is the angle between the wave vectors of signal and pump in the crystal. The bandwidth (FWHM) of signal photons is calculated to be 35.4 nm. Taking single-frequency approximation which can be guaranteed by narrowband interference filters, we can simplify Eq.(2) into

$$|\psi_1\rangle = \frac{1}{\sqrt{2}}(|1,0\rangle + |0,1\rangle), \quad (3)$$

which is a single-photon path-entangled state.

For the two-photon path entanglement indicated in Fig. 1(d), we can deduce the two-photon state to be

$$|\psi\rangle = \Psi_0 \int dv \phi(v) \left[\hat{a}^+_{s_1}\left(\frac{\omega_p}{2}+v\right)\hat{a}^+_{i_1}\left(\frac{\omega_p}{2}-v\right) + \hat{a}^+_{s_2}\left(\frac{\omega_p}{2}+v\right)\hat{a}^+_{i_2}\left(\frac{\omega_p}{2}-v\right) \right]|0\rangle. \quad (4)$$

In this case the phase mismatching is $\Delta k_y = -v^2 \cos\theta_2 \frac{d}{d\omega}\left(\frac{1}{u}\right)|_{\omega=\omega_p/2}$, wherein $\theta_2 = 1.095°$ is the angle between the wave vectors of signal and pump. The theoretical bandwidth (FWHM) of down-converted photons is about 31.4 nm. Using single-frequency approximation, we can simplify Eq.(4) into

$$|\psi_2\rangle = \frac{1}{\sqrt{2}}(|2,0\rangle + |0,2\rangle), \quad (5)$$

which is a two-photon NOON state.

Figures 2(a)–2(d) record the spatial distribution evolution of down-converted photons generated from the HPLT when the temperature increased. Figures 2(b) and 2(d) disclose the spatial distribution of single-photon and two-photon path-entangled sources when the QPM conditions of Figs. 1(c) and 1(d) are satisfied, respectively. Although the coherence between two single- or two-photon modes is guaranteed physically by the coherence of concurrent nonlinear processes sharing the same pump, it is necessary to verify the path entanglement. Here we carry out spatial beating experiments to reveal quantum nature of the states by obtaining the fringe visibility and coherence length, which is equivalent to combining two single-photon or two-photon modes by a beam splitter but more stable since it is an integrated photonic setup.

At 154.6℃, as shown in Fig. 2(b) the photon pair emit into either one of two tangent cones, resulting in the heralded single-photon path entanglement. Theoretically, when two coherent signal modes with wavelength λ intersect at an angle of 2θ, the spatial beating fringes will appear as

$$I_{s.c.} \propto 1 + \cos\left(\frac{2\pi x}{\lambda/(2\sin\theta)}\right). \quad (6)$$

Figure 3(a) shows the experimental setup. The coherent signal modes are picked out by two slits and intersect at an angle of $2\theta = 10.4$ mrad. Their beating fringes are recorded by a CCD camera which is sensitive to a few photons. In Figs. 3(b) and 3(c) we show the interference fringe when two optical paths are precisely controlled to be equal. The experimental data (blue dots) are fitted with a sinusoidal curve weighed by a Gaussian profile. The fitted curve shows a period of 104.4 μm, which agrees with the calculated value. The fringe visibility is 70.9%, which confirms that the two modes sketched in Fig. 1

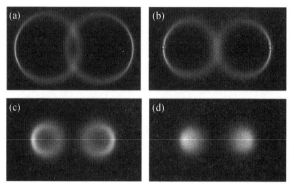

FIG.2. (color online) (a) The spatial Fourier spectrum obtained at the focal plane of a convex lens when the crystal is set at 145℃ by a sensitive CCD camera. (b) The spatial Fourier spectrum at 154.6 ℃. The two small red circles indicate the two modes of the single-photon entangled state. The angle between two signal modes outside of the crystal is measured to be about 9.54° which agrees well with the theoretical design of 9.38°. (c) The spatial Fourier spectrum at 168℃. (d) The spatial Fourier spectrum at 172.3 ℃. The two bright spots correspond to the noncollinear beamlike two-photon modes with the relative emitting angle outside of the crystal 4.88°(theoretically 4.69°). The measured divergence angle of each mode is about 2.26°(FWHM).

(c) are generated in a coherent way. The coherence is due to the indistinguishability of their idler modes. The coherence length, which is estimated by recording the relationship between the fringe visibility and the path difference as shown in Fig. 3(d), is measured to be 23.5 μm(the visibility drops to 1/e), which fits well with theoretical value of 22.6 μm deduced from the theoretical single-photon bandwidth after taking the interference filter's (IF) transmission profile into account. When tuning the temperature away from 154.6℃, two signal modes will share less of the idler mode and tend to be incoherent. At 143℃, we find that the fringe visibility decreases to less than 30% under balanced path lengths of two modes.

As shown in Fig. 2(d), when the crystal is controlled at 172.3℃, the photon pair will emit into either one of the bright well-defined beamlike spots, resulting in the two-photon path entanglement. Theoretically, the two-photon spatial correlation is proportional to

$$R_{c.c.} \propto 1 + \cos\left(\frac{2\pi x}{\lambda/(4\sin\theta)}\right), \tag{7}$$

when two two-photon modes intersect at an angle of 2θ. Two-photon beating fringes should present the period half of the single-photon case with the same wavelength. Figure 4(a) is the corresponding experimental setup for observing two-photon spatial beating fringe. As shown in Fig. 4(b) although the single counts of D_1 and D_2 show a smooth distribution, the coincidence counts reveal an interference fringe with the period of 48 μm, which consists well with Eq.(7) when $2\theta = 10.9$ mrad. The equivalent wavelength is reduced to $\lambda/2 = 532$ nm. The visibility of interference fringe is $68.2 \pm 1.8\%$, and it reaches $82.4 \pm 1.8\%$

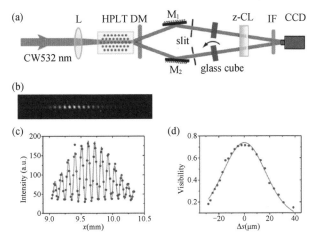

FIG.3. (color online) (a) Experimental setup. The crystal is controlled at 154.6 ℃ and pumped with a cw single longitudinal mode 532 nm laser. After the crystal, the down-converted photons of 1064 nm are first separated from the pump by a dichromatic mirror DM, then two signal modes are both reflected by flat mirrors to ensure a small intersecting angle $2\theta = 10.4$ mrad so that the beating fringes can be resolved by the CCD camera which is sensitive to a few photons. A cylindrical lens z-CL gathers the intensity along the z axis. For each path, a 500 μm width slit(corresponding to an arc of 43 mrad on the ring of down-converted photons) is used to pick out the coherent signal mode. Preceding the CCD is a 40 nm bandwidth IF centered at 1064 nm. The optical path length of each mode is adjusted by rotating a K9 glass cube with 25.4 mm thickness. (b) The interference fringe observed by the CCD when the two modes have no optical path difference. (c) The intensity distribution extracted from Fig. 2(b). The solid red curve is sinusoidal fitting weighed by a Gaussian profile. (d) The relationship between the fringe visibility and the optical path-length difference.

when accidental coincidence counts are excluded from the raw data; therefore, the two-photon modes are coherently superposed. The nonideal visibility is caused by the nonperfect mode overlapping as well as the 1∶2 intensity imbalance between the two modes since they go through different optical elements before the interaction. The intensity imbalance is valued through two-photon coincidence measurement for each mode by blocking the other one. Furthermore, to examine the two-photon coherence length which is theoretically determined by the pump, we increase the path difference by 1 mm which is much larger than the single photon coherence length of 113 μm(estimated by the 10 nm-bandwidth of IF) to find that the two-photon beating fringe [Fig. 4(c)] keeps almost the same visibility as in Fig. 4(b). For each beamlike mode, the conversion efficiency is $\sim 4.8 \times 10^{-10}$ when the pump power is 14 mW.

The spatial form of the down converted beams from the crystal is revealed to be controllable and versatile which exhibits the morphing behavior from two symmetric conical beams into two beamlike spots, resulting in the direct generation and easy switch of two types of path entanglement. It is worth noting that the spatial forms of photon pairs in

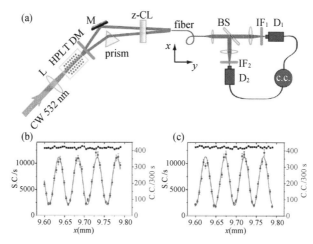

FIG. 4. (color online) (a) Experimental setup. The crystal temperature is controlled at 172.3℃. A prism and a reflecting flat mirror are used to ensure that the two modes intersect at a small angle, and a cylindrical lens z-CL gathers the photons along the z axis into the fiber. The fiber tip is scanned along the x axis and cascaded by a coincidence counting measurement. Specifically, entangled photons are separated by a beam splitter (BS) after collimated by a lens and collected into two single photon detectors D_1 and D_2, respectively, by two lenses. The bandwidth of IF is 10 nm. (b) Measured single (black squares) and coincidence (blue dots) counts versus the transverse position of the fiber tip. The red solid line is sinusoidal fitting. (c) Measured single (black squares) and coincidence (blue dots) counts when one optical path length is extended by about 1 mm. Error bars show $\pm\sqrt{R_{c.c}}$.

Fig. 2 seem to be similar to those from some type-Ⅱ birefringence-phase-matching (BPM) SPDC processes[38−40]; however, there exist intrinsic differences. The photon pair always emit into the same cone or beamlike spot in this work, while for BPM type-Ⅱ cases the signal and idler photons usually emit into different ones. It is worth emphasizing that a recent work aiming to engineer the two-photon NOON state has just been published[18]. The results are similar to the two-photon case in this work; however, because of the parasitical SPDC process participated by the collinear reciprocal vector there exists an unwanted contribution of $|1,1\rangle$ which causes some limitations.

Considering the state-of-the-art domain-engineered technique, such compact path-entangled states can be extended into the high-dimensional fashion and multiphoton level[17]. Figure 5 shows our experimental results in the realization of heralded single-photon high-dimensional path entanglement. Figure 5(a) is the spatial form of photon pairs generated from a single lithium niobate crystal wafer with cascaded tetragonally and hexagonally poled blocks. Each block can emit a heralded single-photon two-mode entangled state as Eq. (3). Since four concurrent SPDC processes share the same idler mode, the signal photon can emit into any one of modes 1, 2, 3, and 4 coherently as shown in Fig. 5(b). The single-photon multi-partite entanglement can be written as

$$|\psi\rangle = \alpha(|1000\rangle + |0100\rangle) + \beta e^{i\varphi}(|0010\rangle + |0001\rangle). \tag{8}$$

α and β are decided by the SPDC efficiencies from each block. φ has a fixed value which relates with the length of each block and the interval between two blocks[41]. Following such a principle, the single-photon N-dimensional ($N \geqslant 5$) multipartite entanglement can further be approached. In addition, other degrees of freedom like polarization, orbital angular momentum, and frequency can also be manipulated inside the same crystal chip, which will result in new types of compact photonic entanglement like hyperentanglement over several degrees of freedom. The integrated and miniaturized quantum light sources based on concurrent multiple QPM processes will act as a key element in exploring the knowledge bounda-ries of quantum mechanics and prompting the developments of practical quantum technologies.

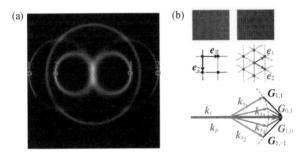

FIG.5. (color online) (a) The spatial Fourier spectrum of 1064 nm photon pairs generated from a single domain-cascaded lithium niobate crystal. (b) Micrograph of the sample and involved four concurrent QPM conditions. Both domain blocks are ~6 mm long with a separation of 2 mm and each can supply a pair of reciprocal vectors ($G_{1,1}$, $G_{1,-1}$) for the tetragonal distribution with poling period of 6.271 μm or ($G_{1,0}$, $G_{0,1}$) for the hexagonal distribution with poling period of 6.445 μm.

References and Notes

[1] A. Politi, M. J. Cryan, J. G. Rarity, S. Yu, and J. L. O'Brien, Science **320**, 646(2008).

[2] J.C.F. Matthews, A. Politi, A. Stefanov, and J.L. O'Brien, Nat. Photonics **3**, 346(2009).

[3] B.J. Smith, D. Kundys, N. Thomas-Peter, P.G.R. Smith, and I. A. Walmsley, Opt. Express **17**, 13516 (2009).

[4] A. Politi, J.C.F. Matthews, and J.L. O'Brien, Science **325**, 1221(2009).

[5] L. Sansoni, F. Sciarrino, G. Vallone, P. Mataloni, A. Crespi, R. Ramponi, and R. Osellame, Phys. Rev. Lett. **105**, 200503(2010).

[6] A. Laing, A. Peruzzo, A. Politi, M. R. Verde, M. Halder, T. C. Ralph, M. G. Thompson, and J. L. O'Brien, Appl. Phys. Lett. **97**, 211109(2010).

[7] M.F. Saleh, G. Di Giuseppe, B.E.A. Saleh, and M.C. Teich, Opt. Express **18**, 20475(2010).

[8] A. Peruzzo, A. Laing, A. Politi, T. Rudolph, and J. L. O'Brien, Nat. Commun. **2**, 224(2011).

[9] P. J. Shadbolt, M. R. Verde, A. Peruzzo, A. Politi, A. Laing, M. Lobino, J.C.F. Matthews, M.G. Thompson, and J. L. O'Brien, Nat. Photonics **6**, 45(2011).

[10] J.C.F. Matthews, A. Politi, D. Bonneau, and J.L. O'Brien, Phys. Rev. Lett. **107**, 163602(2011).

[11] D. Bonneau, M. Lobino, P. Jiang, C. M. Natarajan, M. G. Tanner, R.H. Hadfield, S.N. Dorenbos, V. Zwiller, M.G. Thompson, and J. L. O'Brien, Phys. Rev. Lett. **108**, 053601(2012).

[12] V. Berger, Phys. Rev. Lett. **81**, 4136(1998).

[13] N.G.R. Broderick, G.W. Ross, H.L. Offerhaus, D.J. Richardson, and D. C. Hanna, Phys. Rev. Lett. **84**, 4345(2000).

[14] J.P. Torres, A. Alexandrescu, S. Carrasco, and L. Torner, Opt. Lett. **29**, 376(2004).

[15] X.Q. Yu, P. Xu, Z.D. Xie, J.F. Wang, H.Y. Leng, J. S. Zhao, S. N. Zhu, and N. B. Ming, Phys. Rev. Lett. **101**, 233601(2008).

[16] H.Y. Leng, X.Q. Yu, Y.X. Gong, P. Xu, Z.D. Xie, H. Jin, C. Zhang, and S. N. Zhu, Nat. Commun. **2**, 429(2011).

[17] Y.X. Gong, P. Xu, Y.F. Bai, J. Yang, H. Y. Leng, Z. D. Xie, and S. N. Zhu, Phys. Rev. A **86**, 023835 (2012).

[18] E. Megidish, A. Halevy, H.S. Eisenberg, A. Ganany-Padowicz, N. Habshoosh, and A. Arie, Opt. Express **21**, 6689(2013).

[19] S. Carrasco, J.P. Torres, L. Torner, A. Sergienko, B.E.A. Saleh, and M. C. Teich, Opt. Lett. **29**, 2429 (2004).

[20] S. E. Harris, Phys. Rev. Lett. **98**, 063602(2007).

[21] M.B. Nasr, S. Carrasco, B.E.A. Saleh, A.V. Sergienko, M.C. Teich, J.P. Torres, L. Torner, D.S. Hum, and M.M. Fejer, Phys. Rev. Lett. **100**, 183601(2008).

[22] S. Sensarn, G.Y. Yin, and S.E. Harris, Phys. Rev. Lett. **104**, 253602(2010).

[23] A.M. Brańczyk, A. Fedrizzi, T.M. Stace, T.C. Ralph, and A. G. White, Opt. Express **19**, 55(2010).

[24] T. Nagata, R. Okamoto, J. L. O'Brien, K. Sasaki, and S. Takeuchi, Science **316**, 726(2007).

[25] J.W. Pan, Z.B. Chen, C.Y. Lu, H. Weinfurter, A. Zeilinger, and M. Żukowski, Rev. Mod. Phys. **84**, 777 (2012).

[26] A.N. Boto, P. Kok, D.S. Abrams, S.L. Braunstein, C.P. Williams, and J.P. Dowling, Phys. Rev. Lett. **85**, 2733(2000).

[27] Y. Kawabe, H. Fujiwara, R. Okamoto, K. Sasaki, and S. Takeuchi, Opt. Express **15**, 14244(2007).

[28] S.M. Tan, D.F. Walls, and M.J. Collett, Phys. Rev. Lett. **66**, 252(1991).

[29] L. Hardy, Phys. Rev. Lett. **73**, 2279(1994).

[30] S. J. van Enk, Phys. Rev. A **72**, 064306(2005).

[31] S. B. Papp, K. S. Choi, H. Deng, P. Lougovski, S. J. Van Enk, and H. J. Kimble, Science **324**, 764 (2009).

[32] G. Björk, A. Laghaout, and U.L. Andersen, Phys. Rev. A **85**, 022316(2012).

[33] I. Usmani, C. Clausen, F. Bussières, N. Sangouard, M. Afzelius, and N. Gisin, Nat. Photonics **6**, 234(2012).

[34] C.K. Hong, Z.Y. Ou, and L. Mandel, Phys. Rev. Lett. **59**, 2044(1987).

[35] Y. H. Shih and C. O. Alley, Phys. Rev. Lett. **61**, 2921(1988).

[36] M. H. Rubin, Phys. Rev. A **54**, 5349(1996).

[37] D.V. Strekalov, Y.H. Kim, and Y.H. Shih, Phys. Rev. A **60**, 2685(1999).

[38] P.G. Kwiat, K. Mattle, H. Weinfurter, A. Zeilinger, A.V. Sergienko, and Y.H. Shih, Phys. Rev. Lett. **75**, 4337(1995).

[39] S. Takeuchi, Opt. Lett. **26**, 843(2001).
[40] C. Kurtsiefer, M. Oberparleiter and H. Weinfurter, J. Mod. Opt. **48**, 1997(2001).
[41] W. Ueno, F. Kaneda, H. Suzuki, S. Nagano, A. Syouji, R. Shimizu, K. Suizu, and K. Edamatsu, Opt. Express **20**, 5508(2012).
[42] The authors thank Z. Y. Ou for helpful discussions. This work was supported by the State Key Program for Basic Research in China(No. 2012CB921802 and No. 2011CBA00205), the National Natural Science Foundations of China(Contract No. 91121001, No. 11174121, No. 11021403, and No. 11004096), and the Project Funded by the Priority Academic Program development of Jiangsu Higher Education Institutions(PAPD), the Program for New Century Excellent Talents in University(NCET), and a Foundation for the Author of National Excellent Doctoral Dissertation of People's Republic of China (FANEDD).

On-chip Steering of Entangled Photons in Nonlinear Photonic Crystals[*]

H.Y.Leng[1], X.Q.Yu[1,2], Y.X.Gong[1], P.Xu[1], Z.D.Xie[1], H.Jin[1], C.Zhang[3] and S.N.Zhu[1]

[1] *National Laboratory of Solid State Microstructures and School of Physics, Nanjing University, Nanjing 210093, China*

[2] *Physics Department, Southeast University, Nanjing 211189, China*

[3] *School of Modern Engineering and Applied Science, Nanjing University, Nanjing 210093, China*

One promising technique for working toward practical photonic quantum technologies is to implement multiple operations on a monolithic chip, thereby improving stability, scalability and miniaturization. The on-chip spatial control of entangled photons will certainly benefit numerous applications, including quantum imaging, quantum lithography, quantum metrology and quantum computation. However, external optical elements are usually required to spatially control the entangled photons. Here we present the first experimental demonstration of on-chip spatial control of entangled photons, based on a domain-engineered nonlinear photonic crystal. We manipulate the entangled photons using the inherent properties of the crystal during the parametric downconversion, demonstrating two-photon focusing and beam-splitting from a periodically poled lithium tantalate crystal with a parabolic phase profile. These experimental results indicate that versatile and precise spatial control of entangled photons is achievable. Because they may be operated independent of any bulk optical elements, domain-engineered nonlinear photonic crystals may prove to be a valuable ingredient in on-chip integrated quantum optics.

During parametric downconversion, two lower-frequency photons (usually called the signal and idler) are generated from a pump photon via a nonlinear crystal. Due to the conservation of energy and momentum of the original pump photon, the frequency and momentum of the signal and idler are strongly correlated. In particular, the spatial entanglement of the photon pair has led to interesting research in many fields, including quantum imaging[1-6], quantum lithography[7-9], quantum metrology[10-13] and quantum computation[14]. To prepare specific two-photon states for various applications, the spatial entanglement of the signal and idler is tailored by manipulating the wavefront of the pump beam[15,16] or modifying the entangled photons after their generation[17-19], using various optical elements, such as lenses, multi-slits or spatial light modulators. These bulk optical elements inevitably hinder the performance of the photonic quantum circuits during practical applications, which require more stability, scalability and miniaturization[20-24].

---

[*] Nat.Commun.,2011,2(1):429

The aforementioned difficulty can be overcome with a different strategy that applies domain engineering in quadratic nonlinear photonic crystals, which is widely used in quasi-phase-matching(QPM) nonlinear optics[25—29]. Although the spatial control of entangled photons via domain engineering has been already theoretically proposed[30], few related experiments, other than a recent experiment showing that the amplitude of the entangled photons can be modulated by a multi-stripe nonlinear photonic crystal[31], have been attempted. Comprehensive control of spatial entanglement (particularly the phase of entangled photons) has still not been experimentally demonstrated.

In this work, we experimentally demonstrate the steering of entangled photons (that is, the wavefront shaping of the entangled photons) via domain engineering in nonlinear photonic crystals. By introducing a transverse inhomogeneity into the crystal, the wavefront of the entangled photons can be shaped at will, and the propagation of the entangled photons can be steered. We investigate a periodically poled lithium tantalate (PPLT) crystal with a transverse parabolic phase profile and find that the generated entangled photons are focused at a fixed distance from the crystal. In this case, our engineered crystal is equivalent to a homogeneous nonlinear crystal and a focusing lens[Fig. 1(a)]. Additionally, by translating the crystal, we realize a dual-focusing condition in which the two-photon is focused onto either of two symmetric directions. Under this condition, our engineered crystal serves as the entangled photon source, lens and beam splitter[Fig. 1(b)].

Results

Sample structure. The quadratic nonlinear coefficient of our crystal is expressed as

$$\chi^{(2)}(x,z) = d_{eff} U(x) \sum_m F(G_m) e^{iG_m[z+\phi(x)]}, \tag{1}$$

where d_{eff} is the effective quadratic susceptibility, $G_m = 2m\pi/\Lambda$ is the m th-order reciprocal vector for the longitudinal modulation with period Λ in the beam's propagation direction z, $F(G_m)$ represents the corresponding Fourier coefficient and $\phi(x)$ and $U(x)$ describe the nonuniform transverse phase and amplitude distributions, respectively. Because $\phi(x) = \omega_p x^2/(2cfG_3)$ is parabolic, a two-photon carrying the phase front $\omega_p x^2/(2cf)$ is focused at a distance f from the crystal, where ω_p is the angular pump frequency, c is the speed of light, G_3 is the third-order reciprocal vector that compensates for the longitudinal phase mismatching in the QPM and f represents the equivalent focal length of the crystal for the pump wavelength. The focusing behaviour can be described by the Huygens-Fresnel principle: each point on the primary wavefront acts as a source of secondary wavelets of the pump, as well as a source of the entangled photon pair. Therefore, such an engineered nonlinear photonic crystal is termed a QPM two-photon lens. By following the Huygens-Fresnel principle, we may engineer this crystal for any arbitrary spatial manipulation of

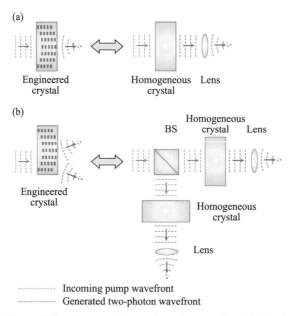

----- Incoming pump wavefront
----- Generated two-photon wavefront

FIG. 1. The multifunction nonlinear photonic crystals. (a) In the single-focusing experiment, the engineered nonlinear photonic crystal serves as the entangled photon source and lens. (b) In the dual-focusing experiment, the engineered nonlinear photonic crystal serves as the entangled photon source, lens and beam splitter(BS).

entangled photons, such as focusing, beam-splitting or even multifunction integration.

Figure 2(a) is a micrograph of our etched crystal. The transverse direction has a multi-stripe pattern that means that the amplitude modulation $U(x)$ in equation (1) takes the form of a grating with a stripe width of 10 μm and a stripe interval $\Lambda_{tr} = 20$ μm. The initial position of each stripe follows the function $\omega_p x^2/(2cfG_3)$, and the equivalent focal length of the crystal at the pump wavelength is designed to be $f = 33.3$ mm.

Single-focusing experiments. The layout of our experiment is shown in Figure 2(b). With the input tip of the fibre scanning in the transverse(x) and longitudinal(z) direction, we obtain the spatial correlation of entangled photon pairs by performing coincidence counting using the two detectors. The pump coincides with the centre of the parabolic PPLT sample. Figure 3(a) shows a simulation of the two-photon propagation dynamics based on the Huy-gens-Fresnel principle. Experimentally we obtained the minimum two-photon spatial correlation at a distance of $z = 33$ mm, as can be seen in Figure 3(b). The full width of the correlation peak, at which the intensity drops to $1/e^2$ of its maximum value, was 28 μm. In contrast, we theoretically expected a 24-μm focusing spot with a full pump width of 0.82 mm in the experiment. The experimental value was slightly larger than the theoretical calculation primarily because the pump has a greater divergence angle than the TEM$_{00}$ mode. In addition to the primary focusing spot along the pump beam direction, which is located at $x = 0$ mm, two other focusing spots were distributed symmetrically at x

= +0.78 mm and −0.76 mm. The minor focusing spots resulted from multi-stripe interference. The theoretical separation between the two adjacent focusing spots was $2\pi c f/(\omega_p \Lambda_{tr})$ = 0.76 mm, which is consistent with the experimental value.

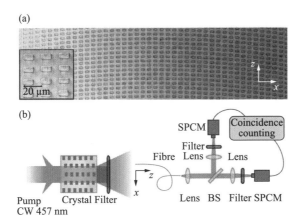

FIG.2. Steering of the entangled photons by domain engineering. (a) Microscope photograph of the etched quadratic nonlinear crystal. The sample was fabricated from a congruent lithium tantalate crystal at room temperature by electric field poling technique. The periodicity in the longitudinal direction is Λ = 13.917 μm, which ensures efficient entangled photon generation with the third-order reciprocal vector $G_3 = 3 \times 2\pi/\Lambda$. The stripe interval is Λ_{tr} = 20 μm, the stripe width is b = 10 μm and the stripe length is L = 6 mm. (b) Experimental setup. The crystal was pumped with a CW single longitudinal 457 nm laser. Degenerate photon pairs of 914 nm were generated at 180.1 ℃ with the polarization configuration of $e \rightarrow e + e$. After separation from the pump beam, the entangled photon pairs were directed into an optical fibre. Finally a 50/50 beam splitter (BS) was used, and photons on each path were collected by a lens and detected by a single-photon counting module (SPCM-AQR-14, PerkinElmer). A 10 nm bandwidth interference filter was placed before each detector to further suppress the pump photons.

Figure 3(c) shows an enlarged view of the simulated two-photon propagation dynamics around the focusing spot. Figure 3(d) shows the corresponding measured results. Whenever the fibre tip deviates from the focal plane, the measured two-photon spatial correlation widens. For example, at distances of z = 29 mm and z = 37 mm, the spatial correlation peak widths are 92 μm and 120 μm, respectively. The Rayleigh range is measured to be 0.9 mm. We used a TEM_{00} Gaussian beam to fit the focusing behaviour and found that the two-photon behaves similar to a Gaussian beam with the pump wavelength (rather than the signal or idler wavelength).

Dual-focusing experiments. It is worth emphasizing that we engineered the parabolic PPLT with a transverse multi-stripe structure rather than a continuous structure. This introduces periodicity into the transverse direction of the crystal. For the structure that we fabricated, in any two stripes separated by a distance of $2\pi c f/(\omega_p \Lambda_{tr})$ = 0.76 mm, the

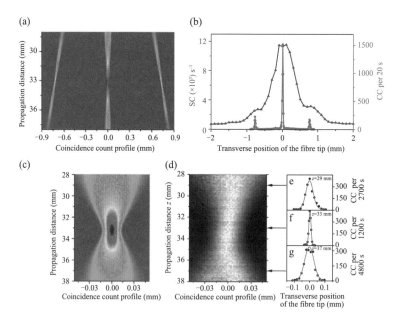

FIG.3. The focused two-photon. (a) Simulation of the coincidence count profile as a function of the propagation distance. (b) Measured single(blue triangles) and coincidence(red dots) counts versus the transverse position of the fibre tip in the focal plane. (c) Simulation of the coincidence count profile around the focusing spot as a function of the propagation distance. (d) Measured coincidence count profile around the focusing spot as a function of the propagation distance z. Coincidence counts at each distance have been normalized respectively. (e-g) Measured transverse correlation at distances of 29 mm, 33 mm and 37 mm, respectively. The solid red curves in (e)-(g) are Gaussian fittings.

added phases of the entangled photon pairs are approximately equivalent. Thus, the stripes at the $2j\pi cf/(\omega_p \Lambda_{tr})$ positions, where j is an integer, are all equivalent principal axes of the QPM two-photon lens, which are denoted by P_j in Figure 4(a). In our experiment, we find that when the crystal is translated in the x direction by multiples of 0.76 mm, the spatial distributions of the single counts and the two-photon coincidence counts remain identical.

A new type of spatially entangled state is generated when the incident pump lies halfway between the two principal axes; this situation is illustrated in Figure 4(a). The two-photon is deflected and focused onto either of the two symmetric directions around the pump beam. Figure 4(b) shows the simulated propagation dynamics of the dual-focused two-photon. Figure 4(c) shows the measured two-photon spatial correlation and single counts in the focal plane. There are two narrow correlation peaks with equivalent intensities, whereas the single counts follow a relatively smooth distribution. The measured peak interval of 0.77 mm agrees well with the theoretical prediction of $2\pi cf/(\omega_p \Lambda_{tr})=0.76$ mm. The phase-matching of the dual focusing is actually achieved using two tilted reciprocal vectors as shown in Figure 4(e). For such concurrent spontaneous parametric downconversion processes, it has been experimentally verified that their

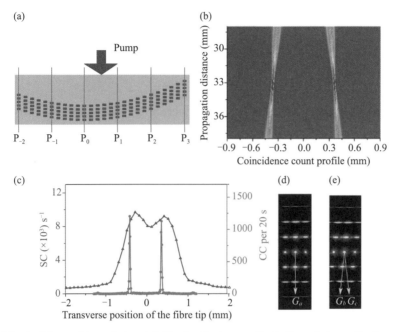

FIG. 4. The dual-focused two-photon. (a) The incident pump lies halfway between the two principal axes. P_j ($j = 0, \pm 1, \pm 2, \cdots$) are equivalent principal axes of the QPM two-photon lens. (b) Simulation of the coincidence count profile as a function of the propagation distance for the dual-focusing case. (c) Measured single (blue triangles) and coincidence (red dots) counts versus the transverse position of the fibre tip in the focal plane for the dual-focusing case. (d), (e) Measured Fourier spectra of the multi-stripe parabolic PPLT for the single-focusing and dual-focusing case, respectively. The third-order reciprocal vectors used in the QPM are marked.

contributions to the two-photon state have a fixed phase relation[32]. Therefore, if we collect the photons in the two foci described here, we obtain a NOON state with $N=2$. As we change the incident position of the pump between the two principal axes, the proportions of the photon pairs focused onto the two directions can be dynamically tuned. In this case, the crystal serves as both the beam splitter (with a tunable splitting ratio) in the pump, and the lens.

Phase matching analysis. To obtain a better understanding of the working principle of the engineered crystal, we analysed the Fourier spectra of the multi-stripe parabolic PPLT, which are the measured far-field diffraction patterns of the crystal illuminated by the pump laser beam. Figure 4(d) corresponds to the single-focusing case, in which the pump coincides with the principal axis, whereas Figure 4(e) corresponds to the dual-focusing case, in which the pump beam lies halfway between the two principal axes. The third-order reciprocal vectors, which are used in the QPM, are marked by G_a in Figure 4(d) and by G_b and G_c in Figure 4(e). In the single-focusing case, G_a ensures an efficient collinear downconversion and the two-photon is focused along the pump direction. In the dual focusing case, as G_b and G_c are not along the pump direction, the downconversion occurs using a non-collinear geometry. Hence, the two-photon is focused onto either of two possible directions, as shown in Figure 4(c).

Discussion

We experimentally realized the on-chip steering of entangled photons, based on a domain-engineered nonlinear photonic crystal. Using a transversely parabolic PPLT, we demonstrated two-photon focusing and beam-splitting. Our measured results agree well with the designed parameters of the domain structure, which shows that accurate spatial control of entangled photons can be achieved via domain engineering. Because this approach enables the control of the amplitude and phase of the two-photon up to lithographic precision, unique optical elements, such as lenses with extremely small focal lengths and high numerical apertures, can be engineered. In our experiment, the crystal served as an entangled photon source, lens and beam splitter, as illustrated in Figure 1. This multi-functionality shows the potential of integrating multiple optical transformations, such as a battery of lenses, into a single crystal. This multifunctional integration, which is inherent to the crystal, is free from any bulk optical elements and, therefore, may be exploited for on-chip integrated quantum optics. However, for a fully integrated device, other necessary functions(such as spectral filtering) should also be realized on the chip. This requires further consideration.

The flexible, state-of-the-art crystal poling technique enabled the fabrication of a wide variety of domain structures, such that we could spatially control the entangled photons at will. Combined with the temporal control of the entangled photons using longitudinal domain engineering[33—38], more interesting two-photon states may be prepared from nonlinear photonic crystals. This may attract interest in both fundamental physics and practical quantum technologies. Our technique might also find applications in research fields, such as quantum walk and continuous-variable encoding.

Methods

Calculation of the two-photon correlation. Assuming a monochromatic plane pump wave, the generated entangled two-photon state is

$$|\Psi\rangle = \Psi_0 \sum_{k_s, k_i} \Phi(\Delta k_z L) H_{tr}(\boldsymbol{\kappa}_s + \boldsymbol{\kappa}_i) \delta(\omega_p - \omega_s - \omega_i) \hat{a}_s^\dagger \hat{a}_i^\dagger |0\rangle, \qquad (2)$$

where Ψ_0 is a normalization constant, $\Phi(\Delta k_z L) = \sin(\Delta k_z L/2)/(\Delta k_z L/2)$ is the longitudinal detuning function(in which $\Delta k_z = k_p - k_{sz} - k_{iz} - G_3$ and L is the stripe length), H_{tr} is the transverse mode function(which is the Fourier transform of the transverse inhomogeneity, including the pump profile and the transverse domain structure), and $\boldsymbol{\kappa}_s$ and $\boldsymbol{\kappa}_i$ are the transverse wave vectors of the signal and idler photons, respectively. For a transversely infinite and homogeneous crystal, the momentum correlation H_{tr} is $\delta(\kappa_s+\kappa_i)$. However, in our experiment, a parabolic phase profile has been introduced, and therefore the momentum correlation of the photon pair is modified. According to Glauber's theory, the spatial correlation as a function of x and z can be derived to be

$$R(x,z) \propto \langle \Psi | E_1^{(-)} E_2^{(-)} E_2^{(+)} E_1^{(+)} | \Psi \rangle \propto \left| \int_A dx' E_p(x') U(x') e^{i\frac{w_p}{2c}(\frac{1}{z}-\frac{1}{f})x'^2} e^{-i\frac{w_p}{cz}xx'} \right|^2, \quad (3)$$

where E_1 and E_2 are the electric fields evaluated at the two detectors and E_p is the pump field. Equation(3) shows that the two-photon exhibits its minimum spatial correlation at a distance of $z=f$; therefore the corresponding plane is termed the two-photon focal plane. For a transversely infinite and homogeneous crystal(that is, $U(x')$ = const.) in the focal plane, the two-photon spatial correlation is expressed as

$$R(x, z=f) \propto \begin{cases} \delta(x) & \text{for } E_p(x') = \text{const.} \\ e^{-2(\frac{\pi w_0}{\lambda_p f}x)^2} & \text{for } E_p(x') = e^{-(x'/w_0)^2} \end{cases}. \quad (4)$$

For a plane pump wave with an infinite beam width, the two-photon is focused onto an infinitesimal point. For a Gaussian pump beam, the two-photon is focused onto a spot, the size of which is determined by the pump beam size w_0 and the pump wavelength(rather than the signal or idler wavelength).

References and Notes

[1] Pittman, T. B., Shih, Y. H., Strekalov, D. V. & Sergienko, A. V. Optical imaging by means of two-photon quantum entanglement. *Phys. Rev. A* **52**, R3429 – R3432(1995).

[2] D'Angelo, M., Kim, Y. H., Kulik, S. P. & Shih, Y. H. Identifying entanglement using quantum ghost interference and imaging. *Phys. Rev. Lett.* **92**, 233601(2004).

[3] Bennink, R. S., Bentley, S. J., Boyd, R. W. & Howell, J. C. Quantum and classical coincidence imaging. *Phys. Rev. Lett.* **92**, 033601(2004).

[4] Gatti, A., Brambilla, E. & Lugiato, L. A. Entangled imaging and wave-particle duality: from the microscopic to the macroscopic realm. *Phys. Rev. Lett.* **90**, 133603(2003).

[5] Blanchet, J.-L., Devaux, F., Furfaro, L. & Lantz, E. Measurement of sub-shotnoise correlations of spatial fluctuations in the photon-counting regime. *Phys. Rev. Lett.* **101**, 233604(2008).

[6] Brida, G., Genovese, M. & Ruo Berchera, I. Experimental realization of sub-shot-noise quantum imaging. *Nature Photon.* **4**, 227 – 230(2010).

[7] Boto, A. N. *et al.* Quantum interferometric optical lithography: exploiting entanglement to beat the diffraction limit. *Phys. Rev. Lett.* **85**, 2733 – 2736(2000).

[8] D'Angelo, M., Chekova, M. V. & Shih, Y. H. Two-photon diffraction and quantum lithography. *Phys. Rev. Lett.* **87**, 013602(2001).

[9] Kawabe, Y., Fujiwara, H., Okamoto, R., Sasaki, K. & Takeuchi, S. Quantum interference fringes beating the diffraction limit. *Opt. Express* **15**, 14244 – 14250(2007).

[10] Nagata, T., Okamoto, R., O'Brien, J. L., Sasaki, K. & Takeuchi, S. Beating the standard quantum limit with four-entangled photons. *Science* **316**, 726 – 729(2007).

[11] Afek, I., Ambar, O. & Silberberg, Y. High-NOON states by mixing quantum and classical light. *Science* **328**, 879–881(2010).

[12] Xiang, G. Y., Higgins, B. L., Berry, D. W., Wiseman, H. M. & Pryde, G. J. Entanglement-enhanced measurement of a completely unknown optical phase. *Nature Photon.* **5**, 43–47(2011).

[13] Giovannetti, V., Lloyd, S. & Maccone, L. Advances in quantum metrology. *Nature Photon.* **5**, 222–229(2011).

[14] Gao, W.-B. *et al*. Experimental realization of a controlled-NOT gate with four-photon six-qubit cluster states. *Phys. Rev. Lett.* **104**, 020501(2010).

[15] Pittman, T. B. *et al*. Two-photon geometric optics. *Phys. Rev. A* **53**, 2804–2815(1996).

[16] Monken, C. H., Souto Ribeiro, P. H. & Pádua, S. Transfer of angular spectrum and image formation in spontaneous parametric down-conversion. *Phys. Rev. A* **57**, 3123–3126(1998).

[17] Vaziri, A., Weihs, G. & Zeilinger, A. Superpositions of the orbital angular momentum for applications in quantum experiments. *J. Opt. B* **4**, S47–S51(2002).

[18] Neves, L. *et al*. Generation of entangled states of qudits using twin photons. *Phys. Rev. Lett.* **94**, 100501(2005).

[19] Lima, G., Vargas, A., Neves, L., Guzmán, R. & Saavedra, C. Manipulating spatial qudit states with programmable optical devices. *Opt. Express* **17**, 10688–10696(2009).

[20] Politi, A., Cryan, M.J., Rarity, J.G., Yu, S. & O'Brien, J.L. Silica-on-silicon waveguide quantum circuits. *Science* **320**, 646–649(2008).

[21] Matthews, J. C. F., Politi, A., Stefanov, A. & O'Brien, J. L. Manipulation of multiphoton entanglement in waveguide quantum circuits. *Nature Photon.* **3**, 346–350(2009).

[22] Smith, B. J., Kundys, D., Thomas-Peter, N., Smith, P. G. R. & Walmsley, I.A. Phase-controlled integrated photonic quantum circuits. *Opt. Express* **17**, 13516–13525(2009).

[23] Sansoni, L. *et al*. Polarization entangled state measurement on a chip. *Phys. Rev. Lett.* **105**, 200503 (2010).

[24] Saleh, M. F., Di Giuseppe, G., Saleh, B. E. A. & Teich, M. C. Modal and polarization qubits in Ti: LiNbO$_3$ photonic circuits for a universal quantum logic gate. *Opt. Express* **18**, 20475–20490(2010).

[25] Berger, V. Nonlinear photonic crystals. *Phys. Rev. Lett.* **81**, 4136–4139(1998).

[26] Broderick, N. G. R., Ross, G. W., Offerhaus, H. L., Richardson, D. J. & Hanna, D. C. Hexagonally poled lithium niobate: a two-dimensional nonlinear photonic crystal. *Phys. Rev. Lett.* **84**, 4345–4348(2000).

[27] Kurz, J. R., Schober, A. M., Hum, D. S., Saltzman, A. J. & Fejer, M. M. Nonlinear physical optics with transversely patterned quasi-phase-matching gratings. *IEEE J. Sel. Top. Quantum Electron.* **8**, 660–664(2002).

[28] Qin, Y. Q., Zhang, C., Zhu, Y. Y., Hu, X. P. & Zhao, G. Wave-front engineering by Huygens-Fresnel principle for nonlinear optical interactions in domain engineered structures. *Phys. Rev. Lett.* **100**, 063902(2008).

[29] Ellenbogen, T., Voloch-Bloch, N., Ganany-Padowicz, A. & Arie, A. Nonlinear generation and manipulation of Airy beams. *Nature Photon.* **3**, 395–398(2009).

[30] Torres, J. P., Alexandrescu, A., Carrasco, S. & Torner, L. Quasi-phase-matching engineering for spatial control of entangled two-photon states. *Opt. Lett.* **29**, 376–378(2004).

[31] Yu, X. Q. *et al*. Transforming spatial entanglement using a domain-engineering technique. *Phys. Rev. Lett.* **101**, 233601(2008).

[32] Guillet de Chatellus, H., Sergienko, A. V., Saleh, B. E. A., Teich, M. C. & Di Giuseppe, G. Non-collinear and non-degenerate polarization-entangled photon generation via concurrent type-I parametric downconversion in PPLN. *Opt. Express* **14**, 10060–10072(2006).

[33] Carrasco, S. *et al*. Enhancing the axial resolution of quantum optical coherence tomography by chirped quasi-phase matching. *Opt. Lett.* **29**, 2429–2431(2004).

[34] Canalias, C. & Pasiskevicius, V. Mirrorless optical parametric oscillator. *Nature Photon.* **1**, 459–462 (2007).

[35] Harris, S. E. Chirp and compress: toward single-cycle biphotons. *Phys. Rev. Lett.* **98**, 063602(2007).

[36] Nasr, M. B. *et al*. Ultrabroadband biphotons generated via chirped quasi-phasematched optical parametric down-conversion. *Phys. Rev. Lett.* **100**, 183601(2008).

[37] Sensarn, S., Yin, G. Y. & Harris, S. E. Generation and compression of chirped biphotons. *Phys. Rev. Lett.* **104**, 253602(2010).

[38] Branczyk, A. M., Fedrizzi, A., Stace, T. M., Ralph, T. C. & White, A. G. Engineered optical nonlinearity for quantum light sources. *Opt. Express* **19**, 55–65(2011).

[39] We thank Y. H. Shih for helpful discussions and comments during the preparation of the manuscript. The authors also thank Y. Yuan for help in fabricating the crystal. This work was supported by the National Natural Science Foundations of China(contract nos 11021403, 10904066 and 11004030), the State Key Program for Basic Research in China(nos 2011CBA00205 and 2010CB630703) and the Project Funded by the Priority Academic Program Development of Jiangsu Higher Education Institutions.

[40] X.Q.Y. conceived the study. X.Q.Y. fabricated the parabolic PPLT sample, with assistance from C.Z. in designing the parabolic pattern of the crystal. P.X. and H.Y.L. completed the theoretical deduction. H.Y.L., P.X. and X.Q.Y. performed the experiment, with assistance from Y.X.G., Z.D.X. and H.J., H.Y.L. and P.X. analysed the results and wrote the paper. S.N.Z supervised the study and commented on the paper.

[41] H.Y.Leng and X.Q.Yu contributed equally to this work.

Lensless Imaging by Entangled Photons from Quadratic Nonlinear Photonic Crystals[*]

P. Xu,[1] H. Y. Leng,[1] Z. H. Zhu,[2] Y. F. Bai,[1] H. Jin,[1] Y. X. Gong,[1,3]
X. Q. Yu,[3] Z. D. Xie,[1] S. Y. Mu,[1] and S. N. Zhu[1]

[1] *National Laboratory of Solid State Microstructures and School of Physics, Nanjing University, Nanjing 210093, China*

[2] *College of Opto-Electronic Engineering, National University of Defense Technology, Changsha 410073, China*

[3] *Physics Department, Southeast University, Nanjing 211189, China*

Lenses play a key role in quantum imaging but inevitably constrain the spatial resolution and working wavelength. In this work we develop and demonstrate a lensless quantum ghost imaging by engineering quadratic nonlinear photonic crystals. With a transverse parabolic domain modulation introduced into the lithium tantalate crystal, the entangled photon pairs generated from parametric down-conversion will self-focus. Therefore we can dispense with additional lenses to construct imaging in a nonlocal way. The lensless imaging is found to follow a specific imaging formula where the effective focal length is determined by the domain modulation and pump wavelength. Additionally, two nonlocal images can be retrieved when the entangled photon pair is generated under two concurrent noncollinear phase-matching geometries. Our work provides a principle and method to realize lensless ghost imaging, which may be extended to other wavelengths and stimulate new types of practical quantum technologies.

1. Introduction

Developing advanced imaging techniques is of essential importance to human daily life and the scientific world. As one of the latest achievements of quantum mechanics, quantum imaging emerges with great advantages over the classical imaging. By using the quantum nature of entangled photons[1—8] or entangled beams[9—12], quantum imaging can surpass classical imaging with higher resolution and better sensitivity. In addition, quantum imaging is nonlocal. The image is constructed by joint detection between a beam that never interacts with the object and one that does. This ghost imaging was first demonstrated by using entangled photons in 1995[1]. Since then, ghost imaging has drawn considerable attention due to its fundamental interest and potential applications. In practical implementations of quantum ghost imaging, the lens plays a key role in conveying the information from objects to images. But lenses severely constrain the spatial resolution and

---

[*] Phys.Rev.A,2012,86(1):013805

present engineering challenges at some wavelengths. Consequently, developing a lensless ghost imaging system is an important goal in quantum imaging science.

Here we develop and demonstrate lensless quantum imaging by engineering quadratic nonlinear photonic crystals(NPCs). The NPC offers high-efficiency nonlinear interactions in a quasi-phase-matching(QPM) way[13,14], and specifically a two-dimensional NPC[15,16] shows particular functionalities in the spatial control of second-harmonic beams[17—19] and entangled photons[20—22]. Very recently, a two-photon lens based on the transverse parabolic NPC was studied theoretically and experimentally[21,22]. In that case, the longitudinally periodic modulation of $\chi^{(2)}$ ensures the generation of entangled photon pairs by QPM spontaneously parametric down-conversion (SPDC), while in the transverse parabolic domain modulation, the crystal can tailor the wave front of the two photons, acting as an equivalent lens. In this work we investigate the two-photon lens and experimentally demonstrate lensless quantum ghost imaging. The schematic layout for this lensless imaging is displayed in Fig. 1. When an object is put into one of the down-converted photon paths, by coincidence measurement the image can be reconstructed in the other down-converted path. No lens is required in this setup and even no beamsplitter is required when the signal and idler photon propagate noncollinearly. Furthermore, we obtain two lensless ghost images simultaneously under concurrent noncollinear QPM geometries in a multistripe parabolic NPC. This lensless quantum imaging presents a way to improve the resolution which is conventionally restricted by the size of lens, and offers a principle for engineering an equivalent lens by nonlinear interaction, hence it may be beneficial at frequencies where we do not have efficient ways of manufacturing lenses, such as X-rays.

The paper is organized as follows. Section 2 includes the mathematical description of the NPC's structure and theoretical calculations on the imaging formula for entangled photons from the NPC. In Sec. 3, the experimental results of the single lensless ghost image and doule lensless ghost images are given. In Sec. 4, the two-color lensless ghost imaging is discussed. The conclusion is drawn is Sec. 5.

2. Theoretical Calculations on the Imaging Formula

2.1 Description of the Parabolic NPC

The quadratic nonlinear coefficient of parabolic NPC is expressed by
$$\chi^{(2)}(x,z) = d_{33} \mathrm{sgn}\{\cos[2\pi(z+\alpha x^2)/\Lambda]\}, \tag{1}$$
which indicates the maximum nonlinear coefficient d_{33} is utilized and its sign changes when the domain is inverted. The longitudinal domain modulation period is Λ and the width of the positive domain is equal to that of the negative domain, while the transverse modulation follows the function $z=-\alpha x^2$. The Fourier expansion of $\chi^{(2)}(x,z)$ is

$d_{33}\sum_n F(g_n)e^{ig_n(z+ax^2)}$, where $g_n = n\dfrac{2\pi}{\Lambda}(n=\pm 1, \pm 2,\cdots)$ represents the nth-order reciprocal vector and $F(g_n)$ is its corresponding Fourier coefficient. The entangled photons are generated under the QPM condition $k_p - k_s - k_i + g_n = 0$ along the propagation direction in which k_p, k_s, and k_i are wave vectors of the pump, signal, and idler, respectively. Along the x axis it is interesting to find that the wave front of entangled photon pairs takes a parabolic profile $e^{ig_nax^2}$. Hence the NPC is equivalent to a two-photon cylindrical lens[21].

FIG.1. (color online) The schematic setup of lensless ghost imaging based on a multistripe parabolic NPC. The solid "face" represents the object and the dotted one represents the image.

In this work we design a multistripe parabolic NPC instead of a continuous one, as shown schematically in Fig. 1. The stripe interval is Λ_{tr} and the stripe width is $\Lambda_{tr}/2$. For each stripe, only its center follows $z=-ax^2$. This design will bring new characters to the two-photon lens, such as the transverse periodicity and two lensless ghost images. The nonlinearity of a multistripe parabolic NPC is

$$\chi^{(2)}(x,z) = d_{33}\sum_n F(g_n)e^{ig_nz}\sum_m \text{rect}\left(\dfrac{x-m\Lambda_{tr}}{\Lambda_{tr}/2}\right)e^{ig_na(m\Lambda_{tr})^2}. \qquad (2)$$

We define

$$U(x) = \sum_m \text{rect}\left(\dfrac{x-m\Lambda_{tr}}{\Lambda_{tr}/2}\right)e^{ig_na(m\Lambda_{tr})^2} \qquad (3)$$

as the transverse structure function which includes both the amplitude and phase modulation. In this experiment we design the third-order reciprocal vector $g_{-3}(-g_3)$ to satisfy the quasi-phase-matching condition $k_p - k_s - k_i - g_3 = 0$. The effective two-photon wave front then takes the form of $e^{-ig_3a(m\Lambda_{tr})^2}$. For deriving an analytical solution of two photons' spatial correlation, we use an approximated description of $U(x)$:

$$U(x) = \sum_m \text{rect}\left(\dfrac{x-m\Lambda_{tr}}{\Lambda_{tr}/2}\right)e^{-ig_3ax^2}, \qquad (4)$$

where within each stripe the phase profile still follows the parabolic function $e^{-ig_3ax^2}$ instead of the constant one $e^{-ig_3a(m\Lambda_{tr})^2}$. It is worth noting that this approximation is reasonable since each stripe's width $\Lambda_{tr}/2$ is small and satisfies the far-field condition of $f_{eff}\gg(\Lambda_{tr}/2)^2/\lambda$ when the measurement is implemented in the two-photon focal plane $z=f_{eff}$[23]. The

Fourier expansion of $U(x)$ is $\frac{1}{2}\sum_m \mathrm{sinc}(m\pi/2)e^{i2\pi mx/\Lambda_{\mathrm{tr}}}e^{-ig_3 ax^2}$ in which $\mathrm{sinc}(x)=\frac{\sin(x)}{x}$. This form of $U(x)$ is convenient to use in the following calculations.

2.2 The Imaging Formula

Based on the first-order perturbation theory[24] and using the Hamiltonian $H_I = \varepsilon_0 \int_V dr \chi^{(2)}(x,z)\{E_p^{(+)}E_s^{(-)}E_i^{(-)}\} + \mathrm{H.c.}$, we obtain the two-photon state:

$$|\psi\rangle \propto \psi_0 \sum_{k_s,k_i} H_{\mathrm{tr}}(\boldsymbol{\kappa}_s,\boldsymbol{\kappa}_i)\hat{a}^{(+)}(\boldsymbol{\kappa}_s)\hat{a}_i^{(+)}(\boldsymbol{\kappa}_i)|0\rangle, \quad (5)$$

in which all the slowly varying terms and constants are absorbed into ψ_0. $\boldsymbol{\kappa}_s, \boldsymbol{\kappa}_i$ are the transverse wave vectors of signal and idler photons, respectively. Here we assume the frequency mode function and spatial mode function of the two-photon state factor and are only concerned with the transverse part[24]. The transverse two-photon mode function is calculated as

$$H_{\mathrm{tr}}(\boldsymbol{\kappa}_s,\boldsymbol{\kappa}_i) = \int d\boldsymbol{\rho}\{U(\boldsymbol{\rho})E(\boldsymbol{\rho})\}e^{-i(\boldsymbol{\kappa}_s+\boldsymbol{\kappa}_i)\cdot\boldsymbol{\rho}}, \quad (6)$$

in which $E(\boldsymbol{\rho})$ is the pump beam's profile and $U(\boldsymbol{\rho})$ represents the two-dimensional structure function of the NPC. Since the state-of-the-art crystal poling technique only enables two-dimensional domain engineering, the domain modulation along the y axis is homogenous. For the plane-wave pump, the mode function along the y axis is given by $H_{\mathrm{tr}}(\kappa_{sy},\kappa_{iy}) = \int dy\, e^{-i(\kappa_{sy}+\kappa_{iy})\cdot y} \propto \delta(\kappa_{sy}+\kappa_{iy})$, while the mode function along the x axis is $H_{\mathrm{tr}}(\kappa_{sx},\kappa_{ix}) = \int dx\, U(x)\, e^{-i(\kappa_{sx}+\kappa_{ix})\cdot x} \propto \sum_m \mathrm{sinc}\left(\frac{m\pi}{2}\right)\int d\kappa\, \delta\left(\kappa_{sx}+\kappa_{ix}-\kappa-\frac{2\pi m}{\Lambda_{\mathrm{tr}}}\right)e^{\frac{\kappa^2}{4g_3 a}}$. In the following part, we will only be concerned with the spatial correlation along the x axis.

Suppose signal and idler photons are captured by detector D_1 and D_2, respectively. $E_j^{(+)}(\boldsymbol{r}_j,t_j)$ is the electromagnetic field evaluated at the two detectors' spatial coordinate $\boldsymbol{r}_j(\boldsymbol{\rho}_j,z_j)(j=s,i)$. z_s and z_i stand for the distance from the crystal to the object and the imaging plane, respectively. The propagation of the two free-space electromagnetic fields is $E_j^{(+)}(\boldsymbol{r}_j,t_j) = \sum_{k_j} E_j e^{-i\omega_j t_j} g(\boldsymbol{\kappa}_j,\omega_j;\boldsymbol{\rho}_j,z_j)\hat{a}_{kj}$, in which Green's function takes the form of $g(\boldsymbol{\kappa}_j,\omega_j;\boldsymbol{\rho}_j,z_j) = \frac{e^{ik_j z_j}}{i\lambda_j z_j}\int d\boldsymbol{\rho}_s e^{i\frac{\omega_j}{2cz_j}|\boldsymbol{\rho}_j-\boldsymbol{\rho}_s|^2}e^{i\boldsymbol{\kappa}_j\cdot\boldsymbol{\rho}_s}$ [24]. Suppose the transverse component is smaller and satisfies $|\boldsymbol{\kappa}_j| \ll |\boldsymbol{k}_j|$, then we have $\boldsymbol{k}_j = |\boldsymbol{k}_j|\hat{e}_z + \boldsymbol{\kappa}_j$. Two-photon amplitude $A(x_s,x_i) = \langle 0|E_2^{(+)}E_1^{(+)}|\psi\rangle$ is calculated to be

$$A(x_s,x_i) \propto \int dx E(x) \sum_m \mathrm{sinc}(m\pi/2)e^{i2\pi mx/\Lambda_{\mathrm{tr}}} \times e^{i\left(\frac{\omega_s}{2cz_s}+\frac{\omega_i}{2cz_i}-g_3 a\right)x^2} e^{-i\left(\frac{\omega_s x_s}{cz_s}+\frac{\omega_i x_i}{cz_i}\right)x}. \quad (7)$$

When taking the pump to be transversely infinite, we obtain a point-to-point correspondence along the x axis between the signal and idler photon:

$$G^{(2)}(x_s,x_i) = |A(x_s,x_i)|^2 \propto \left|\sum_m \mathrm{sinc}\left(\frac{m\pi}{2}\right)\delta\left(\frac{\omega_s x_s}{cz_s}+\frac{\omega_i x_i}{cz_i}-\frac{2\pi m}{\Lambda_{\mathrm{tr}}}\right)\right|^2, \quad (8)$$

under the condition of $\frac{\omega_s}{2cz_s}+\frac{\omega_i}{2cz_i}-g_3\alpha=0$. Thus the imaging formula is obtained and it can be written as

$$\frac{1}{\lambda_s z_s}+\frac{1}{\lambda_i z_i}=\frac{1}{\lambda_p f_{\text{eff}}}, \tag{9}$$

in which f_{eff} equals $\frac{\pi}{g_3\alpha\lambda_p}$. So f_{eff} is relevant to the pump wavelength, the curvature of the parabolic NPC, and the reciprocal vector for the longitudinal quasi-phase-matching condition. The two-photon wave front takes another form of $\varphi(x)=e^{-ig_3\alpha x^2}=e^{-ik_p x^2}/2f_{\text{eff}}$. If one uses a point detector to scan and detect two-photon probability after the crystal, two-photon focusing as well as the far-field diffraction of the multistripe will be observed at the plane $z=f_{\text{eff}}$[22], thus multiple focal spots may be observed. When one puts an object in one of the paths, the image will be recovered in the other path by coincidence measurement. The image is linearly magnified by a factor of $-\frac{\lambda_i z_i}{\lambda_s z_s}$ when the object is put into the signal path. For degenerate photon pair generation, the imaging formula will be simplified to $\frac{1}{z_s}+\frac{1}{z_i}=\frac{2}{f_{\text{eff}}}$. If we set $z_s=z_i=f_{\text{eff}}$, an equal-size reproduction of the object will be retrieved. According to Eq.(8), there should be many equally spaced ghost images and the intensity of high-order images are decided by the sin c function. However, the number of observable ghost images should be further limited by the single-photon distribution.

Consider a pump with a Gaussian spatial profile with beam waist w_0, the spatial correspondence of photon pair will take the form:

$$G^{(2)}(x_s,x_i)\propto\left|\sum_m\text{sinc}\left(\frac{m\pi}{2}\right)e^{-\pi^2\left(\frac{x_s}{\lambda_s z_s}+\frac{x_i}{\lambda_i z_i}-\frac{m}{\Lambda_{\text{tr}}}\right)^2 w_0^2}\right|^2. \tag{10}$$

For a zero-order ghost image, i.e., $m=0$, the imaging resolution is $\lambda_i z_i/w_0\pi$ when the object is put in the signal path.

3. Experimental Results of Lensless Ghost Imaging

3.1 Single Lensless Ghost Image

In this experiment we engineer the multistripe ferroelectric domains in LiTaO$_3$ as Fig. 2(a) shows. The longitudinal periodicity of NPC is $\Lambda=13.917$ μm, which ensures efficient entangled photon generation with the third-order reciprocal vector $g_3=3(2\pi)/\Lambda$. The transverse domain modulation follows $z=-\alpha x^2$ and $\alpha=15.226$ m^{-1}. It indicates that $f_{\text{eff}}=33.3$ mm when the crystal is pumped by 457 nm. The stripe interval Λ_{tr} is 20 μm, stripe width b is 10 μm, and stripe length L is 6 mm. The crystal is pumped with a cw single longitudinal mode 457-nm laser and embedded in an oven for temperature control to produce the degenerate photon pairs at 914 nm. M_1 reflects the pump and keeps the entangled photons transmitted. M_2 is a beamsplitter. A bucket detector D_1 is put in the signal path and collects all the photons after the object which is a double slit aperture with slit width 150 μm and slit separation 300 μm. For the transmitted path, the fiber tip scans along the direction of x axis and it is followed by another single-photon detector D_2. Two bandpass filters IF_1 and IF_2 with 10-nm FWHM are put before two single-photon detectors to further suppress the pump. No additional imaging lens is required in this imaging setup.

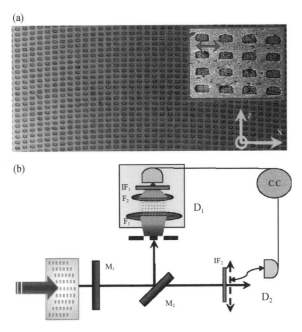

FIG.2. (color online) Photograph of NPC sample(a) and experimental setup (b). The bucket detector D_1 consists of two collection lenses and a single-photon detector. It collects all the photons after the double slit.

When the working temperature of the NPC is set at 180.1℃ and the pump is incident along its center, degenerate entangled photon pairs are generated collinearly; the two-dimensional single-photon counting profile at D_2 is presented in Fig. 3(a). When making coincidence counting between two detectors, a clear image of the double slit appears as in Fig. 3(b). When we set the object distance $z_s = 33.3$ mm, the ghost image plane is found to be located at around $z_i = 32.4$ mm. The measured separation of the double slit is 303 μm, and a Gaussian fitting of the peaks gives slit widths(FWHM) of 146 μm and 142 μm, respectively. Hence a nearly equal-size image is obtained. When the scanning fiber tip locates out-of-focus, the image gets blurred. A preliminary test for the spatial resolution gives 33 μm(FWHM) by using the single-mode fiber tip as the object, while the theoretical value is 28 μm with a full pump width of 0.82 mm. Since no additional imaging lens is used in this experiment, the imaging resolution is determined only by the pump size and wavelength of the entangled photons. As we have calculated, the photon pair emerges from the same point of crystal and each photon pair takes all the possible combinations of momentum with a certain phase distribution which results from the transverse parabolic modulation. This lensless ghost imaging can be explained through analogies to geometric optics. Figure 3(c) is an unfolded layout of the lensless imaging. The geometric ray in Fig. 3(c) actually represents the two-photon amplitude. The superposition of these two-photon amplitudes builds up a nonlocal point-to-point correspondence between the signal and idler path. The parabolic NPC works equivalently as the imaging lens. The spatial resolution of such lensless imaging is the same as the ghost imaging implemented with a homogeneous bulk nonlinear optical crystal and a cylindrical lens with a large enough aperture under the same experiment configuration including the wavelength of entangled photons, the crystal size, and the image distance. Here, we have to emphasize that higher order ghost images are not observed since the spatial interval of two adjacent ghost images is beyond the single-photon distribution as shown in Fig. 3(a).

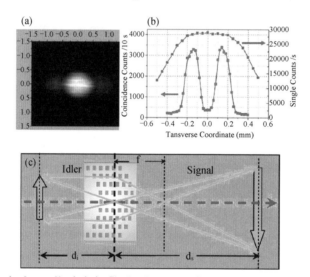

FIG.3. (color online) (a) Single-photon profile at D_2. (b) Single-photon counting distribution is denoted by the upper blue symbols and the coincidence counting by the lower red symbols. The bucket detector D_1 gives the photon counting rate 0.25 Mc/s and the maximum coincidence counting rate is 330 c/s. (c) The unfolded layout of lensless ghost imaging.

3.2 Double Lensless Ghost Images

As first considered in Ref.[21], the alignment of the multiple stripes transforms the two-photon spatial entanglement and brings new characters. Here for the multistripe parabolic NPC sample described by Eq. (3), we find the structure function $U(x)$ reproduces itself after a translation of $d = \dfrac{\lambda_p f_{\text{eff}}}{\Lambda_{\text{tr}}}$ which indicates our crystal structure has a translational periodicity:

$$U\left(x-\frac{\lambda_p f_{\text{eff}}}{\Lambda_{\text{tr}}}\right) = \sum_m \text{rect}\left(\frac{x-\lambda_p f_{\text{eff}}/\Lambda_{\text{tr}}-m\Lambda_{\text{tr}}}{\Lambda_{\text{tr}}/2}\right) e^{-ig_{3\alpha}(m\Lambda_{\text{tr}})^2}$$
$$= \sum_m \text{rect}\left(\frac{x-m'\Lambda_{\text{tr}}}{\Lambda_{\text{tr}}/2}\right) e^{-ig_{3\alpha}(m'\Lambda_{\text{tr}})^2} = U(x),$$
(11)

where we have used relations $g_{3\alpha} = \frac{\pi}{\lambda_p f_{\text{eff}}}$, $\frac{\lambda_p f_{\text{eff}}}{\Lambda_{\text{tr}}} = 38.08 \Lambda_{\text{tr}} \approx 38 \Lambda_{\text{tr}}$, and $m' = m + \frac{\lambda_p f_{\text{eff}}}{\Lambda_{\text{tr}}^2} \approx m + 38$. Therefore when the crystal is translated by a distance of $j\lambda_p f_{\text{eff}}/\Lambda_{\text{tr}}(j=0, \pm 1, \pm 2, \cdots)$, both the image positions and the image intensity remain the same. The crystal works as a two-photon lens which has multiple equally spaced principal axes on it. All these principal axes are equivalent so that when the incident pump lies halfway between two principal axes, the two images near the center will have equivalent intensities.

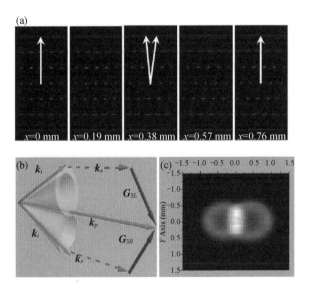

FIG.4. (color online) (a) Theoretical Fourier spectra of NPC when the pump is incident on different positions of NPC. (b) The corresponding QPM geometry for two single-photon cones. Each cone is quasi-phase-matched by a tilted reciprocal vector g_{3L} or g_{3R} and the cone axis is along $k_p - g_{3L}$ or $k_p - g_{3R}$. (c) The measured single-photon distribution.

Experimentally we found that the single-photon distribution and ghost imaging can be recovered when the crystal shifts by multiples of $\lambda_p f_{\text{eff}}/\Lambda_{\text{tr}}$ (0.76 mm), which consists well with the above theoretical calculation. This transverse periodicity can also be understood from the viewpoint of reciprocal space of such NPCs. Figure 4(a) is the theoretical simulation about the reciprocal space of the crystal by Fourier transformation when the pump is incident on different positions of the crystal. The reciprocal space is distorted when the pump moves away from the central position and it will recover when the pump moves 0.76 mm away. When the pump is incident in between, i.e., $x = 0.38$ mm, from the reciprocal space we can see that two noncollinear reciprocal vectors g_{3L} and g_{3R} with mirror symmetry exist, which may work for new types of entangled photon generation under $k_s + k_i + g_{3L(3R)} = k_p$. For a two-dimensional NPC without parabolic modulation, the classical parametric down-conversion process participant by such a pair of reciprocal vectors has been reported[25,26]. For the spontaneous parametric down-conversion in this parabolic NPC sample, the QPM geometry is similar. The only difference is that the photon pair carries a parabolic phase profile. Figure 4(b) shows two concurrent noncollinear QPM geometries which result in two tangent single-photon cones. If one takes the advantage of concurrent QPMs in the NPC sample, multiple images can be obtained.

In the experiment, when the crystal moves 0.38 mm away from the center and the temperature is tuned to 178.5℃, two tangent single-photon rings are captured, as shown

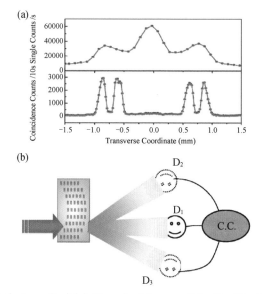

FIG.5. (color online)(a) Two ghost images of double slit whose size is close to that of the object mask. (b) The schematic layout for simultaneous construction of two ghost images. The single photon detector D_1 in the object path is a bucket detector whose combination is similar to that in Fig. 2(b). Single-photon detectors D_2 and D_3 are spatially resolving.

in Fig. 4(c). So when the signal photon travels along the pump direction, the idler will emerge from either ring indistinguishably. If we put the object into the overlap region of the two rings, two ghost images will be expected at the outer sides of two single-photon rings. Figure 5(a) shows the experimental results under the same experimental configuration as in Fig. 2(b). Two equal-size images of double slit are obtained in the transmitted path when the fiber tip is scanning in the in-focus plane. The experimental setup can be further simplified into Fig. 5(b). No beamsplitter is required since the photon pair is naturally separated. Such concurrent ghost images may also be realized by the homogenous bulk nonlinear crystal; however, additional optical elements including a lens and beamsplitter are required.

4. Discussions

Since quasi-phase-matching generation of entangled photons enriches the range of working wavelengths, with nondegenerate photon pairs we may achieve a two-color lensless ghost imaging. The resolving power and field of view can be greatly enhanced[6] when the photons in the spatially resolved path have much higher frequency. Also the wavelengths of the photon pair can be designed according to practical use such as the midinfrared, fiber communication wavelength, or nondestructive wavelength for observing biological samples. Besides, the resolution of the lensless imaging can be further improved through increasing the pump size and engineering a continuous parabolic NPC. But we have to emphasize that the lensless imaging demonstrated in this work is one-dimensional, and the two-dimensional one relies on the fabrication of three-dimensional NPCs.

5. Conclusions

The lensless quantum imaging demonstrated in this work is performed by the momentum correlation manipulation of the entangled two-photon pair by engineering nonlinear photonic crystals. The engineering of spatial entanglement helps us explore fundamental concerns in quantum mechanics and practical quantum technologies such as the lensless quantum imaging described in this work. The lensless imaging offers a principle for engineering equivalent lenses during the nonlinear interaction and opens a door for lensless imaging at specific regions of the electromagnetic spectrum like X-rays[27], terahertz, infrared, microwave radiation, as well as acoustic wave. Preparing other new types of entangled states with tailored mode functions and the extension to entangled bright beams from high-gain parametric down-conversion deserves further investigation.

References and Notes

[1] T. B. Pittman, Y. H. Shih, D. V. Strekalov, and A. V. Sergienko, Phys. Rev. A **52**, R3429(1995).

[2] T. B. Pittman, D. V. Strekalov, D. N. Klyshko, M. H. Rubin, A. V. Sergienko, and Y. H. Shih, Phys. Rev. A **53**, 2804(1996).

[3] A. F. Abouraddy, B. E. A. Saleh, A. V. Sergienko, and M. C. Teich, Phys. Rev. Lett. **87**, 123602 (2001).

[4] A. F. Abouraddy, B. E. A. Saleh, A. V. Sergienko, and M. C. Teich, J. Opt. Soc. Am. B **19**, 1174 (2002).

[5] M. D'Angelo, Y. H. Kim, S. P. Kulik, and Y.H. Shih, Phys. Rev. Lett. **92**, 233601(2004).

[6] M. H. Rubin and Y. H. Shih, Phys. Rev. A **78**, 033836(2008).

[7] M. Malik, H. Shin, M. O'Sullivan, P. Zerom, and R. W. Boyd, Phys. Rev. Lett.**104**, 163602(2010).

[8] B. Jack, J. Leach, J. Romero, S. Franke-Arnold, M. Ritsch-Marte, S. M. Barnett, and M. J. Padgett,

Phys. Rev. Lett. **103**, 083602(2009).

[9] M. I. Kolobov, Rev. Mod. Phys. **71**, 1539(1999).

[10] A. Gatti, E. Brambilla, and L. A. Lugiato, Phys. Rev. Lett. **90**, 133603(2003).

[11] G. Brida, M. Genovese, and I. Ruo Berchera, Nature Photonics **4**, 227(2010).

[12] V. Boyer, A. M. Marino, R. C. Pooser, and P. D. Lett, Science **321**, 544(2008).

[13] J. A. Armstrong, N. Bloembergen, J. Ducuing, and P. S. Pershan, Phys. Rev. **127**, 1918(1962).

[14] P. A. Franken and J. F. Ward, Rev. Mod. Phys. **35**, 23(1963).

[15] V. Berger, Phys. Rev. Lett. **81**, 4136(1998).

[16] N. G. R. Broderick, G. W. Ross, H. L. Offerhaus, D. J. Richardson, and D. C. Hanna, Phys. Rev. Lett. **84**, 4345(2000).

[17] J. R. Kurz, A. M. Schober, D. S. Hum, A. J. Saltzman, and M. M. Fejer, IEEE J. Sel. Top. Quantum Electron. **8**, 660(2002).

[18] Y. Q. Qin, C. Zhang, Y. Y. Zhu, X. P. Hu, and G. Zhao, Phys. Rev. Lett. **100**, 063902(2008).

[19] T. Ellenbogen, N. Voloch-Bloch, A. Ganany-Padowicz, and A. Arie, Nat. Photon. **3**, 395(2009).

[20] X. Q. Yu, P. Xu, Z. D. Xie, J. F. Wang, H. Y. Leng, J. S. Zhao, S. N. Zhu, and N. B. Ming, Phys. Rev. Lett. **101**, 233601(2008).

[21] J. P. Torres, A. Alexandrescu, S. Carrasco, and L. Torner, Opt. Lett. **29**, 376(2004).

[22] H. Y. Leng, X. Q. Yu, Y. X. Gong, P. Xu, Z. D. Xie, H. Jin, C. Zhang, and S. N. Zhu, Nature Commun. **2**, 429(2011).

[23] When this far-field condition is satisfied, we will ignore the propagation phase difference within each stripe and only be concerned with the initial phase of the photon pair then integrate over the stripe. For the stripe centered at $x_0 = m\Lambda_{tr}$, the integration over the mth stripe gives $E_m = \int_{x_0-\Lambda_{tr}/4}^{x_0+\Lambda_{tr}/4} e^{ig_n\alpha x_0^2} dx$ for the flat stripe and $E'_m = \int_{x_0-\Lambda_{tr}/4}^{x_0+\Lambda_{tr}/4} e^{ig_n\alpha x^2} dx$ for the parabolic stripe. Under the condition of $g_n\alpha(\Lambda_{tr}/4)^2 \ll 1$, we will derive $E_m = E'_m/\text{sinc}(g_n\alpha x_0\Lambda_{tr}/2)$ in which $\text{sinc}(g_n\alpha x_0\Lambda_{tr}/2) \approx 1$ for any stripe within the pump beamwidth. So the approximation in Eq.(4) is reasonable and it will not affect the focal spot size and focal length of such two-photon lenses.

[24] M. H. Rubin, Phys. Rev. A **54**, 5349(1996).

[25] H.-C. Liu and A. H. Kung, Opt. Express **16**, 9714(2008).

[26] K. Gallo, M. Levenius, F. Laurell, and V. Pasiskevicius, Appl. Phys. Lett. **98**, 161113(2011).

[27] S. Shwartz and S. E. Harris, Phys. Rev. Lett. **106**, 080501(2011).

[28] The authors thank Y. H. Shih for valuable discussions and Q. J. Wang for evaluating the quality of the sample. This work was supported by the State Key Program for Basic Research of China (No. 2012CB921802 and No. 2011CBA00205), the National Natural Science Foundations of China (No. 11174121, No. 91121001, and No. 11021403), the Project Funded by the Priority Academic Program Development of Jiangsu Higher Education Institutions(PAPD), and the Foundation for the Author of National Excellent Doctoral Dissertation of PR China(FANEDD).

Observation of Quantum Talbot Effect from a Domain-engineered Nonlinear Photonic Crystal *

H. Jin, P. Xu, J. S. Zhao, H. Y. Leng, M. L. Zhong, and S. N. Zhu

National Laboratory of Solid State Microstructures and School of Physics, Nanjing University, Nanjing 210093, China

The quantum Talbot effect is observed from a domain-engineered nonlinear photonic crystal dispensing with a real grating. We deduce and experimentally verify the quantum self-imaging formula which is related to the crystal's structure parameter and working wavelengths. A two-photon Talbot carpet is captured to characterize the Fresnel diffraction dynamics of entangled photons wherein the quantum fractional Talbot effect is specified. The compact and stable quantum Talbot effect can be considered as the contactless diagnosis of domain's homogeneity and developed for new types of entangled photon source and quantum technologies such as quantum lithography with improved performance.

Talbot effect[1,2] is a well-known Fresnel diffraction phenomenon in which a periodic object illuminated with a plane wave replicates itself at multiplies of a certain longitudinal propagation distance. Talbot effect holds on a variety of applications in image processing, photolithography, spectrometry, and so on.[3] It also prompts developments in the array illuminator,[4,5] hard-X-ray dark-field imaging,[6] matter waves,[7,8] and optical traps.[9] Observations of nonlinear Talbot effect[10] and plasmonic Talbot effect[11,12] are also reported, which enriches the research areas and suggests more possible applications.

Since quantum Talbot effect was conceptually proposed[13,14] and experimentally demonstrated,[15] Talbot effect has been endowed with new characters and attracts more interests. In analogy to quantum imaging, quantum Talbot effect can be nonlocal, and the Talbot length with degenerate entangled photon pairs is relevant with the pump wavelength, hence is twice of classical Talbot effect. Here we report the direct observation of quantum Talbot effect from a domain-engineered nonlinear photonic crystal(NPC). With a two-dimensional domain modulation, the NPC works as an efficient platform for entangled photon generation under quasi-phase-matching(QPM) condition; meanwhile, its transverse periodicity engenders quantum Talbot effect directly. So the quantum Talbot effect emerges dispensing with a real grating. The compact and stable quantum Talbot effect can be considered as the contactless diagnosis of domain's homogeneity and developed for new types of entangled photon source and quantum technologies such as

---

* Appl. Phys. Lett., 2012, 101(21): 211115

quantum lithography with improved performance.

Domain-engineered NPCs are active in nonlinear optics for efficient frequency conversion[16] and beam shaping[17] based on the QPM technique. In quantum optics, special attentions are paid because NPCs can be utilized for the generation of bright entangled photon pairs and multifunctional manipulation of two-photon spatial entanglement.[18-20] The NPC structure in this experiment is shown in Fig. 1(a), which is a multi-stripe periodically poled lithium tantalate(MPPLT). The initial position of each stripe is precisely positioned at the same coordinate along the z axis, so that the two-photon generated from each stripe has the same initial phase. The quadratic nonlinear coefficient of the MPPLT crystal can be expressed as

$$\chi^{(2)}(x,z) = d_{33} U(x) \sum_m f(G_m) e^{iG_m z}. \qquad (1)$$

The maximum nonlinear coefficient d_{33} is involved since the pump, signal, and idler are all e-polarized. $U(x) = \sum_{n=-\infty}^{\infty} \text{rect}[(x-nd)/a]$ is a grating function indicating the transverse modulation along the x-axis with period d and stripe width a. $G_m = 2\pi m/\Lambda$ is the m-th order reciprocal vector of the longitudinal domain modulation with period Λ and the corresponding Fourier coefficient is $f(G_m)$. In our experiment the first order reciprocal vector G_1 is designed for ensuring efficient generation of entangled photons under the QPM condition $k_p - k_s - k_i - G_1 = 0$.

Following the previous studies,[21,22] we trivialize the temporal part of the state and write the two-photon state as

$$|\psi\rangle = \psi_0 \int d\boldsymbol{\kappa}_s d\boldsymbol{\kappa}_i \int d\boldsymbol{\rho} \{E_p(\boldsymbol{\rho})U(\boldsymbol{\rho})\} e^{-i(\boldsymbol{\kappa}_s+\boldsymbol{\kappa}_i)\cdot\boldsymbol{\rho}} a^\dagger_{\boldsymbol{\kappa}_s} a^\dagger_{\boldsymbol{\kappa}_i} |0\rangle, \qquad (2)$$

where ψ_0 is a normalization constant and $\boldsymbol{\kappa}_s$ and $\boldsymbol{\kappa}_i$ are the transverse wave vectors of signal and idler photons, respectively. For simplicity, the pump field $E_p(\boldsymbol{\rho})$ is assumed to be even distributed. Since the domain along the y-axis is homogenous, the transverse modulation function $U(\boldsymbol{\rho})$ can be written as $U(x)$ and expanded as the Fourier series $U(x) = \sum_{n=-\infty}^{\infty} c_n e^{i2\pi n x/d}$, where c_n is the Fourier coefficient of the n-th harmonic.

The two-photon spatial correlation for free propagating entangled photons can be derived as[21,22]

$$R_{c.c.}(x_s,z_s;x_i,z_i) \propto |\langle 0 | E_1^{(+)} E_2^{(+)} | \psi \rangle|^2 \propto \left| \int dx_0 e^{i\pi\eta(x_0-\xi)^2} U(x_0) \right|^2 \\ \propto \left| \sum_{n=-\infty}^{\infty} c_n e^{i2\pi n \xi/d} e^{-i\pi n^2/(d^2\eta)} \right|^2, \qquad (3)$$

where $\eta = 1/(\lambda_s z_s) + 1/(\lambda_i z_i)$, $\xi = \left(\dfrac{x_s}{\lambda_s z_s} + \dfrac{x_i}{\lambda_i z_i}\right) / \left(\dfrac{1}{\lambda_s z_s} + \dfrac{1}{\lambda_i z_i}\right)$, $E_1^{(+)}$ and $E_2^{(+)}$ are the electric fields evaluated at the two detectors, λ_s and λ_i are the wavelengths of signal and idler photons, respectively, $z_s(z_i)$ is the distance from the crystal to the plane where the single photon detector $D_1(D_2)$ lies in, x_0 is the transverse coordinate of the output plane of the crystal, and $x_s(x_i)$ is the detector's transverse coordinate. The first exponential term in Eq.(3) is the n-th Fourier component of a periodic function, and the second one is a phase shift dependent on the order of the Fourier component. If the second term equals 1 for all n, that is

$$\frac{1}{\lambda_s z_s} + \frac{1}{\lambda_i z_i} = \frac{1}{2md^2}, \tag{4}$$

where m is an integer indicating the m-th Talbot plane, Eq.(3) turns into a reproduction of grating function. So the quantum Talbot effect is directly observable after the MPPLT crystal. Since ξ is an associate coordinate, the Talbot self-image of the domain structure is magnified, and the exhibited period depends on the way of detection and the wavelengths of the photon pair. In particular, when we set $z_s = z_i = z$ and $x_s = x_i = x$ which is an in-step measurement, Eq.(3) indicates the two-photon amplitude can be equivalently considered as the pump photon's Fresnel diffraction, and the coincidence counting rate exactly replicates the transverse domain function, i.e., $R_{c.c.} \propto U(x)$ at the Talbot plane $z_T = 2d^2/\lambda_p$ and multiples of that. So the two-photon Talbot length is defined by the wavelength of pump wave, which is different from the classical one.

Fig.1(a) shows the micrograph of the ferroelectric domain structure of the etched MPPLT sample used in this experiment. The longitudinal periodicity is $\Lambda = 7.548$ μm for all stripes, which contributes to the QPM condition during the generation of entangled photons when the sample works at 170°C. The transverse stripe interval is $d = 160$ μm with stripe width $a = 20$ μm and stripe length $L = 10$ mm. The two-photon Talbot length is $z_T = 96.2$ mm when the crystal is pumped by 532 nm.

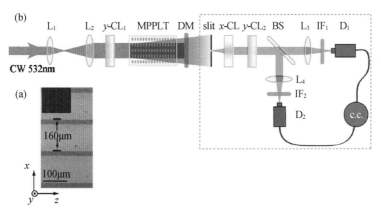

FIG.1. (a) Micrograph of the etched MPPLT. (b) Experimental setup. $L_1 - L_4$ are convex lenses, y-CL$_1$ and y-CL$_2$ are cylindrical lenses focusing along the y-axis, and x-CL is a cylindrical lens focusing along the x-axis.

Experimentally the two-photon Talbot effect is studied under two schemes. In the first one, both detectors are placed at the same distance from the output face of the crystal and scan in step along the same direction. As sketched in Fig. 1(b), in order to produce the degenerate photon pairs at 1064 nm, the crystal is pumped with a CW single longitudinal mode 532 nm laser. The pump is first expanded and collimated by two convex lenses L_1 and L_2 and focused along the y-axis by a cylindrical lens y-CL_1. After the crystal, the entangled photons are first separated from the pump by a dichromatic mirror DM. A 20 μm slit is used to scan along the x-axis and cascaded by a two-photon coincidence counting measurement. Specifically, after the slit, the entangled photons are collimated by two cylindrical lenses x-CL and y-CL_2, separated by a beam splitter and finally collected into two single photon detectors D_1 and D_2, respectively, by two collection lenses. Preceding each detector is a 10 nm bandwidth interference filter centered at 1064 nm to further suppress the pump. The slit and cascaded x-CL are fixed on a motorized precision positioning stage which can scan in both x and z directions. The slit and the following components can be regarded as an assembly two-photon detector denoted by a dashed rectangle in Fig. 1(b), which is capturing two-photon probability at the slit position.

In Fig. 2(a) we show the coincidence counts between two detectors as well as the D_1 single counts versus the transverse position of the slit when located at the first Talbot length $z_T = 96.2$ mm. The single counting rate shows an even distribution, whereas the coincidence counting rate reveals a periodic interference pattern with the periodicity of 160 μm, which is just the same as the transverse separation d of domain stripes; thus, the two-photon Talbot effect is confirmed. The fitted value for the full width at $1/e^2$ of the maximum of each correlation peak is 33.9 ± 1.8 μm which is larger than the theoretical value 24.1 μm due to that the spatial resolution is limited by the 20 μm scanning slit. The theoretical value is given under the 5 mm pump beam size ($1/e^2$), and the Fresnel diffraction broadens the image of each stripe. In order to investigate the two-photon diffraction dynamics, we capture a two-photon Talbot carpet as shown in Fig. 2(b) by measuring two-photon correlation in different detection planes located from $(1/2)z_T$ (48.1 mm) to $(3/2)z_T$ (144.3 mm) by the step size of 1 mm. The measured two-photon

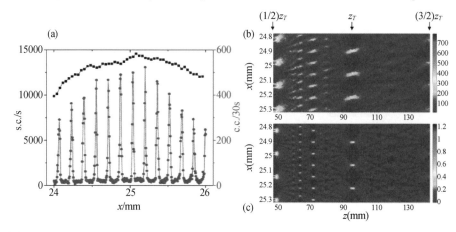

FIG. 2. (a) Measured single counts (black squares) and coincidence counts (blue dots) versus the transverse position of the slit at the first Talbot length $z_T = 96.2$ mm. (b) Experimental and (c) theoretical two-photon Talbot carpet from $(1/2)z_T$ (48.1 mm) to $(3/2)z_T$ (144.3 mm).

carpet is in good agreement with the theoretical one as shown in Fig. 2(c). From the two-photon Talbot carpet, it is convenient for understanding the evolution of the two-photon diffraction and interference when the photon pair propagates in the free space. At distances $(1/2)z_T$ and $(3/2)z_T$, the two-photon interference patterns are the same as that at z_T except with a half-period shift. Besides these integer Talbot images, fractional images[7,23] with a reduced period $(1/n)d$ $(n=2,3,4)$ are picked out and shown in Figs. 3(b)-3(d). The integer images at z_T and $(1/2)z_T$ are also shown for comparison in Figs. 3(a) and 3(e), respectively. For quantum Talbot imaging, there is no net improvement in the spatial resolution[15] due to the doubled Talbot length, which is different from quantum lithography.

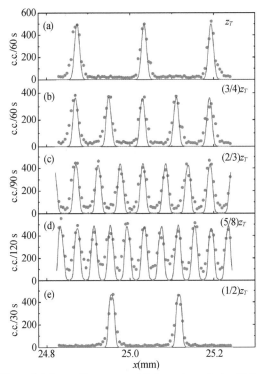

FIG.3. The red dots are the measured integer two-photon Talbot images detected at distances of (a) z_T, (e) $(1/2)z_T$ and fractional images at distances of (b) $(3/4)z_T$, (c) $(2/3)z_T$, and (d) $(5/8)z_T$. The blue lines are theoretical curves.

In order to verify the self-imaging condition $\frac{1}{z_s}+\frac{1}{z_i}=\frac{2}{z_T}$ for degenerate photon pairs indicated by Eq.(4) and disclose the physics of Talbot effect more intuitively, we carry out the two-photon Talbot experiment with another scheme. In this scheme, the two detectors scan independently. As shown in Fig. 4(a), the entangled photon pairs are separated by a 50/50 beam splitter. For each path, the single photon collection scheme is similar with Fig. 1(b), a 20 μm slit cascaded by two cylindrical lens x-CL_1 (x-CL_2) and y-CL_2 (y-CL_3) scans to give the coincidence counts. The slit and the following components can be regarded as an assembly detector D_1 (D_2) whose detection position is determined by the position of the slit. When both detectors are located at z_T, the coincidence counts still presents a grating function as shown in Fig. 4(e) except that the period is $2d$, which is twice of the transverse period of the MPPLT sample. Then we fix the detector D_2 at $z_T \pm 5$ mm and $z_T \pm 10$ mm, respectively, and perform a two-dimensional scanning of D_1 along the x and z directions to find out the corresponding self-imaging distances as shown in the inset in Fig. 4(b), where the values of coincidence counts are represented by brightness. Fig. 4(b) shows the measured self-imaging formula by five sets of z_s and z_i which consists well with Eq.(4). When the detector D_2 is fixed at distances $z_i = z_T - 10$ mm, $z_i = z_T - 5$ mm, $z_i = z_T + 5$ mm, $z_i = z_T + 10$ mm, the self-images are shown in Figs. 4(c), 4(d), 4(f), and 4(g), respectively. The coincidence counts exhibit images with the magnification of $1 + \dfrac{z_s}{z_i}$, which can be derived from Eq.(3). This experiment further differs the Talbot self-imaging from the conventional imaging by a lens. A point to point correspondence between the image plane and object plane is established by a lens, which means that in the imaging process, every point on the object corresponds to only one point, or spot more precisely due to the diffraction effect, on the image plane. However, in the experiment above, the coincidence counts show one-to-many relationship, which intuitively discloses that Talbot self-image is not a real image but a result of optical interference. Quantum Talbot image supplies more insights on the physics of Talbot effect.

It is worth noting that for the direct quantum Talbot imaging after the MPPLT, since the entangled photon pair can be designed at any wavelength within the transmitted window of the material due to the versatility of QPM, two-color Talbot effect is expected to be achieved easily. Furthermore, when implementing the two-color Talbot effect by scanning the two detectors in step but in opposite direction, two-photon interference at the Talbot plane z_T will exhibit the beating pattern with a magnified period $\dfrac{\lambda_s + \lambda_i}{\lambda_s - \lambda_i} d$. The Talbot beating effect may be employed as an effective way to reveal the periodic domain with an extremely small separation.

In conclusion, from a two-dimensional domain engineered lithium tantalate, we observed the direct quantum Talbot effect. With no real grating involved, the direct quantum Talbot effect results from the periodicity of two-photon spatial mode, hence is an inherent phenomenon existing concomitantly within the spontaneous parametric down conversion for entangled photons. By examining the propagation dynamics of two-photon diffraction and interference within one Talbot length, we capture the two-photon Talbot carpet and verify the self-imaging condition. With entangled photon pairs, it is more intuitive to reveal the hypostasis of Talbot effect as a diffraction phenomenon instead of imaging. Since the absence of grating and lens can improve stability and miniaturization, our results may be helpful for developing new types of entangled photon source and quantum technologies such as quantum lithography with improved performance when the flexible QPM technique is involved.

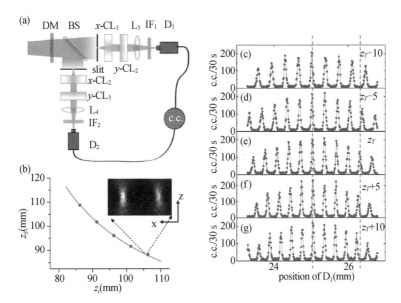

FIG.4. (a) The experimental setup of the detection part for the second scheme. (b) The relationship between two detection distances when Talbot images occur. The blue dots are measured results and the red line is a theoretical curve. The inset shows the measured coincidence counts as detector D_1 scans at different distances while D_2 is fixed at $z_T + 10$ mm. (c)—(g) The measured coincidence counts versus the transverse position of detector D_1 when D_2 is fixed at distances of (c) $z_i = z_T - 10$ mm, (d) $z_i = z_T - 5$ mm, (e) $z_i = z_T$, (f) $z_i = z_T + 5$ mm, and (g) $z_i = z_T + 10$ mm.

References and Notes

[1] H. F. Talbot, Philos. Mag. **9**, 401(1836).
[2] L. Rayleigh, Philos. Mag. **11**, 196(1881).
[3] K. Patorski, in *Progress in Optics*, edited by E. Wolf(North-Holland, Amsterdam, 1989), Vol. 27, pp. 1–108.
[4] A. W. Lohmann and J. A. Thomas, Appl. Opt. **29**, 4337(1990).
[5] P. Maddaloni, M. Paturzo, P. Ferraro, P. Malara, P. De Natale, M. Gioffrè, G. Coppola, and M. Iodice, Appl. Phys. Lett. **94**, 121105(2009).
[6] F. Pfeiffer, M. Bech, O. Bunk, P. Kraft, E. F. Eikenberry, C. H. Brönnimann, C. Grünzweig, and C. David, Nat. Mater. **7**, 134(2008).
[7] S. Nowak, Ch. Kurtsiefer, T. Pfau, and C. David, Opt. Lett. **22**, 1430(1997).
[8] B. Brezger, L. Hackermüller, S. Uttenthaler, J. Petschinka, M. Arndt, and A. Zeilinger, Phys. Rev. Lett. **88**, 100404(2002).
[9] Y. Y. Sun, X.-C. Yuan, L. S. Ong, J. Bu, S. W. Zhu, and R. Liu, Appl. Phys. Lett. **90**, 031107 (2007).

[10] Y. Zhang, J. M. Wen, S. N. Zhu, and M. Xiao, Phys. Rev. Lett. **104**, 183901(2010).

[11] M. R. Dennis, N. I. Zheludev, and F. J. García de Abajo, Opt. Express **15**, 9692(2007).

[12] D. van Oosten, M. Spasenovic, and L. Kuipers, Nano Lett. **10**, 286(2010).

[13] K. H. Luo, J. M. Wen, X. H. Chen, Q. Liu, M. Xiao, and L. A. Wu, Phys. Rev. A 80, 043820 (2009); **83**, 029902(E)(2011).

[14] C. H. R. Ooi and B. L. Lan, Phys. Rev. A **81**, 063832(2010).

[15] X. B. Song, H. B. Wang, J. Xiong, K. G. Wang, X. D. Zhang, K. H. Luo, and L. A. Wu, Phys. Rev. Lett. **107**, 033902(2011).

[16] M. M. Fejer, G. A. Magel, D. H. Jundt, and R. L. Byer, IEEE J. Quantum Electron. **28**, 2631 (1992).

[17] J. R. Kurz, A. M. Schober, D. S. Hum, A. J. Saltzman, and M. M. Fejer, IEEE J. Sel. Top. Quantum Electron. **8**, 660(2002).

[18] J. P. Torres, A. Alexandrescu, S. Carrasco, and L. Torner, Opt. Lett. **29**, 376(2004).

[19] X. Q. Yu, P. Xu, Z. D. Xie, J. F. Wang, H. Y. Leng, J. S. Zhao, S. N. Zhu, and N. B. Ming, Phys. Rev. Lett. **101**, 233601(2008).

[20] H. Y. Leng, X. Q. Yu, Y. X. Gong, P. Xu, Z. D. Xie, H. Jin, C. Zhang, and S. N. Zhu, Nat. Commun. **2**, 429(2011).

[21] M. H. Rubin, Phys. Rev. A **54**, 5349(1996).

[22] P. Xu, H. Y. Leng, Z. H. Zhu, Y. F. Bai, H. Jin, Y. X. Gong, X. Q. Yu, Z. D. Xie, S. Y. Mu, and S. N. Zhu, Phys. Rev. A **86**, 013805(2012).

[23] Z. H. Chen, D. M. Liu, Y. Zhang, J. M. Wen, S. N. Zhu, and M. Xiao, Opt. Lett. **37**, 689(2012).

[24] The authors thank L. A. Wu and J. M. Wen for valuable discussions. This work was supported by the State Key Program for Basic Research of China (Nos. 2012CB921802 and 2011CBA00205), the National Natural Science Foundations of China(Nos. 11174121, 91121001, and 11021403), the Project Funded by the Priority Academic Program Development of Jiangsu Higher Education Institutions (PAPD), and a Foundation for the Author of National Excellent Doctoral Dissertation of PR China (FANEDD).

Mode-locked Biphoton Generation by Concurrent Quasi-phase-matching*

Y. F. Bai,[1] P. Xu,[1] Z. D. Xie,[1] Y. X. Gong,[2] and S. N. Zhu[1]

[1] *National Laboratory of Solid State Microstructures and School of Physics, Nanjing University, Nanjing, 210093, China*

[2] *Physics Department, Southeast University, Nanjing, 211189, China*

We report a scheme for mode-locked biphoton generation by engineering quasi-phase-matching materials. An aperiodically poled lithium tantalate is designed to supply multiple reciprocal vectors for two-photon frequency comb generation. Temporally, such two-photon state exhibits mode-locked pulse train. The two-photon pulse duration reaches 4.2 fs when the frequency comb is centered at 800 nm and covers 442 nm. The HOM interference of the mode-locked biphoton approaches a 1.26-μm dip and exhibits a reappearance of dip and peak, which can be taken as a new scheme for engineering quantum light source. Furthermore, we also design another aperiodically poled crystal to generate two sets of frequency combs. The HOM interference will appear as a revival of spatial beatings. The mode-locked biphotons suggest applications in precise time synchronization, quantum metrology, quantum optical coherence tomography, and quantum computing.

Entangled photons lie at the heart of quantum information science and technology. A widely used method to generate entangled photons is spontaneous parametric down-conversion (SPDC)[1,2] in nonlinear optical crystals. Among them quasi-phase-matching (QPM) crystals play important roles in efficient generation of entangled photons as well as the manipulation of spatial[3-5] and frequency entanglement[6-9]. By designing chirped quasi-phase-matching nonlinear crystals[6-8], several groups focus on broadband two-photon generation and have achieved an ultrashort temporal correlation toward single-cycle limit, which is extraordinarily useful in clock synchronization[10], quantum metrology[11], and quantum optical coherence tomography[12].

An alternative way to achieve the broadband spectrum is engineering the frequency comb, which is the working principle of classical mode-locked laser. The two-photon frequency comb should have equivalent applications in quantum optics, like a continuous broadband spectrum, but the brightness of each mode will be much higher. Ou[13] first reported a two-photon frequency comb by using Fabry-Perot cavity after the entangled photons were generated. In that work, the frequency spacing is concerned with the cavity

---

* Phys.Rev.A,2012,85(5):053807

round trip and the total bandwidth of frequency comb is determined by the spectrum of SPDC process. In this work we aim to engineer the broadband optical frequency comb at the source. The mode-locked two-photon state is directly generated from an aperiodically poled nonlinear crystal(APNC). The mode spacing can be any value as we designed and the time correlation of such mode-locked biphoton exhibits as a pulse train. Each correlation peak reaches single cycle when the bandwidth of frequency comb is close to the central frequency. By designing another different set of reciprocal vectors of APNC, we can obtain two sets of frequency combs and achieve a new type of mode-locked two-photon state. We design the HOM interferometer to investigate the mode-locked biphotons. Serving as a new scheme for single-cycle biphoton preparation, the optical material engineering can be applied for exploiting new types of quantum light source, studying nonlinear optical process at single photon level and playing an important role in quantum computing[14—16].

For a periodically poled nonlinear crystal, usually only a single reciprocal vector is used to support a degenerate or nondegenerate photon pair generation by satisfying the phase matching condition $\Delta k = k_p - k_s - k_i - g = 0$. For frequency comb biphoton generation, we use an aperiodically poled nonlinear crystal, which can supply multiple reciprocal vectors to satisfy multiple spontaneous down-conversions concurrently. The poling function[17] can be written as

$$\chi^{(2)}(z) = \chi^{(2)} \mathrm{sgn}\left[\sum_n a_n \cos(g_n z + \phi_n)\right], \tag{1}$$

where g_n is the nth reciprocal vector, a_n and ϕ_n are adjustable parameters representing amplitude and phase for each g_n, respectively. $\chi^{(2)}$ is the second-order nonlinearity of the material. By our design, all amplitudes a_n are equal and all reciprocal vectors are in phase with $\phi_n = 0$. Theoretically, the cosine poling function can provide the maximum Fourier coefficient for each g_n to ensure a most efficient quasi-phase-matching process. However, from the viewpoint of practical engineering, the crystals such as lithium tantalate and lithium niobate can only be poled via periodic reversion of domains, which will reduce the effective nonlinearity. For a periodically poled nonlinear crystal, the maximum achievable Fourier coefficient for the designed reciprocal vector is $2/\pi$. Figure 1 shows how to design the domain structure where three SPDC processes are mixed.

Here we design N reciprocal vectors for N concurrent SPDC processes. Each generated frequency pair is taken as a mode and all modes are equally spaced. If all the modes are in coherent superposition, mode-locked two-photon will be generated. The interaction Hamilton inside the APNC can be written as

$$\hat{H} = \varepsilon_0 \sum_n \int \mathrm{d}\mathbf{r}\chi_n^{(2)} \mathrm{e}^{-ig_n z} \hat{E}_p^{(+)} \hat{E}_s^{(-)} \hat{E}_i^{(-)} + \mathrm{H.c.}. \tag{2}$$

Considering the first-order perturbation approximation, we can derive the wave function

$$|\psi\rangle_{ML} = \psi_0 \sum_{n=1}^{N} \int \mathrm{d}\nu \,\mathrm{sinc}\left(\frac{1}{2}\nu D_n L\right) \hat{a}_{s_n}^\dagger \hat{a}_{i_n}^\dagger |0\rangle. \tag{3}$$

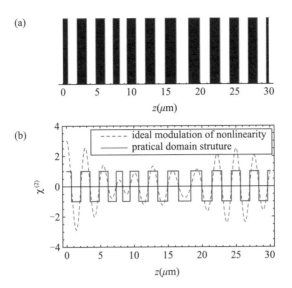

FIG. 1. (color online) Design of the APNC structure. (a) The domain structure of APNC. (b) Comparison between ideal modulation of nonlinearity and practical domain structure.

Since we design a constant amplitude for each g_n, the $\chi_n^{(2)}$ can be considered as a constant $\chi_{\text{eff}}^{(2)}$. In the above equation, $\psi_0 = \frac{\chi_{\text{eff}}^{(2)} E_p L}{2 V_Q n_s n_i} \sqrt{\omega_s \omega_i}$, which includes all the constants and slowly varying terms. N is the total number of frequency pairs. ν is the the detuning frequency around the center frequency. $D_n = \frac{1}{u_{s_n}} - \frac{1}{u_{i_n}}$ is the group velocity dispersion of the nth frequency pair and L is the crystal length. For each frequency pair of the two-photon state, we have $\omega_p = \omega_{s_n} + \omega_{i_n}$. Suppose $\Omega_{s_n}, \Omega_{i_n}$ are the central frequency of each frequency pair, then we have $\omega_{s_n} = \Omega_{s_n} + \nu, \omega_{i_n} = \Omega_{i_n} - \nu$. Suppose $\Delta\omega$ is the mode spacing, then $\Omega_{s_n} = \Omega_s + n\Delta\omega, \Omega_{i_n} = \Omega_i - n\Delta\omega$, in which Ω_s and Ω_i indicate the central frequency of the frequency pair most close to the degenerate frequency. As we only consider the temporal correlation, we neglect the spatial part of the wave function. Thus, the second-order time correlation function is calculated to be

$$G^{(2)}(\tau_1, \tau_2) \propto \left| \sum_{n=1}^{N} e^{-i\omega_p \frac{\tau_1+\tau_2}{2}} e^{-i\frac{\Omega_{d_n}}{2}\tau} \text{rect}\left(\frac{2\tau}{D_n L}\right) \right|^2, \quad (4)$$

where $\Omega_{d_n} = \Omega_{s_n} - \Omega_{i_n}$, $\tau = \tau_1 - \tau_2$. In the ideal case, we take all D_n as the same value D to simplify the two-photon time correlation function:

$$G^{(2)}(\tau) \propto \text{rect}\left(\frac{2\tau}{DL}\right) \left[\frac{\sin\left(\frac{N\Delta\omega}{2}\tau\right)}{\sin\left(\frac{\Delta\omega}{2}\tau\right)}\right]^2. \quad (5)$$

For the photon pair with a single frequency pair, the time correlation function is a

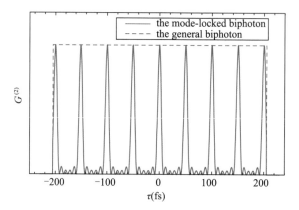

FIG. 2. (color online) Comparison between time correlation of mode-locked biphotons from the APNC and that of the general biphotons from a single plain crystal.

rectangle function in the form of $\text{rect}\left(\frac{2\tau}{DL}\right)$, whose width is determined by the group velocity dispersion and the crystal length. Usually the photon pairs generated from lithium tantalate are well synchronized by picosecond precision. However, in this work, different frequency modes are equally spaced which contributes an equidifferent phase to the time correlation function. Superposition and interference result in a temporally ultrashort pulse train in the form of $\left[\sin\left(\frac{N\Delta\omega}{2}\tau\right)/\sin\left(\frac{\Delta\omega}{2}\tau\right)\right]^2$. This is the physics behind the two-photon mode-locking. The width of two-photon correlation peaks is determined by the mode spacing and the number of frequency pairs. If the frequency comb covers as broadly as the central frequency, it will approach the level of femtosecond, i.e., a single-cycle. However, when discussing this we make an assumption that the group velocity dispersion for each frequency pair is taken to be constant, which is valid for some small dispersion materials, or the frequency comb covering not too broadly. We will give a specific discussion about it later.

In analogy with the mode-locked laser, the physics of mode-locked two-photon lies in that all the frequency pairs are in phase since they come from the same original pump photon and inherit the phase information. Besides, all reciprocal vectors are designed to be in phase. We have to emphasize that although the spectrum of signal or idler photon takes the form of frequency comb, neither the signal nor the idler photon is mode-locked due to the fact that the phase of single photon is random.

In our numerical calculation, we take $\lambda_p = 400$ nm, $\Delta\omega = 1.26 \times 10^{14}$ Hz, $L = 10$ mm, $N = 6$. The overall frequency comb covers 442 nm. The APNC is fabricated from a congruent lithium niobate(CLT) crystal. By using the Sellmeier equation of CLT, we can derive the magnitude of each reciprocal vector at 180℃: 2.1372 μm^{-1}, 2.1269 μm^{-1}, 2.1062 μm^{-1}, 2.0752 μm^{-1}, 2.0337 μm^{-1}, 1.9816 μm^{-1}. The two-photon then will be mode-locked to 4.2 fs as shown in Fig. 2. The rectangle time correlation function in Fig. 2

corresponds to the situation of 400 nm →779 nm +822 nm in a 10-mm-long crystal. We have to emphasize that in practical engineering we may choose the third-order reciprocal vectors working for the quasi-phase-matching conditions so as to ensure the homogeneity of the sample. Up to now, no single-photon detectors can resolve the picosecond time duration. Instead of measuring the time correlation of such two-photon state directly, we can set up a HOM interferometer to reveal the mode-locked character. In Fig. 3, for a collinear type-I SPDC, mode-locked biphotons are separated by a dichroic mirror. Signal photons propagate along the reflected path while the idler photons propagate along the transmitted path. The beam splitter (BS) recombines the photon pair and two-photon interference will take place. The beam splitter is displaced from its symmetry position by changing small distances Δx. Both of the single-photon detectors can detect all the wavelengths of frequency comb. To be mentioned that, the optical path before the BS has to be accurately chosen to ensure that all the modes are in phase when they arrive at the BS.

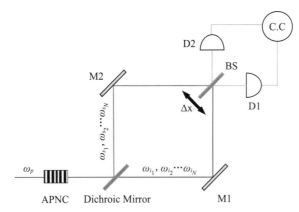

FIG. 3. (Color online) Schematic diagram of the HOM interferometer. A dichroic mirror is used to split the photon pair into two paths. The reflected signal photon and transmitted idler photon will be recombined together after passing through the beam splitter. Here we suppose the two single-photon detectors can respond to all the frequencies.

The electromagnetic field operators at D_1 and D_2 are expressed by

$$\hat{E}_1^{(+)}(t_1,x_1) = \sqrt{T}\,\hat{E}_{20}^{(+)} + i\sqrt{R}\,\hat{E}_{10}^{(+)} e^{i\omega_1 \frac{\Delta x}{c}},$$
$$\hat{E}_2^{(+)}(t_2,x_2) = \sqrt{T}\,\hat{E}_{10}^{(+)} + i\sqrt{R}\,\hat{E}_{20}^{(+)} e^{-i\omega_2 \frac{\Delta x}{c}}. \qquad (6)$$

Then we can calculate the two-photon amplitude,

$$A(\tau_1,\tau_2) \propto e^{-i\omega_p \frac{\tau_1+\tau_2}{2}} e^{-i\Omega_d \frac{\Delta x}{c}} \left[T g(\tau) - R e^{i\Omega_d \frac{\Delta x}{c}} g\!\left(\tau - 2\frac{\Delta x}{c}\right) \right], \qquad (7)$$

where $g(\tau) = \int d\nu \phi(\nu) e^{-i\nu\tau}$, $\phi(\nu)$ is the spectrum function of each mode, and $\Omega_d = \Omega_s - \Omega_i$. In our calculation, we take $T = R = \frac{1}{2}$ and suppose each mode has equal spectrum width σ

= 6.3×10^{12} Hz.

There are N pairs of frequency modes. Each pair provides its two-photon amplitude $A_n(\tau)$. By summing all of them, we get the total two-photon amplitude $A(\tau)$. Then we can work out the coincidence counting rate by integrating $|A(\tau)|^2$ from $-\infty$ to $+\infty$,

$$R_{cc} = 1 - \frac{\sum_n \int d\tau g_n(\tau) g_n\left(\tau - 2\frac{\Delta x}{c}\right) \cos\left(\Omega_{dn}\frac{\Delta x}{c}\right)}{\sum_n \int d\tau g_n(\tau)^2}. \quad (8)$$

where we neglect the small contributions from terms $A_n(\tau)A_m^*(\tau)(m \neq n)$.

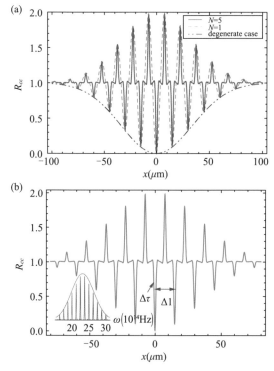

FIG.4. (color online) HOM interference pattern of mode-locked biphoton. (a) HOM interference pattern of mode-locked two-photon state for different number of frequency pairs. (b) HOM interference pattern of mode-locked two-photon state when the number of frequency pairs is 6. The inset is the corresponding frequency spectrum of optical frequency comb with Gaussian envelope.

By designing the APNC structure, we can get any desired spectrum envelope of frequency comb. In the following calculation, we consider all the frequency combs have a Gaussian envelope. By changing the number of frequency pairs, we obtain different results after the HOM interferometer. As shown in Fig. 4(a), the width of HOM dip is 48 μm when only one reciprocal vector takes the role for 800-nm degenerate photon pair generation. When another reciprocal vector contributes to the nondegenerate photon pair at 779 nm and 822 nm, we will see the beating inside the HOM dip as first reported in Ref.

[18]. The beating frequency is 1.26×10^{14} Hz. Furthermore, when five pairs of nondegenerate frequency are taken into account, the HOM interference pattern turns out to be a series of narrow dips and peaks. When the FWHM of the frequency comb envelope is 242 nm($N=6$), each dip or peak covers 1.26 μm. This indicates that temporally the two-photon is 4.2-fs mode-locked pulse. Narrower dip or narrower mode-locked two-photon pulse can be achieved when more frequency pairs are involved.

The width(FWHM) of each dip in Fig. 4(b) is determined by the bandwidth of frequency envelope. The dip width is written as

$$\Delta \tau \sim \frac{\pi}{N \Delta \omega}. \tag{9}$$

It is the equal mode spacing that causes the reappearance of the dip. Here the period for dip reappearance is 15 μm, which is associated with the mode spacing,

$$\Delta l \sim \frac{2c\pi}{\Delta \omega}. \tag{10}$$

For a traditional HOM interference, the dip or minimum coincidence rate can be understood as the results of two-photon interference. Photon pairs are bunched and captured two by two by either single-photon detector. A reversed HOM interferometer[19] will present correlation peak when two paths before the beam splitter have equal optical path. This means photon pair are more likely to be captured separately by two detectors and $|1,1\rangle$ state is dominant. Then what does the reappearance of peak and dip mean? We now analyze the two-photon state after the beam splitter,

$$|\psi\rangle = \sum_n \{(|2_a, 0_b\rangle_n + e^{i\theta_n}|0_a, 2_b\rangle_n) + i(|1_a, 1_b\rangle_n - e^{i\theta_n}|1_b, 1_a\rangle_n)\}, \tag{11}$$

where a, b indicate different optical paths. For each frequency pair, when $\theta_n = 0$, the correlation dip appears, while when $\theta_n = \pi$, the correlation peak emerges. In our case, θ_n can be written as

$$\theta_n = \frac{2n \Delta \omega \Delta x}{c}. \tag{12}$$

The superposition of different frequency pairs will lead to reappearance of the dip and peak when the BS is displaced at definite positions. Since θ changes with the optical path difference of HOM interferometer, dip and peak reappear periodically. By changing the optical path difference of the interferometer, we can engineer different quantum light source. It is worth mentioning that in addition to the HOM interferometer we can also use broadband sum-frequency-generation as an ultrafast correlator[8] to explore the temporal character of mode-locked biphotons.

A more general mode-locked biphoton state is that the photon pair covers two sets of frequency combs. The signal comb and idler comb both carrying Gaussian envelopes are separated away from the degenerate wavelength. We find a new character when observing the HOM interference as shown in Fig. 5. The coincidence counting rate between two detectors exhibits periodic revival, which is not the neat dip or peak but the spatial beating. The beating cycle is concerned with the difference of center mode of two frequency combs. This mode-locked two-photon with two separate frequency combs is first calculated here. In classical laser, no similar concepts can be referred to. Here we take $\lambda_p = 532$ nm, $\lambda_{s0} = 810$ nm, $\lambda_{i0} = 1550$ nm, $N = 9$, $\Delta\omega = 7.85 \times 10^{13}$ Hz and $\sigma = 7.85 \times 10^{12}$ Hz.

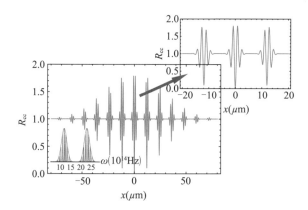

FIG.5. (color online) HOM interference pattern of mode-locked two-photon state, which covers two sets of frequency combs. The inset is the corresponding frequency spectrum when the signal comb and idler comb both carry Gaussian envelopes.

In the above discussions, we take the group velocity dispersion between signal and idler photons almost the same for each frequency pair. It comes to be true when we consider some materials with small group velocity dispersion and also for the case that the frequency comb is not too broad. Experimentally we can use a medium with the negative dispersion to compensate the dispersion. When taking the group velocity dispersion of CLT into the calculation of time correlation, we find the width of peak around $\tau = 0$ does not change and peaks elsewhere are broadened. The middle correlation peak corresponds to the situation when the optical path before BS is equal. The broadening effect cannot be eliminated because it is caused by the nonlinear crystal's intrinsic property. However, experimentally we are only concerned about central correlation peak, which is still well mode-locked. Another problem one may encounter in practical engineering is the poling of tiny domains when several different reciprocal vectors coexist. For a 0.5-mm-thick sample, the domains smaller than 2 μm can hardly be poled. To solve this problem, one may practically neglect the tiny domain and incorporate it with its adjacent domain. It will make the poling function deviate a little from the desired one, which can be considered as a fabrication error. It will broaden each frequency mode but affect little on the shape of the frequency comb. The narrow dip and peak of HOM interference will not change except the envelope of interference pattern will be varied. But such effects are not dominant since the possibility of finding tiny domains is not high.

A well-designed cascaded or multistripe periodically poled nonlinear crystal may be used to generate a two-photon frequency comb. However, extra efforts are required to make all the frequency modes in phase. For the cascaded periodically poled crystal, different frequency pairs are generated at different locations. Different modes propagate and would be out of phase, which will not result in a mode-locked biphoton from such crystals. For the multistripe crystal, we have to make the photon pair from different stripes indistinguishable, such as using single-mode fiber to collect the photon pairs. So the scheme by use of APNC for mode-locked biphoton generation is compact and feasible.

In summary, we have proposed a scheme to generate mode-locked biphoton by designing the aperiodically poled nonlinear crystal with concurrent quasi-phase-matchings. We analyze the temporal correlation of two types of such states in virtue of the HOM interferometer. When the frequency comb is broadband, the ultrashort time correlation can reach single-cycle length, which is quite useful in a wide range of areas. Furthermore, when the signal and idler comb is nondegenerate, we find a reappearance of two-photon optical beating after HOM interferometer. The physics behind the results is the superposition of different frequency modes when they are in phase. The multiple frequency modes may carry diverse information, which would be of great use in quantum wavelength-division multiplexing. Besides, due to the high-dimensionality of frequency degree, shaping the spectrum of biphoton will benefit a wealth of physical phenomena and supply a new scheme for quantum-state engineering. Since optical frequency comb can be engineered as a large cluster state, which is the core of the one-way quantum computation scheme[14—16], the aperiodically poled nonlinear crystal may supply as a crucial component in this field.

References and Notes

[1] Y.H. Shih and C. O. Alley, Phys.Rev.Lett. **61**, 2921(1988).
[2] C. K. Hong, Z. Y. Ou, and L. Mandel, Phys. Rev. Lett. **59**, 2044(1987).
[3] J. P. Torres, A. Alexandrescu, S. Carrasco, and L. Torner, Opt. Lett. **29**, 376(2004).
[4] X. Q. Yu, P. Xu, Z. D. Xie, J. F. Wang, H. Y. Leng, J. S. Zhao, S. N. Zhu, and N. B. Ming, Phys. Rev.Lett. **101**, 233601(2008).
[5] H. Y. Leng, X. Q. Yu, Y. X. Gong, P. Xu, Z. D. Xie, H. Jin, C. Zhang, and S. N. Zhu, Nature Commun. **2**, 429(2011).
[6] S. E. Harris, Phys. Rev. Lett. **98**, 063602(2007).
[7] M. B. Nasr, S. Carrasco, B. E. A. Saleh, A. V. Sergienko, M. C. Teich, J. P. Torres, L. Torner, D.S. Hum, and M. M. Fejer, Phys. Rev. Lett. **100**, 183601(2008).
[8] S. Sensarn, G. Y. Yin, and S. E. Harris, Phys. Rev. Lett. **104**, 253602(2010).
[9] A. M. Braänczyk, A. Fedrizzi, T. M. Stace, T. C. Ralph, and A. G. White, Opt. Express **19**, 55 (2011).
[10] A. Valencia, G. Scarcelli, and Y. Shih, Appl. Phys. Lett. **85**, 2655(2004).
[11] V. Giovannetti, S. Lloyd, and L. Maccone, Science **306**, 1330(2004).
[12] M. B. Nasr, B. E. A. Saleh, A. V. Sergienko, and M. C. Teich, Phys. Rev. Lett. **91**, 083601(2003).
[13] Y. J. Lu, R. L. Campbell, and Z. Y. Ou, Phys. Rev. Lett. **91**, 163602(2003).

[14] N. C. Menicucci, S. T. Flammia, and O. Pfister, Phys. Rev. Lett. **101**, 130501(2008).

[15] M. Pysher, Y. Miwa, R. Shahrokhshahi, R. Bloomer, and O. Pfister, Phys. Rev. Lett. **107**, 030505 (2011).

[16] M. Pysher, A. Bahabad, P. Peng, A. Arie, and O. Pfister, Opt. Lett. **35**, 565(2010).

[17] A. Norton and C. de Sterke, Opt. Express **12**, 841(2004).

[18] Z. Y. Ou and L. Mandel, Phys. Rev. Lett. **61**, 54(1988).

[19] J. Chen, K. F. Lee, and P. Kumar, Phys. Rev. A **76**, 031804(2007).

[20] The authors thank Y. H. Shih and L. Yan for helpful discussions. This work was supported by the State Key Program for Basic Research of China(Grant Nos. 2012CB921802 and 2011CBA00205), the National Natural Science Foundations of China(Grant Nos. 10409066, 91121001, and 11021403), and the Project Funded by the Priority Academic Program Development of Jiangsu Higher Education Institutions(PAPD).

Generation of N00N State with Orbital Angular Momentum in a Twisted Nonlinear Photonic Crystal*

Yang Ming, Jie Tang, Zhao-xian Chen, Fei Xu, Li-jian Zhang, and Yan-qing Lu

National Laboratory of Solid State Microstructures, College of Engineering and Applied Sciences, Collaborative Innovation Center of Advanced Microstructures, Nanjing University, Nanjing 210093, China

We investigate wavefront engineering of photon pairs generated through spontaneous parametric down conversion in lithium niobate-based nonlinear photonic crystals(NPCs). Due to the complexity of domain structures, it is more convenient to describe photon interaction based on the nonlinear Huygens-Fresnel principle than conventional quasiphase matching regime. Analytical expressions are obtained to describe the transverse properties of down-converted photon states. The convenience of domain engineering in $LiNbO_3$ crystals provides a potential platform for flexible wavefront manipulation of multiphoton states. The generation of N00N state with orbital angular momentum in a twisted NPC is studied utilizing this method. The obtained state is of great value in quantum cryptography, metrology, and lithography applications.

1. Introduction

SPONTANEOUS parametric down-conversion (SPDC) is an important approach to obtaining quantum photonic states, such as the heralded single-photon state[1]–[3] and the entangled photon state[4]–[11]. The SPDC process is commonly implemented in $\chi^{(2)}$ nonlinear optical crystals[4],[12], in which a single photon is transformed into a photon pair. In practice, different kinds of complex photon states are required for various specific applications, which need appropriate combinations and modulations of down-converted photons. To perform this task, the so-called nonlinear photonic crystal (NPC)[13] with spatially variable nonlinear coefficient $\chi^{(2)}(r)$ (the corresponding function is usually periodic or of special form) presents an effective platform. One of the most popular kind of NPC is the domain-inverted ferroelectric crystals which are selectively poled to drive the sign of $\chi^{(2)}$ changing between positive and negative with certain patterns[13],[14]. Among such crystals, the domain-engineered lithium niobate crystal is a valuable candidate[13],[15],[16]. Owing to the mature processing technologies[17]–[19], suitable domain structure could be flexibly introduced into the $LiNbO_3$ crystals for SPDC engineering[15],[16],[20]. Moreover,

---

* IEEE J. Sel. Top. Quant.,2015,21(3):1

the multifunctional feature of LiNbO$_3$ crystal also brings opportunities for function-integrated quantum circuits[11], [15], [21].

In this work, we investigate wavefront engineering of down-converted photons based on domain-engineered LiNbO$_3$ crystals. Tailoring the wavefront of entangled photons has been demonstrated to play an important role in spatial entanglement[20], [22] and orbital angular momentum entanglement of photons[23], which are valuable resources for quantum communication[24], [25], metrology[26], [27] and imaging[16], [28], [29]. In these works, the parametric processes for tailoring wavefront are usually described in conventional phase matching regime. In a NPC, the complexity of domain pattern increases and the corresponding symmetry reduces. It is more convenient to describe the photon interactions based on the nonlinear Huygens-Fresnel principle[30]. Analytical expressions are obtained to outline the wavefront of photon pairs generated from arbitrary domain-engineered LiNbO$_3$ crystals. As an illustration, the SPDC process in a twisted LiNbO$_3$ crystal is investigated. It is revealed that helical wavefront known as orbital angular momentum is imprinted on the N00N state[31].

2. Transverse Properties of SPDC in Domain-Engineered Lithium Niobate Crystals

As is known, inverted domains are introduced into a LiNbO$_3$ crystal through poling it along the direction of optical axis(usually marked as z-direction)[17]–[19]. It is relatively convenient to obtain various transverse patterns in the xy plane. Thus, for effective tailoring of the wavefront, we choose to launch the pump light along the z-direction. In a usual situation, the $\chi^{(2)}$ coefficient in a single LiNbO$_3$ slice is z-independent. However, there is still an available way to realizing z-directional modulation of $\chi^{(2)}$ through stacking the slices together[32] (High quality LiNbO$_3$ thin films with a thickness of several hundred nanometers are experimentally achievable through the ion slicing technique[33]). Without loss of generality, we consider a 3-D system, in which the $\chi^{(2)}$ coefficient is x, y, z-dependent.

According to nonlinear Huygens-Fresnel principle[30], [34], in SPDC process, each point on the wavefront of the pump light could act as a source of secondary down-converting wavelets through interacting with NPC. The transverse properties of the output photon pairs are determined by the pump and the structure of NPC together. The influence of each secondary photon pair source on the output photon state could be estimated by the corresponding Hamiltonian density as

$$\mathcal{H}_I(\boldsymbol{r}) = \frac{1}{2} \hat{\boldsymbol{P}}_{NL}(\boldsymbol{r}) \cdot \hat{\boldsymbol{E}}_p(\boldsymbol{r}) = \frac{1}{2} \chi^{(2)}(\boldsymbol{r}) \hat{E}_p^{(+)}(\boldsymbol{r}) \hat{E}_s^{(-)}(\boldsymbol{r}) \hat{E}_i^{(-)}(\boldsymbol{r}), \tag{1}$$

where the subscripts p, s and i correspond to the pump, signal and idler photons, respectively. The pump field E_p is usually treated as a classical field and assumed to be undepleted. The signal and idler fields should be quantized and represented by field operators. Detailed expressions are given as follows:

$$
\begin{aligned}
E_p^{(+)} &= E_{p0} \Psi_{pt}(\boldsymbol{r}) e^{i(\boldsymbol{k}_p \cdot \boldsymbol{r} - \omega_p t)}, \\
E_s^{(-)} &= i \sum_\sigma \sum_{l_s} \int d\omega_s \int d\boldsymbol{k}_s C_{s\sigma} a_{s\sigma}^\dagger e^{-i(\boldsymbol{k}_s \cdot \boldsymbol{r} + l_s \theta - \omega_s t)}, \\
E_i^{(-)} &= i \sum_\sigma \sum_{l_i} \int d\omega_i \int d\boldsymbol{k}_i C_{i\sigma} a_{i\sigma}^\dagger e^{-i(\boldsymbol{k}_i \cdot \boldsymbol{r} + l_i \theta - \omega_i t)},
\end{aligned}
\tag{2}
$$

where E_{p0} is the amplitude of pump light, and $\Psi_{pt}(\widetilde{r})$ represents the transverse shape of the field which includes the wavefront information of the pump. Here we consider the monochromatic pump. The parameter σ corresponds to the polarization state of the down-converted photons, while $C_{j\sigma}(j=s,i)$ is the normalization parameter. The symbols l_s and l_i are defined as topological charge, which indicates that the orbital angular momentum of signal and idler photons are $l_s \hbar$ and $l_i \hbar$, respectively[23]. Moreover, $\theta = \tan^{-1}(y/x)$ is the azimuthal coordinate. Based on these analyses, the contribution of a secondary source in the down-converted photon state could be presented through the evolution equation as

$$
\begin{aligned}
|d\Psi\rangle &= \left[1 - \frac{i}{\hbar} \int_{-\infty}^{\infty} dt\, \mathcal{H}_I(t)\right] |0\rangle \\
&= |0\rangle + \frac{iE_{p0}}{2\hbar} \sum_{\sigma,\sigma'} \int d\boldsymbol{k}_s \int d\boldsymbol{k}_i \int_{-\infty}^{\infty} dt\, C_{s\sigma} C_{i\sigma'} K(\boldsymbol{r}; \boldsymbol{k}_s, \boldsymbol{k}_i) e^{-i(\omega_p - \omega_s(\boldsymbol{k}_s) - \omega_i(\boldsymbol{k}_i))t} a_{s\sigma}^\dagger a_{i\sigma'}^\dagger |0\rangle,
\end{aligned}
\tag{3}
$$

with

$$
K(\boldsymbol{r}; \boldsymbol{k}_s, \boldsymbol{k}_i) = \chi^{(2)}(\boldsymbol{r}) \Psi_{pt}(\boldsymbol{r}) e^{i(\boldsymbol{k}_p - \boldsymbol{k}_s - \boldsymbol{k}_i) \cdot \boldsymbol{r}} e^{-i(l_s + l_i)\theta}. \tag{4}
$$

The transverse properties of the generated two-photon state are connected to the pump light and the NPC through the function $K(\boldsymbol{r}; \boldsymbol{k}_s, \boldsymbol{k}_i)$. Thus the transverse wavefront of the down-converted photons are determined by the integral of function K in the entire quadratic interaction region. The expression could be given as

$$
|\Psi\rangle_f = \int d\boldsymbol{r}\, |d\Psi\rangle = |0\rangle + \frac{iE_{p0}}{2\hbar} \sum_{\sigma,\sigma'} \int d\boldsymbol{k}_s \int d\boldsymbol{k}_i\, C_{s\sigma} C_{i\sigma'} \Omega(\boldsymbol{k}_s, \boldsymbol{k}_i) a_{s\sigma}^\dagger a_{i\sigma'}^\dagger |0\rangle, \tag{5}
$$

with

$$
\Omega(\boldsymbol{k}_s, \boldsymbol{k}_i) = \int d\boldsymbol{r}\, K(\boldsymbol{r}; \boldsymbol{k}_s, \boldsymbol{k}_i) \delta(\omega_p - \omega_s(\boldsymbol{k}_s) - \omega_i(\boldsymbol{k}_i)). \tag{6}
$$

where the Dirac-δ function is given by the temporal integration. The final state is formed through superposing the contributions of all secondary down-converting sources. The spatial integral $\Omega(\boldsymbol{k}_s, \boldsymbol{k}_i)$ represents the correspondence between nonlinear Huygens-Fresnel principle and the superposition principle of quantum states.

Wavefront engineering could be realized through appropriate design of the spatial distribution of the nonlinear coefficient $\chi^{(2)}$, which could be expressed as

$$
\chi^{(2)}(\boldsymbol{r}) = \chi^{(2)}_{\text{eff}} \sum_{Z_n} G(z \mid Z_n) \text{sgn}\{\cos[2\pi F Z_n(\boldsymbol{r}_t)]\}. \tag{7}
$$

In the equation, the function $G(z \mid Z_n)$ represents the z-dependence of $\chi^{(2)}$. As is mentioned above, to date z-directional modulation of $\chi^{(2)}$ could only be achieved through stacking LiNbO$_3$ slices together, so it is treated discretely. $F_{Z_n}(r_t)$ corresponds to the transverse domain pattern of the nth LiNbO$_3$ slice. Complex transverse domain structures can be produced conveniently with current techniques. In theory, $F_{Z_n}(r_t)$ could be arbitrary 1D or 2D function. Reasonable combinations of the functions $G(z|Z_n)$ and $F_{Z_n}(r_t)$ are critical for elaborate wavefront engineering.

3. N00N State With Orbital Angular Momentum

In this section, utilizing the approach discussed above, we investigate wavefront engineering of the well-known N00N state in twisted NPCs. A N00N state is expressed as

$$|N::0\rangle = (|N,0\rangle + |0,N\rangle)/\sqrt{2}, \tag{8}$$

where N is a positive integer which represents the number of photons. Such states are very useful in quantum precise phase measurement[35] and super-resolution quantum lithography[36]. The actual effect of N00N state in these applications increases with the photon number N. As we know, the two-photon N00N state could be easily experimentally generated through various methods[31],[37]-[40]. However, multi-photon N00N state ($N>2$) is hard to produce because of technical difficulties[41]. In the following text, we will show that a two-photon N00N state carrying orbital angular momentum may work like a multi-photon N00N state in certain applications. Through appropriate wavefront engineering based on a NPC with a twisted domain pattern, helical wavefront could be imprinted on the photons of a N00N state with $N=2$. As is known, the helical wavefront corresponds to the orbital angular momentum of a photon[23].

For the transverse domain structure of the NPC, we choose the "Fork Grating"(FG) pattern. In classical linear optics, such patterns are widely used to generate optical vortex with helical wavefront expressed as $\Psi = \exp(im\theta)$[42]. The illustrations of FGs are given in Fig. 1(a) and (b). Second harmonic generation in this kind of NPC with FG has also been well investigated[14],[43]. Besides, the interaction of transverse structured light and nonlinear optical crystals is also a recent research hotspot[44].

In this work, we would focus on the properties of the quantum state generated in such twisted NPC through SPDC process. The nonlinear coefficient $\chi^{(2)}$ could be written as

$$\chi^{(2)}(r) = \chi^{(2)}_{\text{eff}} G[\text{Int}(z/a)]\text{sign}\{\cos[2\pi F(r_t) + l_C \theta]\}, \tag{9}$$

where $\text{Int}(x)$ refers to the floor function and a is the thickness of a LiNbO$_3$ slice. The function $G[\text{Int}(z/a)]$ is used to distinguish which LiNbO$_3$ slice a z-coordinate is located in and describe the influence of the corresponding slice on $\chi^{(2)}$. The value l_C relates to the number of branchings in the central "fork" of the pattern, while θ is the azimuthal coordinate. N_{LN} is the total number of LiNbO$_3$ slices.

The pump light is set to be $E_p(r) = E_{p0}\exp(ik_{pz}z + il_p\theta)$ with the topological charge marked as l_p, thus the function $\Psi_{pt}(r)$ is corresponding to $\exp(il_p\theta)$. Substituting $\Psi_{pt}(r)$

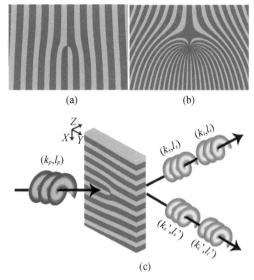

FIG.1. The FG patterns are plotted based on different topological charges as (a) $l_C = 1$ and (b) $l_C = 20$, which also correspond to transverse domain structures of the $LiNbO_3$-based NPCs. The light blue and purple regions are positive and negative domains, respectively. (c) Schematic of wavefront engineering of N00N state in a twisted NPC. Please note that the helical wavefront does not correspond to the actual value of topological charge, which is just a diagram.

and Eq.(9) into Eq.(6), we have

$$\Omega_I = \chi_{\text{eff}}^{(2)} \left(\int dz G[\text{Int}(z/a)] e^{i(k_{pz}-k_{sz}-k_{iz})z} \right) \int d\bm{r}_t \, \text{sign}\{\cos[2\pi F(\bm{r}_t) + l_C\theta]\} e^{i(l_p-l_s-l_i)\theta} e^{-i(\bm{k}_{st}+\bm{k}_{it})\cdot\bm{r}_t}. \tag{10}$$

In the transverse integral, the function $F(\bm{r}_t)$ is a 1-D or 2-D function. For the FG pattern, we have $F(\bm{r}_t) = x/\Lambda$. Λ is defined as the poling period in the x-direction. To obtain a clear physical description, the sign function is analyzed with Fourier transform. The set of Fourier basis is chosen to be $\{\exp[im(2\pi x/\Lambda + l_C\theta)]\}$. The expression of is $\chi^{(2)} = \chi_{\text{eff}} \sum_m F_m \exp(i2\pi mx/\Lambda) \exp(iml_C\theta)$, where F_m is the Fourier coefficient calculated by $F_m = (1/2\pi) \int \text{sign}[\cos(2\pi x/\Lambda + l_C\theta)] \cdot \exp[im(2\pi x/\Lambda + l_C\theta)]$. In all the oscillating terms, the first-order terms with $m = \pm 1$ are dominant[14]. Neglecting high-order terms, the transverse integral could be written as

$$I_t = \sum_{\pm 1} F_{\pm 1} \int d\bm{r}_t \, e^{i(l_p \pm l_C - l_s - l_i)\theta} e^{\pm i\frac{2\pi x}{\Lambda}} e^{-i(\bm{k}_{st}+\bm{k}_{it})\cdot\bm{r}_t}. \tag{11}$$

For the z-directional dependence of $\chi^{(2)}$, we would consider a simple case in which the transverse patterns of all $LiNbO_3$ slices are the same but they are placed parallelly and antiparallelly to the z-axis in turn. That means the function $G[\text{Int}(z/a)]$ is expressed as

$$G[\text{Int}(z/a)] = (-1)^{\text{Int}(z/a)}. \tag{12}$$

Based on the analyses above, the state vector of SPDC could be obtained through substituting Eqs.(10)—(12) into Eq.(5), which is expressed as

$$|\Psi\rangle_f = |0\rangle + \sum_{\pm 1} \frac{iE_{p0}F'_{\pm 1}d_{22}}{2\hbar} \int d\bm{k}_s \int d\bm{k}_i C_{s\sigma}C_{i\sigma'}$$
$$\times \left[\int_0^{N_{LN}a} dz(-1)^{\text{Int}(z/a)} e^{i(k_{pz}-k_{sz}-k_{iz})z}\right]$$
$$\times \left[\int d\bm{r}_t e^{\pm i\frac{2\pi x}{\Lambda}} e^{-i(\bm{k}_{st}+\bm{k}_{it})\cdot \bm{r}_t} e^{i(l_p\pm l_C-l_s-l_i)\theta}\right] a^\dagger_{s\sigma}a^\dagger_{i\sigma'}|0\rangle. \qquad (13)$$

In the equation, only the influence of the first order terms of $\chi^{(2)}$ are included. On the one hand, the transformation efficiencies of the high-order terms are relatively low; on the other hand, as the output angles of down-converted photons are different, their contributions could be eliminated through proper post-selections. The pump light is incident along z-direction and assumed to be y-polarized. In this situation, the only available element in nonlinear coefficient of LiNbO$_3$ is d_{22}, and the down-converted photons should also be y-polarized. In the practical diffraction process, the polarization state of photon may be influenced by non-idealities, and the d_{31}-dominated nonlinear process may contribute to the down-converted photons. However, this proportion of the contribution is quite small and does not have crucial influences on the final conclusion, thus we neglect that in calculations for simplicity. In practical situations, it is also convenient to eliminate the corresponding influence through putting additional polarizers after the LiNbO$_3$ crystal into the optical paths. Based on these analyses, the summation about the polarization state could be simplified. The characteristics of the down-converted state are derived from Eq. (13). In a high efficiency situation, the obtained photons are imprinted with helical wavefront $\exp[i(l_p\pm l_C)\theta]$ corresponding to a total angular momentum $(l_p\pm l_C)\hbar$.

For the integral term $\int d\bm{r}_t \exp(\pm i2\pi x/\Lambda)\times \exp[-i(\bm{k}_{st}+\bm{k}_{it})\cdot \bm{r}_t]$, the generation efficiencies of corresponding spatial modes are determined by its modular square. The modular square is calculated to be proportional to $\text{sinc}^2[(k_{sx}+k_{ix}\pm 2\pi/\Lambda)X_0/2]\cdot \text{sinc}^2[(k_{sy}+k_{iy})Y_0/2]$, where X_0 and Y_0 represent the transverse sizes of the NPC. Just as what is utilized in Ref. [4], two desired modes could be selected with equally high transformation efficiencies, in which we have $k_{sx}=k_{ix}=\pm\pi/\Lambda$ and $k_{sy}=k_{iy}=0$ together with a degenerate frequency of the signal and idler photons. This corresponds to two equal possibilities. In both cases, a pair of photons is emitted from the NPC with an angle of equal degree but opposite direction, as is shown in Fig. 1(c). Thus a N00N state is formed. Unlike conventional N00N states[37]—[41], this N00N state possesses optical angular momentum. In practical applications, band pass filters with certain center frequency could be place at the corresponding angles to improve the purity of the state.

The z-directional integral $\int_0^{N_{LN}a} dz(-1)^{\text{Int}(z/a)} e^{i(k_{pz}-k_{sz}-k_{iz})z}$ also has critical influences on the transformation efficiency of SPDC process. To improve the transformation efficiency, the value of a should be set at the coherent length decided by $L_{coh}=2/(k_{pz}-k_{sz}-k_{iz})$. This point could also be understood through the QPM regime. When N_{LN} becomes a large number, it is also convenient to express the function as $\text{sgn}[\sin(2\pi z/L_{coh})]$. Under this condition, the transformation efficiency increases with N_{LN}. For a detailed understanding, we calculate the values of transformation efficiency for a practical system with $N_{LN}=1,10,50$, respectively. The pump wavelength is set at 0.405 μm, while the corresponding signal and idler wavelengths are 0.81 μm. The coherent length is calculated to be 1.07 μm. The values of l_p and l_C are assumed to be 3 and 1. The poling period in the x-direction is 10.3 μm, which is determined by $\Lambda = 0.405/n_{PDC} \cdot \sin\theta_{PDC}$ (μm) with the propagation angles of down-converted photons $\theta_{PDC}=\pm 1°$. The output angle after exiting NPC will be $\theta_{out} \sim n_{PDC} \cdot \theta_{PDC} = \pm 2.25°$. When the pump power 1 W, the transformation efficiencies are 1.227×10^{-19}, 4.973×10^{-17}, 6.216×10^{-15}, respectively. These values are obtained through the coupled wave equations for quadratic nonlinear interaction[14]. The average generation rates of N00N state are calculated to be 0.25 s^{-1}, 101.28 s^{-1} and 12659.88 s^{-1}, respectively.

To present quantum interference, the N00N state with orbital angular momentum could be adjusted and detected through the combinations of several optical components, such as Dove prism[45],[46], q-plate[47] and the HOM interferometer[31]. The function of a Dove prism is rotating the wavefront[45],[46]. The scheme is shown in Fig. 2(a). Since the down-converted photons pass through the Dove prisms in different arms, the corresponding photonic state could be expressed as

$$|\Psi\rangle \propto \sum_{l_s, l'_s}(e^{i(l_s+l_i)\alpha}|1_{l_s}+1_{l_i},0_{l'_s,l'_i}\rangle + e^{i(l'_s+l'_i)\alpha}|0_{l_s,l_i},1_{l'_s}+1_{l'_i}\rangle)$$

$$\propto \sum_{l_s, l'_s}(e^{i(l_p+l_C)\alpha}|1_{l_s}+1_{l_i},0_{l'_s,l'_i}\rangle + e^{i(l_p-l_C)\alpha}|0_{l_s,l_i},1_{l'_s}+1_{l'_i}\rangle)$$

$$\propto \sum_{l_s, l'_s}(|1_{l_s}+1_{l_i},0_{l'_s,l'_i}\rangle + e^{-i2l_C\alpha}|0_{l_s,l_i},1_{l'_s}+1_{l'_i}\rangle), \qquad (14)$$

where α is the rotating angle of Dove prism. In this equation, the orbital angular momentums of the generated photons are marked as (l_s, l_i) and (l'_s, l'_i). The four parameters are topological charges of corresponding photons. After they output from the NPC, we have $l_s+l_i=l_p+l_C$ and $l'_s+l'_i=l_p-l_C$. The two-photon coincidence could be measured utilizing a HOM interferometer. It is worth mentioning that the azimuthal index l of a reflective photon becomes $-l$. The coincidence rate when APD1 detects a photon with OAM $l_1\hbar$ and APD2 detects a photon with OAM $l_2\hbar$ is expressed as

$$R_C \propto \langle\Psi_f|b^\dagger_{l_1}b^\dagger_{l_2}b_{l_2}b_{l_1}|\Psi_f\rangle. \qquad (15)$$

In this equation, the subscript 1 and 2 correspond to the upper and lower paths after the beam splitter, respectively. The values of l_1 and l_2 are chosen from those of $l_s, -l'_s, l_i$, and $-l'_i$. The vector $|\psi_f\rangle$ corresponds to the state after the beam splitter. For any given value of l_p and l_C, the perfect two-photon coincidence appears only when the orbital angular momentum of down-converted photons satisfies that $l_s+l_i=-l'_s-l'_i$ and either $l_s=-l'_s$ & $l_i=-l'_i$ or $l_s=-l'_i$ & $l_i=-l'_s$. In this case, the terms $|1_{l_1},1_{l_2}\rangle$ and $|1_{l'_1},1_{l'_2}\rangle$ are

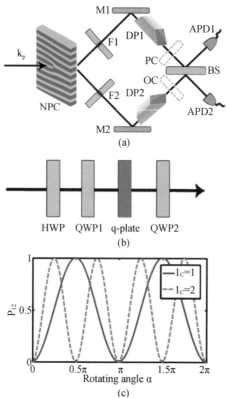

FIG.2. (a) The scheme of quantum coherence based on the N00N state with orbital angular momentum. The meanings of the corresponding symbols are listed as follows, F1 & F2: band-pass filters, M1 & M2: mirrors, DP1 & DP2: Dove prisms, BS: beam splitter, APD1 & APD2: avalanche photodiodes, OC(for $l_p \neq 0$): orbital angular momentum converter, PC(for $l_p \neq 0$): phase compensator. (b) The detailed construction of OC. HWP: half wave plate, QWP: quarter wave plate. (c) The probability of two-photon coincidence against the rotating angle α corresponding to a two-photon N00N state with $l_C = 1$ and $l_C = 2$.

indistinguishable for maximal interference, and it is convenient to calculate the corresponding probability of coincidence for each possible combination of l_s, l_i, l'_s and l'_i. In practical measurement, appropriate post-selections of OAM may be done according to the values of l_s, l_i, l'_s and l'_i to ensure maximal interference before detection. The approach is similar to that in Ref.[23]. For the other cases when the mentioned condition is not satisfied, the corresponding two terms are independent and there is no interference. Based on these analyses, detailed treatments could be divided into two cases:

i) $l_p = 0$, the condition of maximal interference is automatically satisfied. The probability of two-photon coincidence is

$$P_{12} = \frac{1}{2}[1 - \cos(2l_C \alpha)]. \tag{16}$$

ii) $l_p \neq 0$, the total orbital angular momentum of down-converted photons on the two paths is equal to $(l_p + l_C)\hbar$ and $(l_p - l_C)\hbar$, respectively. To achieve the maximal condition, an additional orbital angular momentum converter(OC) has to be used. We choose the well-known q-plate[47]. The typical design of OC is shown in Fig. 2(b). The half wave plate(HWP) and quarter wave plate(QWP) before the q-plate are utilized to transform the linearly polarized photons into right-handed circularly polarized photons. The q value is set at $l_p/2$. After passing through the q-plate, each photon obtains a part of extra orbital angular momentum of $-l_p\hbar$ and further becomes a left-handed circularly polarized photon. The total orbital angular momentum of the down-converted photons thus is changed into $[(l'_s-l_p)+(l'_i-l_p)]\hbar=[(l_p-l_C)-2l_p]\hbar=(-l_C-l_p)\hbar$. The QWP2 after the q-plate makes the photons linearly polarized along y-direction again. Moreover, a phase compensator(PC) should be placed in the upper path to balance the phase shift caused by the OC. The probability of coincidence also has the same expression as Eq.(16).

In Fig. 2(c), the two-photon coincidence probability is plotted against the rotating angle α. It is worth mentioning that the Dove prism influences the polarization state of photon. However, as is shown in Refs.[45], [46], the influence is quite low at short wavelength band. The effect of coherence is nearly unaffected.

In practical applications of N00N state, the photon number is preferred to be large. The precision of phase measurement and the resolution of quantum lithography are both proportional to $1/N$[35], [36]. However, increasing the photon number of N00N state brings great technical challenges in photonic state manipulation and detection. On the contrary, the measurement of angular displacement can be greatly benefited from our two-photon N00N state with orbital angular momentum, which is expressed as $(|2,0\rangle+e^{i2l_C\alpha}|0,2\rangle)/\sqrt{2}$, and the related limit of angular resolution is $1/2l_C$. With this state, increasing the photon number of N00N state can be avoided, while the estimation precision can be improved through increasing the value of l_C. It has been shown that OAM of light can be used to detect the structure and motion of rotating objects. Thus the quantum enhancement provided by the N00N state with OAM may find wide applications in remote sensing[48]–[50].

4. Conclusion

In summary, we investigate wavefront engineering of photon pairs generated through SPDC in domain-engineered $LiNbO_3$ crystals which is a kind of the well-known NPC. The photon interaction processes in $LiNbO_3$ crystals with complex domain patterns are described based on the nonlinear Huygens-Fresnel principle. Analytical expressions about the wavefront of photon pairs generated from arbitrary domain-engineered $LiNbO_3$ crystals are obtained. As a detailed illustration, the generation of N00N state with orbital angular momentum in a twisted $LiNbO_3$ crystal is presented. Such state can be utilized to increase the information capacity of quantum-secured communications[23] as well as to achieve the quantum-enhanced sensing[26]. Besides the azimuthal index(topological charge), the radial index may further expand the degrees of freedom in the system based on nonlinear optical interaction of Laguerre-Gaussian (LG) modes, which has been realized with lithium-niobate-based NPCs[51]. The flexible domain design in LiNbO$_3$ crystals provides a powerful platform for wavefront engineering of two-photon states, which may present a new sight for function-integrated quantum photonic applications.

References and Notes

[1] T. B. Pittman, B. C. Jacobs, and J. D. Franson, "Heralding single photons from pulsed parametric down-conversion," *Opt. Commun.*, vol. 246, pp. 545–550, Feb. 2005.

[2] E. Pomarico, B. Sanguinettil, C. I. Osorio, H. Herrmann, and R. T. Thew, "Engineering integrated pure narrow-band photon sources," *New. J. Phys.*, vol. 14, pp. 033008-1–033008-13, Mar. 2012.

[3] T. Jennewein, M. Barbieri, and A. G. Whiteb, "Single-photon device requirements for operating linear optics quantum computing outside the post-selection basis," *J. Mod. Opt.*, vol. 58, pp. 276–287, Feb. 2011.

[4] P. G. Kwiat, K. Mattle, H. Weinfurter, and A. Zeilinger, "New high-intensity source of polarization-entangled photon pairs," *Phys. Rev. Lett.*, vol. 75, pp. 4337–4341, Dec. 1995.

[5] H. Takesue and K. Inoue, "Generation of polarization-entangled photon pairs and violation of Bell's inequality using spontaneous four-wave mixing in a fiber loop," *Phys. Rev. A*, vol. 70, pp. 031802(R)-1–031802(R)-4, Sep. 2004.

[6] M. Hunault, H. Takeuse, O. Tadanaga, Y. Nishida, and M. Asobe, "Generation of time-bin entangled photon pairs by cascaded second-order non-linearity in a single periodically poled LiNbO$_3$ waveguide," *Opt. Lett.*, vol. 35, pp. 1239–1241, Apr. 2010.

[7] R. T. Horn *et al.*, "Inherent polarization entanglement generated from a monolithic semiconductor chip," *Sci. Rep.*, vol. 3, pp. 2314-1–2314-5, Jul. 2013.

[8] R. T. Thew, A. Acin, H. Zbinden, and N. Gisin, "Bell-type test of energy-time entangled qutrits," *Phys. Rev. Lett.*, vol. 93, pp. 010503-1–010503-4, Jul. 2004.

[9] J. L. Smirr *et al.*, "Optimal photon-pair single-mode coupling in narrow-band spontaneous parametric down-conversion with arbitrary pump profile," *J. Opt. Soc. Amer. B*, vol. 30, pp. 288–301, Feb. 2013.

[10] J. F. Dynes *et al.*, "Efficient entanglement distribution over 200 kilometers," *Opt. Exp.*, vol. 17, pp. 11440–11449, Jul. 2009.

[11] Y. Ming *et al.*, "Integrated source of tunable nonmaximally mode-entangled photons in a domain-engineered lithium niobate waveguide," *Appl. Phys. Lett.*, vol. 104, pp. 171110-1–171110-4, Apr. 2014.

[12] M. Fiorentino *et al.*, "Spontaneous parametric down-conversion in periodically poled KTP waveguides and bulk crystals," *Opt. Exp.*, vol. 15, pp. 7479–7488, Jun. 2007.

[13] V. Berger, "Nonlinear photonic crystals," *Phys. Rev. Lett.*, vol. 81, pp. 4136–4139, Nov. 1998.

[14] N. V. Bloch *et al.*, "Twisting light by nonlinear photonic crystals," *Phys. Rev. Lett.*, vol. 108, pp. 233902-1–233902-5, Jun. 2012.

[15] Y. Ming *et al.*, "Tailoring entanglement through domain engineering in a lithium niobate waveguide," *Sci. Rep.*, vol. 4, pp. 4812-1–4812-9, Apr. 2014.

[16] P. Xu and S. N. Zhu, "Review article: Quasi-phase-matching engineering of entangled photons," *AIP*

Adv., vol. 2, pp. 041401-1-041401-11, Dec. 2012.

[17] R. G. Batchko, V. Y. Shur, M. M. Fejer, and R. L. Byer, "Backswitch poling in lithium niobate for high-fidelity domain patterning and efficient blue light generation," *Appl. Phys. Lett.*, vol. 75, pp. 1673-1675, Sep. 1999.

[18] A. C. Busacca, S. Stivala, L. Curcio, and G. Assanto, "Parametric conver-sion in micrometer and submicrometer structured ferroelectric crystals by surface poling," *Int. J. Opt.*, vol. 2012, pp. 606892-1-606892-11, 2012.

[19] D. Yudistira *et al.*, "UV direct write metal enhanced redox(MER) domain engineering for realization of surface acoustic devices on lithium niobate," *Adv. Mater. Interfaces*, vol. 1, pp. 1400006-1-1400006-7, Jul. 2014.

[20] J. P. Torres, A. Alexandrescu, S. Carrasco, and L. Torner, "Quasi-phase-matching engineering for spatial control of entangled two-photon states," *Opt. Lett.*, vol. 29, pp. 376-378, Feb. 2004.

[21] D. Bonneau *et al.*, "Fast path and polarization manipulation of telecom wavelength single photons in lithium niobate waveguide devices," *Phys. Rev. Lett.*, vol. 108, pp. 053601-1-053601-5, Feb. 2012.

[22] C. H. Monken, P. H. Souto Ribeiro, and S. Padua, "Transfer of an-gular spectrum and image formation in spontaneous parametric down-conversion," *Phys. Rev. A*, vol. 57, pp. 3123-3126, Apr. 1998.

[23] A. Mair, A. Vaziri, G. Weihs, and A. Zeilinger, "Entanglement of the orbital angular momentum states of photons," *Nature*, vol. 412, pp. 313-316, Jul. 2001.

[24] A. Vaziri, G. Weihs, and A. Zeilinger, "Experimental two-photon, three-dimensional entanglement for quantum communication," *Phys. Rev. Lett.*, vol. 89, pp. 240401-1-240401-4, Dec. 2002.

[25] Y. Sun, Q. Y. Wen, and Z. Yuan, "High-efficient quantum key distribution based on hybrid entanglement," *Opt. Commun.*, vol. 284, pp. 527-530, Jan. 2011.

[26] N. Thomas-Peter *et al.*, "Real-world quantum sensors: Evaluating resources for precision measurement," *Phys. Rev. Lett.*, vol. 89, pp. 240401-1-240401-4, Dec. 2002.

[27] V. D'Ambrosio *et al.*, "Photonic polarization gears for ultra-sensitive angular measurements," *Nature Commun.*, vol. 4, pp. 2432-1-2432-8, Sep. 2013.

[28] T. B. Pittman *et al.*, "Two-photon geometric optics," *Phys. Rev. A*, vol. 53, pp. 2804-2815, Apr. 1996.

[29] A. F. Abouraddy, B. E. A. Saleh, A. V. Sergienko, and M. C. Teich, "Entangled-photon fourier optics," *J. Opt. Soc. Amer. B*, vol. 19, pp. 1174-1184, May 2002.

[30] Y. Q. Qin, C. Zhang, Y. Y. Zhu, X. P. Hu, and G. Zhao, "Wave-front engineering by Huygens-Fresnel principle for nonlinear optical interac-tions in domain engineered structures," *Phys. Rev. Lett.*, vol. 100, pp. 063902-1-063902-4, Feb. 2008.

[31] Z. Y. J. Ou, *Multi-Photon Quantum Interference*. New York, NY, USA: Springer, 2007.

[32] A. Bahabad and A. Arie, "Generation of optical vortex beams by nonlinear wave mixing," *Opt. Exp.*, vol. 15, pp. 17619-17624, Dec. 2007.

[33] G. Poberaj *et al.*, "Ion-sliced lithium niobate thin films for active photonic devices," *Opt. Mater.*, vol. 31, pp. 1054-1058, May 2009.

[34] H. Y. Leng *et al.*, "On-chip steering of entangled photons in nonlinear photonic crystals," *Nature Commun.*, vol. 2, pp. 429-1-429-5, Aug. 2011.

[35] M. W. Mitchell, J. S. Lundeen, and A. M. Steinberg, "Super-resolving phase measurements with a

multiphoton entangled state," *Nature*, vol. 429, pp. 161–164, May 2004.

[36] A. N. Boto et al., "Quantum interferometric optical lithography: Exploiting entanglement to beat the diffraction limit," *Phys. Rev. Lett.*, vol. 85, pp. 2733–2736, Sep. 2000.

[37] Z. Y. Ou, X. Y. Zou, L. J. Wang, and L. Mandel, "Experiment on nonclassical fourth-order interference," *Phys. Rev. A*, vol. 42, pp. 2957–2965, Sep. 1990.

[38] J. G. Rarity et al., "Two-photon interference in a Mach–Zehnder interfer-ometer," *Phys. Rev. Lett.*, vol. 65, pp. 1348–1351, Sep. 1990.

[39] E. Megidish et al., "Compact 2D nonlinear photonic crystal source of beamlike path entangled photons," *Opt. Exp.*, vol. 21, pp. 6689–6696, Mar. 2013.

[40] H. Jin et al., "Compact engineering of path-entangled sources from a monolithic quadratic nonlinear photonic crystal," *Phys. Rev. Lett.*, vol. 111, pp. 023603–1–023603–5, Jul. 2013.

[41] H. Kim, H. S. Park, and S. K. Choi, "Three-photon N00N states generated by photon subtraction from double photon pairs," *Opt. Exp.*, vol. 17, pp. 19720–19726, Oct. 2009.

[42] B. Y. Wei et al., "Generating switchable and reconfigurable optical vortices via photopatterning of liquid crystals," *Adv. Mater.*, vol. 26, pp. 1590–1595, Mar. 2014.

[43] S. Sharabi, N. Voloch-Bloch, I. Juwiler, and A. Arie, "Dislocation parity effects in crystals with quadratic nonlinear response," *Phys. Rev. Lett.*, vol. 112, pp. 053901–1–053901–5, Feb. 2014.

[44] G. H. Shao Z. J. Wu, J. H. Chen, F. Xu, and Y. Q Lu, "Nonlinear frequency conversion of fields with orbital angular momentum using quasi-phase-matching," *Phys. Rev. A*, vol. 88, pp. 063827–1–063827–7, Dec. 2013.

[45] M. J. Padgett and J. P. Lesso, "Dove prisms and polarized light," *J. Mod. Opt.*, vol. 46, pp. 175–179, Feb. 1999.

[46] I. Moreno, G. Paez, and M. Strojnik, "Polarization transforming properties of Dove prisms," *Opt. Commun.*, vol. 220, pp. 257–268, May 2003.

[47] L. Marrucci, C. Manzo, and D. Paparo, "Optical spin-to-orbital angular momentum conversion in inhomogeneous anisotropic media," *Phys. Rev. Lett.*, vol. 96, pp. 163905–1–163905–4, Apr. 2006.

[48] M. P. J. Lavery, F. C. Speirits, S. M. Barnett, and M. J. Padgett, "Detection of a spinning object using light's orbital angular momentum," *Science*, vol. 341, pp. 537–540, Aug. 2013.

[49] F. Tamburini, B. Thidé, G. Molina-Terriza, and G. Anzolin, "Twisting of light around rotating black holes," *Nature Phys.*, vol. 7, pp. 195–197, Mar. 2011.

[50] M. Harwit, "Photon orbital angular momentum in astrophysics," *Astro-phys. J.*, vol. 597, pp. 1266–1270, Nov. 2003.

[51] K. Shemer et al., "Azimuthal and radial shaping of vortex beams generated in twisted nonlinear photonic crystals," *Opt. Lett.*, vol. 38, pp. 5470–5473, Dec. 2013.

[52] This work was supported by 973 Programs Nos. 2011CBA00205 and 2012CB921803, by the National Science Fund of China under Grants 61225026, 61322503, and 61490714, and by the Program for Changjiang Scholars and Innovative Research Team at the University under Contract IRT13021.

Tailoring Entanglement Through Domain Engineering in a Lithium Niobate Waveguide[*]

Yang Ming[1,3], Ai-Hong Tan[2], Zi-Jian Wu[1,3], Zhao-Xian Chen[1,3], Fei Xu[1,3] & Yan-Qing Lu[1,3]

[1] *National Laboratory of Solid State Microstructures and College of Engineering and Applied Sciences, Nanjing University, Nanjing 210093, China*

[2] *Laboratory for Quantum Information, China Jiliang University, Hangzhou 310018, China*

[3] *National Center of Microstructures and Quantum Manipulation, Nanjing University, Nanjing 210093, China*

We propose to integrate the electro-optic (EO) tuning function into on-chip domain engineered lithium niobate(LN) waveguide. Due to the versatility of LN, both the spontaneously parametric down conversion(SPDC) and EO interaction could be realized simultaneously. Photon pairs are generated through SPDC, and the formation of entangled state is modulated by EO processes. An EO tunable polarization-entangled photon state is proposed. Orthogonally-polarized and parallel-polarized entanglements of photon pairs are instantly switchable by tuning the applied field. The characteristics of the source are theoretically investigated showing adjustable bandwidths and high entanglement degrees. Moreover, other kinds of reconfigurable entanglement are also achievable based on suitable domain-design. We believe tailoring entanglement based on domain engineering is a very promising solution for next generation function-integrated quantum circuits.

Entanglement is a crucial physical resource for quantum information science and technology. In practical investigations and applications, entanglement systems based on photons are widely used owing to the extra long decoherence time. To perform various specific tasks, different kinds of entanglements are required, such as multi-particle entanglement, nonmaximally entanglement, and mixed state entanglement[1–7]. For the generation of entangled photons, spontaneous parametric down conversion(SPDC) in $\chi^{(2)}$ nonlinear crystals is one of the most powerful tools[8]. In SPDC process, a single photon is transformed into a photon pair. However, in previous works[1–4], it is usually necessary to design complicated optical paths to transform the initially generated states into target entangled states. Numerous large-scale bulk optical elements have to be fixed onto sizeable optical benches in the laboratory, such as lenses, wave plates and filters[9,10]. The physical size and the inherent instability of these components hinder the development of more complex schemes, and bring significant challenges for the integration of quantum circuits,

---

[*] Sci.Rep.,2014,4:4812

which needs to be improved in practical applications[11,12].

In this article, we propose a different strategy for on-chip tailoring entanglement through domain engineering in a lithium niobate (LN) waveguide. LN is a typical ferroelectric nonlinear optical material, which is simultaneously available for SPDC[13] and electro-optic(EO) effect[14—16]. Due to its quite large effective nonlinear coefficient, LN could serve as entangled photon-pair source of high quality[17,18]. What's more, EO modulation is an effective way to manipulate entangled photons. Previous investigations have demonstrated the control of phase[19] or polarization state[14]. An integrated quantum relay operator has been established based on EO effect of single domain LN crystals with some special electrode design[15]. In fact, at the quantum level, EO could be regarded as a two-photon interaction process, in which the quantum states of photons could exchange. In this work, we derive relevant theories to describe this process. Combining the theories for EO and SPDC, entangled states of desired formations could be designed flexibly. Through suitable domain engineering, which is widely used in quasi phase matching (QPM) nonlinear optics[9], the EO and SPDC processes could be effectively combined to generate the needed entangled states. Moreover, LN is also a good platform to realize versatile functions including efficient infrared photon detection[20], negative permittivity[21], EO quantum logic gate[22] and lensless ghost imaging[23]. All of them could be integrated together toward future practical large scale quantum circuit integration.

Following, as an illustration of our entanglement design approach, we establish an EO tunable polarization-entangled photon pair source in a domain engineered LN waveguide with suitable artificial structures. When a suitable voltage is applied, the source produces a pair of entangled photons contains the same polarization, with either o-or e-polarization state. The notations o-and e-represent ordinary and extraordinary light in uniaxial anisotropy crystal, respectively. In contrast, if the voltage is turned off, the entangled photons bear orthogonal polarization states, i.e., one photon is o-polarized while another one has e-polarization. Since we may control the polarization state intentionally, the formations of entangled states thus could be regulated accordingly, which possess considerable potential in modulation of multi-photon entangled states.

Results

Tailoring entanglement in a domain engineered LN waveguide. In recent works[24—26], integrated quantum circuits established on silicon waveguides are adequately investigated. Utilizing the femtosecond laser waveguide writing technology, on-chip devices to support and manuscript polarization-encoded qubits have been fabricated[27,28]. A controlled-NOT gate has been experimental demonstrated with switchable entanglement. It is no doubt that silicon waveguides have excellent capabilities to confine and propagate light. However, limited by the inherent properties of silicon, they are not very convenient for fast manipulating and modulating entangled photons; such as the modulation scheme based on thermo-optical effect[25], the response time still has much room to improve in practical applications. In addition, to realize multifunctional integrated quantum circuits, entanglements of complicated formations are normally required such as the nonmaximally entanglement[3,5] and mixed state entanglement[4,6,7]. Under these circumstances, the LN-based system is considered as an alternative choice. Besides excellent nonlinear optical properties, EO effect in LN material provides an effective way to change the quantum state of photons through the photon interaction process. For a LN crystal, if we set its symmetry axis as the z-axis and apply a voltage along the y-axis, a variation is led into the dielectric constant. It could be written as $\varepsilon = \varepsilon(0) + \Delta\varepsilon$ with $\Delta\varepsilon_{jk} = -\varepsilon_0 \gamma_{51} E_a n_o^2 n_e^2 (jk = 23, 32)$[16]. In the equation, γ_{51} is the effective EO coefficient, and E_a represents the applied electric field. The refractive indices of ordinary and extraordinary light are marked as n_o and n_e. This additional portion of permittivity corresponds to an equivalent polarization item, which is expressed as $P_j = -\sum_k \varepsilon_0 \gamma_{51} E_a n_o^2 n_e^2 \Delta_{jk} E_k$ with $\Delta_{jk} = \begin{cases} 1, jk=(23,32) \\ 0, \text{else} \end{cases}$. From $H_I = \frac{1}{2}\int_V d^3 r P \cdot E$, we obtain the interaction Hamiltonian for EO process as

$$H_I = \frac{\hbar \varepsilon_0 \gamma_{51} E_a}{4}\left[\int_{-L}^{0} dx f(x) e^{i(\beta_e - \beta_o)x} \iint d\omega d\omega' F_{EO} \sqrt{\frac{\omega\omega' n_o^2 n_e^2}{N_o N_e}} a_o^\dagger a_e e^{-i(\omega - \omega')t} + \int_{-L}^{0} dx f(x) e^{i(\beta_o - \beta_e)x} \iint d\omega d\omega' F_{EO} \sqrt{\frac{\omega\omega' n_o^2 n_e^2}{N_o N_e}} a_e^\dagger a_o e^{-i(\omega' - \omega)t}\right] \quad (1)$$

with

$$F_{EO} = \iint dy dz \vartheta_o(r)\vartheta_e(r). \quad (2)$$

In the equation, the quantized expression of electric field is utilized. F_{EO} is the overlap integral of waveguide modes expressed as $\vartheta(r)$. The function $f(x)$ represents the structures of the inverted domains. The EO coefficient is periodically modulated, thus the o- and e-beams are coupled due to the periodic index ellipsoid deformation. The phase matching condition should be satisfied between the ordinary and extraordinary photons to ensure high conversion efficiency. The first item of the equation represents the annihilation of an e-polarized photon and the creation of an o-polarized photon, while the second corresponds to the inverse process. Accompanied by the two photon interaction process, the old photon state is substituted by new state which is desired to form target entangled state. Choosing proper EO processes to modulate entangled photon states supplies an innovative approach for entanglement architectures.

As a simple illustration, the polarization qubit of photon is considered. Assuming H_I is acted on a photon with polarization state $|o\rangle$, it would be transformed into polarization state $|e\rangle$ with a certain possibility. Referring to the corresponding classical situation[16], we could obtain a superposition of $|o\rangle$ and $|e\rangle$ through modulating the applied voltage.

The ratio of these two portions is determined by the value of applied voltage. The overall effect could be equivalent to $H_{EO} = R(\alpha)H_I$, where R is a standard rotation operator, and α is the equivalent rotation angle. For a biphoton state, the modulation process could be written as

$$|\Psi_f\rangle = H_{EO1} \otimes H_{EO2} |\Psi_i\rangle \tag{3}$$

with the matrix expression of the operator as

$$\begin{bmatrix} \sin\theta\sin\phi & -\sin\theta\cos\phi & -\cos\theta\sin\phi & \cos\theta\cos\phi \\ -\sin\theta\cos\phi & -\sin\theta\sin\phi & \cos\theta\cos\phi & \cos\theta\sin\phi \\ -\cos\theta\sin\phi & \cos\theta\cos\phi & -\sin\theta\sin\phi & \sin\theta\cos\phi \\ \cos\theta\cos\phi & \cos\theta\sin\phi & \sin\theta\cos\phi & \sin\theta\sin\phi \end{bmatrix}, \tag{4}$$

where θ and φ are the corresponding equivalent rotation angles to determine the relative ratio of the two photon vector states, which are controlled by the applied voltage. It works similar to a half waveplate (HWP). However, there are still apparent differences. Assuming the input light is a linearly polarized light, a circular polarized beam is obtained if the wave plate thickness is only half. In contrast, the light passing through a PPLN with half-length is still linear-polarized, while the orientation angle is also half if the applied field is the same.

For universal multiphoton entangled state, they could not be generated through SPDC directly, so the postprocessing modulations are quite critical. In multiphoton entanglement situation, only certain formations may lead to potential applications. We need to choose proper photons from the initial state $F\{|\Psi_1\rangle;|\Psi_2\rangle\cdots|\Psi_N\rangle\}$, where $\{|\Psi_n\rangle\}$ represents the photon pairs from SPDC, and the function $F(|\Psi\rangle)$ corresponds to their initial combination formation. After suitable modulations are applied based on the effective H_{EO}, the state thus could be transformed into a valuable formation such as GHZ state or cluster state. This process could be expressed as

$$|\Psi_f\rangle = \bigotimes_{|\Psi_{k_i}\rangle \in \{|\Psi_n\rangle\}}^{kN} H_{EO}^{k_i} F\{|\Psi_1\rangle;|\Psi_2\rangle\cdots|\Psi_N\rangle\}, \tag{5}$$

where $\otimes$ refers to the direct product of all $H_{EO}^{k_i}$.

After the corresponding scheme is decided, the ferroelectric domains of the LN waveguide could be well-designed to satisfy the phase matching conditions of the related SPDC and EO processes. Through analyses of the set of $\{\Delta k_n\}$, the matching vectors could be obtained as $\{2\pi/\Lambda_m\}$. Accordingly, the corresponding Fourier coefficients and the Fourier bases could be derivate. Based on these, the original function is recovered through

$$f(\boldsymbol{r}) = \sum_m F_m e^{iK_m \cdot r}, \tag{6}$$

where $f(r)$ represents the actual domain arrangement of the entangled photon-pair source. It is worth mentioning that the correspondence between $\{\Delta k_n\}$ and $\{2\pi/\Lambda_m\}$ does not have to be one-to-one. The choice of $\{\Lambda_m\}$ is flexible, which is an important part of domain engineering.

EO tunable polarization-entangled photon state. As a detailed illustration, we consider the generation of an EO tunable polarization-entangled photon state with switchable characteristic. We consider the processes of photon interactions in a z-cut, x-propagating titanium in-diffused LN waveguide. There are well defined transverse electric (TE) and transverse magnetic (TM) propagation modes in the waveguide at near-infrared frequencies[29]. In this work, only the fundamental modes of the TE and TM modes are discussed. For convenience, they are marked as o-polarized and e-polarized, respectively[30]. Three photon interaction processes are included. They are two SPDC processes, which are expressed as $o_p \to o_s + e_i, o_p \to e_s + o_i$ (p, s and i represent the pump, signal, and idler waves. o and e correspond to ordinary and extraordinary polarization); and an EO process takes place for the signal wave, namely $o_s \to e_s (e_s \to o_s)$. If there is no voltage applied, only the SPDC processes happen in the waveguide. The power of the pump light is launched into the downconversion photons in corresponding propagation modes of the waveguide, the generated entangled state is type-II-like with the formation $|o_s, e_i\rangle + |e_s, o_i\rangle$. In contrast, if there is an applied voltage, the EO effect is activated. The EO polarization rotation and SPDC processes happen simultaneously and are coupled everywhere in the crystal. In this case, the down-converted photons are equally generated everywhere and affected by the locally EO perturbed refractive index ellipsoid. The final obtained polarization state, including, e_s, o_i, e_i and o_s are the superposition result of all "sub-sources" inside the crystal. Therefore the rotation of polarization is realized through coherent superposition of all kinds of photons randomly collected in the sample. Corresponding to the coupling wave theory, that means the power of the fundamental TE mode(o_s) and the TM mode(e_s) at the signal frequency exchanges with each other. For a given sample length, the efficiency is dependent on the interaction strength, which is proportionate to the applied voltage. We could always find a proper voltage to make the efficiency reach ~100%, then the entangled state is switched to be type-I-like with the formation $|e_s, e_i\rangle + |o_s, o_i\rangle$. Thus the entangled state is switchable through controlling the applied voltage on and off.

To satisfy phase matching conditions of the three interaction processes, the domain structures need to be well designed. General speaking, we may design three sets of domain periods so that their reciprocal vectors are able to compensate these three wave vector mismatches. However, this straightforward scheme normally induces very thin domains and complicated structures. To solve this problem, several approaches have been proposed. For example, for the optimized wavelengths and domain structures, the needed domain periods could be reduced to two and they are of integer ratio, so that the domain structure could be easily fabricated. The schematic is shown in Fig. 1. Correspondingly, the sign of nonlinear coefficient $\chi^{(2)}$ and EO coefficient γ change periodically. The modulation functions are expressed as $d(x) = d_{31} f_1(x) f_2(x)$ and $\gamma(x) = \gamma_{51} f_1(x) f_2(x)$[30], where $f(x) = \text{sign}[\cos(2\pi/\Lambda)] = \sum_m G_m exp(ik_m x)$. In the equations, d is the substitute of nonlinear coefficient $\chi^{(2)}$ with the relationship $d=\chi^{(2)}/2$. The effective component of d and γ in our situation are d_{31} and γ_{51}, respectively. Thus the reciprocal vectors are given as $k_m=2m\pi/\Lambda$, with their corresponding Fourier coefficients as $G_m=(2/m\pi)\sin(m\pi/2)$. As a consequence, the expansion formulation of the coefficients are obtained as[31]

$$d(x)=d_{31}\sum_{m,n}G_{m,n}e^{iK_{m,n}x}$$
$$\gamma(x)=\gamma_{51}\sum_{m,n}G_{m,n}e^{iK_{m,n}x} \quad (7)$$

with

$$G_{m,n}=\frac{4}{mn\pi^2}\sin\left(\frac{m\pi}{2}\right)\sin\left(\frac{n\pi}{2}\right)$$
$$K_{m,n}=\frac{2m\pi}{\Lambda_1}+\frac{2n\pi}{\Lambda_2}, \quad (8)$$

where Λ_1 and Λ_2 represents the corresponding two modulation periods. Based on these analyses, the equations of QPM conditions could be expressed as

$$\Delta\beta_1=\beta_{p,o}-\beta_{s,o}-\beta_{i,e}=K_{m_1,n_1}$$
$$\Delta\beta_2=\beta_{p,o}-\beta_{s,e}-\beta_{i,o}=K_{m_2,n_2}.$$
$$\Delta\beta_3=\beta_{s,o}-\beta_{s,e}=K_{m_3,n_3} \quad (9)$$

In these equations, $\beta_{j,\sigma}$ ($j=p,s,i;\sigma=o,e$) refers to the propagation constant of the corresponding waveguide mode, which is obtained based on the Hermite-Gauss formulations[29]. $\Delta\beta_1$ and $\Delta\beta_2$ correspond to two SPDC processes, while $\Delta\beta_3$ corresponds to the EO interaction. K_{m_i,n_i} ($i=1, 2, 3$) are the required reciprocal vectors to compensate the phase mismatches.

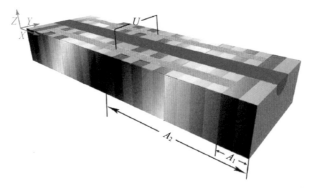

FIG.1. Schematic of a dual-PPLN waveguide. The light and dark blue portions represent the positive and negative domains of the PPLN, respectively; while the purple portion is the core of the waveguide. The two golden strips correspond to the electrodes. Λ_1 and Λ_2 are the two periods of the structure. Moreover, the polarization directions of o-and e-polarized photons are corresponding to y-and z-directions, respectively.

Through detailed calculations, we find an appropriate set of solutions for Eq.(9). The width and depth of the waveguide core are both set at 10 μm, and the maximum of index difference is set at 0.003. The pump, signal and idler wavelengths are 0.7335 μm, 1.6568 μm and 1.3162 μm, respectively. As room temperature operation is always more appreciated for future quantum circuits, the simulation temperature is chosen at 25 ℃. The photorefractive effect might bring some influences. However, the photorefractive damage of LN is not a simple process. It depends on many factors such as laser pulse width, operation wavelength and surface quality. Through recent technologies[32], the damage threshold has been increased to an acceptable degree for many applications. Moreover, the MgO doping is also an effective way to further avoid crystal damage[33,34]. In this situation, the corresponding values for the (m, n) series are $\{(m_1, n_1)=(3, 1), (m_2, n_2)=(3, -1), (m_3, n_3)=(1, 1)\}$. The dual-modulation periods are obtained at $\Lambda_1 = 25.84$ μm and $\Lambda_2 = 154.96$ μm, respectively. The ratio of these two periods is $\Lambda_2/\Lambda_1 = 6$ and the duty cycle is 0.5, as is shown in Fig. 1. (More details about the selecting process could be referred to the Methods section).

To give a quantum mechanics description of our tunable entangled photon-pair source, we derive the state vector of entangled photons through the effective Hamiltonian. The total interaction Hamiltonian consists of two parts, which is expressed as $H_I = H_{SPDC} + H_{EO}$. H_{SPDC} corresponds to the SPDC processes, while H_{EO} refers to the EO interaction. Both of them arise from polarization, namely, P_{NL} and P_{EO}. In detailed treatments, the pump field is usually treated as an undepleted classical wave. The signal and idler fields are quantized and represented by field operators. Thus the formulations of the pump, signal and idler fields could be written as

$$E_p^{(+)} = E_{po} \vartheta_{po}(\boldsymbol{r}) e^{i(\beta_{po} x - \omega_p t)}$$

$$E_s^{(-)} = i \sum_\sigma \int d\omega_s \sqrt{\frac{\hbar \omega_s}{2\varepsilon_0 n_{s\sigma}^2 N_{s\sigma}}} \vartheta_{s\sigma}(\boldsymbol{r}) a_{s\sigma}^\dagger e^{-i(\beta_{s\sigma} x - \omega_s t)},$$

$$E_i^{(-)} = i \sum_\sigma \int d\omega_i \sqrt{\frac{\hbar \omega_i}{2\varepsilon_0 n_{i\sigma}^2 N_{i\sigma}}} \vartheta_{i\sigma}(\boldsymbol{r}) a_{i\sigma}^\dagger e^{-i(\beta_{i\sigma} x - \omega_i t)} \quad (10)$$

where σ represents the polarization state o or e, and $n_{j\sigma}$ ($j = s, i$) is the corresponding refractive index. $N_{s\sigma}$ and $N_{i\sigma}$ are normalization parameters. $\vartheta_{po}(\boldsymbol{r})$, $\vartheta_{s\sigma}(\boldsymbol{r})$ and $\vartheta_{i\sigma}(\boldsymbol{r})$ correspond to the transverse mode profiles of pump, signal and idler field, respectively.

For SPDC processes, if using the rotating wave approximation, H_{SPDC} is derived as

$$H_{SPDC} = \frac{1}{2} \int_V d^3r P_{NL} \cdot E = \varepsilon_0 \int_V d^3r d(x) E_p^{(+)} E_s^{(-)} E_i^{(-)} + \mathrm{H.C.}$$

$$= -\frac{\hbar E_{po}}{2} \sum_{m,n} d_{31} G_{m,n} \left[\int_{-L}^{0} dx\, e^{i(\beta_{po} - \beta_{so} - \beta_{ie} + K_{m,n})x} \right.$$

$$\times \iint d\omega_s d\omega_i F_{oe} \sqrt{\frac{\omega_s \omega_i}{n_{so}^2 n_{ie}^2 N_{so} N_{ie}}} a_{so}^\dagger a_{ie}^\dagger e^{-i(\omega_p - \omega_s - \omega_i)t}$$

$$+ \int_{-L}^{0} dx\, e^{i(\beta_{po} - \beta_{se} - \beta_{io} + K_{m,n})x} \iint d\omega_s d\omega_i F_{eo} \sqrt{\frac{\omega_s \omega_i}{n_{se}^2 n_{io}^2 N_{se} N_{io}}} a_{se}^\dagger a_{io}^\dagger e^{-i(\omega_p - \omega_s - \omega_i)t} + \mathrm{H.C.} \Bigg]$$

$$(11)$$

with
$$F_{oe} = \iint dy\,dz\,\vartheta_{po}(\mathbf{r})\vartheta_{so}(\mathbf{r})\vartheta_{ie}(\mathbf{r})$$
$$F_{eo} = \iint dy\,dz\,\vartheta_{po}(\mathbf{r})\vartheta_{se}(\mathbf{r})\vartheta_{io}(\mathbf{r}) \tag{12}$$

For the EO process, the interaction Hamiltonian is derivate as above. H_{EO} is expressed as

$$H_{EO} = \frac{\hbar \varepsilon_0}{4} \sum_{m,n} \gamma_{51} E_a G_{m,n} \left[\int_{-L}^{0} dx\, e^{i(\beta_{s'e}-\beta_{so}+K_{m,n})x} \times \right.$$
$$\iint d\omega_s\,d\omega_{s'} F_{EO} \sqrt{\frac{\omega_s \omega_{s'} n_{so}^2 n_{s'e}^2}{N_{so} N_{s'e}}} a_{so}^\dagger a_{s'e} e^{-i(\omega_s - \omega_{s'})t} +$$
$$\int_{-L}^{0} dx\, e^{i(\beta_{so}-\beta_{s'e}+K_{m,n})x} \times$$
$$\left. \iint d\omega_s\,d\omega_{s'} F_{EO} \sqrt{\frac{\omega_s \omega_{s'} n_{so}^2 n_{s'e}^2}{N_{so} N_{s'e}}} a_{s'e}^\dagger a_{so} e^{-i(\omega_{s'}-\omega_s)t} \right] \tag{13}$$

with
$$F_{EO} = \iint dy\,dz\,\vartheta_{so}(\mathbf{r})\vartheta_{se}(\mathbf{r}). \tag{14}$$

The entangled state vector could be obtained through $|\Psi(t)\rangle = \exp[(-i/\hbar)\int dt(H_{SPDC} + H_{EO}(E_a))]|0\rangle$. Detailed treatments thus could be divided into two cases:

i) when there is no applied voltage, we expand the evolution operator to the first-order perturbation, so the state vector is expressed as

$$|\Psi\rangle = \left[1 - \frac{i}{\hbar}\int_{-\infty}^{\infty} dt\, H_{SPDC}(t)\right]|0\rangle$$
$$= |0\rangle + \frac{iE_{po}}{2}\sum_{m,n} d_{31} G_{m,n} \iint d\omega_s\,d\omega_i \left[F_{oe}\sqrt{\frac{\omega_s \omega_i}{n_{so}^2 n_{ie}^2 N_{so} N_{ie}}} \int_{-L}^{0} dx\, e^{i(\beta_{po}-\beta_{so}-\beta_{ie}+K_{m,n})x} \times \right.$$
$$\int_{-\infty}^{\infty} dt\, e^{-i(\omega_p - \omega_s - \omega_i)t} a_{so}^\dagger a_{ie}^\dagger + F_{eo}\sqrt{\frac{\omega_s \omega_i}{n_{se}^2 n_{io}^2 N_{se} N_{io}}} \int_{-L}^{0} dx\, e^{i(\beta_{po}-\beta_{se}-\beta_{io}+K_{m,n})x} \times$$
$$\left. \int_{-\infty}^{\infty} dt\, e^{-i(\omega_p - \omega_s - \omega_i)t} a_{se}^\dagger a_{io}^\dagger \right]|0\rangle \tag{15}$$

In the equation, we have $\int dt\, e^{-i(\omega_p-\omega_s-\omega_i)t} = 2\pi\delta(\omega_p - \omega_s - \omega_i)$ and $\int dx\, e^{i\Delta\beta x} = h(L\Delta\beta)$ with the formation $h(x) = e^{ix/2}\,\text{sinc}(x/2)$. Moreover, the actual SPDC process doesn't merely arise at the perfect phase-matching frequencies Ω_s and Ω_i. As the natural bandwidth is ν, we set $\omega_s = \Omega_s + \nu$ and $\omega_i = \Omega_i - \nu$, then the state vector is simplified as

$$|\Psi\rangle = |0\rangle + \int d\nu P_{oe} h(L\Delta\beta_{ooe}) a_{so}^\dagger a_{ie}^\dagger |0\rangle + \int d\nu P_{eo} h(L\Delta\beta_{oeo}) a_{se}^\dagger a_{io}^\dagger |0\rangle \tag{16}$$

with

$$P_{oe} = i\pi L E_{po} d_{31} G_{3,1} F_{oe} \sqrt{\frac{\Omega_s \Omega_i}{n_{so}^2 n_{ie}^2 N_{so} N_{ie}}}$$
$$P_{eo} = i\pi L E_{po} d_{31} G_{3,-1} F_{eo} \sqrt{\frac{\Omega_s \Omega_i}{n_{se}^2 n_{io}^2 N_{se} N_{io}}}. \tag{17}$$

In the equation, the phase mismatch terms could be expressed as $\Delta\beta_{ooe} = \beta_{p,o} - \beta_{s,o} - \beta_{i,e}$ and $\Delta\beta_{oeo} = \beta_{p,o} - \beta_{s,e} - \beta_{i,o}$. As $|\nu| \ll \Omega_s, \Omega_i$, we have substituted $\omega_s(\omega_i)$ by $\Omega_s(\Omega_i)$ in P_{oe} and P_{eo}. For the second case, the voltage E_a is applied. The evolution operator should be expanded to the second-order term. The formulation of the state vector could be expressed as

$$|\Psi\rangle = \left[1 + \left(-\frac{i}{\hbar}\right)^2 \int_{-\infty}^{\infty} dt_2 \int_{-\infty}^{\infty} dt_1 T(H_{EO}(t_2) H_{SPDC}(t_1))\right]|0\rangle$$

$$= |0\rangle + \frac{\varepsilon_0 d_{31}\gamma_{51} E_{po} E_a L^2}{8} \left\{ \iint d\omega_s d\omega_{s'} \int_{-\infty}^{\infty} dt_2 e^{-i(\omega_s - \omega_{s'})t_2} F_{EO} \left[G_{1,1} h(L\Delta\beta_{eo}) \sqrt{\frac{\omega_s \omega_{s'} n_{so}^2 n_{s'e}^2}{N_{so} N_{s'e}}} a_{so}^\dagger a_{s'e} \right. \right.$$

$$\left. + G_{1,1} h(L\Delta\beta_{oe}) \sqrt{\frac{\omega_s \omega_{s'} n_{so}^2 n_{s'e}^2}{N_{so} N_{s'e}}} a_{s'e}^\dagger a_{so} \right] \right\} \left\{ \iint d\omega_s d\omega_i \int_{-\infty}^{\infty} dt_1 e^{-i(\omega_p - \omega_s - \omega_i)t_1} \right.$$

$$\left. \times \left[G_{3,1} h(L\Delta\beta_{oee}) F_{oe} \sqrt{\frac{\omega_s \omega_i}{n_{so}^2 n_{ie}^2 N_{so} N_{ie}}} a_{so}^\dagger a_{ie}^\dagger + G_{3,-1} h(L\Delta\beta_{oeo}) F_{eo} \sqrt{\frac{\omega_s \omega_i}{n_{se}^2 n_{io}^2 N_{se} N_{io}}} a_{se}^\dagger a_{io}^\dagger \right] \right\} |0\rangle. \tag{18}$$

In the equation, T is the time-ordering operator. The notation $\Delta\beta_{oe}$ corresponds to the mismatch of $\beta_{s,o}$ and $\beta_{s,e}$. The functions $h(L\Delta\beta_{ooe})$ and $h(L\Delta\beta_{oeo})$ are spatial integration for SPDC, while $h(L\Delta\beta_{oe})$ and $h(L\Delta\beta_{eo})$ corresond to the EO processes. There are four product terms of effective operators, including $a_{so}^\dagger a_{se} a_{se}^\dagger a_{io}^\dagger, a_{se}^\dagger a_{so} a_{so}^\dagger a_{ie}^\dagger, a_{so}^\dagger a_{se} a_{so}^\dagger a_{ie}^\dagger$ and $a_{se}^\dagger a_{so} a_{se}^\dagger a_{io}^\dagger$. The first two correspond to actual processes. If they act on the vacuum state, the results show $a_{so}^\dagger a_{se} a_{se}^\dagger a_{io}^\dagger |0\rangle = (a_{so}^\dagger|0_{so}\rangle)(a_{se}a_{se}^\dagger|0_{se}\rangle)(a_{io}^\dagger|0_{io}\rangle) = |1_{so}, 0_{se}, 1_{io}\rangle$ and $a_{se}^\dagger a_{so} a_{so}^\dagger a_{ie}^\dagger|0\rangle = (a_{se}^\dagger|0_{se}\rangle)(a_{so}a_{so}^\dagger|0_{so}\rangle)(a_{ie}^\dagger|0_{ie}\rangle) = |1_{se}, 0_{so}, 1_{ie}\rangle$. For the last two terms, they don't affect the final entangled state because as $a_{se}|0\rangle = 0$ and $a_{so}|0\rangle = 0$. Therefore the parallel-polarization entangled state is written as

$$|\Psi\rangle = |0\rangle + \int d\nu \int d\nu' P_{oo} h(L\Delta\beta_{eo}) h(L\Delta\beta_{oeo}) a_{so}^\dagger a_{io}^\dagger |0\rangle$$

$$+ \int d\nu \int d\nu' P_{ee} h(L\Delta\beta_{oe}) h(L\Delta\beta_{ooe}) a_{se}^\dagger a_{ie}^\dagger |0\rangle \tag{19}$$

with

$$P_{oo} = \frac{\pi^2 \varepsilon_0^2}{2} d_{31} \gamma_{51} \Omega_s E_{po} E_a G_{1,1} G_{3,-1} F_{EO} F_{eo} L^2 \sqrt{\frac{n_{so}^2 \Omega_s \Omega_i}{n_{io}^2 N_{se}^2 N_{io} N_{so}}}$$

$$P_{ee} = \frac{\pi^2 \varepsilon_0^2}{2} d_{31} \gamma_{51} \Omega_s E_{po} E_a G_{1,1} G_{3,1} F_{EO} F_{oe} L^2 \sqrt{\frac{n_{se}^2 \Omega_s \Omega_i}{n_{ie}^2 N_{so}^2 N_{ie} N_{se}}} \tag{20}$$

Similarly, we also set $\omega_s = \Omega_s + \nu$ and $\omega_i = \Omega_i - \nu$, while $\omega_s(\omega_i)$ is substituted by $\Omega_s(\Omega_i)$ in P_{oo} and P_{ee}. Summarizing the two situations above, we rewrite the entangled state vector in a terser formation:

$$|\Psi\rangle = |0\rangle + \left(\int_{0^-}^{0^+} \delta(E_a) dE_a\right) \left[\int d\nu P_{oe} h(L\Delta\beta_{ooe}) a_{so}^\dagger a_{ie}^\dagger |0\rangle + \int d\nu P_{eo} h(L\Delta\beta_{oeo}) a_{se}^\dagger a_{io}^\dagger |0\rangle\right]$$

$$+ \left(1 - \int_{0^-}^{0^+} \delta(E_a) dE_a\right) \times \left[\int d\nu \int d\nu' P_{oo} h(L\Delta\beta_{eo}) h(L\Delta\beta_{oeo}) a_{so}^\dagger a_{io}^\dagger |0\rangle\right.$$

$$\left. + \int d\nu \int d\nu' P_{ee} h(L\Delta\beta_{oe}) h(L\Delta\beta_{ooe}) a_{se}^\dagger a_{ie}^\dagger |0\rangle\right]. \tag{21}$$

In the equation, $\delta(E_a)$ represents the Dirac-δ function. When $E_a = 0$, we have $\int_{0-}^{0+} \delta(E_a) dE_a = 1$. The generated entangled state corresponds to $C_{oe}|o\rangle_s|e\rangle_i + C_{eo}|e\rangle_s|o\rangle_i$. If E_a is an appropriate nonzero value, the value of $\left(1 - \int_{0-}^{0+} \delta(E_a) dE_a\right)$ should be 1, the entangled state vector is $C_{oo}|o\rangle_s|o\rangle_i + C_{ee}|e\rangle_s|e\rangle_i$. To ensure the EO transformation of the downconverted photons, several calculations have been done. The crystal length is set at 3 cm, and the average pump power is set at 1 W. Based on the coupled wave equations for EO interaction[16] and those for downconversion process, we obtain that when E_a is 3.8 $\times 10^5$ V/m (Usually, the distance between the electrodes used for the LN waveguide is 10~1000 μm, the corresponding applied voltage is about 1 ~ 100 V)[35], the transformation efficiencies are $P_{se}/(P_{se} + P_{so}) = 0.9927$ (for $|o\rangle_p \to |o\rangle_s|e\rangle_i \to |e\rangle_s|e\rangle_i$) and $P_{so}/(P_{se} + P_{so}) = 0.9932$ (for $|o\rangle_p \to |e\rangle_s|o\rangle_i \to |o\rangle_s|o\rangle_i$), respectively. As for the idler photon, the polarization state is hardly rotated. That's because the PPLN's EO effect is quite special with wavelength selectivity. Only lights at the designed discrete wavelengths are affected. The corresponding wavelength bandwidth also could be predesigned. For a direct verification, the transformation efficiency of polarization of the idler photon is calculated to be about 1%. Therefore the corresponding influence could be neglected. The orthogonally polarized entangled pair could be transformed into parallel polarized entangled pair with a near 100% efficiency. Therefore EO tunable entangled states are realized. Furthermore, due to the intrinsic properties of EO modulation, the entangled state is fast switchable. That could be realized in both CW and pulsed pump regimes. In the CW pump regime, we could just set the voltage at a certain value until different output photon states are requested. For a 3-cm long sample, the switching time is limited by the propagation time from the entry to the exit surface, which is about 0.2 ns. Being different from the CW pump, the pulsed regime may give some more constrains. We should avoid changing the voltage when the pumped pulses propagate right in the sample. Since the highest repetition rate of commercial lasers are only around 100 MHz, the time gap between cascading pulses should be long enough for EO switching.

Characterization of the tunable entangled photon-pair source. To ensure the quality of our tunable entangled photon-pair source, we investigate the corresponding properties. Firstly, the spectrum character is discussed, which is represented by the modulus squares of h-functions. For type-II-like polarization entangled photons, the corresponding expressions are $|h(L\Delta\beta_{ooe})|^2$ and $|h(L\Delta\beta_{oeo})|^2$; while those for the type-I-like polarization entangled photons are simultaneously affected by SPDC process and EO interaction, so the formulations are $|h(L\Delta\beta_{eo})h(L\Delta\beta_{ooeo})|^2$ and $|h(L\Delta\beta_{oe})h(L\Delta\beta_{ooe})|^2$, respectively. Calculation results are presented in Fig. 2. The natural spectral bandwidth of entangled photon-pair source is mainly influenced by the group velocities of the signal and idler wavelengths. In our situation, these two wavelengths are close with each other, so the differences between the bandwidths for the two SPDC processes, i. e., $o_p \to o_s + e_i$, $o_p \to e_s + o_i$ are relatively small. From Fig. 2(a) and Fig. 2(b), when the waveguide length L is 3 cm, these two quantities are 0.35 nm and 0.27 nm, respectively. Compared with the previous report[30], these values are relatively smaller, which is beneficial to the improvement of entanglement degree. On the other hand, the natural bandwidth increases with the decrease of the waveguide length L. For instance, we could see that the value of bandwidth increase to~0.9 nm for $o_p \to o_s + e_i$ process when the crystal length reduces to 1 cm. For $C_{oo}|o\rangle_s|o\rangle_i + C_{ee}|e\rangle_s|e\rangle_i$ pairs, as the EO process is integrated together, the natural bandwidths are further restricted through the corresponding h-functions. If the waveguide length L is 3 cm, the bandwidths for ordinary and extraordinary photon pairs are 0.20 nm and 0.23 nm, respectively. The difference between these two values is smaller, and the spectrum is almost symmetrical.

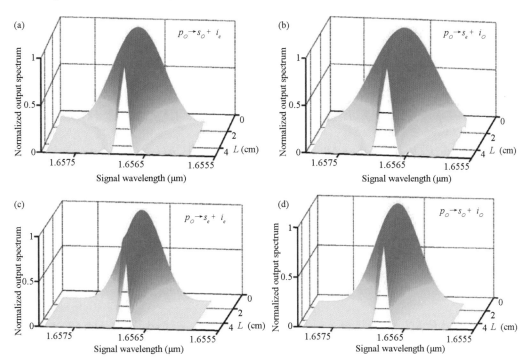

FIG.2. Normalized output signal spectra corresponding to the two type of entangled states, respectively. (a) $o_p \to o_s + e_i$; (b) $o_p \to e_s + o_i$; (c) $o_p \to o_s + e_i \to e_s + e_i$; (d) $o_p \to e_s + o_i \to o_s + o_i$. The corresponding natural bandwidths are 0.27 nm, 0.35 nm, 0.20 nm and 0.23 nm, when the waveguide length L is 3 cm.

The anticorrelation dip could be calculated based on the spectrum function, which is usually different for type-I and type-II SPDC[36]. It is the Fourier transform of the SPDC spectrum. The corresponding formulation is $R_C(\tau) \sim 1 - (1/2\pi)\int |h(\nu)|^2 \cos(\nu\tau)d\nu$. A narrower dip corresponds to a broader spectrum, so the width of anticorrelation dip is mainly determined by the natural bandwidth of the SPDC spectrum. For the orthogonally polarized entangled pair, $|h(\nu)|^2$ corresponding to $\text{sinc}^2(L\Delta\beta_{ooe}/2)$ and $\text{sinc}^2(L\Delta\beta_{oeo}/2)$.

We expand the phase mismatching to the first nonzero order of ν. The corresponding formulations are $\Delta\beta_{ooe} \approx -(\beta'_{so}-\beta'_{ie})\nu$ and $\Delta\beta_{oeo} \approx -(\beta'_{se}-\beta'_{io})\nu$. $\beta'_{j\sigma}(j=p,s,i;\sigma=o,e)$ represents the dispersion parameter. For the parallel polarized entangled pair, as its spectrum is further confined by EO process, the spectrum property is naturally different from those of the traditional cases. The $|h(\nu)|^2$ functions are written as $\text{sinc}^2(L\Delta\beta_{ooe}/2) \cdot \text{sinc}^2(L\Delta\beta_{oe}/2)$ and $\text{sinc}^2(L\Delta\beta_{oeo}/2) \cdot \text{sinc}^2(L\Delta\beta_{eo}/2)$. Similarly, we have $\Delta\beta_{oe} \approx (\beta'_{so}-\beta'_{se})\nu$ and $\Delta\beta_{eo} \approx (\beta'_{se}-\beta'_{so})\nu$. As an illustration, the anticorrelation dips corresponding to $\text{sinc}^2(L\Delta\beta_{ooe}/2)$ and $\text{sinc}^2(L\Delta\beta_{ooe}/2) \cdot \text{sinc}^2(L\Delta\beta_{oe}/2)$ are plotted in Fig.3

Based on the discussions above, we calculate the von Neumann entropy of our source, in which the influences of $P_{k\delta}(k,\delta=o,e)$ are included. It is defined as $S = -tr(\rho_{sub}\log_2\rho_{sub})$[37]. ρ_{sub} represents the reduced density operator for the subsystems. For a product state, the quantity vanishes. If the state is maximally entangled, the value of the entropy is 1. To describe tunable entangled photon-pair source, the density operator of the entangled state is expressed as follows

$$\hat{\rho}_{si} = \left(\int_{0-}^{0+}\delta(E_a)dE_a\right)|\varphi_O\rangle\langle\varphi_O| + \left(1-\int_{0-}^{0+}\delta(E_a)dE_a\right)|\varphi_P\rangle\langle\varphi_P| \qquad (22)$$

with

$$|\varphi_O\rangle = \int d\nu P_{oe}h(L\Delta\beta_{ooe})a_{so}^\dagger a_{ie}^\dagger|0\rangle + \int d\nu P_{eo}h(L\Delta\beta_{oeo})a_{se}^\dagger a_{io}^\dagger|0\rangle$$

$$|\varphi_P\rangle = \int d\nu\int d\nu' P_{oo}h(L\Delta\beta_{eo})h(L\Delta\beta_{oeo})a_{so}^\dagger a_{io}^\dagger|0\rangle + \int d\nu\int d\nu' P_{ee}h(L\Delta\beta_{oe})h(L\Delta\beta_{ooe})a_{se}^\dagger a_{ie}^\dagger|0\rangle.$$

(23)

The reduced density operator could be obtained through $\hat{\rho}_s = -tr_i(\hat{\rho}_{si})$. Therefore the quantity S is dervied as

$$S = -\left(\int_{0-}^{0+}\delta(E_a)dE_a\right)\left[\frac{\Theta_{oe}}{\Theta_O}\log_2\left(\frac{\Theta_{oe}}{\Theta_O}\right) + \frac{\Theta_{eo}}{\Theta_O}\log_2\left(\frac{\Theta_{eo}}{\Theta_O}\right)\right]$$

$$- \left(1-\int_{0-}^{0+}\delta(E_a)dE_a\right)\left[\frac{\Theta_{oo}}{\Theta_P}\log_2\left(\frac{\Theta_{oo}}{\Theta_P}\right) + \frac{\Theta_{ee}}{\Theta_P}\log_2\left(\frac{\Theta_{ee}}{\Theta_P}\right)\right] \qquad (24)$$

where $\Theta_{oe} = |\int d\nu P_{oe}h(L\Delta\beta_{ooe})|^2$, $\Theta_{eo} = |\int d\nu P_{eo}h(L\Delta\beta_{oeo})|^2$, $\Theta_{oo} = |\int d\nu\int d\nu' P_{oo}h(L\Delta\beta_{eo})h(L\Delta\beta_{oeo})|^2$ and $\Theta_{ee} = |\int d\nu\int d\nu' P_{ee}h(L\Delta\beta_{oe})h(L\Delta\beta_{ooe})|^2$. Besides, $\Theta_O = \Theta_{oe}+\Theta_{eo}$ and $\Theta_P = \Theta_{oo}+\Theta_{ee}$.

From the equation above, we could see that the value of S is mainly determined by Θ_{ij} ($i,j=o,e$). In an ideal situation, i.e., $\Theta_{oe}=\Theta_{eo}$ and $\Theta_{oo}=\Theta_{ee}$, S would reach its maximal value 1, which corresponds to the maximal entanglement. If the natural bandwidth is narrow, and the ratio $P_{oe}/P_{eo}, P_{oo}/P_{ee}$ is close to 1, Θ_{oe} and Θ_{oo} thus are almost equal to Θ_{eo} and Θ_{ee}, which makes $S\approx 1$ and a very high degree of entanglement. On the contrary, wider bandwidth would affect the entanglement accordingly. As is discussed above, the natural bandwidth in our case is quite small. Following the treatments in Ref. [38], we simplify the calculation of S into the perfect phase matching situation. Thus S has the same formulation for two types of entangled states. This result verifies that EO process

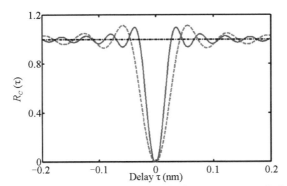

FIG. 3. The anticorrelation dips $R_c(\tau)$ are calculated based on $\text{sinc}^2(L\Delta\beta_{ooe}/2)$ and $\text{sinc}^2(L\Delta\beta_{ooe}/2) \times \text{sinc}^2(L\Delta\beta_{oe}/2)$, respectively. The blue solid line corresponds to the former, while the red dashed line represents the latter one. The crystal length is set at 3 cm.

does not affect the entanglement degree. Besides, we could see that $P_{k\delta}$ is mainly influenced by geometry and material parameters from Eq.(21). We find that the source has high tolerance for the geometrical variations. Until the waveguide modes of λ_s and λ_i are nearly cutoff, S still stays above 0.99. The degree of entanglement maintains in a high level.

Discussion

What we have presented above is merely a simplest case for the two-photon entangled state, where the entangled states are generated and manipulated through EO modulation. Moreover, this type of system also could be used to generate multiphoton entanglement. A straightforward proposal is considered based on the switchable two-photon entanglement source proposed above. For our device, if the voltage is turned on, the generated entangled state would be $|ee\rangle+|oo\rangle$. We may increase the pump power to improve the simultaneous generation possibility of double photon pairs in SPDC processes. In this situation, we may obtain four states with the same possibility, namely, $|ee\rangle|ee\rangle$, $|oo\rangle|oo\rangle$, $|ee\rangle|oo\rangle$ and $|oo\rangle|ee\rangle$. To demonstrate the entanglement, we'd better erase influences of $|ee\rangle|oo\rangle$ and $|oo\rangle|ee\rangle$. That could be realized through a post-selected detection scheme, which is widely used in energy-time entanglement. After the generated photon pairs pass through a single domain LN waveguide, the arrival time of o-polarized and e-polarized photons would be different due to the birefringence. If the frequencies of the downconverted photons are close and chosen around telecom band, the frequency dispersion would be small enough to be neglected[39]. Therefore $|ee\rangle|ee\rangle$ and $|oo\rangle|oo\rangle$ could be detected the same time, while there would be time difference in detection of photons in $|ee\rangle|oo\rangle$ and $|oo\rangle|ee\rangle$ states. With a narrow time window that is usually used in energy-time entanglement[40], a four-photon GHZ state could be selected and detected. Moreover, with the help of on-chip optical elements, such as, mode analyzer, directional coupler, Mach-Zehnder interferometer and Y-coupler[19,22,39], entangled state of more than four photons and preselected entanglement scheme are designable.

Moreover, with the help of artificially designed domain structures, the nonmaximally entangled state[3,5] could also be easily realized, which can find applications in the loophole-free tests of Bell inequalities[41] and quantum metrology with the presence of photon loss[42]. The LN waveguide could be divided into several functional portions along the propagation direction[19,22]. We may set the first portion to regulate the amplitude ratio of o-polarized and e-polarized pump light through EO interaction. Since both the o-polarized and e-polarized pump lights are involved, two corresponding SPDC processes, namely, $o_p \to o_s + o_i$ and $e_p \to e_s + e_i$ may happen simultaneously in the second portion. They are dominated by the nonlinear coefficients d_{22} and d_{33}, respectively. In this case, the QPM conditions also could be satisfied for these two processes through suitable domain design. The non-maximally entangled state thus could be obtained that is expressed as $|o_s,o_i\rangle + \varepsilon |e_s,e_i\rangle$. The relative ratio ε of amplitude may be controlled by the voltage applied on the first portion. As a consequence, an EO tunable non-maximally entanglement source is realized. The mixed state entanglement is not hard to be generated based on nonmaximal entanglement. A usual procedure is to create an entangled state with desired degree of entanglement, modulate the density matrix and then introduce decoherence[6,7]. The first step could be realized through non-maximal entanglement with a suitable entanglement degree, and then we could utilize the EO polarization rotation effect to modulate the non-maximally entangled state to generate off-diagonal terms in the corresponding density matrix. To introduce decoherence into the system, experiments using bulk optical elements have been demonstrated based on the spacial properties of the SPDC emission cone[6,7]. However, in waveguide systems, it's not convenient to use spatial decoherence directly. We could consider the temporal decoherence. Through introducing birefringent retardation, the vector $|oe\rangle$ and $|eo\rangle$ could be distinguished from $|oo\rangle$ and $|ee\rangle$ via the relative arrival time difference between two photons. Tracing over time, coherence between distinguishable terms in the density matrix could be erased, thus specific off-diagonal terms become smaller to form the target mixed entangled state.

It's worth mentioning that besides the uniform domain periods we used all above, the domain structure of LN or similar materials could be elaborately prepared in more complicated patterns, such as quasiperiodic structures, which could provide complex reciprocal vectors to compensate the multi-process phase mismatching with high efficiency[43]. We believe the domain engineered nonlinear waveguide is really a very promising platform for quantum integration circuits, which deserves more in-depth studies in both theories and experiments[44].

As for practical applications of our device, the quantum cryptography protocol may be a potential area, which is the most realizable quantum information application nowadays[45]. Based on the fast switching entangled state, our device provides a different strategy for quantum cryptography. Instead of switching the complementary analysis basis at the users' (Alice and Bob) places, we could switch between the states directly at the source while keeping the analysis basis always oriented along the same direction. Although this strategy may cause certain constrains, an advantage is that the operation conveniences for users may be improved. Another possible application example would be the classical and quantum communication without a shared reference frame[46]. Alice may encode a logical qubit as $|0_L\rangle=(|HH\rangle+|VV\rangle)/\sqrt{2}$, $|1_L\rangle=(|HV\rangle+|VH\rangle)/\sqrt{2}$, and use it to communicate with Bob. If one applies a rotation to both of the photons, $|0_L\rangle$ is unchanged, i.e., it is invariant under arbitrary rotations of the reference frame around the propagation axis. Although $|1_L\rangle$ will be changed, it is still orthogonal to $|0_L\rangle$. Therefore without a shared reference frame or the pre-alignment of Alice and Bob's polarization basis, a projective measurement on $|0_L\rangle$ by Bob will distinguish the two logical bases. Furthermore, with small postprocessing modifications based on on-chip optical elements mentioned above[19,22,39], our source can switch between the generation of two entangled states with different symmetries $(|HH\rangle+|VV\rangle)/\sqrt{2}$ and $(|HV\rangle-|VH\rangle)/\sqrt{2}$, both of which are invariant under the rotation of the reference frame[47]. Thus a symmetry measurement by Bob will distinguish between the two states. The similar idea can also be applied to the error correction protocol in quantum computations. Moreover, in most of the integrated platforms to date, the quantum information is encoded in the spatial modes[24—26]. Manipulating the polarization of photons on chip allows us to encode and decode classical and quantum information in the polarization degree of freedom. Fast switching means the shorter operation time of qubit, which is a desirable feature in practical applications. We believe over 1 GHz modulation bandwidth would be easily achieved in this case.

In summary, we theoretically propose an effective approach to tailoring entangled photons through the combination of SPDC and EO effects. Based on the domain-engineered LN waveguide, an EO tunable polarization entangled photon-pair source is realized. It could selectively generate orthogonal-polarization or parallel-polarization photon by adjusting the applied voltage. The bandwidths and entanglement degrees of both types of entangled states are of excellent performances. In comparison with the bulk or in-fiber beam splitting, polarization handling, and wave plate elements, we proposed an integrated solution possibly on a single monolithic LN chip. This on-chip integration would be very desired in future quantum information related applications since it could be more compact, low loss and stable. In addition, the EO tunability is another attractive feature that supplies extremely fast manipulation on entangled photons, which is superior to the fixed wave plate solution. Moreover, the manipulation of some other entangled states are also discussed based on our proposals, including the multi-photon entangled state, the mixed entangled state and the nonmaximally entangled state, which show some attractive advantages. Introducing EO polarization modulation into domain engineered nonlinear waveguides offers opportunities for on-chip manipulation of quantum states and fast reconfigurability of the dedicated quantum chips, which is very desired for practical applications in future quantum circuits.

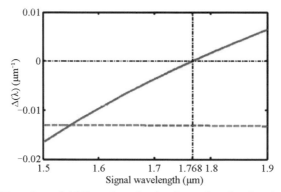

FIG.4. The values of $\Delta(\lambda)$ against the signal wavelengths when λ_p is equal to 775 nm. The blue solid line corresponds to $(m_3, n_3) = (1, 1)$, while the red dash line corresponds to $(m_3, n_3) = (1, -1)$. Only the blue line has crossover point with the zero line, so $(1, 1)$ is possible solution for the equation $\Delta(\lambda) = \Delta\beta_3(\lambda) - G_{1,\pm 1} = 0$.

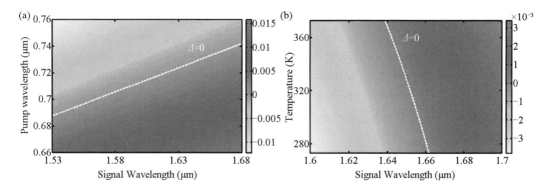

FIG. 5. In the situation where the subscript of $\Delta\beta_3$ corresponds to $(m_3, n_3) = (1, 1)$, The function $\Delta(\lambda)$ is tuned by the pump wavelength and the operation temperature, respectively. The green dashed line marks $\Delta(\lambda) = 0$. (a) Values of $\Delta(\lambda)$ change with the signal wavelengths and the pump wavelengths. (b) Values of $\Delta(\lambda)$ changes with the signal wavelengths and the operation temperature. From the $\Delta(\lambda) = 0$ lines, suitable pairs of pump, signal and idler wavelengths could be chosen for specific situations.

References and Notes

[1] Eibl, M. *et al.* Experimental observation of four-photon entanglement from parametric down-conversion. *Phys. Rev. Lett.* **90**, 200403(2003).

[2] Zou, X. B. & Mathis, W. Generating a four-photon polarization-entangled cluster state. *Phys. Rev. A* **71**, 032308(2005).

[3] White, A. G., James, D. F., Eberhard, P. H. & Kwiat, P. G. Nonmaximally entangled states: Production, characterization, and utilization. *Phys. Rev. Lett.* **83**, 3103–3107(1999).

[4] Bennett, C. H., DiVincenzo, D. P., Smolin, J. A. & Wootters, W. K. Mixed-state entanglement and quantum error correction. *Phys. Rev. A* **54**, 3824–3851(1996).

[5] Thew, R. T., Tanzilli, S., Tittel, W., Zbinden, H. & Gisin, N. Experimental investigation of the robustness of partially entangled qubits over 11 km. *Phys. Rev. A* **66**, 062304(2002).

[6] Cinelli, C., di Nepi, G., de Martini, F., Barbieri, M. & Mataloni, P. Parametric source of two-photon states with a tunable degree of entanglement and mixing: Experimental preparation of Werner states and maximally entangled mixed states. *Phys. Rev. A* **70**, 022321(2004).

[7] Barbieri, M., de Martini, F., di Nepi, G. & Mataloni, P. Generation and characterization of Werner states and maximally entangled mixed states by a universal source of entanglement. *Phys. Rev. Lett.* **92**, 177901(2004).

[8] Kwiat, P. G., Waks, E., White, A. G., Appelbaum, I. & Eberhard, P. H. Ultrabright source of polarization-entangled photons. *Phys. Rev. A* **60**, R773–R776(1999).

[9] Leng, H. Y. *et al*. On-chip steering of entangled photons in nonlinear photonic crystals. *Nat. Commun.* **2**, 429(2011).

[10] Politi, A. *et al*. Integrated quantum photonics. *IEEE J. Sel. Top. Quantum Electron.* **15**, 1673–1684 (2009).

[11] Tanzilli, S. *et al*. On the genesis and evolution of integrated quantum optics. *Laser & Photon. Rev.* **6**, 115–143(2012).

[12] Suhara, T. Generation of quantum-entangled twin photons by waveguide nonlinear-optic devices. *Laser & Photon. Rev.* **3**, 370–393(2009).

[13] de Chatellus, H. G., Sergienko, A. V., Saleh, B. E., Teich, M. C. & Di Giuseppe, G. Non-collinear and non-degenerate polarization-entangled photon generation via concurrent type-I parametric downconversion in PPLN. *Opt. Express* **14**, 10060–10072(2006).

[14] Bonneau, D. *et al*. Fast path and polarization manipulation of telecom wavelength single photons in lithium niobate waveguide devices. *Phys. Rev. Lett.* **108**, 053601(2012).

[15] Martin, A., Alibart, O., de Micheli, M. P., Ostrowsky, D. B. & Tanzilli, S. A quantum relay chip based on telecommunication integrated optics technology. *New. J. Phys.* **14**, 025002(2012).

[16] Lu, Y. Q., Wan, Z. L., Wang, Q., Xi, Y. X. & Ming, N. B. Electro-optic effect of periodically poled optical superlattice $LiNbO_3$ and its applications. *Appl. Phys. Lett.* **77**, 3719–3721(2000).

[17] Kaiser, F. *et al*. High-quality polarization entanglement state preparation and manipulation in standard telecommunication channels. *New. J. Phys.* **14**, 085015(2012).

[18] Herrmann, H. *et al*. Post-selection free, integrated optical source of nondegenerate, polarization entangled photon pairs. *Opt. Express* **21**, 27981–27991(2013).

[19] Lugani, J., Ghosh, S. & Thyagarajan, K. Electro-optically switchable spatial-mode entangled photon pairs using a modified Mach-Zehnder interferometer. *Opt. Lett.* **37**, 3729–3731(2012).

[20] Song, X. S., Yu, Z. Y., Wang, Q., Xu, F. & Lu, Y. Q. Polarization independent quasi-phase-matched sum frequency generation for single photon detection. *Opt. Express* **19**, 380–386(2011).

[21] Lu, Y. Q. *et al*. Optical properties of an ionic-type phononic crystal. *Science* **284**, 1822–1824(1999).

[22] Saleh, M. F., Di Giuseppe, G., Saleh, B. E. & Teich, M. C. Modal and polarization qubits in Ti: $LiNbO_3$ photonic circuits for a universal quantum logic gate. *Opt. Express* **18**, 20475–20490(2010).

[23] Xu, P. *et al*. Lensless imaging by entangled photons from quadratic nonlinear photonic crystals. *Phys.*

Rev. A **86**, 013805(2012).

[24] Politi, A., Cryan, M. J., Rarity, J. G., Yu, S. & O'Brien, J. L. Silica-on-silicon waveguide quantum circuits. *Science* **320**, 646 – 649(2008).

[25] Matthews, J. C., Politi, A., Stefanov, A. & O'Brien, J. L. Manipulation of multiphoton entanglement in waveguide quantum circuits. *Nat. photon.* **3**, 346 – 350(2009).

[26] Metcalf, B. J. *et al.* Multiphoton quantum interference in a multiport integrated photonic device. *Nat. commun.* **4**, 1356(2013).

[27] Sansoni, L. *et al.* Polarization entangled state measurement on a chip. *Phys. Rev. Lett.* **105**, 200503 (2010).

[28] Crespi, A. *et al.* Integrated photonic quantum gates for polarization qubits. *Nat. Commun.* **2**, 566 (2011).

[29] Sharma, A. & Bindal, P. Analysis of diffused planar and channel waveguides. *IEEE J. Quantum Electron.* **29**, 150 – 153(1993).

[30] Thyagarajan, K. *et al.* Generation of polarization-entangled photons using type-II doubly periodically poled lithium niobate waveguides. *Phys. Rev. A* **80**, 052321(2009).

[31] Gong, Y. X. *et al.* Compact source of narrow-band counterpropagating polarization-entangled photon pairs using a single dual-periodically-poled crystal. *Phys. Rev. A* **84**, 053825(2011).

[32] Lambda Photometrics Ltd., http://www.lambdaphoto.co.uk/pdfs/Inrad_datasheet_LNB.pdf(accessed 09/02/2014).

[33] Volk, T. R. *et al.* Optical and non-linear optical investigations in $LiNbO_3$: Mg and $LiNbO_3$: Zn. *Ferroelectrics* **109**, 345(1990).

[34] Kumar, R. M. *et al.* SIMs-depth profile and microstructure studies of Ti-diffused Mg-doped near-stoichiometric lithium niobate waveguide. *J. Cryst. Growth* **287**, 472 – 477(2006).

[35] Huang, C. Y., Lin, C. H., Chen, Y. H. & Huang, Y. C. Electro-optic Ti:PPLN waveguide as efficient optical wavelength filter and polarization mode converter. *Opt. Express* **15**, 2548 – 2554 (2007).

[36] Burlakov, A. V., Chekhova, M. V., Karabutova, O. A. & Kulik, S. P. Collinear two-photon state with spectral properties of type-I and polarization properties of type-II spontaneous parametric down-conversion: preparation and testing. *Phys. Rev. A* **64**, 041803(R)(2001).

[37] Nielson, M. A. & Chuang, I. L. Quantum computation and quantum information. (Cambridge University Press, 2006).

[38] Lugani, J. *et al.* Generation of modal-and path-entangled photons using a domain-engineered integrated optical waveguide device. *Phys. Rev. A* **83**, 062333(2011).

[39] Saleh, M. F., di Giuseppe, G., Saleh, B. E. A. & Teich, M. C. Photonic circuits for generating modal, spectral, and polarization entanglement. *IEEE Photon. J.* **2**, 736 – 752(2010).

[40] Tanzilli, S. *et al.* PPLN waveguide for quantum communication. *Eur. Phys. J. D* **18**, 155 – 160 (2002).

[41] Giustina, M. *et al.* Bell violation using entangled photons without the fairsampling assumption. *Nature* **497**, 227 – 230(2013).

[42] Dorner, U. *et al.* Optimal quantum phase estimation. *Phys. Rev. Lett.* **102**, 040403(2009).

[43] Zhu, S. N., Zhu, Y. Y. & Ming, N. B. Quasi-phase-matched third-harmonic generation in a quasi-periodic optical superlattice. *Science* **278**, 843 – 846(1997).

[44] Ming, Y. *et al.* Quantum entanglement based on surface phonon polaritons in condensed matter

systems. *AIP Advances* **3**, 042122(2013).

[45] Martin, A. *et al*. Cross time-bin photonic entanglement for quantum key distribution. *Phys. Rev. A* **87**, 020301(R)(2013).

[46] Bartlett, S. D. *et al*. Classical and quantum communication without a shared reference frame. *Phys. Rev. Lett.* **91**, 027901(2003).

[47] Aolita, L. *et al*. Quantum communication without alignment using multiple-qubit single-photon states. *Phys. Rev. Lett.* **98**, 100501(2007).

[48] The authors thank the helpful discussions with Dr. Li-jian Zhang for preparing the revised manuscript. This work is sponsored by 973 programs with contract No. 2011CBA00205 and 2012CB921803, and the National Science Fund for Distinguished Young Scholars with contract No. 61225026. The authors also thank the supports from PAPD and Fundamental Research Funds for the Central Universities. The technical support from Miss Ting-ting Xu is also acknowledged.

[49] Y.Q.L. and Y.M. contributed to the original idea. Y.M., Z.J.W. and Z.X.C. did theoretical analysis, calculations and interpretations. A.H.T. provided theoretical guidance. Y.M. and Y.Q.L. wrote the manuscript together. Y.Q.L. and F.X. supervised the project. All authors reviewed the manuscript.

[50] Methods: To find a proper set of pump, signal and idler wavelengths, they are scanned carefully. Firstly, when the pump wavelength is fixed, calculation results reveal that $\Delta\beta_1$ and $\Delta\beta_2$ (for SPDC processes) are close but much greater than $\Delta\beta_3$ (for EO interaction). Thus they are assumed to correspond to the solutions $(3, \pm 1)$ for (m, n), respectively. For certain pairs of signal and idler frequencies, the values of $\Lambda_1(\lambda)$ and $\Lambda_2(\lambda)$ could be calculated through linear superposition of $\Delta\beta_1(\lambda)$ and $\Delta\beta_2(\lambda)$, and then we compare the linear combination $(1, \pm 1)$ of $2\pi/\Lambda_1$ and $2\pi/\Lambda_2$ with $\Delta\beta_3$ to see if they are matched. The corresponding expression is $\Delta(\lambda) = \Delta\beta_3(\lambda) - G_{1,\pm 1}$. For the wavelength satisfying $\Delta(\lambda)=0$, three phase mismatches could be compensated simultaneously.

In this study, we choose the pump wavelength at 775 nm. The values of $\Delta(\lambda)$ against the signal wavelengths are plotted in Fig. 4 ($\lambda_p = 775$ nm). It is seen from the figure that only the linear combination $(1, 1)$ of $2\pi/\Lambda_1$ and $2\pi/\Lambda_2$ provides a solution for the equation $\Delta(\lambda)=0$. Thus $G_{3,1}$, $G_{3,-1}$, $G_{1,1}$ are selected as the compensations of $\Delta\beta_1$, $\Delta\beta_2$ and $\Delta\beta_3$, respectively. Moreover, the phase matching conditions could be adjusted by the pump wavelength and the operation temperature, which are correspondingly shown in Fig. 5(a) and Fig. 5(b). To ensure the ratio between Λ_1 and Λ_2 is an integer, we finally choose the pump, signal and idler wavelengths at 0.7335 μm, 1.6568 μm and 1.3162 μm, respectively. In this situation, $\Lambda_2 = 6\Lambda_1$, where the operating temperature is set at 25℃. For quantum communication applications, the operation wavelengths might be preferred to be chosen from the telecom-band. That could be easily realized by finding the solutions of Eq.(9) accordingly, as is shown in Fig. 5(a). It is clearly from the figure that the whole telecom C band(1.53 μm - 1.57 μm) and L band(1.57 μm - 1.61 μm) could be selected for operation. Furthermore, the operating temperature could also be conveniently adjusted according the simulation results as shown in Fig. 5(b).

On-Chip Generation and Manipulation of Entangled Photons Based on Reconfigurable Lithium-Niobate Waveguide Circuits[*]

H. Jin,[1] F. M. Liu,[2] P. Xu,[1] J. L. Xia,[2] M. L. Zhong,[1] Y. Yuan,[1] J. W. Zhou,[2]
Y. X. Gong,[3] W. Wang,[2] and S. N. Zhu[1]

[1] *National Laboratory of Solid State Microstructures and College of Physics, Collaborative Innovation Center of Advanced Microstructures, Nanjing University, Nanjing 210093, China*

[2] *Beijing Institute of Aerospace Control Devices, Beijing 100094, China*

[3] *Department of Physics, Southeast University, Nanjing 211189, China*

A consequent tendency toward high-performance quantum information processing is to develop the fully integrated photonic chip. Here, we report the on-chip generation and manipulation of entangled photons based on reconfigurable lithium-niobate waveguide circuits. By introducing a periodically poled structure into the waveguide circuits, two individual photon-pair sources with a controllable electro-optic phase shift are produced within a Hong-Ou-Mandel interferometer, resulting in a deterministically separated identical photon pair. The state is characterized by 92.9 ± 0.9% visibility Hong-Ou-Mandel interference. The photon flux reaches ~1.4×10^7 pairs nm^{-1} mW^{-1}. The whole chip is designed to contain nine similar units to produce identical photon pairs spanning the telecom C and L band by the flexible engineering of nonlinearity. Our work presents a scenario for on-chip engineering of different photon sources and paves the way to fully integrated quantum technologies.

Tremendous progress has been achieved in integrated photonic circuits[1—10], providing a solid strategy for high-performance quantum information processing. To increase the integration complexity, further integration of photon sources together with the photonic circuits are of essential importance[11—13]. For practical applications, strategic effort should be devoted to developing the photonic chip's characteristics of low energy cost, high-efficiency internal photon sources, and fast and convenient phase modulation for enabling multiple high-fidelity quantum operations on a single chip. Therefore, integrated optical materials capable of producing and manipulating entangled photons are highly desired, and these will become pivotal ingredients for fully integrated quantum technologies.

The lithium-niobate(LN) crystal, also called the "silicon of photonics"[14,15], belongs to one of the valuable materials for integrated quantum optics, owing to strong $\chi^{(2)}$ nonlinearity, large piezoelectric, acousto-optic and electro-optic coefficient features, as

---

[*] Phys.Rev.Lett.,2014,113(10):103601

well as the well-developed waveguide fabrication technique using either the proton-exchange or the titanium-indiffusion method. In telecommunications applications, the LN modulators based on electro-optic effect have reached 40 GHz (100 GHz in the laboratory[16]) and have recently been demonstrated for fast path and polarization control of single photons[17]. More specifically, a domain-engineering technique[18,19] can be introduced to LN, leading to the desirable quasi-phase-matching (QPM) spontaneous parametric downconversion (SPDC) process for efficient generation of entangled photons. When complex designs of the poling structure are considered, multiple operations on entangled sources can be achieved on a single bulk crystal platform[20,21]. When the bulk LN is further fabricated into the waveguide structure, the photon flux will be greatly enhanced, as was first demonstrated from a single periodically poled LN (PPLN) waveguide[22,23]. Considering the recently developed quantum memory[24] in a LN waveguide, the LN is promising for photonic chips with a complete family of robust, integrated quantum devices.

All the aforementioned features allow complex waveguide circuits embedded with high-flux photon sources and fast phase shifters on a single LN chip, which makes LN qualified for more complex quantum tasks. Here, based on a single LN photonic chip we demonstrate the generation and deterministic separation of an identical photon pair at telecom wavelengths. The deterministically separated identical photon pair is characterized by $92.9\pm0.9\%$ visibility Hong-Ou-Mandel (HOM) interference. The photon flux reaches $\sim 1.4\times10^7$ pairs nm^{-1} mW^{-1}. Continuous morphing from a two-photon separated state to a bunched state is further demonstrated by on-chip fast control of electro-optic phase shift (EOPS). The whole chip is designed to contain nine similar units to produce identical photon pairs spanning the telecom C and L band by the flexible engineering of nonlinearity. These results denote that the LN photonic chip is competent for practical applications of fully integrated quantum technologies with low energy cost, high-efficiency and flexible internal photon sources, fast and convenient phase modulators, and reconfigurable waveguide circuits.

Figure 1(a) shows the main structure of this chip. Basically, it is composed of annealed proton exchanged channel waveguides integrated on a Z-cut PPLN crystal. This design enables the generation, interference, and filtering of entangled photons from separate regions of PPLN waveguides, leading to reconfigurable on-chip quantum light sources. The chip can be characterized by three sections. Section I is designed to deal with the classical pump light. A 780 nm pump is coupled into waveguide L_0 and equally distributed by a Y-branch single mode beam splitter at the wavelength of 780 nm. After the Y branch, the electro-optic effect is considered to control the phase shift between two paths. Electrodes are fabricated above two pump waveguides for applying a voltage. Transition tapers then follow to connect the 780 nm Y branch with the 1560 nm single mode waveguides. Section II is the PPLN region, in which degenerate photon pairs at

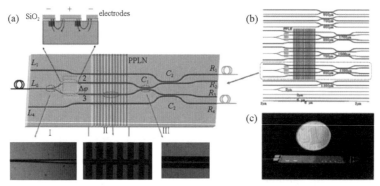

FIG.1. (color) (a) The photonic chip with dimensions of 50 mm×5 mm×0.5 mm. The widths of single mode waveguides are 2 μm for the pump and 6 μm for entangled photons. The periodically poled section is 10 mm long with a period of 15.32 μm. The interaction length of C_1/C_2 is 650 μm/1300 μm at a gap of 4 μm. The upper inset is the structure of the electrooptic modulator. The electrode pairs are 8.35 mm long with a separation of 6 μm. The buffer layer(SiO$_2$) is etched along the gap between the electrodes to suppress dc drift. The lower three insets are the micrographs of the Y branch, the PPLN waveguide, and the directional coupler, respectively. (b) The whole structure of the chip. (c) Photograph of the chip with fixed fiber pigtails.

1560 nm are generated indistinguishably from either one of the two PPLN waveguides, yielding a path-entangled state $(|2,0\rangle + e^{i\Delta\varphi}|0,2\rangle)/\sqrt{2}$. The relative phase $\Delta\varphi$ is transferred from the phase difference of pump modes. In section III, quantum interference is realized by a 2×2 directional coupler(C_1). The on-chip phase is adjusted by changing the voltage applied in section I. The entangled photons are separated from the pump by on-chip wavelength filters (C_2). Entangled photons are transferred to the neighboring waveguides R_1 and R_4. The pump is left in R_2 and R_3. The input waveguide L_0 and output waveguides R_1 and R_4 are directly connected with optical fiber tips, which are fixed with the chip by using UV-curing adhesive after reaching a high coupling efficiency. This removes the need for optimizing the coupling during the observation of quantum interference.

When the phase shift $\Delta\varphi$ between pump modes in waveguide 2 and 3 is zero, the chip worked as a balanced time-reversed Hong-Ou-Mandel(BRHOM) interferometer, realizing a deterministic separation(sep) of an identical 1560 nm photon pair,

$$\frac{1}{\sqrt{2}}(|2,0\rangle + |0,2\rangle) \xrightarrow{\text{BRHOM}} |1,1\rangle. \qquad (1)$$

In this case, the photon pair always emits from different waveguides(one from R_1 and the other from R_4), marked as $|\Psi\rangle_{\text{sep}} = |1,1\rangle$.

In a more general case, when the electro-optic phase shift $\Delta\phi$ is introduced, the state evolution in the chip can be written as

$$\frac{1}{\sqrt{2}}(|2,0\rangle + e^{i\Delta\phi}|0,2\rangle) \xrightarrow{\text{EOPS}} \frac{1}{\sqrt{2}}(|2,0\rangle - |0,2\rangle)\sin(\Delta\phi/2) + |1,1\rangle\cos(\Delta\phi/2). \qquad (2)$$

The output state is a superposition of the two-photon separated state $|\Psi\rangle_{\text{sep}} = |1,$

1⟩ ($\Delta\phi=0$) and the bunched state $|\Psi\rangle_{bunch}=1/\sqrt{2}(|2,0\rangle-|0,2\rangle)$ ($\Delta\varphi=\pi$, the photon pair always emits together from either R_1 or R_4). By varying the bias voltage U, and thus the phase difference, we will observe a continuous evolution from the two-photon separated state to the bunched state. Specifically, the photon-pair rate from R_1 and R_4 follows

$$P_{sep} \propto \cos^2(\Delta\phi/2) = \cos^2(\pi U/2V_\pi), \tag{3}$$

while the photon-pair rate from R_1 or R_4 gives

$$P_{bunch} \propto \sin^2(\Delta\phi/2) = \sin^2(\pi U/2V_\pi). \tag{4}$$

$V_\pi = \lambda_p d/2\Gamma\gamma_{33} n_p^3 L$ is a half-wave voltage, in which Γ is a fill factor reflecting the overlap between the electric field and the optical mode field[25], γ_{33} is the electro-optic coefficient, n_p and λ_p are the refractive index and wavelength of the pump, and L and d are the length and separation of electrode pairs. Here we design the electrodes as the push-pull configuration; i.e., two waveguides both experience the electric field but with different sign, so the effective electrode length should be $2L$.

Actually, as shown in Fig. 1(b), the LN chip contains three main photonic circuits, and each one has a structure similar to Fig. 1(a). The difference lying among three main circuits is the interaction lengths of C_1 and C_2. Different interaction lengths, including 650 and 1300 μm, 750 and 1500 μm, and 850 and 1700 μm, are designed to ensure a good splitting ratio for photon pairs. The dashed rectangle region indicates the one we use in this experiment, with the interaction lengths of 650 μm for C_1 and 1300 μm for C_2. Additionally, we also design the involved waveguide elements like single mode waveguides for 780 and 1560 nm, directional couplers, etc. around the three main photonic circuits for testing and optimizing the waveguide fabrication technology. Figure 1(c) shows a photograph of the chip with fixed fiber pigtails.

Experimentally, the transformation from a two-photon bunched state to a separated one is observed by making coincidence counting measurements on R_1 and R_4 under different bias voltage. The experimental setup is sketched in Fig. 2(a). The photonic chip is temperature controlled at 25.5 ℃. The 780 nm pump is a continuous-wave fiber laser that is polarization controlled and connected with input fiber L_0. We make coincidence counting measurement between R_1 and R_4, R_1 and R_1, and R_4 and R_4. The measured pump suppression is 29.2 dB for R_1 and 31.4 dB for R_4, which mainly arises from the integrated filters and the propagation losses in the waveguide. The pump is further suppressed by interference filters centered at 1560 nm which are inserted into the fiber delay lines[as shown in Fig. 2(a)]. The coincidence counts are recorded in Figs. 2(b)-2(d), respectively, exhibiting interference fringes with the visibilities of 98.9±0.3%, 97.5±0.9%, and 88.4±2.1%, respectively. The visibility is calculated by $V=(R_{max}-R_{min})/(R_{max}+R_{min})$, in which R_{max} and R_{min} are the maximum and minimum values of fitting. The accidental coincidence counts have been excluded. The period of these fringes is 7.1 V, giving the half-wave voltage $V_\pi = 3.55$ V. These fringes agree with Eq. (2) except

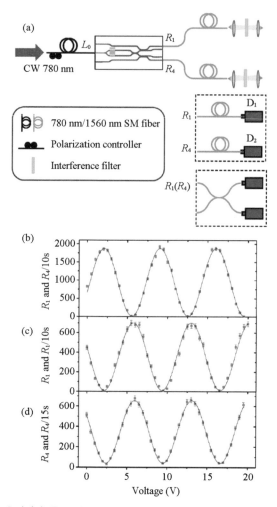

FIG.2. (color) (a) Experimental setup of on-chip two-photon interference. The fiber laser is polarization controlled and connected with the input fiber L_0. Before single-photon detectors, 14 nm bandwidth interference filters centered at 1560 nm are used to further eliminate the pump. The coincidence counting measurements are made between R_1 and R_4, R_1 and R_1, and R_4 and R_4, respectively. The results are recorded in(b)-(d). The solid lines are sinusoidal fitting, and error bars show $\pm\sqrt{\text{counts}}$.

that it is not a clean separated state $|\Psi\rangle_{\text{sep}}=|1,1\rangle$ when $U=0$. From Fig. 2(b), the $|1,1\rangle$ state is obtained when $U_{\text{offset}}=2.3$ V, namely, the offset bias voltage. This results from certain optical path difference between two PPLN waveguides. Here the difference between the visibilities of R_1 and R_1 and R_4 and R_4 is mainly caused by the imperfect coupling efficiency of $C_1(T\sim54\%)$ and different losses in waveguides 2 and 3. The different losses make the two-photon state $(|2,0\rangle+\alpha e^{i\Delta\phi}|0,2\rangle)/\sqrt{2}$ before C_1, in which α stands for relative loss. When the photons interfere on C_1, the visibilities for the separated and bunched states are $V_{\text{sep}}^{(1,4)}=2\alpha/(1+\alpha^2)$, $V_{\text{bunch}}^{(1)}=2\alpha T(1-T)/[T^2+\alpha^2(1-T)^2]$, $V_{\text{bunch}}^{(4)}=$

$2\alpha T(1-T)/[\alpha^2 T^2+(1-T^2)]$. This imperfect coupling efficiency can be further optimized when applying a voltage to C_1. The small deviation between the positions of peaks in Fig. 2(b) and the dips in Figs. 2(c) and 2(d) is attributed to dc drift associated with the buffer layer(SiO_2)[26,27]. Here the dc drift has been depressed a lot by etching a groove on the SiO_2 buffer layer as shown in the inset of Fig. 1(a). The dc drift can be further minimized by replacing SiO_2 with a transparent conductive buffer layer such as indium tin oxide[27].

According to the above measurement, we configured the chip to emit a $|1,1\rangle$ state by controlling the voltage at 2.3 V. To evaluate the quality of such a state, we performed a HOM interference experiment by an external fiber beam splitter. The experimental setup is sketched in Fig. 3(a). Photons from R_1 and R_4 are sent into two fibers, one with a variable delay line in free space and the other with a polarization controller. Therefore, the indistinguishability of two photons in the arrival time and polarization can be ensured when they interfere on the fiber beam splitter. Figure 3(b) shows the coincidence counts that we recorded while varying the displacement of the free-space delay line. We observed a HOM dip with a visibility of $(92.9\pm0.9)\%$ after excluding the accidental coincidence counts. The nonideal visibility indicates that $|\Psi\rangle_{sep}$ is not absolutely clean, containing some bunched portion [as shown in Fig. 2(d)] as well as some polarization rotations in the external single mode fibers which are not compensated perfectly.

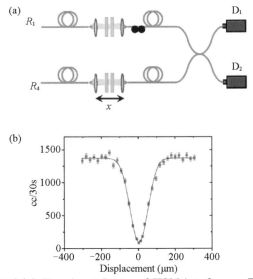

FIG.3. (color)(a) Experimental setup of HOM interference. The photonic chip is controlled to emit separated photon pairs. In each arm there are two interference filters of 14 nm bandwidth. (b) Coincidence counts(cc) versus the displacement of the free-space delay line. Error bars show $\pm\sqrt{counts}$.

To evaluate the conversion efficiency of our device, we configure the chip to emit a separated state and measure the photon pairs picked out by an interference filter with 14 nm bandwidth. Two single counting rates were $R_1=40000$ Hz and $R_4=29000$ Hz, and the coincidence counting rate was $R_{cc}=186$ Hz when the pump power coupled into the waveguide L_0 is 31 μW. The production rate of photon pairs in the PPLN waveguides reached ~6.2×10^6 Hz(calculated by $N=R_1R_4/R_{cc}$). The production rate of photon pairs per units bandwidth and pump power was ~1.4×10^7 Hz nm^{-1} mW^{-1} for the PPLN waveguide source of 10 mm length. The conversion efficiency reaches $N/N_{pump}=5\times10^{-8}$. The total loss experienced by a photon exiting from R_1 or R_4 is 21.9 and 23.3 dB[calculated by $-10\log(R_{cc}/R_{1(4)})$], respectively, including the propagation loss in waveguides, the coupling efficiency from the chip to fiber, the filtering efficiencies, and the detector efficiencies. The coupling efficiency from the chip to the fiber is 1 dB. The total transmission efficiency of the fiber-free space-fiber filtering system(as shown in Fig. 2)is 30%(5.2 dB loss) including the 60% collection efficiency and 50% transmission efficiency of the 1560 nm interference filter. The detectors are set to give 9.2 and 8.5 dB loss, respectively. Thus, we can calculate the propagation loss of entangled photons in waveguides to be 6.5 dB for R_1 and 8.6 dB for R_4. The low propagation efficiency (compared with Refs. [28,29]) may result from the random scratches or other defects during the sample fabrication. However, the propagation efficiency as well as the efficiencies of detectors and filtering systems can be further optimized.

The PPLN is the key section of this photonic chip, providing a flexible and feasible on-chip photon source. Here, by varying the poling period Λ, we actually design nine units on a 3-in. LN wafer as shown in Fig. 4(a), intending to generate identical photon pairs whose wavelengths cover the C band(1530 - 1565 nm) and L band(1565 - 1625 nm) in fiber communication. Each channel contains the same waveguide circuits as shown in Fig. 1(b) but different poling period, ranging from 14.36 μm to 16.28 μm in a step of 0.24 μm(the corresponding SPDC process at room temperature will produce identical photon pairs around 1560 nm by ~12 nm separation). In this measurement, we verify the poling period and corresponding photon-pair wavelength by second harmonic generation(SHG), the inversed one of spontaneous parametric down-conversion in the assistant waveguide(L_1 or L_4). Figure 4(b) shows the wavelengths for efficient SHG in different channels at room temperature(25.5℃). In this work we choose the fifth unit with the poling period of 15.32 μm. From this chip, the two-color photon pair is also accessible by increasing the working temperature due to the QPM condition of PPLN.

Recently, a related work was published[11]. It makes a great step toward fully integrated quantum optics by integrating entangled photons by four-wave mixing processes together with waveguide circuits on a silicon-on-insulator photonic chip. Silicon materials are considered to be competitive for quantum photonic devices for the mature fabrication technologies and accessible four-wave mixing photon sources[30—35]. But at the present stage, the nonlinearity and phase modulator still need to be improved. In contrast, although the LN circuits contain larger footprints and the fabrication technology is not compatible with CMOS electronics, it contains efficient PPLN waveguide photon sources

Chapter 6 Review Article: Quasi-phase-matching Engineering of Entangled Photons

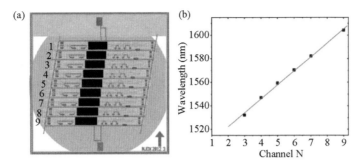

FIG. 4. (color) (a) The layout of nine channels with the same waveguide circuits but different poling periods on a LN wafer. (b) Efficient SHG wavelengths in different channels at 25.5℃. The red line is a linear fitting.

and fast electro-optic modulators, which makes it quite competent for on-chip engineering of quantum light sources. By a careful design of QPM sections, the polarization, frequency, spatial mode, and path degrees of freedom can be engineered during the SPDC processes. The efficient identical photon pair demonstrated here will stimulate the heralded generation of single photon sources[36] as well as the extension to the multiphoton fashion. When associated with phase-controlled circuits, more types of quantum light sources can be exploited in addition to the on-chip quantum information processing. This will pave a new way for the fully integrated quantum optics. Additionally, the LN photonic chip can act as an alternative platform for quantum walk, especially when the nonlinear PPLN section is embedded[37,38], which may bring new characteristics to quantum walk and stimulate further applications in quantum computing.

References and Notes

[1] A. Politi, M. J. Cryan, J. G. Rarity, S. Yu, and J. L. O'Brien, Science **320**, 646(2008).

[2] J.L.O'Brien, A. Furusawa, and J. Vuckovic, Nat. Photonics **3**, 687(2009).

[3] J.C. F. Matthews, A. Politi, A. Stefanov, and J. L. O'Brien, Nat. Photonics **3**, 346(2009).

[4] A. Politi, J. C. F. Matthews, and J. L. O'Brien, Science **325**, 1221(2009).

[5] B. J. Smith, D. Kundys, N. Thomas-Peter, P. G. R. Smith, and I. A. Walmsley, Opt. Express **17**, 13516(2009).

[6] S. Tanzilli, A. Martin, F. Kaiser, M. P. De Micheli, O. Alibart, and D. B. Ostrowsky, Laser Photonics Rev. **6**, 115(2012).

[7] P. J. Shadbolt, M. R. Verde, A. Peruzzo, A. Politi, A. Laing, M.Lobino, J. C. F. Matthews, M. G. Thompson, and J.L. O'Brien, Nat. Photonics **6**, 45(2011).

[8] B. J. Metcalf *et al.*, Nat. Commun. **4**, 1356(2013).

[9] M. Tillmann, B. Dakić, R. Heilmann, S. Nolte, A. Szameit, and P. Walther, Nat. Photonics **7**, 540 (2013).

[10] A. Crespi, R. Osellame, R. Ramponi, D. J. Brod, E. F. Galvão, N. Spagnolo, C. Vitelli, E. Maiorino, P. Mataloni, and F. Sciarrino, Nat. Photonics **7**, 545(2013).

[11] J. W. Silverstone *et al.*, Nat. Photonics **8**, 104(2014).

[12] A. Martin, O. Alibart, M. P. De Micheli, D. B. Ostrowsky, and S. Tanzilli, New J. Phys. **14**, 025002 (2012).

[13] N. Matsuda, H. L. Jeannic, H. Fukuda, T. Tsuchizawa, W. J. Munro, K. Shimizu, K. Yamada, Y. Tokura, and H. Takesue, Sci. Rep. **2**, 817(2012).

[14] M. Kösters, B. Sturman, P. Werheit, D. Haertle, and K. Buse, Nat. Photonics **3**, 510(2009).

[15] R. S. Weis and T. K. Gaylord, Appl. Phys. A **37**, 191(1985).

[16] A. Kanno, T. Sakamoto, A. Chiba, T. Kawanishi, K. Higuma, M. Sudou, and J. Ichikawa, IEICE Electron. Express **7**, 817(2010).

[17] D. Bonneau, M. Lobino, P. Jiang, C. M. Natarajan, M. G. Tanner, R. H. Hadfield, S. N. Dorenbos, V. Zwiller, M. G. Thompson, and J. L. O'Brien, Phys. Rev. Lett. **108**, 053601(2012).

[18] J. A. Armstrong, N. Bloembergen, J. Ducuing, and P. S. Pershan, Phys. Rev. **127**, 1918(1962).

[19] P. A. Franken and J. F. Ward, Rev. Mod. Phys. **35**, 23(1963).

[20] H. Y. Leng, X. Q. Yu, Y. X. Gong, P. Xu, Z. D. Xie, H. Jin, C. Zhang, and S. N. Zhu, Nat. Commun. **2**, 429(2011).

[21] H. Jin, P. Xu, X. W. Luo, H. Y. Leng, Y. X. Gong, W. J. Yu, M. L. Zhong, G. Zhao, and S. N. Zhu, Phys. Rev. Lett. **111**, 023603(2013).

[22] S. Tanzilli, H. De Riedmatten, H. Tittel, H. Zbinden, P. Baldi, M. De Micheli, D. B. Ostrowsky, and N. Gisin, Electron. Lett. **37**, 26(2001).

[23] K. Sanaka, K. Kawahara, and T. Kuga, Phys. Rev. Lett. **86**, 5620(2001).

[24] E. Saglamyurek, N. Sinclair, J. Jin, J. A. Slater, D. Oblak, F. Bussières, M. George, R. Ricken, W. Sohler, and W. Tittel, Nature(London) **469**, 512(2011).

[25] T. Fujiwara, A. Watanabe, and H. Mori, IEEE Photonics Technol. Lett. **2**, 260(1990).

[26] S. Yamada and M. Minakata, Jpn. J. Appl. Phys. **20**, 733(1981).

[27] C. M. Gee, G. D. Thurmond, H. Blauvelt, and H. W. Yen, Appl. Phys. Lett. **47**, 211(1985).

[28] K. R. Parameswaran, M. Fujimura, M. H. Chou, and M. M. Fejer, IEEE Photonics Technol. Lett. **12**, 654(2000).

[29] C. Langrock, S. Kumar, J. E. McGeehan, A. E. Willner, and M. M. Feijer, J. Lightwave Technol. **24**, 2579(2006).

[30] J. E. Sharping, K. F. Lee, M. A. Foster, A. C. Turner, B. S. Schmidt, M. Lipson, A. L. Gaeta, and P. Kumar, Opt. Express **14**, 12388(2006).

[31] S. Clemmen, K. Phan Huy, W. Bogaerts, R. G. Baets, Ph. Emplit, and S. Massar, Opt. Express **17**, 16558(2009).

[32] H. Takesue, H. Fukuda, T. Tsuchizawa, T. Watanabe, K. Yamada, Y. Tokura, and S.-i. Itabashi, Opt. Express **16**, 5721(2008).

[33] S. Azzini, D. Grassani, M. J. Strain, M. Sorel, L. G. Helt, J. E. Sipe, M. Liscidini, M. Galli, and D. Bajoni, Opt. Express **20**, 23100(2012).

[34] C. Xiong et al., Opt. Lett. **36**, 3413(2011).

[35] M. J. Collins et al., Nat. Commun. **4**, 2582(2013).

[36] T. Meany, L. A. Ngah, M. J. Collins, A. S. Clark, R. J. Williams, B. J. Eggleton, M. J. Steel, M. J. Withford, O. Alibart, and S. Tanzilli, Laser Photonics Rev. **8**, L42(2014).

[37] A. S. Solntsev, A. A. Sukhorukov, D. N. Neshev, and Y. S. Kivshar, Phys. Rev. Lett. **108**, 023601 (2012).

[38] M. Gräfe, A. S. Solntsev, R. Keil, A. A. Sukhorukov, M. Heinrich, A. Tünnermann, S. Nolte, A.

Szameit, and Yu. S. Kivshar, Sci. Rep. **2**, 562(2012).

[39] This work was supported by the State Key Program for Basic Research in China(No. 2012CB921802 and No. 2011CBA00205), the National Natural Science Foundation of China(Contracts No. 91321312, No. 91121001, No. 11321063, and No. 11422438), and the project funded by the Priority Academic Program Development of Jiangsu Higher Education Institutions (PAPD). P. X. acknowledges the program for New Century Excellent Talents in university(NCET), the Foundation for the Author of National Excellent Doctoral Dissertation of the People's Republic of China(FANEDD), and Deng Feng Scholars Program of Nanjing University.

Generation of Three-mode Continuous-variable Entanglement by Cascaded Nonlinear Interactions in a Quasiperiodic Superlattice[*]

Y. B. Yu, Z. D. Xie, X. Q. Yu, H. X. Li, P. Xu, H. M. Yao, and S. N. Zhu

National Laboratory of Solid State Microstructures, Department of Physics, Nanjing University, Nanjing 210093, People's Republic of China

The generation of three-mode continuous-variable entanglement in a quasiperiodically optical superlattice is studied theoretically in this paper. This work is based on the previous experiment result in which three-color light generated from a quasiperiodically optical superlattice through a stimulated parametric down-conversion cascaded with a sum-frequency process. The degree of quadrature phase amplitude correlations, a nonclassical characteristic, among the three mode was discussed by a sufficient inseparability criterion for continuous-variable entanglement, which was proposed by van Loock and Furusawa.

1. Introduction

The generation of quantum entanglement has attracted much interest, as it is the key resource in applications such as information science, quantum communication, and quantum computation. As is well known, in 1935, Einstein, Podolsky, and Rosen (EPR) proposed a quantum entanglement state in which the momentum and position are continuous variable(CV) and entangled[1]. Later Bohm extended the CV entangled state to a singlet state discrete variable system in which two spin-1/2 particles consist of so-called biparticles EPR states[2]. A lot of theoretical and experimental work has been concentrated on various discrete variable systems, such as trapped ions[3], spontaneously parametric downconversion[4], and nuclear magnetic resonance[5]. Furthermore, many experiments on quantum teleportation[6,7] and quantum entanglement swapping[8] have been successfully realized based on the biphoton EPR state. Recently, it seems that people are putting more effort into CV entanglement. A valuable feature for implementations of quantum optics based on CV is its unconditionalness because of its high efficiency[9]. Ou *et al.*, experimentally demonstrated CV entanglement using an optical parametric oscillator operating below threshold[10]. Later the CV entanglement above the threshold from an optical parametric oscillator was predicted[11] and experimentally confirmed[12]. Up to now,

---

[*] Phys.Rev.A,2006,74(4):042332

the CV entanglement state has been successfully used in unconditional quantum teleportation[13,14], quantum dense coding[15], quantum error correction[16,17], entanglement swapping[18,19], and universal quantum computation[20].

With this progress in CV entanglement research, the generation of more than two partite entanglements has attracted much attention. A truly N-partite entangled state generated by one single mode squeezed state and linear optics was theoretically studied[21]. Subsequently, a CV tripartite entangled state was experimental realized by combining three independent squeezed vacuum states[22]. Recently, generation of CV tripartite entanglement using cascaded nonlinear interaction in an optical cavity without linear optics has been studied theoretically[23]. The scheme mentioned two pump beams and a degenerately parametric down-conversion cascaded by a sum-frequency process. In our former work[24], we experimentally demonstrated that three-color light generated from a quasiperiodical superlattice through quasi-phase-matched (QPM)[25,26] parametric down-conversion cascaded by a QPM frequency adding. The signal at 666 nm, the idler at 2644 nm, and the sum frequency at 443 nm by the frequency adding of idler and pump were simultaneously achieved from a two-component quasiperiodic superlattice using a 532 nm laser as a pump. Based on this result, we propose a scheme to realize three-mode CV entanglement from an optical parametric oscillator cavity containing such a quasiperiodic superlattice as the nonlinear medium. In this scheme, only one pump beam is needed and the system can work in nondegeneracy, thus a three-mode CV entanglement state with three separate frequencies may be achieved. In this paper, we shall make a theoretical analysis of the generation of CV tripartite entanglement in this system.

The paper is structured as follows. A physical system for the cascaded nonlinear interaction in an optical cavity with a quasiperiodic superlattice as nonlinear medium is introduced in the following section. The motion equation and output field are solved in the third section. The three-mode CV entanglement characteristics are discussed in the fourth section. Finally, a brief conclusion is given in the fifth section.

2. The Cascaded Nonlinear Interaction In An Optical Oscillator Cavity

In this work, we only consider one incidence pump beam in an one-sided optical oscillator cavity with three output modes [Fig. 1(a)]. The schematic QPM diagram of the two nonlinear interaction processes is plotted in Fig. 1(b). In the first step, two beams of frequency ω_1 and ω_3 were generated by a spontaneous parametric down-conversion in the quasiperiodic $LiTaO_3$ superlattice. In the second step, the third beam was produced by a sum-frequency generation of pump and the beam ω_3. The energy-matching conditions are $\omega_p = \omega_1 + \omega_3$ and $\omega_2 = \omega_p + \omega_3$. Phase mismatching in the first step $\Delta k = k_p - k_1 - k_3$ and in the second step $\Delta k' = k_2 - k_p - k_3$ were compensated by the two predesigned reciprocals G_1

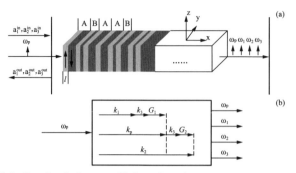

FIG.1. (a) Sketch of the one-sided cavity. A two-component quasiperiodic optical superlattice, which constructed from two building blocks *A* and *B*, is used to realize cascaded nonlinear process. Assume that the widths of positive domains in all *A* and *B* blocks are the same, represented with l. (b) The QPM schematic diagram for the cascaded nonlinear interatcion process.

and G_2, respectively. In this case, we think the bandwidth of pump is very small and approaching zero. k_p, k_1, k_3, and k_2 are the wave vectors of pump, signal, idler, and sum-frequency beam, respectively. G_1 and G_2 are determined by the parameters of superlattice. The first QPM condition for the parametric down-conversion process is $\Delta k = k_p - k_1 - k_3 - G_1 = 0$. The second QPM condition for the sum-frequency generation is $\Delta k' = k_2 - k_p - k_3 - G_2 = 0$. According to these two QPM conditions, one can design a quasiperiodic superlattice with appropriate structure units *A* and *B*, and arranged a sequence of theirs. Such two QPM processes can be simultaneous realized in the same superlattice; therefore, three cavity modes couple with each other. The interaction Hamiltonian for this coupled process can be written as[27]

$$\hat{H}_I = i\hbar e^{-i\omega_p t}(\kappa_1 \hat{a}_1^\dagger \hat{a}_3^\dagger + \kappa_2 \hat{a}_3 \hat{a}_2^\dagger) - i\hbar e_p^{i\omega t}(\kappa_1 \hat{a}_1 \hat{a}_3 + \kappa_2 \hat{a}_3^\dagger \hat{a}_2), \tag{1}$$

where $\hat{a}_i (i=1,2,3)$ are annihilation operators of the cavity modes which frequency are ω_i ($i=1, 2, 3$), respectively. κ_1 and κ_2 are the effective coupling constants, which are proportional to the nonlinear susceptibility, structure parameters, and pump intensity and taken to be real without loss of generality[28,29].

3. The Motion Equation And Output Field

Following the formalism which was obtained by Gardiner and Collett, the quantum Langevin equations for the three cavity modes $\hat{a}_1$, $\hat{a}_2$, and $\hat{a}_3$ can be written as[30]

$$\tau \dot{\hat{a}}_1^\dagger = i\omega_1 \tau \hat{a}_1^\dagger + \kappa_1 \hat{a}_3 e^{i\omega_p t} - \frac{\gamma_1}{2}\hat{a}_1^\dagger + \sqrt{\gamma_1}\,\hat{a}_1^{\dagger\text{in}}, \tag{2}$$

$$\tau \dot{\hat{a}}_2 = -i\omega_2 \tau \hat{a}_2 + \kappa_2 \hat{a}_3 e^{-i\omega_p t} - \frac{\gamma_2}{2}\hat{a}_2 + \sqrt{\gamma_1}\,\hat{a}_2^{\text{in}}, \tag{3}$$

$$\tau \dot{\hat{a}}_3 = -i\omega_3 \tau \hat{a}_3 + \kappa_1 \hat{a}_1^\dagger e^{-i\omega_p t} - \kappa_2 \hat{a}_2 e^{i\omega_p t} - \frac{\gamma_3}{2}\hat{a}_2 + \sqrt{\gamma_3}\,\hat{a}_3^{\text{in}} \tag{4}$$

Where τ is the round-trip time of light in the cavity and it is assumed to be the same value for all the three fields. γ_1 and γ_2 are damping rates which relate to the amplitude reflection and the transmission coefficients of the input and the output couplers of the optical cavity. $\hat{a}_i^{in}(i=1,2,3)$ and input fields operators of the cavity.

From Fourier transformation in rotating frame and the bounday condition $\hat{A}_i^{out} = \sqrt{\gamma_i}\,\hat{A}_i - \hat{A}_i^{in}(i=1,2,3)$ on the coupling mirror[27,29], we can have

$$\hat{A}_1^{\dagger out}(-\omega) = \frac{2\gamma_1\left\{\kappa_2^2 + \frac{1}{4}\left[\gamma_3 + 2i\tau\left(\omega - \frac{\Delta\omega}{2}\right)\right]\left[\gamma_2 + 2i\tau\left(3\omega - \frac{\Delta\omega}{2}\right)\right]\right\}}{M}\hat{A}_1^{\dagger in} -$$

$$\frac{2\kappa_1\kappa_2\sqrt{\gamma_1\gamma_2}}{M}\hat{A}_2^{in} + \frac{\kappa_1\sqrt{\gamma_1\gamma_3}\left[\gamma_2 + 2i\tau\left(3\omega - \frac{\Delta\omega}{2}\right)\right]}{M}\hat{A}_3^{in}, \qquad (5)$$

$$\hat{A}_2^{\dagger out}(3\omega) = \frac{2\gamma_2\left\{-\kappa_1^2 + \frac{1}{4}\left[\gamma_3 + 2i\tau\left(\omega - \frac{\Delta\omega}{2}\right)\right]\left[\gamma_2 + 2i\tau\left(\omega + \frac{\Delta\omega}{2}\right)\right]\right\}}{M}\hat{A}_2^{\dagger in} +$$

$$\frac{2\kappa_1\kappa_2\sqrt{\gamma_1\gamma_2}}{M}\hat{A}_1^{\dagger in} + \frac{\kappa_2\sqrt{\gamma_2\gamma_3}\left[\gamma_1 + 2i\tau\left(\omega + \frac{\Delta\omega}{2}\right)\right]}{M}\hat{A}_3^{in}, \qquad (6)$$

$$\hat{A}_3^{out}(\omega) = \frac{\gamma_3\left[\gamma_2 + 2i\tau\left(3\omega - \frac{\Delta\omega}{2}\right)\right]\left[\frac{\gamma_1}{2} + i\tau\left(\omega + \frac{\Delta\omega}{2}\right)\right]}{M}\hat{A}_3^{in} +$$

$$\frac{\kappa_1\sqrt{\gamma_1\gamma_3}\left[\gamma_2 + 2i\tau\left(3\omega - \frac{\Delta\omega}{2}\right)\right]}{M}\hat{A}_1^{\dagger in} - \frac{\kappa_2\sqrt{\gamma_2\gamma_3}\left[\gamma_1 + 2i\tau\left(\omega + \frac{\Delta\omega}{2}\right)\right]}{M}\hat{A}_2^{\dagger in} \qquad (7)$$

with

$$M = \kappa_2^2\left[\gamma_1 + 2i\tau\left(\omega + \frac{\Delta\omega}{2}\right)\right] + \left[\gamma_2 + 2i\tau\left(3\omega - \frac{\Delta\omega}{2}\right)\right] \\ \times \left\{-\kappa_1^2 + \frac{1}{4}\left[\gamma_3 + 2i\tau\left(\omega - \frac{\Delta\omega}{2}\right)\right]\left[\gamma_1 + 2i\tau\left(\omega + \frac{\Delta\omega}{2}\right)\right]\right\}, \qquad (8)$$

where ω is the analysis frequency. $\Delta\omega = \omega_1 - \omega_3$. When $\Delta\omega = 0$ down-conversion is degenerate.

4. Three-mode CV Entanglement Characteristics

Van Loock and Furusawa have proposed the sufficient inseparability criterion for CV multimode entanglement[31]. The sufficient inseparability criterion for CV three-mode entanglement is[31]

$$\langle \delta^2(X_2 - X_3)\rangle + \langle \delta^2(g_1Y_1 + Y_2 + Y_3)\rangle < 1, \qquad (9)$$

$$\langle \delta^2(X_1 - X_3)\rangle + \langle \delta^2(Y_1 + g_2Y_2 + Y_3)\rangle < 1, \qquad (10)$$

$$\langle \delta^2(X_1-X_2)\rangle + \langle \delta^2(Y_1+Y_2+g_3Y_3)\rangle < 1, \tag{11}$$

where g_1, g_2, and g_3 are adjustable factors. The minimize value in the left of the inequalities can be obtained by choosing an appropriate g. The satisfaction of any pair of above inequalities in sufficient for full inseparability of three-mode entanglement[31].

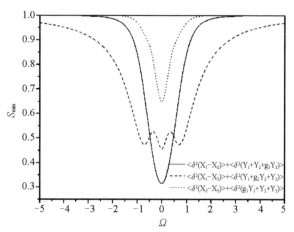

FIG.2. The quantum correlation spectra versus normalized analyzing frequency $\Omega = \omega\tau/\gamma_1$

From the definition of $X_i = \frac{1}{2}[\hat{A}_i^{\text{out}}(\omega) + \hat{A}_i^{\dagger\text{out}}(-\omega)]$ and $Y_i = \frac{1}{2i}[\hat{A}_i^{\text{out}}(\omega) - \hat{A}_i^{\dagger\text{out}}(-\omega)]$, the fluctuations of quadrature amplitude and phase components can be caculated as

$$\begin{pmatrix} X_1^{\text{out}} \\ X_2^{\text{out}} \\ X_3^{\text{out}} \end{pmatrix} = \begin{pmatrix} A & B & C \\ -B & D & E \\ C & -E & F \end{pmatrix} \begin{pmatrix} X_1^{\text{in}} \\ X_2^{\text{in}} \\ X_3^{\text{in}} \end{pmatrix} \tag{12}$$

and

$$\begin{pmatrix} Y_1^{\text{out}} \\ Y_2^{\text{out}} \\ Y_3^{\text{out}} \end{pmatrix} = \begin{pmatrix} A & -B & -C \\ B & D & E \\ -C & -E & F \end{pmatrix} \begin{pmatrix} Y_1^{\text{in}} \\ Y_2^{\text{in}} \\ Y_3^{\text{in}} \end{pmatrix} \tag{13}$$

with

$$A = -1 + 2\gamma_1 \left\{ \kappa_2^2 + \frac{1}{4}\left[\gamma_3 + 2i\tau\left(\omega - \frac{\Delta\omega}{2}\right)\right]\left[\gamma_2 + 2i\tau\left(3\omega - \frac{\Delta\omega}{2}\right)\right]\right\}/M,$$

$$B = 2\kappa_1\kappa_2\sqrt{\gamma_1\gamma_2}/M,$$

$$C = \kappa_1\sqrt{\gamma_1\gamma_3}\left[\gamma_2 + 2i\tau\left(3\omega - \frac{\Delta\omega}{2}\right)\right]/M,$$

$$D = \frac{-1 + 2\gamma_2\left\{-\kappa_1^2 + \frac{1}{4}\left[\gamma_3 + 2i\tau\left(\omega - \frac{\Delta\omega}{2}\right)\right]\left[\gamma_1 + 2i\tau\left(\omega + \frac{\Delta\omega}{2}\right)\right]\right\}}{M}$$

$$E = \kappa_2\sqrt{\gamma_2\gamma_3}\left[\gamma_1 + 2i\tau\left(\omega + \frac{\Delta\omega}{2}\right)\right],$$

$$F = -1 + \gamma_3\left[\gamma_2 + 2i\tau\left(3\omega - \frac{\Delta\omega}{2}\right)\right]\left[\frac{\gamma_1}{2} + i\tau\left(\omega + \frac{\Delta\omega}{2}\right)\right]/M. \tag{14}$$

From the above calculation, the correlation spectra of the total phase quadratures of three-mode and relative amplitude quadratures can be obtained as

$$\langle \delta^2 (X_1 - X_2) \rangle + \langle \delta^2 (Y_1 + Y_2 + g_3 Y_3) \rangle$$
$$= \langle (X_1^{\text{out}} - X_2^{\text{out}})^\dagger (X_1^{\text{out}} - X_2^{\text{out}}) \rangle + \langle (Y_1^{\text{out}} + Y_2^{\text{out}} + g_3 Y_3^{\text{out}})^\dagger (Y_1^{\text{out}} + Y_2^{\text{out}} + g_3 Y_3^{\text{out}}) \rangle$$
$$= \frac{1}{4}(|A+B|^2 + |B-D|^2 + |C-E|^2) + \frac{1}{4}(|A+B-g_3 C|^2 + |-B+D-g_3 E|^2$$
$$+ |-C+E+g_3 F|^2), \tag{15}$$

$$\langle \delta^2 (X_1 - X_3) \rangle + \langle \delta^2 (Y_1 + g_2 Y_2 + Y_3) \rangle$$
$$= \langle (X_1^{\text{out}} - X_3^{\text{out}})^\dagger (X_1^{\text{out}} - X_3^{\text{out}}) \rangle + \langle (Y_1^{\text{out}} + g_2 Y_2^{\text{out}} + Y_3^{\text{out}})^\dagger (Y_1^{\text{out}} + g_2 Y_2^{\text{out}} + Y_3^{\text{out}}) \rangle$$
$$= \frac{1}{4}(|A-C|^2 + |B+E|^2 + |C-F|^2) + \frac{1}{4}(|A+g_2 B-C|^2 + |-B+g_2 D-E|^2$$
$$+ |-C+g_2 E+F|^2), \tag{16}$$

$$\langle \delta^2 (X_2 - X_3) \rangle + \langle \delta^2 (g_1 Y_1 + Y_2 + Y_3) \rangle$$
$$= \langle (X_2^{\text{out}} - X_3^{\text{out}})^\dagger (X_2^{\text{out}} - X_3^{\text{out}}) \rangle + \langle (g_1 Y_1^{\text{out}} + Y_2^{\text{out}} + Y_3^{\text{out}})^\dagger (g_1 Y_1^{\text{out}} + Y_2^{\text{out}} + Y_3^{\text{out}}) \rangle$$
$$= \frac{1}{4}(|B+C|^2 + |D+E|^2 + |E-F|^2) + \frac{1}{4}(|g_1 A+B-C|^2 + |-g_1 B+D-E|^2$$
$$+ |-g_1 C+E+F|^2). \tag{17}$$

The quantum correlation spectra $\langle \delta^2 (X_1 - X_2) \rangle + \langle \delta^2 (Y_1 + Y_2 + g_3 Y_3) \rangle$, $\langle \delta^2 (X_1 - X_3) \rangle + \langle \delta^2 (Y_1 + g_2 Y_2 + Y_3) \rangle$, and $\langle \delta^2 (X_2 - X_3) \rangle + \langle \delta^2 (g_1 Y_1 + Y_2 + Y_3) \rangle$ versus normalized analyzing frequency $\Omega = \omega\tau/\gamma_1$ are plotted in Fig. 2 with $\gamma_1 = \gamma_3 = 0.02$, $\gamma_2 = 0.06$, $\kappa_1/\gamma_1 = 0.4$, $\kappa_2/\gamma_1 = 1$, and $\Delta\omega = 0$ by solid, dashed, and dotted line, respectively. From Fig. 2 we can see that all three inequalities are satisfied in a wide frequency range which is enough to assert that the three modes are entangled. In addition, the smaller the values of quantum correlation spectra are, the larger the correlation degree that can be obtained. From Fig. 2, we can see that the large correlation degree can be obtained at a low analyzing frequency. The correlation degree between a_1 and a_3 which is generated from parametric down-onversion is also larger than that of the other two pairs a_1, a_2 and a_3, a_2 at the large analyzing frequency. However, with the decrease of the analyzing frequency, the correlation degree between a_1 and a_2 is larger than that of a_1 and a_3.

The quantum correlation spectra $\langle \delta^2 (X_1 - X_2) \rangle + \langle \delta^2 (Y_1 + Y_2 + g_3 Y_3) \rangle$, $\langle \delta^2 (X_1 - X_3) \rangle + \langle \delta^2 (Y_1 + g_2 Y_2 + Y_3) \rangle$, and $\langle \delta^2 (X_2 - X_3) \rangle + \langle \delta^2 (g_1 Y_1 + Y_2 + Y_3) \rangle$ versus κ_1/γ_1 are plotted in Fig.3 with $\gamma_1 = \gamma_3 = 0.02$, $\gamma_2 = 0.06$, $\Omega = 0.8$, $\kappa_2/\gamma_1 = 1$, and $\Delta\omega = 0$ by solid, dashed, and dotted lines, respectively. The κ_1/γ_1 includes pump power, structure parameters of the superlattice, and nonlinear coefficient for down-conversion and the threshold is at $\kappa_1 = \gamma_1$. From Fig. 3 we can see that three modes are entangled both below and above the threshold. CV entanglement is superior to discrete variable entanglement for CV entanglement can generate above the threshold. The best correlation degree for all three modes can be achieved at about $\kappa_1/\gamma_1 = 0.5$. The correlation degree of a_1 and a_3 which was generated from the parametric down-conversion is larger than that of the other two

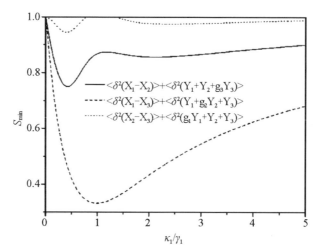

FIG.3. The quantum correlation spectra versus κ_1/γ_1.

pairs a_1, a_2 and a_3, a_2, in turn, the correlation degree between a_1 and a_2 is larger than that between a_2 and a_3. The best correlation of a_1 and a_3 can be obtained at the threshold $\kappa_1 = \gamma_1$. However, a_2 field and a_3 field are not inseparable near the threshold. The three quantum correlation spectra $\langle \delta^2(X_1-X_2)\rangle + \langle \delta^2(Y_1+Y_2+g_3Y_3)\rangle$, $\langle \delta^2(X_1-X_3)\rangle + \langle \delta^2(Y_1+g_2Y_2+Y_3)\rangle$, and $\langle \delta^2(X_2-X_3)\rangle + \langle \delta^2(g_1Y_1+Y_2+Y_3)\rangle$ versus κ_2/γ_1 are plotted in Fig. 4 with $\gamma_1=\gamma_3=0.02$, $\gamma_2=0.06$, $\Omega=0.8$, $\kappa_1/\gamma_1=0.3$, and $\Delta\omega=0$ by solid, dashed, and dotted lines, respectively. The κ_2/γ_1 includes pump power, structure parameters of the superlattice, and the nonlinear coefficient for the sum-frequency process. From Fig. 4 one can see only the correlation of a_1 and a_3 which is generated by parametric down-conversion below 1 at $\kappa_2/\gamma_1 = 0$ as no sum-frequency process takes place under this condition and only a_1 and a_3 generate through parametric down-conversion. If $\kappa_2 \neq 0$, the a_2 field will generate by the sum-frequency process and the partial energy of the pump field and the a_3 field will be transferred to the a_2 field. Then the three-mode entanglement will come about. From Fig. 4 one can see the correlation between a_1 and a_2 increase and the correlation of a_1 and a_3 decrease with the increase of κ_2/γ_1.

Figures 3 and 4 show that the mode entanglement is critically associated with the ratio of the effective coupling constant κ_i to damping rate when the phase matching condition is satisfied ($\Delta k = \Delta k' = 0$). In a quasiperiodic superlattice, κ_1 and κ_2 depend not only on the nonlinear susceptibility of the crystal but also on the structure parameters of quasiperiodic superlattices, that is, the ratio of the positive and negative domain in the structure units A and B[26]. So, in addition to the damping rate determined by the reflectivity of cavity mirrors and nonlinear susceptibility of the crystal, the structural parameter of the superlattice is also an adjustable parameter for the study of quantum correlation. One can obtain stronger three-mode CV entanglement by optimizing the duty cycle of modulation of nonlinear susceptibility in a quasiperiodic superlattice.

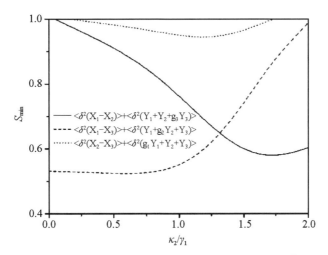

FIG. 4. The three quantum correlation spectra versus κ_2/γ_1.

In Fig. 5, the quantum correlation spectra $\langle\delta^2(X_1-X_2)\rangle+\langle\delta^2(Y_1+Y_2+g_3Y_3)\rangle$, $\langle\delta^2(X_1-X_3)\rangle+\langle\delta^2(Y_1+g_2Y_2+Y_3)\rangle$, and $\langle\delta^2(X_2-X_3)\rangle+\langle\delta^2(g_1Y_1+Y_2+Y_3)\rangle$ versus relative detuning of down-conversion $\Delta\omega/\omega$ with $\gamma_1=0.02$, $\gamma_2=0.06$, $\gamma_3=0.04$, $\Omega=0.2$, $\kappa_1/\gamma_1=0.5$, and $\kappa_2/\gamma_1=1$ are plotted by solid, dashed, and dotted lines, respectively. Their values are notably below 1 within a wide detuning range. The CV entanglement was produced by cascaded nonlinear interactions which can be realized within a wide detuning range by choosing appropriate superlattice structure parameters. However, the best correlations for three modes are not at $\Delta\omega=0$. A better quantum correlation can be realized at $\Delta\omega<0(\omega_1<\omega_3)$ for the energy of the a_3 field partly transferred to the a_2 field after the process of down-conversion.

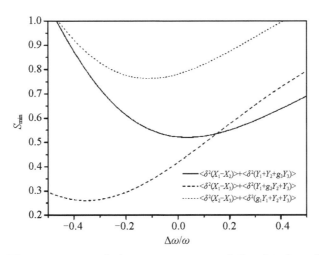

FIG. 5. The quantum correlation spectra versus relative detuning of down-conversion $\Delta\omega/\omega$.

5. Conclusions

We proposed a scheme to generate three-mode CV entanglement by QPM cascaded nonlinear process in a quasiperiodic superlattice. The CV entanglement of three modes is demonstrated theoretically by a sufficient inseparability criterion, and the entanglement characteristics are discussed as well. This cascaded nonlinear process in a quasiperiodic superlattice has been realized by a single-pass QPM process in our laboratory[24], which means that the three-mode CV entanglement is experimentally feasible when the superlattice is set into an optical oscillator cavity. This scheme of three-mode CV entanglement generation by only one pump and a single superlattice is very significant for the application in quantum information and quantum communication.

References and Notes

[1] A. Einstein, B. Podolsky, and N. Rosen, Phys. Rev. **47**, 777(1937).
[2] D. Bohm, *Quantum Theory*,(Prentice Hall, New York, 1951).
[3] Q. A. Turchette *et al.*, Phys. Rev. Lett. **81**, 3631(1998).
[4] Y. H. Shih and C. O. Alley, Phys. Rev. Lett. 61, 2921(1988).
[5] N. Gershenfeld and I. L. Chuang, Science **275**, 350(1997).
[6] D. Bouwmeester *et al.*, Nature(London) **390**, 575(1997).
[7] M. A. Nielsen, E. Knill, and R. Laflamme, Nature(London) **396**,52(1998).
[8] J. W. Pan, D. Bouwmeester, H. Weinfurter, and A. Zeilinger, Phys. Rev. Lett. **80**, 3891(1998).
[9] S. L. Braunstein and P. van Loock, Rev. Mod. Phys. **77**, 513(2005).
[10] Z. Y. Ou, S. F. Pereira, H. J. Kimble, and K. C. Peng, Phys. Rev. Lett. **68**, 3663(1992).
[11] M. D. Reid and P. D. Drummond, Phys. Rev. Lett. **60**, 2731(1988).
[12] A. S. Villar, L. S. Cruz, K. N. Cassemiro, M. Martinelli, and P. Nussenzveig, Phys. Rev. Lett. **95**, 243603(2005).
[13] A. Furusawa *et al.*, Science **282**, 706(1998).
[14] Samuel L. Braunstein, and H. J. Kimble, Phys. Rev. Lett. **80**, 869(1998).
[15] X.Y. Li *et al.*, Phys. Rev. Lett. **88**, 047904(2002).
[16] Samuel L. Braunstein, Nature(London) 394,47(1998); Phys. Rev. Lett. **80**, 4084(1998).
[17] S. Lloyd and Jean-Jacques E. Slotine, Phys. Rev. Lett. **80**, 4088(1998).
[18] R. E. S. Polkinghorne and T. C. Ralph, Phys. Rev. Lett. **83**, 2095(1999).
[19] X. J. Jia *et al.*, Phys. Rev. Lett. **93**, 250503(2004).
[20] S. Lloyd and S. L. Braunstein, Phys. Rev. Lett. **82**, 1784(1999).
[21] P. van Loock and S. L. Braunstein, Phys. Rev. Lett. **84**, 3482(2000).
[22] T. Aoki *et al.*, Phys. Rev. Lett. **91**, 080404(2003).
[23] J. Guo, H. X. Zou, Z. H. Zhai, J. X. Zhang, and J. R. Gao, Phys. Rev. A **71**, 034305(2005).
[24] Y. Du *et al.*, Appl. Phys. Lett. **81**, 1573(2002).
[25] S. N. Zhu, Y. Y. Zhu, and N. B. Ming, Science **278**, 843(1997).
[26] S. N. Zhu *et al.*, Phys. Rev. Lett. **78**, 2752(1997).

[27] M. J. Collett and C. W. Gardiner, Phys. Rev. A **30**, 1386(1984).

[28] A. Ferraro *et al.*, J. Opt. Soc. Am. B **21**, 1241(2004).

[29] Z. Y. Ou, S. F. Pereira, and H. J. Kimble, Appl. Phys. B **55**, 265(1992).

[30] C. W. Gardiner and M. J. Collett, Phys. Rev. A **31**, 3761(1985).

[31] P. van Loock and A. Furusawa, Phys. Rev. A **67**, 052315(2003).

[32] This work is supported by the National Natural Science Foundations of China under Contract Nos. 10534042 and 60578034 and the Sate Key Program for Basic Research of China(2004CB619003).

第七章 介电体超晶格与畴工程学
Chapter 7 Domain Engineering for Dielectric Superlattice

7.1 The Growth Striations and Ferroelectric Domain Structures in Czochralski-grown $LiNbO_3$ Single Crystals

7.2 Growth of Optical Superlattice $LiNbO_3$ with Different Modulating Periods and Its Applications in Second-harmonic Generation

7.3 Growth of Nd^{3+}-doped $LiNbO_3$ Optical Superlattice Crystals and Its Potential Applications in Self-frequency Doubling

7.4 Fabrication of Acoustic Superlattice $LiNbO_3$ by Pulsed Current Induction and Its Application for Crossed Field Ultrasonic Excitation

7.5 $LiTaO_3$ Crystal Periodically Poled by Applying an External Pulsed Field

7.6 Poling Quality Evaluation of Optical Superlattice Using 2D Fourier Transform Method

7.7 Frequency Self-doubling Optical Parametric Amplification: Noncollinear Red-green-blue Lightsource Generation based on a Hexagonally Poled Lithium Tantalate

7.8 Direct Observation of Ferroelectric Domains in $LiTaO_3$ Using Environmental Scanning Electron Microscopy

7.9 Nondestructive Imaging of Dielectric-Constant Profiles and Ferroelectric Domains with a Scanning-Tip Microwave Near-Field Microscope

第七章　介电体超晶格与畴工程学

祝世宁

介电体中的超晶格可以用不同的介电微结构来构建,如相结构、畴结构、组分和异质结构、孪晶结构等。在本书中所涉及的介电体超晶格无论是光学超晶格、声学超晶格还是离子型声子晶体大都是使用了铁电晶体作为基质材料,以铁电畴作为构建超晶格的基本单元。研究铁电晶体中铁电畴的可控制备,畴结构的设计、表征及其与晶体宏观物理性能之间关系导致了畴工程学(Domain Engineering)的诞生。畴工程的概念是由著名的材料物理学家 L.E.Cross 和 J.Fousek 于上世纪九十年代提出来的。他们跟据铁电畴对晶体物性的影响将畴工程分成两类:一类叫畴几何工程(Domain-Geometry-Engineering,简称 DGE),另一类叫畴平均工程(Domain-Average-Engineering,简称 DAE)。研究介电体超晶格中铁电畴的构建属于畴工程的第一类。三十年来,我们重点研究了铁电畴有序分布对铁电晶体中光、声等物理效应的影响,开发出了如光学超晶格、声学超晶格和离子型声子晶体三种新功能晶体。这些超晶格晶体展示了均匀晶体所不具有的独特性能,大大扩展和提升了晶体原有的功能。畴平均工程则是研究驰豫铁电体晶体(如铌镁酸铅(PMN),铌锌酸铅(PZN)等)中微畴及其组态导致的压电异常而归纳出来的,它解释了这类材料优异的压电和机电耦合性能的起源。2003 年 J.Fousek 在英国利兹举行的铁电学研究 55 周年大会发言中指出:"畴工程这一概念已引起极大关注,两种畴工程大致同时呈现了。第一种是指在晶体中将畴排成所需的几何结构,这种可能性被中国和以色列科学家发现。他们在铌酸锂中制备出周期畴结构并证实它的非线性光学效应,之后又在此基础上做了关键性推进,即制备出 Fibonacci 序列的铁电畴超晶格。"

广义地讲畴工程学应该包括超晶格的设计、制备、检测、性能表征和应用等,并涉及其中的物理基础及技术。在前面各章中我们已着重讨论了超晶格结构设计和功能之间的联系,介绍了声学超晶格、光学超晶格和离子型声子晶体的功能及其在高频、超高频声学器件、非线性光学与全固态激光器件以及量子光源与光量子芯片方面重要应用。在本章中我们挑选的论文侧重介绍工程畴的制备、检测和表征技术及其科学原理。我们精选了 9 篇论文,这 9 篇论文大致可分成三类。第一类论文介绍介电体超晶格制备的晶体生长条纹法,它的发明、发展及其应用;第二类论文介绍了介电体超晶格制备的室温电场极化技术和它的应用;第三类论文介绍了我们发展的两种铁电畴无损探测和成像新技术。

7.1~7.4 的 4 篇论文介绍了介电体超晶格的晶体生长条纹方法。这种方法提出于上世纪七十年代末,成熟于上世纪九十年代中。首次实验验证准相位匹配倍频增强效应的周期极化 $LiNbO_3$(LN)晶体就是用这种 Czochralski 技术生长出来的。这种技术能在铌酸锂晶体生长过程中控制晶体内部杂质元素(通常是钇,yttrium)浓度的起伏,通过周期性的浓度起伏在 LN 晶体内建立起周期性的内建电场,在晶体冷却通过相变居里点时这内建电场能诱导出周期极化的电畴。为了满足上述条件，发展了两种方法，一种是让拉晶杆转轴略为偏离温场的中心点(7.1～7.3)，这能引起晶体旋转提拉生长过程中固液界面处温度的涨落和杂质浓度的起伏，产生生长条纹，条纹的周期由拉杆的提拉速度及转速之比决定，而条纹的周期与畴周期一一对应；另一种是通过在金属提拉杆上周期性地提供正反向电流，也可达到同样的效果(7.4)。生长法的最大优点是能制备大截面的超晶格晶片，这种晶片在研制高功率激光变频器、高功率超声器件方面具有优势。生长法还可以将有用的元素作为杂质掺进超晶格晶体，借以改善和提高晶体的光学和热学性能。例如掺镁(Mg)、掺锌(Zn)可以提高晶体的抗损伤阈值，保证LN、$LiTaO_3$(LT)光学超晶格能在高功率激光条件下使用；掺稀土离子如钕(Nd)、铒(Er)则可使光学超晶格也同时具有激光活性，可以用于研制结构紧凑的自倍频激光器(7.3)。生长法的缺点是对生长技术和生长条件要求很高，很难长出光学均匀性很高、周期严格保证的超晶格晶体。

1992年Yamada等人报道了LN晶体在室温实现极性反转的实验结果，利用这一技术，他们在一块0.2 mm厚的LN晶片上实现了周期畴极化反转，制备出蓝光倍频光波导。1994年，我们研究组在LT晶体上也实现了室温极化的突破，先后制备出0.3～0.5 mm厚的LT超晶格晶片，实现了蓝光与绿光的直接倍频(7.5)。在发展室温极化这一超晶格制备关键技术方面，我们的主要贡献是：1. 通过研究畴反转过程中极化电流随时间变化规律，发现其中有两个不同的时间常数，分别对应于高、低场下极化反转所遵循的不同规律；2. 提出了用单脉冲极化方法代替当时普遍采用的多脉冲法，通过控制极化反转电量即极化电流大小和单脉长短来控制畴的反转量，以获得正负畴宽度比的最优值。上述两点后来成为进一步优化极化技术的基础。尽管如此，实际采用极化方法制备出的超晶格晶体中正负畴宽度与设计值仍会有一定的偏离，这一偏离会带来晶体中有效非线性系数下降，影响晶体的使役性能和一致性。7.6介绍了我们发展的根据晶体极性面上正负畴的实际图形推算超晶格晶体有效非线性系数的傅里叶变换方法，用这种方法推算出的超晶格晶体有效二阶非线性光学系数与光学实验实际测量值非常接近。

7.7一文介绍了利用畴工程原理实现红绿蓝三色激光器的设计方案及实验结果。该方案中非线性晶体采用的是六角极化的LT晶片，基波光(绿光，波长532 nm)沿轴x方向入射，由非共线匹配出射的红光与蓝光则是对应于沿这两个方向的两束参量光的共线直接倍频，用一个基波光源和一块光学超晶格获得了红绿蓝三基色激光的同时非共线输出。此方案中这块六角极化的LT超晶格晶体的功能相当于集成了一块光参量放大晶体，两块倍频晶体以及两块激光分束器。

畴工程研究急需要发展新的铁电畴的观察和成像技术，特别是无损技术。长期以来铁电畴的观察一直是采用光学方法即光学显微术，样品观察前要经预处理，要放入酸溶液中进行表面腐蚀，由于畴的正负极性表面腐蚀速率不一样，通常可以通过观察晶体表面腐蚀形貌，即可获得畴的形貌及其分布。这种技术要损坏样品表面，不能用于已完成加工的超晶格样品，因此发展铁电畴的无损观察和成像十分重要，我们针对两种主要制备超晶格的技术发展了两种新的铁电畴无损表征和成像方法：环境扫描电镜二次电子成像和扫描微波近场显微成像。

7.8报告了将环境扫描二次电子成像(ESEM)技术应用于铁电畴的无损观察与成像的原理和实验技术。普通SEM(CSEM)也常被用来观察铁电材料表面畴的结构和形貌，不过观察前样品表面也须先经腐蚀处理产生表面形貌,然后喷镀一层金属薄膜,以避免电子在材料表面过度积累干扰成像。与光学成像相比,CSEM 的分辨更高,可达纳米量级。但腐蚀和镀膜对样品仍是破坏性的。Bihan 和 Sogr 等曾尝试使用 CSEM 技术不取形貌像而直接利用不同极性表面二次电子发射率的差别来形成畴的衬度像,他们在铁电晶体 TGS 的未经处理的极性面上获得了正负畴的衬度像。为了避免表面电荷过度积累,实验中他们使用了低电子加速电压(0.2 - 4 kV)和小束电流($\sim 10^{-8}$ A)等观察条件。不过即使这样,畴所成的衬度像的时间也很短,通常只有几秒钟到几分钟。为了能形成高稳定的畴衬度像,我们在 7.9 中引进了环境扫描电子(ESEM)技术,ESEM 能在低气压($p = 1.0 - 3.0$ Torr)环境下工作,电子束与水分子碰撞所产生的自由离子为沉积在样品表面的电荷提供了转移通道,因而畴的二次电子成像可以更持久。因为电子束工作电压仍可高达 30 kV,所以成像的分辨率要远高于低电压模式,高的分辨率和稳定的成像是这种技术的优势。

ESEM 技术主要用于观察电场极化的超晶格样品,对于晶体生长方法的超晶格样品,我们发展了一种微波近场扫描(SNMM)技术,利用晶体表面对微波的共振响应分别对表面介电常数分布的实部和虚部成像。共振频率的变化对应的是介电常数实部,它反映的是晶体中杂质浓度的变化,损耗即 Q 因子成像描述的是介电常数虚部,畴壁附近损耗最大。7.9 介绍了利用 SNMM 技术对掺钇 LN 晶体中周期结构进行观测和成像的研究结果。

我们在介电体超晶格方面的系统研究工作主要是建立在铁电畴工程基础之上的,该工作同时也催生了铁电材料领域畴工程这一分支学科。

Chapter 7 Domain Engineering for Dielectric Superlattice

Shining Zhu

The superlattice in dielectric crystal can be constructed by various dielectric microstructures, such as the phase structures, domain structures, composition and heterostructures etc. Ferroelectric materials represent a family of dielectric materials. The superlattice in ferroelectric crystal is mainly composed of 180° anti-parallel laminar domains arranged according to some order. Most of the dielectric superlattices we develop are ferroelectric superlattices, which are prepared by domain engineering. The concept of domain engineering was originally proposed by L. E. Cross and J. Fousek in the nineties of the last century. In their original paper, domain engineering is divided into two types: one is called domain-geometry-engineering (DGE), the other is called domain-average-engineering (DAE). The construction of ferroelectric superlattice belongs to the first category. We developed three kinds of superlattice crystals by DGE in terms of their functions: optical superlattice, acoustic superlattice and ionic-type phononic crystal. They exhibit unique properties in contrast to the uniform crystals. DAE mentions the configurations of domain in a relaxed ferroelectric crystal, such as PMN, PZN etc. There is a very large number of micro-domains inside these crystals. The response of DAE crystal to external field is roughly described by tensorial properties averaged over all of the domain states in crystal. J. Fousek addressed in the 55th annual meeting of ferroelectric research (2003): "This possibility was rediscovered especially by researchers in China and Israel. Feng et al. succeeded in producing a periodic domain pattern in LN and proved its efficiency in nonlinear optics. Later Zhu et al. initiated key progress in the field by producing a domain pattern whose geometry corresponds to the Fibonacci superlattice".

In general, domain engineering should include how to prepare the orderly domains in ferroelectric superlattice and how to design, test, characterize and utilize various engineered domain structures. In our previous chapters we have extensively introduced many important applications of engineered domains on acoustic, nonlinear optics and quantum optics. In this chapter, we focus on the preparation, detection and characterization of engineering domain, and the understanding of the scientific principle behind them. Since the 1980s, some techniques for controlling domains in superlattice crystals, either during or after growth, have been developed. Among them, the growth striation technique and external electric field poling technique are used mainly. We select 9 papers in this chapter from several dozen papers we have published. These 9 papers can be divided into three parts. In the first one, from 7.1 to 7.4, we introduce the growth striation technique, and its applications in laser frequency doubling and acoustic devices. The second one, from 7.5 to 7.7, is about external electric field poling technique (poling technique, for short). We give a representative example to show how to design a two-dimension (2D) optical superlattice for RGB laser. In the third part, which consists of 7.8 and 7.9, we introduce two kinds of the new technologies we developed for the ferroelectric domain nondestructive characteristic and imaging.

The growth striation technique was proposed in the late seventies of the last century and matured in the nineties of the last century. The superlattice for the first experimental demonstration of the quasi-phase-matched SHG enhancement effect was prepared by use of the so-called Czochralski growth technology. In this technique the melt is doped with a solute to control domain structure, such as yttrium (Y) for LN, with concentration about 0.1% - 0.5%. In the process of crystal growth, a temperature fluctuation may be introduced into the solid-liquid interface, either via an eccentric rotation (7.1~7.3) or by means of applying an alternating electric current (7.4). This leads to growth striation, i.e. regular solute concentration fluctuation in the crystal along the growth direction. The solutes in the crystal are generally ionized but not completely shielded. A non-uniform solute distribution is equivalent to a non-uniform space-charge distribution in crystal, which can induce local electric field. When cooling through the Curie point, the modulated domain, or superlattice, may be automatically induced by the growth striation. The advantage for the growth striation technique is that it can make a large cross section of the superlattice sample, which allows the use of larger optical and acoustic aperture. The high power laser frequency converter and high power ultrasonic devices can benefit from this advantage (7.4). Another advantage of the growth method is that it can be used to improve the optical and thermal properties of the crystal. For example, Mg and Zn doped LN and LT can improve the anti-damage threshold of the crystals, which can ensure the use of the optical superlattice in high power laser conditions. The rare-earth ions such as Nd and Er can be doped into LN and LT crystal as well, which can make the optical superlattice as a laser gain medium besides nonlinear properties. 7.3 reports the result of the laser self frequency doubling in an Nd doped LN optical superlattice. The growth method, however, has its drawback. It is difficult to guarantee the optical uniformity and the strict period of superlattice.

A definite goal for application is to find a practical technique that can mass-produce ferroelectric superlattice at low cost. In 1993, Yamada *et al.* developed a new technique: poling technique. They successfully fabricated a periodic domain structure in a ~ 0.2 mm thick LN thin wafer by applying a pulsed field at room temperature. In 1994, we also made such breakthrough in LT crystals. We prepared periodic superlattice in 0.3 – 0.5 mm thick LT wafers by using a single electric pulse, and obtained blue and green lights by direct frequency doubling. Our main contributions (7.5) for the room temperature poling technique are: Firstly we found that the domain reversion showed different features in high poling field and low poling field, respectively. The two features correspond to two different mechanisms, hence, have two different time constants. Secondly we improved the poling technique using only single pulse instead of the multi-pulse introduced previously. The method reported in 7.5 eventually becomes the basis for further optimization of poling technology. In the poling technique, the location of the reversed domain is defined by the electrode and the domain duty cycle is controlled by the correct choice of width to spacing ratio of electrode and pulse duration. Nevertheless, both uniformity of domain reversal and control of duty cycle are still somewhat difficult for all ferroelectric crystals. The differences appear from crystal to crystal, from sample to sample. The deviation from the designed values would lead to the decrease of the effective nonlinear coefficient of sample, and influence the crystal's performance. We proposed a Fourier transform method in 7.6. This method can be used to calculate the effective nonlinear coefficient of superlattice according to the etching pattern on the surface of superlattice crystal, hence, is very useful for the evaluation of quality of superlattice samples.

7.7 gives an example for the generation of noncollinear RGB three-color laser from a hexagonally poled LT superlattice crystal using the domain engineering principle. The pump source is a 532 nm green laser, and the red and blue lights result from two OPG process cascading by two SHG processes of the signal and idler in a single-pass setup, respectively. Owing to the noncollinear reciprocal vectors provided by this 2D hexagonally poled optical superlattice, the RGB lights were autoseparated without using any optical separation elements. In fact, the 2D superlattice integrates the functions of one OPG crystal, two pieces of SHG crystal and two pieces of beam splitter together only by the design of domains.

With the development of domain engineering research, new techniques to visualize and image domains are still urgently needed, preferably those that are nondestructive. Great progress has been made by us along this direction. In this chapter, we introduce two novel methods we developed to characterize non-destructively domain structures in crystals. These two methods are the environmental scanning electron microscopy (ESEM) imaging of ferroelectric domains and the scanning-tip near-field microwave microscopy (SNMM) imaging of ferroelectric domains, receptively.

7.8 introduces the ESEM imaging of ferroelectric domain. The conventional scanning electron microscopy (CSEM) is often used to observe the domain pattern on the etched surface of ferroelectric crystals with higher resolution than an optical microscopy. In order to avoid the charge accumulation on the surface of an insulator, the sample surface is to be coated with a layer of thin metal to conduct the deposited electrons to ground during operation. This method cannot be used to observe an uncoated insulator surface in conventional operation condition. In 1989, Bihan and Sogr, using low accelerating voltage, realized the direct imaging of ferroelectric domains on the untreated polar surfaces of TGS crystal and a few other ferroelectric crystals in secondary electron emission mode, respectively. In order to avoid excessive charge accumulation on surface, experiments were performed in the conditions of TV scanning mode, under a low electron acceleration voltage of 0.2 - 4 kV and a beam current of ~ 10-8 A. Even though in such conditions, the domain imaging is short lived, remaining only for a few seconds to a couple of minutes, due to the fact that surface charge accumulation can not be effectively conducted away. In order to overcome this obstacle, we observed the domains in water vapor atmosphere using ESEM. Because ESEM can operate in a gas atmosphere of low pressure ($p = 1.0 - 3.0$ Torr), and the free ions generated by collisions between beam electrons and water molecules provide a conductive path for deposited charges. Therefore, the sample surface is free from charging and a stable contrast image of anti-parallel domains is obtained in the case of LT crystals. The image has high resolution due to the high beam voltage (up to 30 kV) used.

The ESEM technique is used to observe the samples prepared by poling technique. A new technique introduced in 7.9 is developed to observe the domain structure in the superlattice crystal prepared by growth striation technique. Two kinds of images on the surface of a doped LN superlattice crystal were observed using such a SNMM technique. One is the image of resonant frequency f_0 that reflects variations in the dielectric constant associated with changes in dopant concentrations, the other is the image of Q that corresponds to the loss of microwave energy. The losses are large at domain walls as a result of domain wall movement under the influence of the microwave field. Dopant, defects and domain boundary profiles can be imaged simultaneously by this technique. Analyzing these images, one is able to understanding of the formation mechanisms of both the domain structure and defects, which is helpful in optimizing superlattice preparation.

The Growth Striations and Ferroelectric Domain Structures in Czochralski-grown LiNbO₃ Single Crystals[*]

Nai-Ben Ming, Jing-Fen Hong, Duan Feng

Department of Physics and Institute of Solid State Physics, Nanking University, Nanking, People's Republic of China

The rotational striations and power striations are studied in LiNbO$_3$ crystals and one-to-one correspondence between the striations and temperature fluctuations is demonstrated. The ferroelectric domain structures related to the rotational striations and the power striations have been observed. The distribution of solute concentration in rotational striations is measured by means of energy dispersive X-ray analysis in the scanning electron microscope, and it has been concluded that the ferroelectric domain structures depend on the solute concentration gradient.

1. Introduction

It is commonly observed that there are growth striations in most melt-grown single crystals. In order to obtain a crystal with a high degree of homogeneity it is necessary to eliminate growth striations. On the other hand, striations give a useful built-in record of the interface shape at any point within crystal and thus are widely employed in the study of the microscopic growth rate and the morphology of solid-liquid interfaces[1, 2]. Systematic studies of growth striations in melt-grown single crystals have been conducted in semiconductors[3] and, also, many results have been obtained in oxide crystals.

It has been found that the structure of the ferroelectric domain in LiNbO$_3$ single crystals could be changed by doping the crystal with certain solutes[4, 5]. It has been reported by Rauber[6] and Peuzin and Tasson[7] that LiNbO$_3$ single crystals with periodic ferroelectric domain structures have been produced and used as a very efficient hypersonic transducer, and this kind of crystal has also been used to realize quasi-phase matching of second harmonic generation of light by the authors of this paper[8]. In this paper, an account will be given of our experimental results on the growth striations and ferroelectric domain structure investigations in LiNbO$_3$ single crystal.

2. Experimental Methods

We intentionally displace the rotational axis of the growing crystal from the symmetry

---

[*] J.Mater.Sci.,1982,17(6):1663

axis of the temperature field by means of a fine screw to adjust the position of the heater relative to the rotation axis. In so doing, periodic fluctuations of temperature, growth rate and solute concentrations within the crystal are produced during Czochralski growth. The interior rotational striations are compositional variations induced by the growth-rate fluctuations, but the surface rotational striations are directly attributed to variation of the radial component of growth velocity. The spacing of the rotational striations may be changed at will by means of adjusting the ratio of pulling rate to rotation rate. The amplitude of both the temperature fluctuations and growth-rate fluctuations can be changed by means of adjusting the displacement of rotation axis from the symmetry axis of the temperature field.

The proportional integral differential (PID) variables in the automatic diameter control system having been so selected, such that the heating power is periodically fluctuated, and power striations are produced in the growing crystal[9]. The microscopic growth rates in forming the power striations are measured by using the periodic rotational striations as time markers, since the time interval forming any two adjacent rotational striations is equal to the rotational period at that time. In this way the transient microscopic growth rate at any point of crystal can be measured, providing the pulling rate and rotation rate are known and are constant over the time between the two striations being measured.

In order to reveal the structures of ferroelectric domains with etching, the $LiNbO_3$ specimens with surfaces parallel to $\{0\ 0\ 0\ 1\}$ or $\{1\ 0\ \bar{1}\ 0\}$ planes have been prepared. After the determination of crystal orientation, the crystal wafers were prepared by usual methods of cutting, grading and polishing, and then the crystal wafers were immersed in a mixture of 1 part HF and 2 parts HNO_3 (by vol) for 10min at 100℃ and etch patterns showing ferroelectric domain structures were thus obtained.

The distribution of solute concentration in the rotational striations was measured by means of energy dispersive X-ray analysis in the scanning electron microscope (Philips, PSEM-500X).

3. Experimental Results

3.1 The Rotational Striations

When the rotation axis of crystal does not coincide with the symmetry axis of temperature field and the measuring junction of a Pt-10% Rh thermo-couple rotating with the growing crystal is put into the meniscus, the temperature signal caused by crystal rotation is passed through a rotational slipring to a chart-recorder callibrated to record the temperature. Under such a condition, the temperature fluctuations related to crystal rotation[10] were measured. When the rotational rate is suddenly changed in the course of crystal growth, for example, from 4 to 13 rpm, the resulting temperature fluctuations are

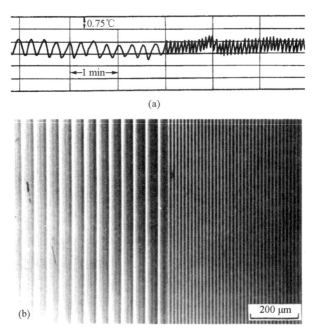

FIG. 1. The rotational rate changed suddenly from 4 to 13 rpm. (a) Temperature fluctuation of meniscus and (b) surface growth striations.

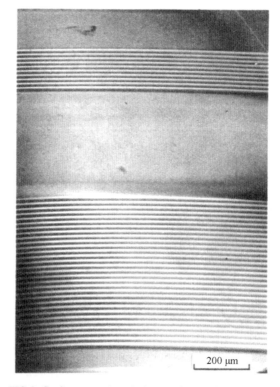

FIG. 2. Surface growth striations and crystal revolutions.

shown in Fig.1(a). Fig. 1(b) shows surface growth striations which correspond to the temperature fluctuations, and surface striations can be observed clearly using a metallographic microscope with oblique illumination. In Fig. 1 it is found that crystal rotational rate, frequency of temperature fluctuations and the number of surface growth striations not only quantitatively agree, but also display the same variation when the rotation rate suddenly changes. We also carried out the following procedure: stop crystal rotation, turn 10 revolutions, stop again, turn 40 revolutions and finally stop the crystal rotation altogether. The result of this procedure is shown in Fig. 2. When the crystal is stopped, the striations are eliminated, and when the crystal begins to rotate, the striations are produced immediately; it can be seen also that the number of striations corresponds completely to the revolutions of crystal. Thus, it is clearly demonstrated that the surface-growth striations are indeed an aspect of the rotational striations themselves.

3.2 The power striations

Adjusting the PID variables in the automatic control system in such manner that the heating power periodically fluctuates, produces power striations such as those shown in Fig. 3(a). In order to measure the transient microscopic growth rate during the power-striation formation, we have used periodic rotational striations as time markers, and the variation of transient microscopic growth rate within the power striations has been measured. The result is shown in Fig. 3(b). The variation of crystal diameter is shown in Fig. 3(c). The relative changes of heating power, indicated by the temperature measured using a Pt-10% Rh thermo-couple near the heating element, are shown in Fig. 3(d). The relationship between the space coordinates and time co-ordinates in Fig. 3 can be established because the pulling rate is a known constant. As shown in Fig. 3, it is clear that the power fluctuations, crystal diameter variations and microscopic growth-rate fluctuations correlate quite well and display the same period. By comparing Fig. 3(a) to (d), the adjusting action in the automatic diameter control system may be inferred, i.e., when the crystal diameter increases (or decreases), the error signal is measured by electric weighting and the signal then triggers a "feed-back" into the input command to increase (or decrease) heating power, so the transient microscopic growth rate decreases (or increases) and thus the increase (or the decrease) of crystal diameter is held down, and the crystal diameter could be kept constant. In general, when power striations formed in the automatic diameter control system, the microscopic growth rate decreases as the diameter increases and vice versa. In the case of the rotational striation formation, the situation is reversed, the microscopic growth rate and crystal diameter both increasing within the half-period of low temperature, and decreasing within the half-period of high temperature.

3.3 Growth striations and ferroelectric domain structure

In $LiNbO_3$ single crystal grown from congruent melt composed of 48.6 mol% Li_2O, we have not observed any evidence showing the influence of temperature fluctuations caused by crystal rotations or power fluctuations on the structures of ferroelectric domains. In LiNbO$_3$ single crystals grown from a stoichiometric melt composed of 50 mol% Li$_2$O, which is equivalent to doping the congruent melt with 2.72 mol% Li$_2$O, the corresponding ferroelectric domain structures in the power striation within the crystal have been observed, as shown in Fig. 4. Although there is no apparent variation effect of rotational striations on the ferroelectric domain structures, however, in the region where the diameter of crystal increases and the microscopic growth rate decreases, as shown in Fig. 3, a positive domain appears, shown in Fig. 4(a). In the region where the diameter of crystal decreases and the microscopic growth rate increases, it is found that domain bands composed of clusters of fine negative domains appear. It is possible to distinguish between positive and negative domains according to their different rates of etching: the etching rate of negative domains is larger than that of positive domains on the (0 0 0 1) plane.

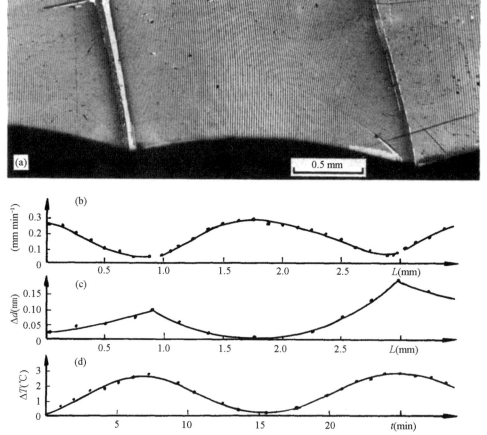

FIG.3. The power striations showing (a) peripheral section of LiNbO$_3$ single crystal, (b) microscopic growth-rate variation of the section depicted in (a), (c) variation of crystal diameter of the section depicted in (a) and (d) relative power fluctuation indicated by the temperature of the heating element.

In the case of a congruent melt doped with 1 wt% yttrium, the ferroelectric domain structures have been effectively changed such that, within the rotational striations, periodic ferroelectric domain structures have been induced, confirming the results reported by Peuzin and Tasson[7]. The one-to-one correspondence between the rotational striations and the ferroelectric domain structures have been demonstrated with the following procedure: stop crystal rotation, turn 10 revolutions and stop crystal rotation again; the result is shown in Fig. 5. Fig. 5(a) shows the surface striations and Fig. 5(b) shows the interior domain structures. Evidently, the surface striations and interior domain structures correspond to each other and to the crystal revolutions. The morphology of the surface striations and interior domain structure having been carefully studied, the surface striation is found to be a spiral winding of 10 turns and the interior periodic laminar domain structure is strictly not composed of parallel lamellae one above the other; rather it is a single lamella in the form of helicoid, or spiral ramp. So Fig. 5(a) is a cross-section pattern of the helicoid and the end of the helicoid can be clearly seen at the lower left corner in Fig. 5(b). Although both the power striations and the rotational striations can change the interior domain structures, the negative region in power striations is a band composed of clusters of fine negative island-like domains (see Fig. 4) and a single lamella negative domain in rotational striation (Fig. 5b). The above observations show that the gradient of yttrium concentration can change the ferroelectric domain structure more efficiently than

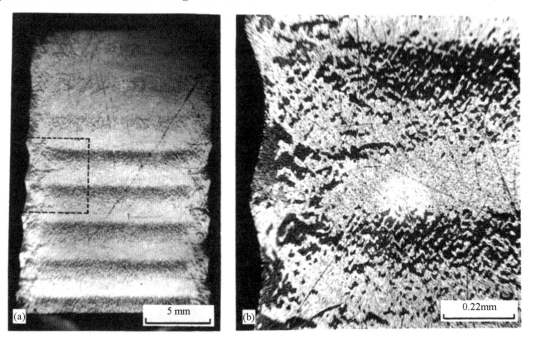

FIG.4. The power striations and ferroelectric domain structures. The area outlined with a broken line in (a) is shown at higher magnification in (b).

can that of lithium concentration. Perhaps it is because the concentration gradient in rotational striations is larger than that in power striations as the period of the rotational striation is shorter and the valence of the yttrium ion is larger than that of the lithium ion.

In order to unravel the relationship between solute concentration distribution and the ferroelectric domain structures, we have used energy dispersive X-ray analysis in the scanning electron microscope to measure the solute concentration distribution in the rotational striations and to study the relationship between the solute distribution and the domain structures. The intensity of the $L\alpha$ spectral line of yttrium has been measured, at distances, d, point by point, along a line normal to the domain boundaries and it has been found to be proportional to the mean solute concentration of a selected region. The diameter of the selected region being measured is 0.125 μm. The result is shown in Fig. 6, in which the curve represents the distribution of yttrium concentration and the shaded area represents the positive domain. Domain boundaries are found near the maxima and minima of the curve and, in the positive domain area, the concentration decreases and its gradient is negative; in the negative domain area, however, the concentration increases and its gradient is positive. This directly confirms that the gradient of the solute concentrations

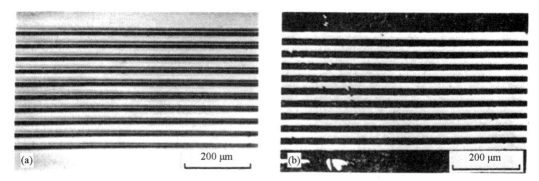

FIG.5. The rotational striations and ferroelectric domain structures. (a) Surface rotational striations and (b) interior ferroelectric domain structures.

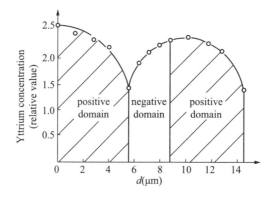

FIG.6. The yttrium concentration distribution and ferroelectric domain structures in rotational striations measured at distances, d, point by point, along a line normal to the domain boundaries.

determines the structures of the ferroelectric domains, which conforms with the conclusion of Tasson et al.[11, 12]. It is worth noting that there are two types of sign inversion of concentration gradient, i.e., at the minima the change of gradient takes place abruptly; at the maxima, the change of gradient takes place gradually. The difference is also shown in the morphology of the domain boundaries observed with the scanning electron microscope (see Fig. 7). An abrupt change of concentration gradient corresponds to a smooth boundary in Fig. 7, while a gradual change of concentration gradient corresponds to a rough boundary in Fig. 7. This effect on domain boundary morphology have also been displayed in the power striations, shown in Fig. 4. There are also two kinds of boundaries of the domain bands composed of clusters of fine negative domains, the diffuse boundaries at the maxima of solute concentration distribution, and the sharp boundaries at its minima.

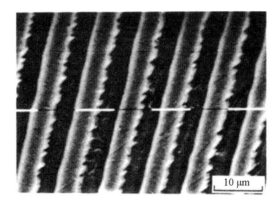

FIG.7. Morphology of ferroelectric domain boundaries.

4. Discussion

4.1 The Production of LiNbO$_3$ Single Crystal with Periodic Laminar Domain Structures

We have established the correspondence between temperature fluctuations, growth-rate fluctuations, solute concentration fluctuations and the ferroelectric domain structure, and reached the conclusion that the ferroelectric domain structures are determined by the gradient of the solute concentration. Therefore, if it is desired to change the domain structure in ferroelectrics, besides varying the applied electric field, we can also use the inhomogeneous solute concentration distribution in the crystal. We have already intentionally produced a crystal with rotational striations and obtained a periodic concentration distribution in the crystal, producing LiNbO$_3$ single crystals with periodic laminar domain structure. The layer thickness of laminar domain structures can be adjusted by changing the ratio of pulling rate to rotation rate.

4.2 The Solute Distribution in Rotational Striation

The measured solute (Y) distribution (Fig. 6) in the rotational striation of the LiNbO$_3$ single crystal resembles the curve of antimony concentration in the rotational striation of silicon crystal measured by Muargai et al.[13], and also similar to the theoretical curve obtained recently by Wilson[14] in the improved version of the Burton-Prim-Slichter theory[15]. It should be noticed that the temperature fluctuation with sinusoidal form caused by the crystal rotation has been measured under our experimental condition[10], but the solute distribution resulting from this form of the temperature fluctuation, deviates from the sinusoidal form as shown in Fig. 6. This phenomenon can be explained as follows. In general, assuming the growth kinetics of a non-facetted interface is linear, the sinusoidal temperature fluctuation will cause a sinusoidal growth rate fluctuation; as a result, a thicker layer of crystal is grown within the half-period of the growth rate larger than the average rate and a thinner layer of crystal is grown within the half-period of growth rate smaller than the average rate. Therefore, the solute distribution in the crystal (space distribution) is distorted from a sine curve. Using the measured value of the average growth rate, the period of rotation and the amplitude of the growth-rate fluctuation, the space solute distribution against distance relation (solid line in Fig. 8) can be transformed into a space solute distribution against time plot (dotted line in Fig. 8). Obviously, the solute distribution against time plot is approximately a sine curve. So far it has been assumed that the validity of the theory of segregation is based on the assumption that the growth interface is at equilibrium, and so the solute concentration of solidifying crystal, C_s, is given by

$$C_s = kC_L, \qquad (1)$$

where k is the equilibrium segregation coefficient and C_L is the concentration of the bulk liquid. Consequently, the dotted curve in Fig. 8 also represents a plot of the variation of solute concentration of liquid at the growth interface against time, caused by crystal rotation, which roughly agrees with the theoretical curve estimated by Wilson[14].

4.3 The Mechanism of Polarization and the Non-uniform Distribution of Solute

The fact that the non-uniform distribution of solutes can change the ferroelectric domain structures may be explained as follows. The solutes in a crystal are generally ionized but not completely shielded, especially when the temperature is decreasing and passing through the ferroelectric phase transition; thus, the non-uniform solute distribution is equivalent to the non-uniform space-charge distribution in the crystal and a non-uniform internal field is produced in it. Although the field is comparatively small, it can induce the ions of lithium and niobium within the lattice to displace preferentially at a temperature close to the Curie point. If we express the solute distribution, related to the periodic rotational striation shown in Fig. 6, as an approximate sine function, the internal field within the crystal can be derived as the same function with a phase difference of π/2. Thus, it can be clearly seen that the domain structures and the solute distribution have the same period and it is the gradient of the solute concentration which determines the domain structure.

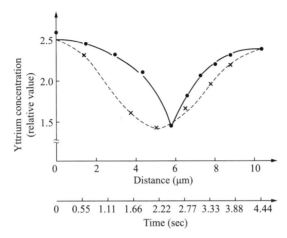

FIG.8. Solute distribution plotted against distance (dotted line) and time (solid line).

References and Notes

[1] N. -B. MING, J. -F. HONG and D. FENG, *Kexue Tongbao* **25**(1980) 256.
[2] J. R. CURRUTHERS and A. F. WITT, in "Crystal Growth and Characterization" edited by R. Ueda and J. B. Mullin (North-Holland, Amsterdam and New York, 1975) p. 107.
[3] H.C. GATOS, J. *Electrochem. Soc.* **122**(1975)287c.
[4] K. NASSN, H. J. LEVINSTON and G. H. LOICONO, *J. Phys. Chem. Sol.* **27**(1966) 983.
[5] H.T. PARFITT and D.S. ROBERTSEN, *Brit. J. Appl. Phys.* **8** (1967) 1709.
[6] A. RAUBER, in "Current Topics in Materials Science" edited by E. Kaldis and H. J. Scheel (North-Holland, Amsterdam and New York, 1977)p. 481.
[7] J.C. PEUZIN and M. TASSON, *Phys. Stat. Sol. a* **37**(1976) 119.
[8] D. FENG, N. -B. MING, J. -F. HONG, Y. -S. YANG, J. -S. ZHU, Z. YANG and Y. -N. WANG, *Appl. Phys. Lett.* **37**(1980) 607.
[9] J.-F. HONG, Z.-M. SUN, Y.-S. YANG and M.-B. MING, *Wuli* **9** (1980) 5.
[10] N.-B. MING, J.-F. HONG, Z. -M. SUN and Y.-S. YANG, *Acta Physica Sinica*, to be published.
[11] M. TASSON, H. LEGAL, J. C. GAY, J. C. PEUZIN and F. C. LISSALDE, *Ferroelectrics* **13**(1976) 479.
[12] *Idem*, *Phys. Stat. Sol. a* **31** (1975) 729.
[13] A. MUARGAI, H.C. GATOS and A. F. WITT, *J. Electronchem. Soc.* **123**(1976) 224.
[14] L.O. WILSON, *J. Crystal Growth* **48**(1980) 435.
[15] J.A. BURTON, R. C. PRIM and W. P. SLICHTER, *J. Chem. Phys.* **21**(1953) 1987.
[16] The authors are indebted to Mr Hai-zhou Guo and Mr Hao-ying Shen of Nanjing Solid State Device Research Institute for their help in the work using the scanning electron microscope.

Growth of Optical Superlattice LiNbO$_3$ with Different Modulating Periods and Its Applications in Second-harmonic Generation[*]

Ya-lin Lu, Yan-qing Lu, Xiang-fei Cheng, and Cheng-cheng Xue

National Laboratory of Solid State Microstructures, Nanjing University, Nanjing 210093, People's Republic of China, and Center for Advanced Studies in Science and Technology of Microstructures, Nanjing 210093, People's Republic of China

Nai-ben Ming

CCAST (World Laboratory), P.O. Box 8370, Beijing 100080, People's Republic of China, and National Laboratory of Solid State Microstructures, Nanjing University, Nanjing 210093, People's Republic of China

Optical superlattice LiNbO$_3$ crystals with a modulation period number over 200 and a modulation period from 2.0 μm to over 15.0 μm were grown by the Czochralski method in a carefully designed asymmetric temperature field system. Measurements of the frequency doubling efficiency for generating light from blue to green were performed. The largest efficiency obtained was 24.0% for frequency doubling of a picosecond 980 nm fundamental light. The angle tuning tolerance of a LiNbO$_3$ optical superlattice was measured to be over 10° of the incidence angle.

Quasiphase matching (QPM) was proposed independently by Bloembergen[1] and Franken and Ward[2] for obtaining efficient frequency conversion in optical second-harmonic generation(SHG) and other nonlinear optical processes. SHG by QPM in a LiNbO$_3$ optical superlattice (OSL), in both bulk[3-5] and waveguide forms,[6-8] has attracted a great deal of attention for its potential applications in constructing compact short-wavelength lasers, and in constructing a QPM optical parametric oscillator(OPO).[9] A LiNbO$_3$ OSL in bulk form offers the following advantages: direct frequency doubling of a laser diode without the need for careful light coupling which exists in waveguide devices, and ease of phase matching and angle tuning in QPM OPO. The latter is difficult to achieve in waveguide form. Several methods have been developed for obtaining a bulk LiNbO$_3$ OSL. Alternating stacks of thin plates of LiNbO$_3$ were constructed for QPM SHG experiments.[10] Single-crystal fibers with a diameter of 250 μm and having periodically alternating ferroelectric domain structures have been applicated to QPM SHG.[11] With the development of integrated-optics technology in LiNbO$_3$[7,8,12,13] some field-induced methods present the

---

[*] Appl. Phys. Lett., 1996, 68(20):2781

potential for preparation of a bulk LiNbO$_3$ OSL.[9,14,15] Using an electric-field-poling method, a LiNbO$_3$ OSL with a domain-inversed depth of 0.5 mm has been successfully prepared and has been performed in QPM SHG or OPO.[9] However, the limitation in the surface area for light-transmitting electric-field-poled samples makes angle tuning difficult in QPM OPO applications. It is also difficult to prepare a LiNbO$_3$ OSL with a small modulating period, which is desirable for SHG to generate short wavelength light using first-order QPM, by field-induced methods. Compared with field-induced methods, the Czochralski method for bulk LiNbO$_3$ growth[3-5,16] grows a LiNbO$_3$ OSL with a practical dimension, with a modulating period as small as 2.0 μm, and with its period controllable by adjusting growing parameters. However, there are some disadvantages to this method, including the formation of various islandlike domains during growth and period variation along the growing direction. The islandlike domains occur easily during crystal growth and destroy the periodicity of the LiNbO$_3$ OSL. The growth of a LiNbO$_3$ OSL with a period below 4.0 μm by the Czochralski method is also relatively difficult due to the troublesome response of periodic ferroelectric domains to periodic temperature fluctuations during crystal growth.

In this letter, with a carefully designed temperature field system, we report the success in growing a LiNbO$_3$ OSL with a modulation period ranging from 2.0 to over 15.0 μm. The formation of various islandlike domains was kept at its lowest level. The period fluctuation along the growing direction was limited to 5%. The continuous period number of the LiNbO$_3$ OSL with a stable modulation period (period fluctuation of less than 5%) can be over 200 when the modulation period is less than about 4.0 μm and easily over 300 when the period is larger than about 4.0 μm. Efficient SHG to generate light from blue to green has been performed with the usage of an OPO as a fundamental source.

Crystals of LiNbO$_3$ doped with 0.5 mol % yttrium were grown along the crystal's a axis by the Czochralski method.[3-5] In designing the asymmetric temperature field system, three parameters-axial temperature gradient above the melt surface, axial temperature gradient below the melt surface, and the radial temperature gradient in the melt surface—were defined as adjusting parameters. In our growing system, the axial temperature gradients above and below the melt surface were kept in the range of 30–40℃/cm and 6–8℃/cm, respectively, when the melt depth was about 30 mm in a 40 mm high platinum crucible. The formation of various islandlike domains, which are caused by solute aggregations in solute boundary layers due to solid-liquid interface instabilities, can be effectively suppressed by adjusting these two axial temperature gradients. The measurement of radial temperature gradient is difficult since the temperature system is asymmetric. But the gradient can be adjusted by changing the thickness of the ZrO$_2$ thermal-insulating layer outside the Pt crucible. Careful adjustment of this thickness can obviously change the shape of the solid-liquid interface, and thus affect the response depth of periodic ferroelectric domain structures in the a-axis as-grown LiNbO$_3$ OSL. The response depth is defined as the width of periodic domain structures along the crystal's z axis on one side of the main x-z plane in an a-axis grown crystal. One of the most important factors determining this response depth is the shape of the solid-liquid interface. If the shape of the solid-liquid is exactly plane (i.e., exactly in the crystal's y-z plane), the periodic domain structures cannot occur due to the fact that the polarization is along the crystal's z axis. A convex solid-liquid interface is beneficial for generating periodic domain structures.

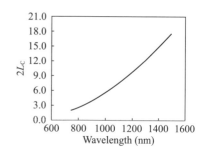

FIG.1. The relationship between fundamental wavelength and double coherence length of the LiNbO$_3$ crystal.

By carefully adjusting these three parameters, we can grow a LiNbO$_3$ OSL with a modulating period from 2.0 to over 15.0 μm. The difficulty in LiNbO$_3$ OSL growth will increase as the modulation period decreases. For a modulation period below about 3.0 μm, which is near the limiting response frequency of periodic domain structures to the temperature fluctuations, the periodic response of the ferroelectric domain not only relies on the properties of the material itself and the temperature system, but also depends on the stability of the pulling and rotating systems of the crystal growth unit.

The relationship between the fundamental wavelength in the range of 0.75 – 1.5 μm and the double coherence length of LiNbO$_3$, which was calculated from the crystal's Sellmier's equation, is shown in Fig. 1, where the fundamental light propagates along the crystal's a axis with its polarization along the crystal's optical axis. In this wavelength range of fundamental light, the most practical fundamental sources for constructing an all-solid short-wavelength laser and a QPM OPO are a Nd:YAG or Nd:VO$_4$ laser at 1.064 μm and a near-infrared laser diode with an output in the range from 800 to 860 nm or from 950 to 980 nm. Thus it requires that the LiNbO$_3$ OSL has a modulation period in the range from about 2.7 to 6.4 μm. For constructing a QPM OPO with 1.064 μm pumping, it is also desirable that the LiNbO$_3$ OSL has a modulation period of over 6.4 μm. Figure 2 shows the photographs of a LiNbO$_3$ OSL with the following modulation periods: (a) 2.7 μm, for frequency doubling of an 809 nm laser diode, (b)

FIG.2. Photographs of a LiNbO$_3$ OSL with modulation period: (a) 2.7 μm, (b) 5.2 μm, (c) 15.0 μm.

5.2 μm, for frequency doubling of a 980 nm laser diode, and (c) 15.0 μm, which will be used in the QPM OPO. These photographs were taken on the crystal's acid-etched y surface. The thickness of the positive domain laminae and that of the negative domain laminae are nearly equal, and the continuity of modulation period and periodicity is very good. SHG experiments were performed by using a picosecond automatic tunable OPO as a fundamental light. The laser has a pulse rate of 1 Hz and a pulse duration of 30 ps. The linewidth of the output light pulse is less than 1 nm at all wavelengths except that at the OPO's degeneracy point of 1064 nm, when the linewidth is about 10 nm. Table 1 lists the results of SHG efficiency and some parameters of five measured samples with different modulation periods. The SHG efficiencies listed in Table 1 are difficult to compare, because SHG efficiency is mainly relevant to fundamental power density, fundamental wavelength, modulation period number, and period stability. These factors are different in these five samples. For comparison, we used a commercial 10-mm-long single-domain LiNbO$_3$ crystal to measure SHG efficiency under similar experimental conditions (the fundamental wavelength is 1064 nm). A SHG efficiency of 18.5% has been obtained under 90° phase-matching conditions. The result shows that sample 4 in Table 1 (the fundamental wavelength is 1026 nm) has a SHG enhancement of about 18.9.[17] It is near the theoretical SHG enhancement of $(d_{33}/d_{31})^2(2/\pi)^2 \approx 23$. The SHG enhancement defined here means that the efficiency comparison is between two crystals (OSL LiNbO$_3$ and single-domain LiNbO$_3$) with the same crystal length and the same experimental conditions.

TABLE 1. Parameters of LiNbO$_3$ OSL samples and results of SHG experiments by using an OPO as a fundamental source.

Sample no.	d_{thick}(mm)	Period(μm)	N_{period}	Fluctuation ($\leqslant$%)	λ_{fun}(nm)	η_{SHG}(%)
1	0.62	2.8	220	5	815	3.0
2	0.78	3.4	230	5	860	4.2
3	1.56	5.2	300	2	980	24.0
4	2.20	6.4	310	5	1026	17.0
5	1.50	8.3	180	2	1130	19.8

For measuring the LiNbO$_3$ OSL's angle tunability, we have measured the curve of SHG efficiency with fundamental wavelength at different incidence angles of sample 4 in Table 1. The sample was rotated in the crystal x-y plane. The rotation angle is 0° when at normal incidence, then is 5°, 10°, and 15°, respectively. These curves are shown in Fig. 3. The peak SHG wavelength was found to increase as the incidence angle was increased, in agreement with the notion that the effective period will increase when the incidence angle increases. However, the peak SHG efficiency decreases and the full width at half-maximum increases with the incidence angle, due to the increase of reflection of fundamental light at the sample's surface and the relative increase of period fluctuation. The peak SHG efficiency decreases sharply when the incidence angle exceeds 10°. The results show that nearly 10° of the incidence angle tuning tolerance can be achieved in this uncoated sample, without the evident decrease of SHG efficiency. This result is useful for tuning in LiNbO$_3$ OSL's applications in SHG or in QPM OPO.

In conclusion, we have grown a LiNbO$_3$ OSL with a modulation period from 2.0 to over 15.0 μm. Frequency doubling efficiencies for generating blue to green light were measured by using an OPO as a fundamental source. The largest SHG efficiency for doubling 980 nm is 24.0%. The LiNbO$_3$ OSL's angle tuning tolerance was measured to be over 10° of the incidence angle.

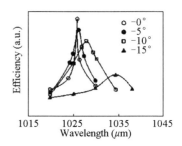

FIG.3. SHG efficiency vs fundamental wavelength of sample 4 at different incidence angles: (○)0°, (●)5°, (□)10°, (▲)15°.

References and Notes

[1] J. A. Armstrong, N. Bloembergen, J. Ducuing, and P. S. Pershan, Phys. Rev. **127**, 1918(1962).
[2] P. A. Franken and J. F. Ward, Rev. Mod. Phys. **35**, 23(1963).
[3] Y. L. Lu, L. Mao, and N. B. Ming, Appl. Phys. Lett. **59**, 516(1991).
[4] Y. L. Lu, L. Mao, and N. B. Ming, Appl. Phys. Lett. **64**, 3092(1994).
[5] Y. L. Lu, L. Mao, and N. B. Ming, Opt. Lett. **19**, 1037(1994).
[6] D. Delacourt, F. Armani, and M. Papuchon, IEEE J. Quantum Electron. **30**, 1090(1994).
[7] E. J. Lim, M. M. Fejer, R. L. Byer, and W. J. Kozlowsky, Electron. Lett. **25**, 731(1989).
[8] J. Webjörn, F. Laurell, and G. Arvidsson, IEEE Photonics Technol. Lett. **1**, 316(1989).
[9] L. E. Myers, G. D. Miller, M. L. Bortz, R. C. Eckart, M. M. Fejer, and R. L. Byer, *1994 Nonlinear Optics Conference* (IEEE LEOS, Piscataway, NJ, 1994), Paper PD8.
[10] M. Okada, K. Takizawa, and S. Ieiri, Opt. Commun. **18**, 331(1976).
[11] G. A. Magel, M. M. Fejer, and R. L. Byer, Appl. Phys. Lett. **56**, 108(1990).
[12] Y. Y. Zhu, S. N. Zhu, and N. B. Ming, Appl. Phys. Lett. **65**, 558(1994).
[13] Y. Y. Zhu, S. N. Zhu, and N. B. Ming, Appl. Phys. Lett. **66**, 408(1995).
[14] M. Yamada, N. Nada, and K. Watanabe, Appl. Phys. Lett. **62**, 435(1993).
[15] W. K. Burns and L. Goldberg, IEEE Photonics Technol. Lett. **6**, 252(1994).
[16] N. B. Ming, J. F. Hong, and D. Feng, J. Mater. Sci. **17**, 1663(1982).
[17] Y. H. Xue, N. B. Ming, and D. Feng, Acta Phys. Sin. **32**, 1515(1983).
[18] This work is supported by a Ke-Li fellowship.

Growth of Nd^{3+}-doped LiNbO$_3$ Optical Superlattice Crystals and Its Potential Applications in Self-frequency Doubling[*]

Ya-lin Lu, Yan-qing Lu, and Cheng-cheng Xue

National Laboratory of Solid State Microstructures, Nanjing University, Nanjing 210093, People's Republic of China and Center for Advanced Studies in Science and Technology of Microstructures, Nanjing University, Nanjing 210093, People's Republic of China

Nai-ben Ming

CCAST (World Laboratory), P.O. 8730, Beijing 100080, People's Republic of China and National Laboratory of Solid State Microstructures, Nanjing University, Nanjing 210093, People's Republic of China

We have grown Nd:MgO:LiNbO$_3$ crystals with periodic ferroelectric domain structures. Absorption and fluorescence spectra measured on these crystals showed little difference from those from Nd:MgO:LiNbO$_3$ with uniform domain structures. Green fluorescence was generated by self-frequency doubling in a cavity having great losses and pumped by a pulsed dye laser.

Self-frequency doubling (SFD) is one of the most important techniques for constructing compact and reliable short wavelength light sources. The combination of laser oscillation of active ions with nonlinear optical properties of the host material offers an opportunity for producing self-frequency doubling, self-modulated, and self-Q-switched lasers as well as miniature waveguide lasers and amplifiers.[1] Nd$_x$Y$_{1-x}$Al$_3$(BO$_3$)[4] (NYAB) has been shown to be a promising material for SFD, but difficulties in crystal growth make it unfavorable for practical applications. Nd laser oscillation in LiNbO$_3$ was achieved in 1967, and self-frequency doubling was achieved in 1979.[3,4] LiNbO$_3$ is uniaxially negative. Therefore Nd ion produces radiation polarized perpendicular(σ) and parallel(π) to the c axis. The π-polarized (σ-polarized) output occurs at 1084 nm (1092 nm) and is the high (low) gain output.[1] Two problems were encountered in the SFD of Nd:LiNbO$_3$; photorefractive damage and high cavity losses associated with the difficulty in producing low-gain σ-polarized emission required for type I phase matching.[3,5] One can minimize photorefractive damage by maintaining the crystal at elevated temperatures, codoping with MgO, and increasing the wavelength of the pump source from the visible($\approx$ 600 nm) to the near IR($\approx$810 nm). Codoping with MgO reduces the photorefractive damage, but stable cw oscillation in Nd:MgO:LiNbO$_3$ requires operation above 100 ℃.[1,5]

---

[*] Appl.Phys.Lett., 1996,68(11):1467

Cavity losses are increased, because a Brewster window is needed to force the low-gain σ-polarized output to satisfy the noncritical phase-matching conditions. A reduction of operation temperature has been achieved in a $Nd:Sc_2O_3:LiNbO_3$ crystal recently. But the critical requirements for temperature stabilization and type I phase matching condition make the SFD laser difficult to miniaturize.[6] In other aspects, all SFD materials mentioned above can only be used to generate green light. One cannot extend the output wavelength to the blue range. Recent interest in all-solid-state blue and green lasers for optical storage applications has intensified the search for new frequency doubling materials, especially those materials which have high nonlinearity, good laser operation at room temperature, and the ability to generate blue light.

Second-harmonic generation (SHG) in quasiphase matched (QPM)[7] in $LiNbO_3$[8-12] and $LiTaO_3$[13] crystal has recently attracted a great deal of attention. In $LiNbO_3$, by modulating the nonlinear susceptibility with periodic ferroelectric domain structures which have an appropriate period ($LiNbO_3$ optical superlattice), it is possible to quasiphase match at an arbitrary temperature over the crystal's entire transparency range and to use the highest nonlinear coefficient d_{33}. So the combination of laser oscillation and SFD through quasiphase matching technique in rare-earth-doped $LiNbO_3$ crystals with periodic ferroelectric domain structures is possible. It has the advantages of high nonlinearity, easy room temperature quasiphase matching (only π-polarization of the fundamental and harmonic light is required in the process of quasiphase matching), and the ability for generating light from blue (corresponding to SFD of 946 nm line of Nd^{3+} ion) to near infrared (corresponding to SFD of 1.53 μm line of Er^{3+}, for example).

In this letter, we report the growth of Nd_2O_3 and MgO codoped $LiNbO_3$ crystals with periodic ferroelectric domain structures. The absorption spectra and fluorescence spectra were recorded and compared with that of $Nd:MgO:LiNbO_3$ with uniform domain structures. SFD phenomenon were observed in a $Nd:MgO:LiNbO_3$ optical superlattice (OSL) crystal.

The single crystals were grown by the conventional Czochralski method using a Pt crucible in a carefully designed asymmetric temperature system.[9-12] Starting materials were prepared by mixing Nd_2O_3 and MgO with a congruent melt of $LiNbO_3$. Pulling and rotating rates were 3 mm/h and 11 rpm, respectively. The Nd_2O_3 concentration was 0.2 wt ‰ in the crystal and the MgO concentration was 4.5 mol ‰. The as-grown a-axis crystals were transparent and bluish in color. The periodic domain structures were revealed by acid etching the crystal's polished b face. The continuous modulation period number reaches over 700, in which over 300 periods are invariable (period fluctuation was kept below 2‰). The modulation period measured by an optical microscope is 7.0 μm, as shown in Fig. 1. Sample was cut out parallel to the laminar domains. The sample size is about $3\times6\times2(b\times c\times a)mm^3$. The light-transmitting length (a-axis direction) is 2 mm. The sample includes about 570 domain laminas.

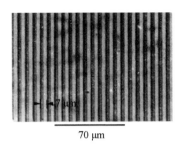

FIG.1. Photograph of a Nd:MgO:LiNbO₃ optical superlattice with a modulation period of 7.0 μm.

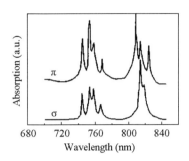

FIG.2. Absorption spectra (σ and π polarized) of a Nd:MgO:LiNbO₃ optical superlattice.

Absorption spectra under $\sigma(E \perp c)$ and $(E \parallel c)$ polarized incident light are shown in Fig. 2. In the 750 nm region, four absorption peaks were observed in both the σ and the π polarization cases, though the relative peak intensity is slightly different. The peak position is almost the same. These four peaks are at 743, 752, 757, and 766 nm. From the spectra, only one peak exists in 810 nm region in σ polarization and the maximal absorption is at 814 nm. Whereas in the case of π polarization, there are four peaks in the same region. They are at 805, 809, 814, and 824.6 nm. The maximal absorption is at 809 nm. The absorption spectra of a Nd:MgO:LiNbO₃ OSL were similar to those of Nd:MgO:LiNbO₃ with a uniform domain structure. It indicates little effect of the periodic domain structures on the Nd absorption characteristics.

Fluorescence spectra were measured by using tunable cw Ti:sapphire laser excitation near the peak absorption. Figure 3 shows the π and σ polarized components of the radiation from the sample obtained with a polarizer. The pumping beam was σ polarized (perpendicular to the crystal's c-axis). The spectral resolution was 0.5 nm in this measurement. The two maximal fluorescence peaks of the π component occurred at 940 and 1084 nm, which is almost the same as that in Nd:MgO:LiNbO₃ with a uniform domain structure. In the case of the σ component, the two main absorption peaks also occurred at 940 and 1084 nm. This differs from that in Nd:MgO:LiNbO₃ with a uniform domain structure, in which the latter main absorption peak is at 1094 nm as shown in Fig. 4. The 1094 nm peak is clearly seen in Fig. 3, but the peak is not at its highest. This is probably caused by the existence of periodic domain structures, but the reason it still unknown. Both these two lines, 940 and 1084 nm, corresponds to a transition of the Nd ion from state $^4F_{3/2}$ to $^4I_{9/2}$ and from state $^4F_{3/2}$ to $^4I_{11/2}$, respectively.

For characterizing the periodicity of the Nd:MgO:LiNbO₃ optical superlattice sample, second-harmonic generation through using third-order QPM technique in the sample has been performed by using a dye laser pumped by a tunable Nd:YAG laser as the fundamental source. The laser has a repetition rate of 10 Hz with a pulsewidth of 7 ns and a linewidth of 0.2 cm^{-1}. The maximal SHG efficiency, which is over 15%, occurred at the fundamental wavelength of about 788 nm. The input energy is about 5 mJ/pulse, and no focusing was used. The full width at half-maximum(FWHM) is less than 0.3 nm. It shows that the periodicity of the sample is very good.

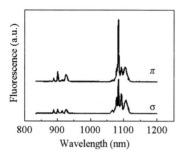

FIG.3. Fluorescence spectra(σ and π polarized) of a Nd:MgO:LiNbO$_3$ optical superlattice.

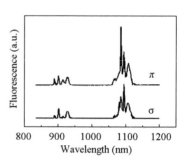

FIG.4. Fluorescence spectra (σ and π polarized) of a Nd:MgO:LiNbO$_3$ with a uniform domain structure.

Using our uncoated sample and with no cavity used, no phenomenon could be observed under 809 nm pumping of the above laser. When a cavity was used and with an input pulse energy over 50 mJ/pulse and no focusing, green fluorescence was generated in the sample, but there was no observable green laser output from the output mirror. This green fluorescence may show that some weak laser oscillation of 1084 nm existed in the cavity and then self-frequency doubled by the optical superlattice through QPM. The cavity consisted of an input plate mirror which has high reflectivity at 1084 nm and high transmitance at 809 nm, an output mirror with a radius of 50 mm which is highly reflective for 1084 nm and highly transmitive for 809 nm. The uncoated sample introduced a great loss in the cavity. Thus the difficulties in laser oscillation may be caused by the large loss.

In summary, we have grown a Nd:MgO:LiNbO$_3$ OSL with a modulating period of 7.0 μm. Absorption and fluorescence spectra of the crystal were measured, and the results from the crystal show little difference from those from a Nd:MgO:LiNbO$_3$ crystal with a uniform domain structure. SFD green fluorescence has been observed by using a pulsed laser as pumping source in a cavity with great loss.

References and Notes

[1] T. Y. Fan, A. Cordova-plaza, M. J. F. Digonnet, R. L. Byer, and H. J. Shaw, J. Opt. Soc. Am. B **3**, 140(1986).

[2] S. Amano, S. Yokoyama, and H. Koyoma, Rev. Laser Eng. **17**, 895(1989).

[3] L. F. Johnson and A. A. Ballman, J. Appl. Phys. **40**, 297(1969).

[4] V. G. Dmitriev, E. V. Raevskii, and A. A. Fomichev, Sov. Tech. Phys. Lett. **5**, 590(1979).

[5] I. P. Kaminow and L. W. Stulz, IEEE J. Quantum Electron. **QE-11**, 306(1975).

[6] J. K Yamamoto, A. Sugimoto, and K. Yamagishi, Opt. Lett. **17**, 1311(1994).

[7] J. A. Armstrong, N. Bloembergen, J. Ducuing, and P. S. Pershan, Phys. Rev. **127**, 1918(1962).

[8] G. A. Magel, M. M. Fejer, and R. L. Byer, Appl. Phys. Lett. **56**, 108(1990).
[9] N. B. Ming, J. F. Hong, and D. Feng, J. Mater. Sci. **17**, 1663(1982).
[10] Y. L. Lu, L. Mao, and N. B. Ming, Appl. Phys. Lett. **59**, 516(1991).
[11] Y. L. Lu, L. Mao, and N. B. Ming, Appl. Phys. Lett. **64**, 3209(1994).
[12] Y. L. Lu, L. Mao, and N. B. Ming, Opt. Lett. **19**, 1037(1994).
[13] W. S. Wang, Q. Zhou, and D. Deng, J. Cryst. Growth **79**, 706(1986).
[14] This work is supported by a Ke-Li fellowship.

Fabrication of Acoustic Superlattice LiNbO₃ by Pulsed Current Induction and Its Application for Crossed Field Ultrasonic Excitation*

Zhi-liang Wan, Quan Wang, Yuan-xin Xi, Yan-qing Lu, Yong-yuan Zhu, and Nai-ben Ming

National Laboratory of Solid State Microstructures, Nanjing University, Nanjing 210093, People's Republic of China

An acoustic superlattice LiNbO₃ crystal with periodic ferroelectric domain structure was fabricated by introducing a periodic electric current through the solid-liquid interface during the crystal growing process. The domain morphology of an as-grown crystal was observed with a scanning electron microscope, and was found to be of good periodicity. A light diffraction experiment indicated that there was a periodic fluctuation of the dielectric constant along the crystal's growing direction. Using the "crossed field" scheme, a 340 MHz ultrasonic was excited in the crystal, which means that the acoustic superlattice is suitable for constructing high-frequency bulk-wave acoustic devices.

Lithium niobate (LN) has long been a research topic because of its outstanding nonlinear optic, electro-optic, and acoustic properties. In recent years, the LN single crystal with periodic ferroelectric domain structure (PFDS) has attracted great research interest due to its applications in quasiphase-matched (QPM) nonlinear optical frequency converter[1—4] and high frequency bulk-wave acoustic devices.[5—7] According to its different usage, the PFDS may be termed as optical superlattice (OSL) or acoustic superlattice (ASL). In this work, we will focus on the latter one, i.e., ASL LN. In order to fabricate the PFDS, some effective techniques have been developed, which may be sorted into two major kinds. The first one is fabricating the PFDS in an as-grown single domain crystal. The patterned electric-field poling technique[8] is an example. This technique is now widely used because of its accurate control of domain period. However, the sample thickness of about 0.5 mm might limit its application in some practical devices.[9] Furthermore, this technique can only fabricate the PFDS with the spontaneous polarization P_s parallel to the domain walls, which is commonly used as OSL. Another PFDS configuration is characterized by P_s perpendicular to the domain walls and has important acoustic applications, but it is impossible to be fabricated by this technique. For LN, fabricating the PFDS directly during the crystal growing process is another major effective way. Up to

---

* Appl.Phys.Lett.,2000,77(12):1891

date, LN crystals with different PFDS configurations and with various dopants have been fabricated with the growth striation technique, which is realized by designing a special asymmetrical temperature field. Because of practical dimension of the sample, this technique is attracting more and more attentions.[4,9,10] However, one disadvantage of this method is that the periodic domain is worse formed in the center of crystal because of the smaller temperature variation, which affects the quality of domain structure. Thus it is very beneficial to find a technique that may fabricate the PFDS structure in LN with both large size and good quality.

In this letter, a current induction technique was proposed for fabricating the ASL LN with PFDS during the Czochralski crystal growing process by applying a periodic pulsed current between the crystal seed and the crucible. Large size ASL LN crystals with the P_s perpendicular to the domain walls were successfully fabricated. The microstructure and acoustic properties of an as-grown crystal were characterized.

As we know, if an electric current is applied through the solid-liquid interface (SLI) during the crystal growing process, the segregation coefficient of impurity in the melt will be affected.[11] Thus if the applied current is periodically varied, a periodic distribution of dopant along the growth direction should be induced in the crystal. This periodic impurity fluctuation is similar to the growth striation by crystal rotation, but their origins are different. Furthermore, the concentration fluctuation of the impurity exists even in the center of the crystal while there is no change of impurity concentration in the crystal's center area for the growth striation technique. Since the growth striation may cause a periodic space-charge field and then makes the periodic ferroelectric domain be written in the crystal when the ferroelectric phase transition takes place.[12] The current-induced periodic impurity distribution should also be able to make the PFDS be produced. In our experiment, 0.5 wt ‰ yttrium was selected as the dopant. The crystal seed was used as the positive pole and the crucible as the negative pole for applying the periodic electric current. The period of the current pulse is 10 s, with 5 s duration of positive pulse, and 5 s of zero current. The current density in the SLI for the positive current is about 15 mA/cm^2. The crystal was grown along the z direction with a pulling rate of 3.5 mm/h. In order to avoid the influence of growth striation due to crystal rotation, we kept the rotation axis static in the experiment. The crystal seed was also put at the center of the temperature field to keep the uniformity of the crystal quality. Using this method, LN crystals with the dimension of 30 mm in diameter and 35 mm in length were successfully grown. Figure 1 is a scanning electron microscope (SEM) photograph of the y face of an as-grown crystal after being etched in a HF:HNO$_3$ mixture to reveal its domain morphology. In this figure of PFDS, the positive domain thickness is almost equal to that of the negative domain. The PFDS was finely built both in the outer region and in the center through out the bulk crystal. The continuous period number of the PFDS is over 400, with a period fluctuation of domain less than 4%.

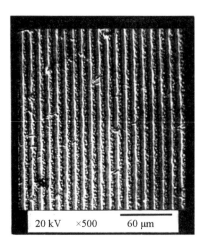

FIG. 1. SEM photograph of the etched y face of a LN crystal with PFDS structure fabricated by pulsed current induction technique.

As we know, the nonlinear optical coefficient and piezoelectric coefficient were modulated from positive to negative domain periodically in LN with PFDS, thus the QPM and high-frequency ultrasonic excitation may be realized. In fact, other third-rank tensors such as the electro-optic coefficient[13] are also modulated periodically, while the even-rank tensor and corresponding physical properties are homogeneous in the crystal. Since the refractive index is equal to the square root of the dielectric constant that is a second-rank tensor, the linear optical properties of a LN with PFDS should be uniform throughout the crystal. However, for a Czochralski grown LN with PFDS, the impurity concentration is not uniform, thus the dielectric constant will also fluctuate along the crystal growing direction. Although this fluctuation is not a consequence of the PFDS, the period is equal to the modulation period of the ferroelectric domain structure. For the PFDS that was fabricated by the growth striation technique, Lu $et\ al.$ demonstrated that there was really a periodic dielectric constant distribution associated with the periodic impurity fluctuation and the periodic domains.[14] For the current-induced PFDS, investigating the relation between the domain structure and the dielectric constant distribution is also interesting and necessary.

To study the dielectric constant distribution, a simple light diffraction experiment was employed. An ASL LN crystal fabricated by the current induction technique with the modulation period of 10.3 μm was selected for the experiment. The crystal was cut into an $8 \times 10 \times 2$ mm^3 ($z \times x \times y$) sample. A He-Ne laser with the wavelength of 6328 Å was shot into the sample along its y axis and a white screen was put behind the sample to record the light spots. After passing through the sample, the light was diffracted into several beams and then several diffraction light spots were observed on the screen. The light diffraction picture shown in Fig. 2 indicates that a periodic dielectric constant fluctuation exists in the sample. The first-order diffraction angle was measured to be 3° 24′, thus the calculated modulation period of the dielectric constant is 10.6 μm, which agrees well with the period of the PFDS. Since the dielectric constant fluctuation is caused by the periodic impurity, we could conclude that there is also a periodic yttrium distribution in the crystal, which results in the PFDS.

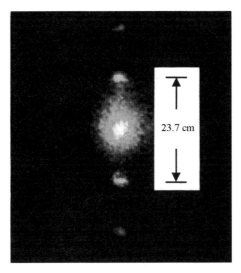

FIG. 2. The diffraction picture of a He-Ne laser passing through an ASL LN with PFDS structure. The distance between the sample and the screen is 200 cm and two first-order diffraction spots are 23.7 cm apart.

As an ASL, the ultrasonic excitation effect was also studied in the same sample. Because of the different signs of piezoelectric coefficient in positive domains and negative domains, the domain boundaries could be viewed as sound sources under the excitation of an alternating external electric field.[15] The ultrasonic waves excited in these sound sources will interfere with each other. As a result, those that satisfy constructive interference will lead to the appearance of resonance and thus the energy of electric field is converted to the elastic energy. The excitation of ultrasonic can also be treated as coupling between vibration of the superlattice and the electromagnetic waves, in which the ASL is considered as a 1D ionic-type phononic crystal.[16] The unique features of ASL for ultrasonic applications are low insertion loss and high working frequency that is determined by the period of domain.[15] There are two different schemes for the excitation of ultrasonic. One is "in-lined field" and the other is "crossed field". Figure 3 shows the diagram of the crossed field excitation scheme. The electric field is applied on the y face of ASL LN to excite the ultrasonic propagating along the z direction. Theoretically there are two types of resonance, the main resonance and the satellite-like resonance.[7] The main resonance is given by

$$f_n = nf_0 \quad (n=1,2,3\cdots) \quad (1)$$

with the fundamental frequency

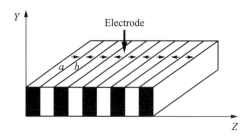

FIG. 3. Schematic of crossed field excitation of ultrasonic propagating along z axis. The arrows represent the orientation of the spontaneous polarization of LN and the alternating electric field is applied along y axis.

$$f_0 = \frac{v}{a+b}, \qquad (2)$$

where a and b is the thickness of positive domain and negative domain, respectively, and v is the velocity of shear wave propagating along z axis. Satellite-like resonance is

$$f_m = f_n \pm \frac{m}{2N} f_0 \ (m = \pm 1, \pm 3, \pm 5 \cdots), \qquad (3)$$

where N is the number of domain periods. For testifying the theoretical prediction above, a pair of Ag electrodes were deposited on the y faces of the ASL LN sample for the experiment. Using a HP8510 network analyzer, the reflection coefficient of the sample was measured and was shown in Fig. 4. The resonant frequency locates at 340 MHz, which is very close to the theoretical value 346 MHz. The slight difference between them might result from the measurement error of the modulation period. The domain walls not being exactly perpendicular to the z axis may influence its resonance frequency as well. From this resonance testing, no satellite-like resonance is observed, which shows that the positive domain thickness is almost equal to thickness of the negative domains.[7] The insertion loss of a transducer based on this sample is determined by[17]

$$IL = -20\log(1-R^2). \qquad (4)$$

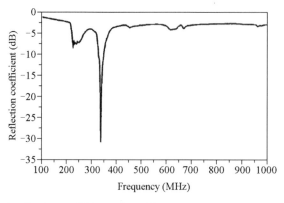

FIG.4. The measured reflection coefficient of an ASL LN with the modulation period of 10.3 μm.

If we ignore the dissipative losses in measurement, the value of insertion loss is 0 dB near the resonant peak and is very advantageous. As we know, the elastic properties, which are related to the even ranks, should be identical in positive domain and in negative domain. As a consequence, ASL LN is acoustically homogenous and there is no additional propagation loss resulted from the periodic domain structure. In addition, a contiguous piece of LN that integrates ASL LN for wave-exciting and single domain LN for wave traveling can be fabricated by our current-induction technique. Thus, by adopting the crossed field scheme, one can devise an attractive kind of acoustic device in which the path of acoustic wave is not obstructed by bonds or electrodes.[18]

In conclusion, the ASL LN with PFDS was successfully fabricated by periodic pulsed current induction during the crystal growth process. Through a simple light diffraction experiment, the periodic variation of the dielectric constant along the growth axis caused by concentration variation of impurities was revealed. A 340 MHz ultrasonic was excited by an applied radio frequency electric field on the y face of the sample, which implies that the ASL LN fabricated by this technique is suitable for constructing the high frequency bulk-wave acoustic devices.

References and Notes

[1] G. A. Magel, M. M. Fejer, and R. L. Byer, Appl. Phys. Lett. **56**, 108(1990).
[2] M. Houé, and P. D. Townsend, J. Phys. D: Appl. Phys. **28**, 1747(1995).
[3] V. Bermúdez, J. Capmany, J. Garía Solé, and E. Diéguez, Appl. Phys. Lett. **73**, 593(1998).
[4] J. J. Zheng, Y. Q. Lu, G. P. Luo, J. Ma, Y. L. Lu, N. B. Ming, J. L. He, and Z. Y. Xu, Appl. Phys. Lett. **72**, 1808(1998).
[5] Y. Y. Zhu and N. B. Ming, Appl. Phys. Lett. **53**, 1381(1988).
[6] S. D. Cheng, Y. Y. Zhu, Y. L. Lu, and N. B. Ming, Appl. Phys. Lett. **66**, 291(1995).
[7] Y. Y. Zhu, S. N. Zhu, and N. B. Ming, J. Phys. D: Appl. Phys. **29**, 185(1996).
[8] L. E. Myers, R. C. Eckardt, M. M. Fejer, R. L. Byer, W. R. Bosenberg, and J. W. Pierce, J. Opt. Soc. Am. B **12**, 2102(1995).
[9] J. Capmany, V. Bermúdez, and E. Diéguez, Appl. Phys. Lett. **74**, 1534(1999).
[10] V. Bermúdez, M. D. Serrano, P. S. Dutta, and E. Diéguez, J. Cryst. Growth **203**, 179(1999).
[11] A. Räuber, Mater. Res. Bull. **11**, 497(1976).
[12] J. Chen, Q. Zhou, J. F. Hong, W. S. Wang, N. B. Ming, D. Feng, and C. G. Fang, J. Appl. Phys. **66**, 336(1989).
[13] Y. Q. Lu, J. J. Zheng, Y. L. Lu, N. B. Ming, and Z. Y. Xu, Appl. Phys.Lett. **74**, 123(1999).
[14] Y. L. Lu, T. Wei, F. Duewer, Y. Q. Lu, N. B. Ming, P. G. Schultz, and X. D. Xiang, Science **276**, 2004(1997).
[15] H. E. Bommel and K. Dransfeld, Phys. Rev. **117**, 1245(1960).
[16] Y. Q. Lu, Y. Y. Zhu, Y. F. Chen, S. N. Zhu, N. B. Ming, and Y. Y. Feng, Science **284**, 1822(1999).
[17] E. K. Sittig, in *Progress in Optics*, edited by E. Wolf(North-Holland, Amsterdam, 1972), Vol. 10, p. 231.

[18] H. Gnewuch, N. K. Zayer, C. N. Pannell, G. W. Ross, and P. G. R. Smith, Opt. Lett. **25**, 305 (2000).

[19] This work is supported by the State Key Program for Basic Research of China, the National Natural Science Foundation Project of China(Contract No. 69708007), and the National Advanced Materials Committee of China. The technical support from Dr. Yi-jun Feng is also acknowledged.

LiTaO₃ Crystal Periodically Poled by Applying an External Pulsed Field[*]

Shi-ning Zhu, Yong-yuan Zhu, Zhi-yong Zhang, Hong Shu, Hai-feng Wang,
Jing-fen Hong, and Chuan-zhen Ge

National Laboratory of Solid State Microstructures, Nanjing University, and Center for Advanced Studies in Science Technology of Microstructures, Nanjing 210093, People's Republic of China

Nai-ben Ming

CCAST(World Laboratory), P.O. Box 8730, Beijing 100080, People's Republic of China and National Laboratory of Solid State Microstructures, Nanjing University, Nanjing 210093, People's Republic of China

A method of periodically poling LiTaO₃ single crystal at room temperature by applying an external pulsed field is proposed. The relationship between the growth of inverted domains and switching current as well as switching time has been studied. The growth of inverted domains can be controlled by the duration of the pulsed field. The domain structure with period $\Lambda > 8$ μm has been fabricated in a 0.3-mm-thick plate of LiTaO₃ by partial switching.

Periodic domain structures in LiNbO₃, LiTaO₃, and KTiOPO₄ have been obtained by various bulk[1-5] and waveguide methods.[6-12] These methods are attracting great interest because quasi-phase-matching(QPM) technique achieved by periodic domain structure has wide potential applications,[13] not only for second harmonic generation (SHG) of coherent blue-green radiation, the difference frequency generation (DFG) of infrared, and far infrared, but also for new fields such as squeezed light generation for optical communication and information processing. Approaches to achieve the required periodically inverted domain structure in LiNbO₃ and LiTaO₃ include periodic modulation of the dopant concentration during crystal growth,[2] surface impurity diffusion,[10-12] and electron-beam injection,[14] and so on.

It was believed that LiNbO₃ and LiTaO₃ are "frozen ferroelectrics" based on the inability to observe polarization reversal using conventional hysteresis-loop techniques,[15] so the domain inversion of LiNbO₃ and LiTaO₃ are difficult at room temperature. Haycock et al.[16] poled LiNbO₃ and LiTaO₃ by an energetic beam of electrons while applying an external field along the c axis. The temperature and fields were 620 ℃ and 10 V/cm for LiNbO₃, and 400 ℃ and 900 V/cm for LiTaO₃, respectively. Camlibel[17] measured the spontaneous polarization of LiNbO₃ and LiTaO₃ using a pulsed-field method, however he

---

[*] J.Appl.Phys.,1995,77(10):5481

did not give detailed records of polarization reversal. Recently, Yamada et al.[18] reported the fabrication of a first-order QPM LiNbO$_3$ waveguide periodically poled by applying an external electric field for blue light generation. And Burns et al.[19] obtained 10 – 13 μm periodic structures in 250-μm-thick slabs of LiNbO$_3$, and second harmonic blue light generation has also been demonstrated on bulk operation.

LiTaO$_3$ has a higher optical damage threshold than LiNbO$_3$, and a large nonlinear coefficient. Much interest has been aroused in LiTaO$_3$ since a several mW power of blue light was obtained using the QPM technique.[20] Here, we report domain inversion induced by applying a pulsed field on LiTaO$_3$ at room temperature. We studied the dependence of domain inversion on the external field and the duration of pulsed field. We found that if the pulsed field has enough amplitude and suitable duration, the polarization of LiTaO$_3$ crystal can be inverted either completely or partially. Using the technique we successfully fabricated periodic domain inversion structures on 0.3-mm-thick LiTaO$_3$ samples with period $\Lambda > 8$ μm.

Shown in Fig. 1 is a schematic experimental setup for inverting domain with the pulsed-field method, which is similar to the one used earlier by Merz[21] and Camlibel.[17] We used z-cut LiTaO$_3$ single crystal as samples. Al thin films 200 nm thick were deposited on both c faces as electrodes. Electrodes with periodic structures can be fabricated by photolithography.

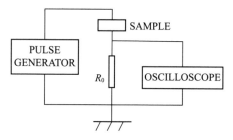

FIG.1. Schematic diagram of the experimental setup.

We studied the relationship between the growth of the inverted domains and the pulsed field. For this purpose, plane electrodes were first used. Voltage treatment was carried out at room temperature in an oil bath. A single positive pulse is applied to the plane electrode on the $+c$ face of the sample, so that the direction of the applied field is opposite to that of the spontaneous polarization P_s. The sample thickness is $d = 0.51$ mm, and the electrode area $A = 5$ mm^2. The amplitude of the rectangular pulse is adjustable from 7 to 10 kV and its width is from 10 ms to 1.5 s. The rise time of the pulse has to be short enough and its duration longer than the maintaining time in order to achieve the reversal of polarization. A reversal of P_s occurs when the applied field E exceeds the coercive field E_c of the sample, which can provide a small switching current i. To check whether the sample was completely switched or not, we observed the current developed by means of the measurement of the voltage across R_o.

Typical experimental results are shown in Figs. 2(a)–2(c). The trace in Fig. 2(a) is the output voltage waveform of the pulse generator. Fig. 2(b) shows the trace of switching current i corresponding to the first pulse. The narrow peak indicated with an arrow is attributed to the sample capacity. In Fig. 2(c), the trace shows the wave form when a second pulse with the same polarity and amplitude was applied to the same sample. The peak is due to the sample capacity, not due to the polarization switching. This means that the sample was completely switched after the first pulse was applied. The observation of the etched cross section of the sample further proves this conclusion.

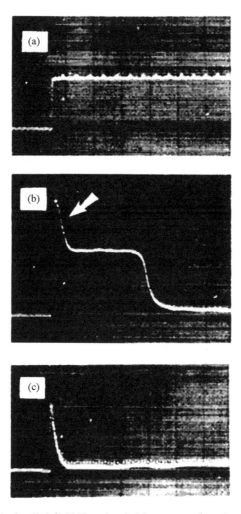

FIG.2. Applied field E and switching current i vs time. The horizontal scale is t. In (b), the time during which the switching current maintains is defined as switching time t_s.

From the measurement, the magnitude of the spontaneous polarization P_s can be calculated using the expression

$$Q = \int i\,dt = 2P_s A, \tag{1}$$

where Q is the total charge during domain reversal. Our measured value of $P_s = 0.51 \pm 0.01$ C/m^2 is in good agreement with the values given by other authors.[17]

In Fig. 3 we plotted the switching time t_s and its reciprocal $1/t_s$ versus the applied field E. The higher the applied field E, the shorter the switching time t_s. The plot of $1/t_s$ versus E can be divided into two parts: a curved part corresponding to low-field strength and a linear part corresponding to high-field strength. This result conforms to the result obtained on BaTiO$_3$ crystal.[21] At the low-field part ($E < 20$ kV/mm) the switching time t_s can be best expressed in the following way

$$t_s \propto e^{-\alpha(T)/E}, \tag{2}$$

where $\alpha(T)$ is a function of temperature T. In Fig. 4, a logarithmic plot of t_s versus $1/E$ is shown. We got a straight line over 3 decades. At the high field part ($E > 20$ kV/mm), t_s is proportional to $1/(E - E')$, thus we can write that

$$t_s \propto \beta(T)/(E - E'), \tag{3}$$

with $\beta(T)$ is related to temperature T. E' is a kind of coercive field strength, which can be determined by the intersection of the extension of the straight in Fig. 3 with the x axis. Here $E' = 16.5$ kV/mm.

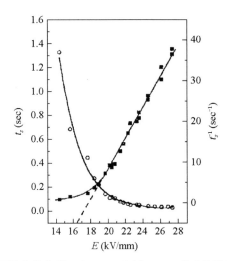 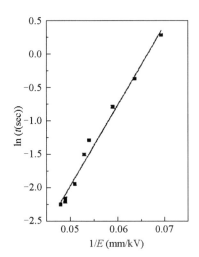

FIG. 3. Switching time t_s and $1/t_s$ vs applied field E. FIG. 4. Logarithmic plot of switching time t_s vs $1/E$.

Based on the above experiments, we studied the periodic electrode sample. The relationship between t_s and E is the same as the plane electrode sample's. However, for the sample completely switched, the domain is wholly inverted, i.e., the periodicity disappears. Therefore, in order to fabricate LiTaO$_3$ samples with periodic domain structures, the nucleation of domain inversion and its development under external pulsed field should be studied. We found that the polarization reversal in a z-cut LiTaO$_3$ crystal is as follows. For a plane electrode sample, the inverted domain first randomly nucleates at the surface and imperfections in the interior in the form of needles, while for a sample with periodic electrode on its surface, the nucleation first occurs under the periodic electrode. Then the inverted domains in these two kinds of samples both grow in the forward direction through the crystal. Finally, the domains spread out and come into contact with each other. Thus, it is very important that the external pulsed field be timely shut off to achieve the required periodically reversed domain structure. To see this more clearly, we prepared two z-cut 0.3-mm-thick $LiTaO_3$ samples A and B. For sample A, plane electrodes were deposited on the both c faces. For sample B, plane electrode was deposited only on its negative c face, while on the positive c face was deposited an electrode with periodic structures ($\Lambda = 8.5$ μm). The pulse duration is $t = 1/2(t_s)$ and may be called partial switching. After that, we cut both samples, and polished and etched the y faces to identify their domain structure. Figures 5(a) and 5(b) are the photographs of the etched cross sections of samples A and B, respectively. In both samples, about 50% of regions are occupied by the inverted domains, however, their distributions are very different. In sample A the inverted domains are randomly distributed, while in sample B, the inverted domain is periodically distributed, essentially across the whole sample with a narrow width. For a sample with a periodic electrode and a pulse width of $t > 1/2(t_s)$, gap regions between the electrodes will be filled by the broadening of the inverted domains. This will reduce the nonlinear conversion efficiency and applicability.

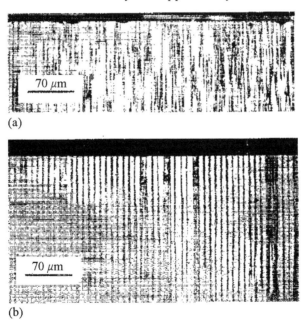

FIG.5. The domain structure in $LiTaO_3$ crystal ($d = 0.3$ mm) by partial switching, (a) for plane electrodes on both c faces, (b) for a periodic electrode on $+c$ face and a plane electrode on $-c$ face. ($E = 21$ kV/mm; $t = 0.246$ s).

In conclusion, we have shown that periodic domain reversal can be achieved by applying a pulsed field across sufficiently thick LiTaO$_3$ samples. By suitably adjusting the pulse duration the growth of inverted domains can be controlled. Further studies on the dependence of the switching current on the applied electrical field, the thickness of sample and temperature, the behavior of the nucleation, and growth of inverted domains during switching and QPM-SHG experiment in bulk and waveguide will be reported later.

References and Notes

[1] D. Feng, N. B. Ming, J. F. Hong, Y. S. Yang, J. S. Zhu, Z. Yang, and Y. N. Wang, Appl. Phys. Lett. **37**, 607 (1980).

[2] N. B. Ming, J. F Hong, and D. Feng, J. Mater. Sci. **17**, 1663 (1982).

[3] Y. H. Xue, N. B. Ming, J. S. Zhu, and D. Feng, Chin. J. Phys. **4**, 554(1984).

[4] A. Feisst and P. Koidl, Appl. Phys. Lett. **47**, 1125 (1985).

[5] G. A. Magel, M. M. Fejer, and R. L. Byer, Appl. Phys. Lett. **56**, 108(1990).

[6] J. A. Armstrong, N. Bloembergen, J. Ducuing, and P. S. Pershan, Phys. Rev. **127**, 1918 (1962).

[7] E. J. Lim, M. M. Fejer, R. L. Byer, and W. J. Kozlovsky, Electron. Lett. **25**, 731 (1989).

[8] Y. Ishigame, T. Suhara, and H. Nishihara, Opt. Lett. **16**, 375 (1991).

[9] X. Cao, B. Rose, R. V. Ramaswamy, and R. Srivastava, Opt. Lett. **17**, 795(1992).

[10] S. Miyazawa, J. Appl. Phys. **50**, 4599 (1979).

[11] J. Webjorn, F. Laurell, and G. Arvidsson, IEEE Photon. Technol. Lett. **1**, 316 (1989).

[12] M. Fujimura, T. Suhara, and H. Nishihara, Electron. Lett. **27**, 1207 (1991).

[13] K. Nakamura, H. Ando, and H. Shimizu, Appl. Phys. Lett. **50**, 1413(1987).

[14] H. Ito, C. Takyu, and H. Inaba, Electron. Lett. **27**, 1221 (1991).

[15] K. Nassau and H. J. Levinstein, Appl. Phys. Lett. **7**, 69 (1965).

[16] P. W. Haycock and P. D. Townsend, Appl. Phys. Lett. **48**, 698 (1986).

[17] I. Camlibel, J. Appl. Phys. **40**, 1690 (1969).

[18] M. Yamada, N. Nada, M. Saitoh, and K. Watanabe, Appl. Phys. Lett. **62**, 435 (1993).

[19] W. K. Burns, W. McElhanon, and L. Goldberg, IEEE Photon. Technol.Lett. **6**, 252 (1994).

[20] K. Mizuuchi, K. Yamamoto, and T. Taniuchi, Appl. Phys. Lett. **58**, 2732(1991).

[21] W. J. Merz, Phys. Rev. **95**, 690 (1954).

[22] This work is supported by a grant for the Key Research Project in Climbing Program from the National Science and Technology Commission of China.

Poling Quality Evaluation of Optical Superlattice Using 2D Fourier Transform Method[*]

X. J. Lv, L. N. Zhao, J. Lu, G. Zhao, H. Liu, Y. Q. Qin and S. N. Zhu

National Laboratory of Solid State Microstructure, Nanjing University, Nanjing 210093, People's Republic of China

In this article we develop a method to evaluate the poling quality of optical superlattice (OSL) based on two-dimensional (2D) Fourier transform. To demonstrate this method, $-Z$ or $+Z$ face etched OSL samples with desired patterns are fabricated by standard electric field poling technique. By analyzing the processed micrograph of the etched surfaces, the magnitude of the reciprocal vectors of the OSL are calculated directly and rapidly. Second harmonic generation (SHG) experiment is performed to validate the evaluation result.

1. Introduction

The development of domain-engineered crystals has gained significant importance not only in nonlinear optical frequency conversion process[1], but also in linear polarization control[2] or modulation[3] devices. Quasi-phase-matching (QPM) devices have been successfully realized in ferroelectric crystals such as lithium niobate ($LiNbO_3$), lithium tantalate ($LiTaO_3$) and potassium titanyl phosphate ($KTiOPO_4$) by the process of electric field poling. Utilizing this approach, QPM pattern with desired reciprocal vectors is able to be fabricated. Due to substrate inhomogeneity and imperfect photolithography process, two kinds of error are inherent: period error and duty cycle error[4]. There are several methods to visualize the resulting domain structures such as second harmonic generation microscopy[5], confocal luminescence microscopy[6], optical near field microscopy[7], etc. However, in these techniques only a small area of the poled pattern can be visualized. An alternative approach to investigate the poling quality of QPM device using diffraction method has been described by Krishnamoorthy et al, in which the duty cycle error is quantified[8].

Effective nonlinear coefficient (d_{eff}) is proportional to the magnitude of the reciprocal vector. Period and duty cycle error influence the magnitude of the reciprocal vector and finally reduce the nonlinear conversion efficiency. In this article, we present a technique to quantify the magnitude of the reciprocal vectors using 2D Fourier transform method. By comparing the measured result with the ideal or designed magnitude of the reciprocal

---

[*] Opt.Express, 2009, 17(20):18241

vector, the poling quality of an OSL can be evaluated.

2. Theory

The second-order nonlinear polarization is given by

$$\begin{pmatrix} P_1 \\ P_2 \\ P_3 \end{pmatrix} = \begin{pmatrix} d_{11} & d_{12} & d_{13} & d_{14} & d_{15} & d_{16} \\ d_{21} & d_{22} & d_{23} & d_{24} & d_{25} & d_{26} \\ d_{31} & d_{32} & d_{33} & d_{34} & d_{35} & d_{36} \end{pmatrix} \begin{pmatrix} E_1^2 \\ E_2^2 \\ E_3^2 \\ 2E_2 E_3 \\ 2E_3 E_1 \\ 2E_1 E_2 \end{pmatrix}. \quad (1)$$

For QPM materials, e.g. periodically poled LiTaO$_3$ (PPLT), the maximum second-order nonlinear optical susceptibility is $d_{33} = 13.8$ pm/V@1064 nm[9]. Define rect function as

$$\text{rect}(t) = \begin{cases} 0, & \text{if } |t| > \frac{1}{2} \\ \frac{1}{2}, & \text{if } |t| = \frac{1}{2} \\ 1, & \text{if } |t| < \frac{1}{2} \end{cases}. \quad (2)$$

We assume the propagation to be along the x-axis. In order to compensate phase mismatch, the modulation function of d_{33} should be

$$f(x) = \sum_{j=-m}^{m} \left[\text{rect}\left(\frac{j + \frac{x}{\Lambda}}{D} \right) - \text{rect}\left(\frac{j - D + \frac{x}{\Lambda}}{1-D} \right) \right], \quad (3)$$

where Λ is the period and D is the duty cycle, satisfying $\frac{1}{\Lambda} = \frac{n(\lambda_1)}{\lambda_1} - \frac{n(\lambda_2)}{\lambda_2} - \frac{n(\lambda_3)}{\lambda_3}$ ($\lambda_1 \leqslant \lambda_2, \lambda_1 < \lambda_3$). The length of the superlattice is $a = (2m+1)\Lambda$. Effective nonlinear coefficient is given by

$$d_{\text{eff}} = d_{33} \cdot |g(f)|, \quad (4)$$

$$g(f) = \frac{1}{a} \int_{-\frac{a}{2}}^{\frac{a}{2}} f(x) \exp(2\pi i f x) \, dx. \quad (5)$$

At the phase-matching points, $f_n = \frac{n}{\Lambda}$ ($n = 1, 2, 3, \cdots$),

$$g(f_n) = \frac{1}{a} \int_{-\frac{a}{2}}^{\frac{a}{2}} f(x) \exp(2\pi i f_n x) \, dx = \frac{\exp(2\pi i n D) - 1}{i n \pi}. \quad (6)$$

If the superlattice is poled perfectly $\left(D = \frac{1}{2} \right)$, $|g(f_n)| = 0$ when n is even, and $|g(f_n)| = \frac{2}{\pi n}$ when n is odd. The most commonly used one is the first order reciprocal vector: when $n = 1$, $|g_1| = 0.6366$.

The ferroelectric domain boundaries of a perfectly poled one-dimensional (1D) OSL parallel to each other. The substrate inhomogeneity and imperfect photolithography result in the distorted domain boundaries. In order to evaluate these error, two-dimensional (2D) evaluation method should be introduced. Similarly, we extend the $g(f_x)$ to the 2D case,

$$g(f_x,f_y) = \frac{1}{a \cdot b} \int_{-\frac{b}{2}}^{\frac{b}{2}} \int_{-\frac{a}{2}}^{\frac{a}{2}} f(x,y) \exp[2\pi i(xf_x + yf_y)] dx dy, \qquad (7)$$

and

$$d_{eff} = d_{33} \cdot |g(f_x,f_y)|, \qquad (8)$$

where a is the length and b is the width of the sampling area. Generally the sampling area should contain enough periods, so the condition of $a, b \gg \Lambda$ should be satisfied. To explore the domain walls, the $-Z$ and $+Z$ surfaces of the same OSL are etched with hydro-fluoric solution. It etches smoothly at $-Z$ face whereas the inverted domain etching rate is negligible. Using polarizing microscope or phase contrast microscope, we can distinguish the $+Z$ domain from the $-Z$ domain. The domain wall in the micrograph usually has a certain width. In order to find the accurate position of the domain walls, image processing operations[10] including sharpening and thinning methods are preformed. After image processing operations, the $+Z$ domains are assigned $+1$ and $-Z$ domains are assigned -1, by which we can calculate the reciprocal vectors of the superlattice.

However, the view-field of the microscope is limited to a small region. Now we consider how to extrapolate the $g(f_x, f_y)$ of the OSL from the region we observed. A limited region $f(x,y) = u(x,y) \cdot \text{rect}\left(\frac{x}{a}\right) \cdot \text{rect}\left(\frac{y}{b}\right)$ is intercepted from the whole region $u(x,y)$. Periodic prolongation of $f(x,y)$ will rebuild $U(x,y)$ as

$$U(x,y) = f(x,y) * \sum_{j=-m}^{m} \sum_{k=-n}^{n} [\delta(ja+x) \cdot \delta(kb+y)]. \qquad (9)$$

m, n are integers and $\delta(x)$ is Dirac function. If the sampling area is representative, we have $U(x,y) \approx u(x,y)$. Then the reciprocal vector of $U(x,y)$ is

$$\begin{aligned}
G(f_x,f_y) &= \frac{1}{S} \mathscr{F}(U(x,y)) \\
&= \frac{1}{(2m+1) \cdot (2n+1) \cdot a \cdot b} \mathscr{F}(f(x,y)) \\
&\quad \cdot \mathscr{F}\left\{ \sum_{j=-m}^{m} \sum_{k=-n}^{n} [\delta(ja+x) \cdot \delta(kb+y)] \right\} \\
&= g(f_x,f_y) \cdot \frac{1}{(2m+1) \cdot (2n+1)} \sum_{j=-m}^{m} \sum_{k=-n}^{n} \int_{-\infty}^{\infty} \int_{-\infty}^{\infty} \exp[2\pi i(xf_x + yf_y)] \\
&\quad \cdot \delta(ja+x) \cdot \delta(kb+y) dx dy \\
&= g(f_x,f_y) \cdot \frac{1}{(2m+1) \cdot (2n+1)} \sum_{j=-m}^{m} \sum_{k=-n}^{n} \exp[-2\pi i(jaf_x + kbf_y)] \\
&= \begin{cases} g(f_x,f_y), f_x = \frac{M}{a} \text{ and } f_y = \frac{N}{b}, M \text{ and } N \text{ are integers} \\ 0, \text{other when } m, n \to \infty \end{cases} \qquad (10)
\end{aligned}$$

This means that at $\left(f_x=\dfrac{M}{a}, f_y=\dfrac{N}{b}\right)$, the Fourier transform result of a limited region is equal to that of the expanded region. If the observed region is representative, by calculating the reciprocal vector at these points, we can obtain the $g\left(\dfrac{M}{a},\dfrac{N}{b}\right)$ of the whole OSL. The fast Fourier transform (FFT) is an efficient algorithm to compute the discrete Fourier transform (DFT). In 1D case, the separated frequency components of FFT are at $f_N=\dfrac{N}{a}$ (assume the sample length in the real space to be a and $N=0,1,2,\cdots$). The first order reciprocal vector (g_1) of the superlattice with a period of Λ is at $f=\dfrac{1}{\Lambda}$. In order to ensure the peak of the first order reciprocal vector coincides with the discrete frequency components of FFT, the condition of $\dfrac{N}{a}=\dfrac{1}{\Lambda}$ should be satisfied, which means that $\dfrac{a}{\Lambda}$ should be integer and the sample region should contain integer periods.

An alternative method is high resolution Fourier transform (HRFT)[11], which is capable of improving the frequency resolution. But the sample window will impact on the reciprocal vectors. If the periodic structure is along x-axis,

$$\begin{aligned}
g(f_x,f_y) &= \dfrac{1}{a\cdot b}\mathscr{F}f(x,y) \\
&= \dfrac{1}{a\cdot b}\cdot\mathscr{F}\left[u(x,y)\cdot\mathrm{rect}\left(\dfrac{x}{a}\right)\cdot\mathrm{rect}\left(\dfrac{y}{b}\right)\right] \\
&= \mathscr{F}[u(x,y)]*[\mathrm{Sinc}(af_x)\cdot\mathrm{Sinc}(bf_y)] \\
&= \left[\delta(f_y)\cdot\sum_{m=-\infty}^{\infty}\dfrac{\delta\left(\dfrac{m}{\Lambda}-f_x\right)-(-1)^m\delta\left(\dfrac{m}{\Lambda}-f_x\right)}{\mathrm{i}2\pi m}\right]*[\mathrm{sinc}(af_x)\cdot\mathrm{sinc}(bf_y)].
\end{aligned}$$

(11)

Theoretically, the magnitude of the even order reciprocal vectors are zero, and the maximum influence of the odd order vectors (3th order) to the first order vector is $\dfrac{\Lambda}{3\pi a}$. If the sample region contains enough period $\left(\dfrac{\Lambda}{3\pi a}\ll 1\right)$, this influence is negligible $\left(\text{e.g., when }\dfrac{a}{\Lambda}=50,\dfrac{\Lambda}{3\pi a}=0.002\right)$. So it is capable to obtain the reciprocal vector by HRFT with very small error. But HRFT algorithm is much slower than FFT. In practice, we use FFT to obtain the approximate position of the reciprocal vector and then use HRFT to calculate the peak value of it. The calculation speed and precision can both be satisfied.

In addition, the sampling interval will also cause calculation error. Assume the sampling interval to be d. The sign of the whole interval is decided by the value of the point at the center. So the sampled duty cycle is $D_s=\dfrac{p}{\Lambda}+D$, where D is the actual duty cycle and p is random number uniformly distributed over the interval $\left(-\dfrac{d}{2},\dfrac{d}{2}\right)$. Mathematical expectation of random variable X is noted as $E(X)$. Considering $E(X_1+X_2+\cdots+X_n)=E(X_1)+E(X_2)+\cdots+E(X_n)$, the Mathematical expectation of Eq. (5) is given by

$$E(g(f)D_s) = \frac{1}{a}E\left\{\int_{-\frac{a}{2}}^{\frac{a}{2}}\sum_{j=-m}^{m}\left[\mathrm{rect}\left[\dfrac{j+\dfrac{x}{\Lambda}}{D_j}\right] - \mathrm{rect}\left[\dfrac{j-D_j+\dfrac{x}{\Lambda}}{1-D_j}\right]\right]\exp(2\pi ifx)\mathrm{d}x\right\}$$

$$= \frac{1}{a}\int_{-\frac{a}{2}}^{\frac{a}{2}}\sum_{j=-m}^{m}\frac{1}{d}\int_{-\frac{d}{2}}^{\frac{d}{2}}\left[\mathrm{rect}\left[\dfrac{j+\dfrac{x}{\Lambda}}{D_j}\right] - \mathrm{rect}\left[\dfrac{j-D_j+\dfrac{x}{\Lambda}}{1-D_j}\right]\right]\mathrm{d}p_j$$

$$\cdot \exp(2\pi ifx)\mathrm{d}x, \tag{12}$$

where $D_j = \dfrac{p_j}{\Lambda} + D$. Considering that $\{p_j\}$ are independent with each other and under the same distribution,

$$E(g(f)D_s) = \frac{1}{ad}\int_{-\frac{d}{2}}^{\frac{d}{2}}\int_{-\frac{a}{2}}^{\frac{a}{2}}\sum_{j=-m}^{m}\left[\mathrm{rect}\left[\dfrac{j+\dfrac{x}{\Lambda}}{D_j}\right] - \mathrm{rect}\left[\dfrac{j-D_j+\dfrac{x}{\Lambda}}{1-D_j}\right]\right]\cdot\exp(2\pi ifx)\mathrm{d}x\mathrm{d}p$$

$$= \frac{1}{d}\int_{-\frac{d}{2}}^{\frac{d}{2}}\frac{\exp(2\pi iD_s)-1}{i\pi}\mathrm{d}p = \frac{\Lambda\exp(2\pi iD)\sin\left(\dfrac{\pi d}{\Lambda}\right)-\pi d}{i\pi^2 d}. \tag{13}$$

When $\dfrac{a}{\Lambda} = 50$ (it means that the sampling area contains 50 periods) and the sampling number is 1024,

$$|E(g_1)| - |E(g_1D_s)| = \begin{cases} 0.0013, D=0.5 \\ 0.0011, D=0.3 \end{cases}. \tag{14}$$

The measured result is always smaller than the actual value. This is easy to understand because the sampling error always decreases the vector magnitude. If the sampling number of each domain is large enough, this error is negligible.

3. Experiment and Discussion

The periodic poled LiTaO$_3$ (PPLT) and LiNbO$_3$ (PPLN) samples are fabricated by electric field poling technique[12]. To explore the superlattice, the $-Z$ and $+Z$ surfaces of the same OSL are etched with hydro-fluoric solution. The original micrography of the $+Z$ surface of sample 1(PPLT, $\Lambda = 28.6\ \mu\mathrm{m}$) is shown in Fig. 1(a). Figure 1(b) shows the image processing result of Fig. 1(a). Then fill the positive domains with black(+1) and the negative domains with white(−1) as shown in Fig. 1(c). The 1024×1024 pixels image is sampled point-by-point and FFT result is shown in Fig. 2(a) and 2(b). Due to the limited resolution of FFT, the peak amplitude of the reciprocal vectors in Fig. 2(b) are not accurate. But we can find the approximate positions of these peaks. Figure 2(c) shows the

HRFT result of Fig. 1(c) at the first order reciprocal vector position. The peak value is $|g_1| = 0.625$. This means that the poling quality of sample1's $+Z$ surface is very high (the perfect value is $2/\pi = 0.6366$). If the d_{33} is given as $d_{33} = 13.8$ pm/V, $d_{eff} = d_{33} \cdot |g_1| = 8.63$ pm/V.

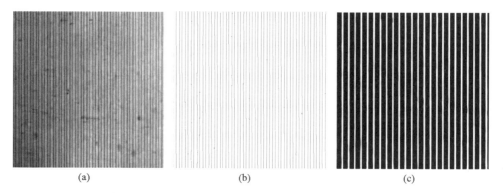

FIG.1. Sample 1: $+Z$ surface ($\Lambda = 28.6$ μm). (a) The original micrograph; (b) The processed image; (c) The final image before analysis.

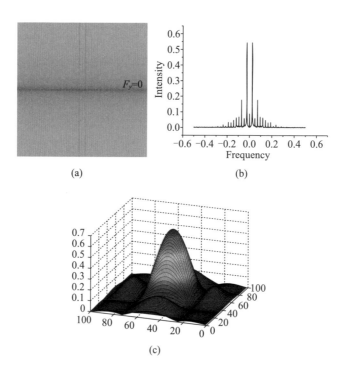

FIG.2. Fourier transform result of Fig. 1(c). (a) 2D FFT result; (b) 2D FFT result at $f_y = 0$; (c) HRFT result is $|g_1| = 0.625$.

Figure 3 shows the $-Z$ surface of sample 1 and the FFT result at $f_y = 0$. The HRFT result of the first order reciprocal vector is $|g_1| = 0.592$. From Fig. 3(b) we can see that due to the decreased poling quality, the second-order reciprocal vector is higher than the third one.

Figure 4 shows the $-Z$ surface of sample 2 (PPLT, $\Lambda = 7$ μm) and it's FFT result. The HRFT result of the first order reciprocal vector is $|g_1| = 0.187$. From Fig. 4(a) we can see that due to the poor poling quality, the one-dimensional superlattice nearly turns in to a two-dimensional one. The nonuniformity of the 2D pattern introduced by the poling process greatly decreases the quality of the superlattice.

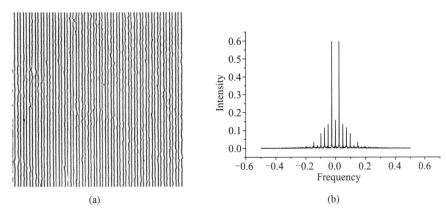

(a) (b)

FIG.3. Sample 1: $-Z$ surface. (a) The processed micrograph; (b) 2D FFT result at $f_y = 0$. The HRFT result is $|g_1| = 0.592$.

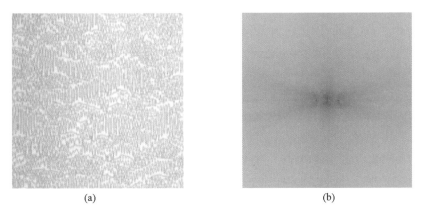

(a) (b)

FIG.4. Sample 2: $-Z$ surface ($\Lambda = 7$ μm). (a) The original micrograph; (b) 2D FFT result. The HRFT result is $|g_1| = 0.187$.

Figure 5(a) shows the $+Z$ surface of a square poled lithium tantalate (Squ-PLT, sample 3, square lattice of circular patterns) with structure parameter $a = 9.05$ μm. The FFT result is shown in Fig. 5(b). The HRFT result of the maximal reciprocal vector is $|g_{01}| = 0.293$, which is designed to be 0.4. Now we derive the analytical expression of the reciprocal vector of the 2D OSL with rectangle lattice. The OSL can be expressed as

$$f(x,y) = \left\{ 2P(x,y) * \left[\text{comb}\left(\frac{x}{\Lambda_x}\right) \text{comb}\left(\frac{y}{\Lambda_y}\right) \right] - 1 \right\} \cdot \text{rect}\left(\frac{x}{a}\right) \cdot \text{rect}\left(\frac{y}{b}\right), \quad (15)$$

where $P(x,y)$ is the function of the positive domain in a period, Λ_x and Λ_y are the period of lattice along x-axis and y-axis, a and b are the width and length of the OSL,

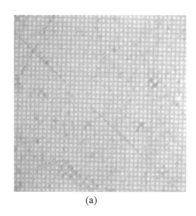 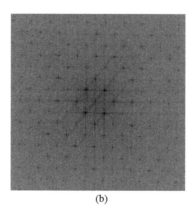

FIG.5. Sample 3: +Z surface (Squ-PLT, $a = 9.05$ μm). (a) The original micrograph; (b) 2D FFT result. The HRFT result is $|g_{01}| = 0.293$.

respectively. The reciprocal vector of 2D OSL is given as

$$g(f_x, f_y) = \{2\Lambda_x\Lambda_y\mathscr{F}[P(x,y)] \cdot \text{comb}(f_x\Lambda_x)\text{comb}(f_y\Lambda_y) - \delta(f_x)\delta(f_y)\} * [\text{sinc}(af_x)\text{sinc}(bf_y)]. \tag{16}$$

For square poled OSL, if we assume the positive domain to be circular and $\Lambda_x = \Lambda_y = a$, the maximum g_{01} is about 0.4 (which is the reciprocal vector we desire) when the radius of the circle is about $0.39a$. From the evaluation we can see that due to poling error, the g_{01} of the fabricated 2D SQL-PLT is smaller than the perfect value. If the pattern is irregular, it is difficult to measure the duty cycle and period error. In this case, our method is still valid, which indicates that it is quite fit for the evaluation of 2D OSL.

In addition, the process of 1024×1024 FFT and HRFT is very fast (several seconds using a computer with 2GHz CUP clock speed). The necessities of this method are only microscope with CCD camera and computer. The simplicity and rapidity of this method make it promising in poling quality evaluation.

In order to validate the evaluation, SHG experiment is performed to inspect the poling quality. Under small-signal condition, the generated second harmonic intensity is

$$I_2 = \frac{2\omega_1^2 d_{\text{eff}}^2 L^2 I_1^2}{c^3 n_1^2 n_2 \varepsilon_0} \frac{\sin^2(\Delta kL/2)}{(\Delta kL/2)^2} = C|g|^2, \tag{17}$$

where I_1 and I_2 are the optical power density of fundamental and second-harmonic wave, respectively. The direct measurement of $|g|$ is difficult because the above equation is derived using the approximation of plane wave, single-longitudinal mode and continuous wave conditions. What's more, the d_{33} of LiNbO$_3$ varies from 25 pm/V to 42 pm/V at different references [13], which will have great impact on the calculation of $|g|$. The relative $|g|^2$ is easier to be achieved. Assuming that the fundamental beam propagates along Y-axis, when scan the fundamental beam along X-axis, we will achieve $I_2(x) = C|g(x)^2|$, where C is a constant if we keep the other parameters (I_1, oven temperature, etc.) constant. In our experiment, the fundamental wave is outputted from a Q-switched

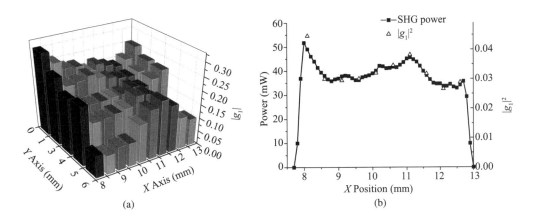

FIG. 6. (a) Evaluated $|g_1|$ at the $+Z$ surface of sample 4 (PPLN, $\Lambda = 6.6$ μm); (b) SHG power and evaluated $|g_1|^2$ at different X-axis position.

1064 nm Nd:YAG laser with pulse duration of 50 ns and repetition rate of 0.2 kHz. The average power is 560 mw. Then it is focused to about 0.1 mm diameter spot with a $f = 200$ mm lens. The fluctuation of fundamental power is less than 2% within 10 minutes. The PPLN oven is fixed at (146.1 ± 0.1) °C to ensure negligible photorefractive effects. The beam is propagating along Y-axis and moving along X-axis. The measured second harmonic power is shown in Fig. 6(b). The maximum output power is about 50 mW and the small-signal approximation is still valid. We divide the $+Z$ surface of the 6.14 mm long (Y-axis, periodically poled), 5mm wide (X-axis) PPLN (sample 4) into 5×10 areas. The poling quality of the $-Z$ surface approximately equals to that of the $+Z$ surface. The period of the OSL is about 6.6 μm and the first reciprocal vector is used for SHG process. We calculate the $|g_1|$ of each area separately (Fig. 6(a)). Each g_1 represents the average intensity of the reciprocal vector at the single area. The average $|g_1|$ of the whole OSL (50 areas) is 0.181 and the standard deviation (SD) is 0.052. In general, the higher of $|g_1|$ and the smaller of SD, the higher quality of the whole OSL. Due to $\mathscr{F}(f_1 + f_2) = \mathscr{F}(f_1) + \mathscr{F}(f_2)$, the $|g_1(x)|$ equals to the average $|g_1|$ of each area along Y-axis at x position. Fig. 6(b) shows the comparison of the calculated $|g_1(x)|^2$ and the measured output SHG power. The fundamental beam is influenced by the edge of the OSL, so the output power is smaller than it should be at the edge of the OSL. At other points, the proportional relation between output power and the calculated $|g_1|^2$ is good. If we define $C(x) = I_2(x)/|g_1(x)|^2$, the relative deviation can be expressed as $|C(x) - \overline{C(x)}|/\overline{C(x)}$ which is smaller than 5% except for the two points at the edge of the OSL. Considering the fluctuation of fundamental power and temperature, we think this error is acceptable.

4. Conclusion

In conclusion, we fabricate PPLT and PPLN samples using standard electric field poling technique. A method is proposed to evaluate the poling quality by analyzing the processed micrograph of the etched surface. The error caused by sampling is derived and estimated. Utilizing this method, the amplitude of the reciprocal vectors are able to be calculated directly. Poling quality of the whole superlattice is evaluated and average quality and deviation are given. SHG experiment is performed to validate the evaluated result, which demonstrates that this method is reliable. In principle, this technique is appropriate for evaluating the poling quality of the periodically poled 1D or 2D superlattice.

References and Notes

[1] Z. D. Gao, S. N. Zhu, S. Y. Tu, and A. H. Kung, "Monolithic red-green-blue laser light source based on cascaded wavelength conversion in periodically-poled stoichiometric lithium tantalate," Appl. Phys. Lett. **89**, 181101(2006).

[2] Q. Chen, Y. Chiu, D. N. Lambeth, T. E. Schlesinger, and D. D. Stancil, "Guided-wave electro-optic beam deflector using domain reversal in $LiTaO_3$," IEEE J. Lightwave Technol. **12**, 1401–1404 (1994).

[3] S. Kumar, D. Gurkan, A. E. Willner, K. Parameswaran, and M. Fejer, "All-optical half adder using a PPLN waveguide and an SOA," OFC 2004 **1**, 23–27 (2004).

[4] M. M. Fejer, G. A. Magel, D. H. Jundt, and R. L. Byer, "Quasi-phase-matched second harmonic generation: tuning and tolerances," IEEE J. Quantum. Electron. **28**, 2631–2654 (1992).

[5] A. Rosenfeldt and M. Florsheimer, "Nondestructive remote imaging of ferroelectric domain distributions with high three-dimensional resolution," Appl. Phys. B. **73**, 523–529 (2001).

[6] V. Dierolf and C. Sandmann, "Inspection of periodically poled waveguide devices by confocal luminescence microscopy," Appl. Phys. B. **78**, 363–366 (2004).

[7] T. J. Yang, V. Gopalan, P. J. Swart, and U. Mohideen, "Direct Observation of Pinning and Bowing of a Single Ferroelectric Domain Wall," Phys. Rev. Lett. **82**, 4106–4109 (1999).

[8] K. Pandiyan, Y. S. Kang, H. H. Lim, B. J. Kim, O. Prakash, and M. Cha, "Poling Quality Evaluation of Periodically Poled Lithium Niobate Using Diffraction Method," J. Opt. Soc. Korea. **12**, 205–209 (2008).

[9] I. Shoji, T. Kondo, A. Kitamoto, M. Shirane, and R. Ito, "Absolute scale of second-order nonlinear-optical coeffients," J. Opt. Soc. Am. B **14**, 2268–2294 (1997).

[10] A. K. Jain, *Fundamentals of Digital Image Processing* (Prentice-Hall, Englewood Cliffs, 1989).

[11] M. D. Sacchi, T. J. Ulrych, and C. J. Walker, "Interpolation and extrapolation using a high-resolution discrete Fourier transform," IEEE Trans. Signal Processing. **46**, 31–38(1998).

[12] S. N. Zhu, Y. Y. Zhu, Z. Y. Zhang, H. Shu, H. F. Wang, J. F. Hong, C. Z. Ge, and N. B. Ming, "$LiTaO_3$ crystal periodically poled by applying an external pulsed field," J. Appl. Phys. **77**, 5481–5483 (1995).

[13] J. Yan, H.W. Li, X. L. Yang, S.W. Xie, and Z. R. Sun, "Second-order nonlinear optical coefficients measurement of $LiNbO_3$ by non-phase-matched second-harmonic generation in uniaxial crystal sphere," Chinese Laser J. **21**, 14–16 (2000).

[14] This work is supported by the National Natural Science Foundation of China Grant (Nos. 10776011, Nos. 10534020) and by the National Key Projects for Basic Research of China (No. 2006CB921804).

Frequency Self-doubling Optical Parametric Amplification: Noncollinear Red-green-blue Lightsource Generation Based on a Hexagonally Poled Lithium Tantalate*

P. Xu, Z. D. Xie, H. Y. Leng, J. S. Zhao, J. F. Wang, X. Q. Yu, Y. Q. Qin, and S. N. Zhu

National Laboratory of Solid State Microstructures, Nanjing University, Nanjing, 210093, China

Simultaneous generation of noncollinear red, green, and blue light from a single hexagonally poled lithium tantalate is reported. It results from the frequency self-doubling optical parametric amplification process, a process of second-order harmonic generation cascaded optical parametric amplification in a single-pass setup. The temperature and spectrum detuning characters of each cascaded quasi-phase-matching process are studied. This unique red-green-blue light source has potential applications in laser display and other laser industries.

The cascaded nonlinear process has attracted great interest in recent years, since it can extend light wavelength to shorter or longer extents[1—20]. It usually requires more than one second-order nonlinear crystal or relies on coincidental crossing of several phase-matching curves inside one nonlinear crystal in which birefringence phase matching (BPM) is utilized[1—7]. However, for this type of cascaded optical parametric process, the $\chi^{(2)}$-modulated nonlinear crystal, also called quasi-phase-matching (QPM) material, should have more advantages. People can modulate $\chi^{(2)}$ to realize intense coupling between arbitrary parametric processes inside one single crystal. Pioneer work including second-order harmonic generation (SHG) or sum frequency generation (SFG) cascaded optical parametric oscillation (OPO), etc., have been accomplished through QPM inside a one-dimensional (1D) periodically, quasi-periodically, or aperiodically poled crystal[8—20]. Some high-order harmonic generations even in a two-dimensional (2D) QPM crystal are achieved[21,22].

In this letter we report on the experimental realization of self-doubling optical parametric amplification (OPA) process, a process of the signal or the idler doubling cascaded OPA inside a hexagonally poled lithium tantalate (HexPLT) crystal. The cascaded SHG of the signal or the idler together with the OPA process both reach a high efficiency in a single-pass setup. The simultaneous happening SHGs of the signal and the

---

* Opt. Lett., 2008, 33(23): 2791

idler together with the residual pump light generate a set of red-green-blue (RGB) primary colors in vision. The special tuning characters on temperature, wavelength, angle, etc. of this HexPLT imply that it is a novel candidate for a high-gain RGB light source and other multiple-wavelengths generators.

In the experiment the HexPLT sample is fabricated by the electric field poling technique[23,24]. Figure 1(a) shows its domain structure after being slightly etched in acids. The near circularly inverted domains (with radius r) distribute hexagonally in the $+\chi^{(2)}$ background with structure parameter $a = 9.05$ μm. We define the reversal parameter $r/a = 28\%$, and the corresponding duty cycle $f = 2\pi r^2/\sqrt{3}a^2$ is then 28%. No domain merging is found across the whole sample dimensions of 15 mm(x)×5 mm (y) with a thickness of 0.5 mm. For each reciprocal lattice vector (RLV) there are six equivalent ones as shown in Fig. 1(b) owing to the sixfold degeneration of this structure.

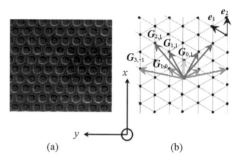

FIG.1. (color online) (a) Micrograph of the HexPLT and (b) its reciprocal space.

The general phase-matching condition for nonlinear interaction in the QPM material is $k_3 - k_1 - k_2 - G_{m,n} = 0$[25,26], where $G_{m,n} = (4\pi/\sqrt{3}a) \times (\sqrt{m^2+n^2+mn})$ is the RLV of the HexPLT with lattice parameter a and the subscripts m and n are integers, representing the order of RLV; k_1, k_2, and k_3 represent the interacted three waves. The pump beam is a 10 Hz 532 nm laser with a pulse width of 3.5 ns and a linewidth of 0.1 nm. We use an $f = 150$ mm lens to focus this $\hat{z}$-polarized pump beam into the HexPLT crystal, getting the beam waist to be around 120 μm.

When the pump beam incidence is along the direction of $G_{0,1}$ ($\hat{x}$ axis) and we tune the crystal temperature during 20℃ - 200℃, a QPM collinear parametric downconversion process happens and the momentum conservation is ensured by $k_p - k_s - k_i - G_{0,1} = 0$. The measured wavelength of the signal and the idler varying on temperature is shown in Fig. 2, which consists well with the theoretical calculation. A wide spectrum about 500 nm is covered when the temperature varies from 20℃ to 200℃.

Two incidental SHG processes occur simultaneously when the temperature is set around 188℃. The QPM condition for the signal and the idler doublings are $k_{ui} - 2k_i - G_{1,0} = 0$ and $k_{us} - 2k_s - G_{2,1} = 0$, respectively. Coupling these two noncollinear SHG pro-cesses with the OPA process is called dual-frequency self-doubling OPA. At 188℃, the signal and the idler are 868.3 and 1373.6 nm and both are collinear with the pump beam. As shown in the insets of Fig. 2, the harmonics of the signal and the idler both shape a pair of spots on the screen behind the crystal. The upper pair is the harmonics of the idler with the central frequency of 686.8 nm, while the lower pair is the harmonics of the signal with the central frequency of 434.2 nm. The mirror symmetry of these two pair spots about the pump beam is due to the mirror symmetry of involved RLVs $G_{2,1}$ and $G_{1,0}$. The energy of two red spots and two blue spots are 11 and 5.2 μJ, respectively, under the pump energy of 1.17 mJ. Adding the remaining pump at 532 nm, we get a set of three primary colors. According to the Commission Internationale de l'Eclairage (CIE) chromaticity dia-gram[27], this set of RGB light can cover most of the area in the CIE chromaticity diagram, which can be used uniquely in laser displays, laser TV, etc.

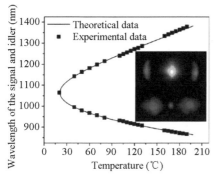

FIG. 2. (color online) Temperature-dependent curve for the OPA process. The insets are the photographs of the signal harmonics participant by $G_{1,1}$ or $G_{2,1}$ and the idler harmonics participant by $G_{3,-1}$ or $G_{1,0}$ when filtering out most of the pump.

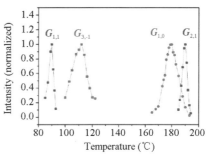

FIF. 3. (color online) Intensity versus temperature for each frequency self-doubling process.

Actually the temperature of 188℃ is deviated from the maximum output temperature for both the red and the blue generation. When lowering the temperature down to 179.6℃, the idler doubling process approaches its maximum efficiency, and the external angle is getting larger toward 5.5° as shown in Fig. 3. The fitted temperature FWHM is around 16℃, which is rather big when compared with 1D periodically poled lithium tantalate (PPLT). It mostly results from the noncollinear phase-matching character in this HexPLT, which has been verified by some earlier work[21]. When the temperature is at 179.6℃, the efficiency of the idler into its harmonics reaching its maximum is about 30% while the efficiency of the idler over the pump is about 10.8% under a pump energy of 1.17 mJ.

The pair of blue spots, i.e., the frequency doubling of the signal, approaches its maximum efficiency at 189.5℃ while the bandwidth is 6℃. It can be seem from Fig. 3 that owing to the wide temperature bandwidth of the signal and the idler doubling processes, it is not hard to reach simultaneously red and blue beams within some range of temperature.

Another dual-frequency self-doubling OPA process is observed around 100℃ in this HexPLT. The signal and idler wavelength are 932.33 and 1240.68 nm, hence the corresponding harmonic waves are at 466.17 and 620.34 nm. Therefore we can get another set of three elementary colors when taking the 532 nm pump into account. Momentum conservations for the idler and signal doubling are $\boldsymbol{k}_{ui} - 2\boldsymbol{k}_i - \boldsymbol{G}_{3,-1} = 0$ and $\boldsymbol{k}_{us} - 2\boldsymbol{k}_s - \boldsymbol{G}_{1,1} = 0$, respectively. The idler and the signal doubling shape another pair of red spots and blue spots with mirror symmetry on the screen behind the crystal. At 110.3℃ the idler doubling reaches its maximum and the temperature detuning bandwidth is about 16.6℃, while the signal doubling reaches it maximum at 89.4℃ and the bandwidth is about 5.3℃.

In Table 1, we list the detailed detuning characters of four frequency self-doubling processes when the temperature changes from room temperature to 200℃. For each signal or idler doubling process, the linewidth $\Delta\lambda$, the external angle α_{ext}, and the relative intensity of the harmonics (λ_d) are listed. We investigate the intensities of four self-doubling processes under the condition of low pump power, which ensures the nondepleted pumping approximation. The experimental and the theoretical results are shown in the last two columns and consist well. The idler doubling process happening around the 179.6℃ participant by $\boldsymbol{G}_{0,1}$ has the maximum output intensity.

It is worth noting here that when we choose the hexagonal structure only one freedom, i.e., the lattice parameter a, can be adjusted to meet the phase-matching requirements. So it is natural that the red and blue are generated not exactly simultaneously. In principle for exact simultaneous phase-matching of the signal and idler doubling, additional structure freedoms should be introduced. Other structures, such as a 2D structure with a parallelogram unit or a 2D fractal structure, can be adopted. In addition, to further enhance the red-blue intensity from this HexPLT, we can put it into an optical cavity that will give the signal or idler high feedback.

TABLE 1. Detailed Detuning Characters of Four Frequency Self-Doubling Processes[a]

$T(\Delta T)$ (℃)	$\lambda_s(\Delta\lambda_s)$ (nm)	$\lambda_i(\Delta\lambda_i)$ (nm)	$\lambda_d(\Delta\lambda_d)$ (nm)	α_{ext} (°)	$G_{m,n}$	$f_{m,n}$	I_d(exp.) (a.u.)	I_d(theo.) (a.u.)
89.4(5.3)	937.5(6.0)	1230(12)	468.7(0.5)	3	$G_{1,1}$	0.04	1	1
110.3(16.6)	920.8(5.5)	1260(10)	630(5.0)	12	$G_{3,-1}$	0.07	3.3	2.3
179.6(15.9)	873(4.0)	1362(7.3)	681.2(3.3)	5.5	$G_{1,0}$	0.32	40	40
189.5(6.0)	867(2.6)	1379(7.7)	433(1)	5	$G_{2,1}$	0.07	8	3

[a] The theoretical calculation including the Fourier coefficients f_{mn} and the intensity of the harmonics I_d (theo.) are based on the real duty cycle about 28% of the HexPLT.

In conclusion, we designed and observed two sets of dual-frequency self-doubling processes in an HexPLT crystal, two sets of signal and idler SHGs cascaded an OPA process. QPM ensures such multiple processes happening inside a single crystal and also enables a high-gain output in this single-pass setup. The generation of the red-blue light together with the remaining green pump from the HexPLT is a new method for noncollinear three primary colors output without any optical separation elements. This unique light source has potential applications in laser display, multicolor laser generation, integrated optics etc.

References and Notes

[1] R. A. Andrews, H. Rabin, and C. L. Tang, Phys. Rev. Lett. **25**, 605 (1970).

[2] J. M. Yarborough and E. O. Ammann, Appl. Phys. Lett. **18**, 145 (1971).

[3] V. Petrov and F. Noack, Opt. Lett. **20**, 2171 (1995).

[4] T. Kartaloğlu, K. G. Köprülü, and O. Aytür, Opt. Lett. **22**, 280 (1997).

[5] K. G. Köprülü, T. Kartaloğlu, Y. Dikmelik, and O. Aytür, J. Opt. Soc. Am. B **16**, 1546 (1999).

[6] E. C. Cheung, K. Koch, and G. T. Moore, Opt. Lett. **19**, 1967 (1994).

[7] R. J. Ellingson and C. L. Tang, Opt. Lett. **18**, 438 (1993).

[8] G. T. Moore, K. Koch, M. E. Dearborn, and M. Vaidyanathan, IEEE J. Quantum Electron. **34**, 803 (1998).

[9] O. Pfister, J. S. Wells, L. Hollberg, L. Zink, D. A. Van Baak, M. D. Levenson, and W. R. Bosenberg, Opt. Lett. **22**, 1211 (1997).

[10] C. McGowan, D. T. Reid, Z. E. Penman, M. Ebrahimzadeh, W. Sibbett, and D. H. Jundt, J. Opt. Soc. Am. B **15**, 694 (1998).

[11] G. Z. Luo, S. N. Zhu, J. L. He, Y. Y. Zhu, H. T. Wang, Z. W. Liu, C. Zhang, and N. B. Ming, Appl. Phys. Lett. **78**, 3006 (2001).

[12] X. P. Zhang, J. Hebling, J. Kuhl, W. W. Rühle, and H. Giessen, Opt. Lett. **26**, 2005 (2001).

[13] W. R. Bosenberg, J. I. Alexander, L. E. Myers, and R. W. Wallace, Opt. Lett. **23**, 207 (1998).

[14] S. N. Zhu, Y. Y. Zhu, and N. B. Ming, Science **278**, 843 (1997).

[15] Y. B. Chen, C. Zhang, Y. Y. Zhu, S. N. Zhu, H. T. Wang, and N. B. Ming, Appl. Phys. Lett. **78**, 577 (2001).

[16] K. F. Kashi, A. Arie, P. Urenski, and G. Rosenman, Phys. Rev. Lett. **88**, 023903 (2002).

[17] Y. Zhang and B. Y. Gu, Opt. Commun. **192**, 417 (2001).

[18] M. M. Fejer, G. A. Magel, D. H. Jundt, and R. L. Byer, IEEE J. Quantum Electron. **28**, 2631 (1992).

[19] Z. D. Gao, S. Y. Tu, S. N. Zhu, and A. H. Kung, Appl. Phys. Lett. **89**, 181101 (2006).

[20] X. P. Hu, G. Zhao, Z. Yan, X. Wang, Z. D. Gao, H. Liu, J. L. He, and S. N. Zhu, Opt. Lett. **33**, 408 (2008).

[21] N. G. R. Broderick, R. T. Bratfalean, T. M. Monro, D. J. Richardson, and C. M. de Sterke, J. Opt. Soc. Am. B **19**, 2263 (2002).

[22] N. Fujioka, S. Ashihara, H. Ono, T. Shimura, and K. Kuroda, J. Opt. Soc. Am. B **24**, 2394 (2007).

[23] S. N. Zhu, Y. Y. Zhu, Z. Y. Zhang, H. Shu, H. F. Wang, J. F. Hong, C. Z. Ge, and N. B. Ming, J. Appl. Phys. **77**, 5481 (1995).

[24] P. Xu, S. H. Ji, S. N. Zhu, X. Q. Yu, J. Sun, H. T. Wang, J. L. He, Y. Y. Zhu, and N. B. Ming, Phys. Rev. Lett. **93**, 133904 (2004).

[25] V. Berger, Phys. Rev. Lett. **81**, 4136 (1998).

[26] N. G. R. Broderick, H. L. Offerhaus, G. W. Ross, D. J. Richardson, and D. C. Hanna, Phys. Rev. Lett. **84**, 4345 (2000).

[27] http://www.nd.edu/~sboker/ColorVision2/CIEColorSpace2.gif.

[28] This work was supported by the National Natural Science Foundation of China (NNSFC) (60578034, 10776011, and 10534020), and by the National Key Projects for Basic Researches of China (2006CB921804 and 2004CB619003).

Direct Observation of Ferroelectric Domains in LiTaO₃ Using Environmental Scanning Electron Microscopy[*]

Shining Zhu and Wenwu Cao

Intercollege Materials Research Laboratory, The Pennsylvania State University, University Park, Pennsylvania 16802

Direct observation of ferroelectric domain structures in LiTaO₃ crystal, without etching or surface coating, has been realized by using environmental scanning electron microscopy in secondary electron emission mode. The new method can nondestructively provide domain contrast image at submicron resolution, and the domain contrast image is very stable. Conditions for best domain contrast of LiTaO₃ crystals have been established.

In recent years, scanning electron microscopy (SEM) has found applications in the study of ferroelectric domain structures[1—4]. Using low accelerating voltage a contrast between antiparallel ferroelectric domains can be imaged on an unmetalized polar crystal surface in secondary electron emission mode. The technique does not need special treatment of the sample surface and is nondestructive. It has been successfully used to study domain structures of several ferroelectric crystals[5—7].

In the SEM method, the key is to avoid surface charge accumulation onto an insulating crystal. The experimental conditions are very strict; particularly, the choice of probe current and accelerating voltage is very critical, and no contrast can be found if the parameters are not properly chosen. Moreover, the contrast image is short lived; i.e., it disappears under the influence of electron beam after a few seconds to a few minutes. The observation time for the contrast shortens with the increase of accelerating voltage and magnification. Consequently, it is not possible to obtain a stable contrast image with high resolution and magnification[1,2].

In this letter we report a new secondary electron image (SEI) technique of observing ferroelectric domains by using environmental scanning electron microscopy (ESEM). Because the ESEM[8,9] can operate at pressures 10000 times higher than that of standard SEM, the free ions which are created by collisions between moving electrons and neutral gas molecules can be used to provide a conducting path for beam-deposited surface charges, which allows one to examine unprepared, uncoated insulating samples. The results show

---

[*] Phys.Rev.Lett., 1997, 79(13):2558

that one can obtain stable domain contrast under relatively high accelerating voltage, large beam current, and slow scan rate using ESEM. Because higher accelerating voltage can be used, higher resolution than the conventional SEM is expected. The technique has been applied to a LiTaO$_3$ crystal with uncoated surface. A clear stable secondary electron image of the domain structures was obtained.

The spontaneous polarization P_s of LiTaO$_3$ can be reversed by a pulse electric field at room temperature[10]. Using different electrode pattern LiTaO$_3$ crystals can be poled into periodic and quasiperiodic domain structures which have been widely used in acoustics[11], nonlinear optics, and solid state laser devices[12—16]. Preliminary examination of the domain structure in a poled LiTaO$_3$ sample was made by optical microscopy on an etched surface. The destructive method of etching often affects further experiments on the sample. Therefore, it is desirable to develop new convenient nondestructive methods for the observation of domain patterns on an unprepared surface of a poled LiTaO$_3$ crystal.

In this work we used c-cut LiTaO$_3$ crystal wafers. The domain structures in the wafers were engineered by field poling method at room temperature. Two kinds of aluminum electrodes were fabricated: one is periodic and the other is uniform plane. During pulse poling using the above two kinds of electrodes two types of domain morphology, stripe and triangular shape, were produced.

The domain observations were performed using an ESEM (ElectroScan Model E-3, Wilmination, MA) in secondary electron emission mode. Figure 1 is a schematic diagram of the experimental arrangement and crystal orientation. A periodically poled sample was etched in a solution containing 2 parts HNO$_3$ and 1 part HF for 15 h at room temperature, and was attached to an aluminum sample mount using double faced tape for observation. The sample mount was grounded. The electric potential of the detector was set at $+0.5$ kV. Water vapor was chosen as the imaging gas. Under normal observation condition of wet mode, chamber pressure 3 – 3.5 Torr, accelerating voltage 15 kV, condenser 40, scan rate 8 – 30 sec/

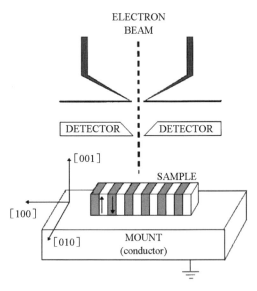

FIG. 1. Schematic diagram of the experimental arrangement.

frame, and aperture size 30 μm, we got a topographic image of the periodic domain pattern on the $+c$ surface, which is [001] orientation, as shown in Fig. 2(a). In the image the surfaces of negative domains are lower than those of positive domains. The formation of the surface steps is due to different etching rates of positive and negative domains in the acid. A wider stripe and some island domains shown in the figure are attributed to imperfect poling. One can see that the contrast only appears in domain boundaries which are brighter than the interior of domains, reflecting the fact that more secondary electrons are emitted at the edges of the etching steps. No contrast was revealed between the positive and negative domains due to the screening of the polarization P_s by the probing current. In other words, the positive and negative domains have nearly equal secondary electron emission rate.

After the first observation, the etched sample was taken out of the chamber and the [001] surface was polished. When the polished sample was reexamined using ESEM under the same condition, nothing interesting was seen except some polishing scratches on the surface. However, after the experiment parameters were changed by decreasing the chamber pressure, increasing the probe current by decreasing the condenser current and increasing the aperture, and increasing the accelerating voltage, contrast stripes began to appear [shown in Fig. 2(b)]. The locations and shapes of the contrast patterns in Fig. 2(b) exactly match the etched patterns in Fig. 2(a). This confirms that the contrast stripes originated from the antiparallel domains.

Optimum contrast was obtained under the following conditions: accelerating voltage, 30 kV; chamber pressure, 2 Torr; aperture size, 50 μm (maximum for the E-3 system); condenser, 10 (lowest of the E-3 system); and scan rate, 17–30 sec/frame.

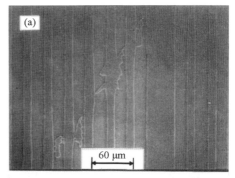
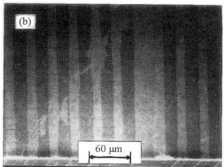

FIG. 2. Comparison of ESEM images of a surface portion in a c-cut LiTaO₃ with antiparallel domains. (a) Etched surface: voltage=15 kV, condenser=40, pressure=3.0 Torr, and aperture=30 μm. (b) Polished surface: voltage=30 kV, condenser=10, pressure=2.0 Torr, and aperture=50 μm.

The positive domains were sufficiently brighter than the negative domains, indicating that the secondary electron emission from positive domains exceeds that from negative ones. The contrast is very stable in time, no change was observed over a period of 4 h which was the longest observation time used in our experiments. This is a marked advance from the conventional SEM method in which the contrast can only be kept for a few seconds to a few minutes[1,2].

With a slight tilt of the sample, we could simultaneously observe the previously etched patterns of the periodic domain structure on the [010] surface of the sample which was not polished [see the bottom of Fig. 2(b)]. The image patterns on the two adjacent

[001] and [010] faces are consistent. However, if we tilt the sample to the [0 $\bar{1}$0] surface which was polished, no contrast stripes were found. This means that only the polar surfaces can show the domain contrast images.

We found that although higher contrast domain images could be obtained under the optimum conditions mentioned above, the experimental conditions are not very critical. With an aperture of 50 μm, domain contrast could appear in the range of accelerating voltages from 20 - 30 kV, condenser current setting from 10 - 30, scan rates from 4 - 60 sec/frame, and chamber pressure from 1.0 - 3.0 Torr. In all cases the positive domains were always brighter than the negative ones.

The visualization of domain contrast is also independent of surface roughness. We could clearly observe the contrast image of domains on a crack surface with appropriate conditions as shown in Fig. 3. The sample in Fig. 3 was partially poled under a pair of plane aluminum electrodes. After poling, the electrodes were removed in NaOH solution which only resolves the aluminum but not the $LiTaO_3$ crystal. In Fig. 3 some triangular domains with positive polarity are shown as lighter regions on the dark background of negative polarity. Generally speaking, the morphology of the inverted domains depends on the symmetry of the crystal as well as the intensity of the poling field. The polarization direction of $LiTaO_3$ is parallel to its triad axis.

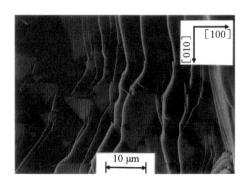

FIG.3. ESEM image of a crack surface of $LiTaO_3$ crystal with nearly [00 $\bar{1}$] orientation. The sample was partially poled by a poling field of E =21 kV/cm. The triangular antiparallel domains with positive polarity are shown as brighter regions on the darker background. Image conditions: voltage = 30 kV, condenser = 10, pressure=2.0 Torr, and aperture=50 μm.

Using a moderate field of $E = 21$ kV/mm, the triangular shape inverted domains can be formed under plane electrode in a partial poling process. The sides of the triangle corresponding to domain walls are parallel to crystallographic x directions. Domains grow by gradually extending the domain walls along crystallographic y directions during poling. This is a typical growth morphology of the antiparallel domains of $LiTaO_3$ in a quasiequilibrium process through two-dimensional nucleation[17]. The domains in Fig. 3 have different sizes, which reveals that the polarization reversal in the crystal is through multisite nucleation and growth; i.e., a large number of inverted domains are nucleated first, then they grow and coalesce to complete the switching process.

The domain contrast in Fig. 3 is better than that in Fig. 2(b) which has to be contrast enhanced by computer. This is because a depolarization layer is produced on the polished surface, while the cracked surface has much less mechanical artifact; thus a better contrast image can be obtained. This phenomenon is also found in the domain contrast image observation using conventional SEM[1,2].

Le Bihan[1] proposed a theoretical model to explain the domain contrast of triglycine sulfate (TGS) crystal using conventional SEM under low voltage condition. It was assumed that the contrast of ferroelectric domains is caused by polarization charges, hence the surface of positive domain should be darker than that of a negative domain. However, this is in contradiction with our results here. Generally speaking, for an insulating sample, surface electric charge equilibrium is the fundamental principle for producing these contrast images. In conventional SEM technique, the charge neutralization relies on the exact balance of surface input and output currents. The samples need to be slightly conductive or coated with a conducting layer in order to accomplish this task under vacuum. For the ESEM this charge balance is achieved by neutralizing surface electron buildup with positive ions under finite vapor pressure.

The secondary electron emission rate depends on the level of interaction between the electron beam and the surface layer of ferroelectric crystal. Although it is not clear if the probing current is sufficient to instantly screen the charge on the sample surface at each frame, there is definitely a charge balance after a short period of time. Therefore, the polarization charges cannot be the main cause of contrast due to the large amount of ionic composition under finite vapor pressure. Because the technique is new, the exact mechanism for the image formation is still under investigation.

In conclusion, for the first time we have successfully observed stable contrast image of antiparallel domains in poled $LiTaO_3$ crystals using ESEM technique. This technique is very promising since it allows us to directly observe antiparallel domains in an unetched and uncoated insulating sample surface. The condition for obtaining the contrast image is not very strict and the domain image for $LiTaO_3$ crystal is stable for several hours. From our experience, better image can be produced with higher voltage, lower condenser current, and larger aperture (more electrons). Because the ions produce a conducting path to neutralize the surface charge accumulation, the technique allows the use of higher accelerating voltage with higher magnification; therefore, submicron resolution can be obtained.

References and Notes

[1] R. Le Bihan, Ferroelectrics **97**, 19 (1989).
[2] A. A. Sogr, Ferroelectrics **97**, 47 (1989).
[3] T. Ozaki, K. Fujii, and S. Aoyagi, J. Appl. Phys. **80**, 1697 (1996).
[4] G. Rosenman et al., J. Appl. Phys. **80**, 7166 (1996).
[5] B. Hilczer, L. Szczesniak, and K. P. Meyer, Ferroelectrics **97**, 59 (1989).
[6] D. V. Roshchupkin and M. Brunel, Scanning Microsc. **7**, 543 (1993).
[7] A. S. Oleinik and V. A. Bokov, Sov. Phys. Solid State **17**, 560 (1975).
[8] G. D. Danilatos, Adv. Electron. Electron Phys. **71**, 109 (1988).
[9] G. D. Danilatos, Adv. Electron. Electron Phys. **78**, 1 (1989).

[10] S. N. Zhu *et al.*, J. Appl. Phys. **77**, 5481 (1995).
[11] Y. F. Chen, S. N. Zhu, Y. Y. Zhu, and N. B. Ming, Appl. Phys. Lett. **70**, 592 (1997).
[12] C. Baron, H. Cheng, and M. C. Gupta, Appl. Phys. Lett. **68**, 481 (1996).
[13] R. L. Byer, Nonlinear Opt. **7**, 234 (1994).
[14] S. N. Zhu *et al.*, Appl. Phys. Lett. **67**, 320 (1995).
[15] K. S. Abedin, T. Tsuritani, M. Sato, and H. Ito, Appl. Phys. Lett. **70**, 10 (1997).
[16] S. N. Zhu, Y. Y. Zhu, Y. Q. Qin, H. F. Wang, C. Z. Ge, and N. B. Ming, Phys. Rev. Lett. **78**, 2752 (1997).
[17] V. Y. Shur and E. L. Rumyantsev, Ferroelectrics **151**, 171 (1994).
[18] This research was sponsored by the NSF and the Office of Naval Research.

Nondestructive Imaging of Dielectric-Constant Profiles and Ferroelectric Domains with a Scanning-Tip Microwave Near-Field Microscope[*]

Y. Lu, F. Duewer, X.-D. Xiang

Molecular Design Institute, Lawrence Berkeley National Laboratory, Berkeley, CA 94720, USA

T. Wei and P. G. Schultz

Howard Hughes Medical Institute and Molecular Design Institute, Lawrence Berkeley National Laboratory, Berkeley, CA 94720, USA

Y. Lu and N.-B. Ming

National Laboratory of Solid State Microstructures, NanJing University, NanJing 210093, China

Variations in dielectric constant and patterns of microwave loss have been imaged in a yttrium-doped $LiNbO_3$ crystal with periodic ferroelectric domains with the use of a scanning-tip near-field microwave microscope. Periodic profiles of dielectric constant and images of ferroelectric domain boundaries were observed at submicrometer resolution. The combination of these images showed a growth-instability-induced defect of periodic domain structure. Evidence of a lattice-edge dislocation has also been observed through a stress-induced variation in dielectric constant.

Crystals with periodic and quasi-periodic ferroelectric domain structures, such as $LiNbO_3$, $LiTaO_3$, and $KTiOPO_4$ superlattice crystals[1,2] (either in bulk form or as thin-film waveguides), have attracted considerable interest and found important applications in quasi-phase matched nonlinear optics[3] and in acoustics[4]. Currently, a destructive method involving optical imaging of differentially etched surfaces is commonly used to characterize the domain structures. Several other (mainly charge-or polarization-sensitive) techniques have also been developed[5], but none allows nondestructive, high-resolution imaging of ferroelectric domains over a large area. Moreover, there is no effective technique to analyze the variations in dielectric constant corresponding to variations in dopant concentration, which could give rise to domain formation, in these materials and related devices.

We have previously developed a scanning-tip microwave near-field microscope with 5-μm resolution[6] and have now improved its performance to nondestructively image surface dielectric constant and microwave loss with submicrometer spatial resolution and a scanning range of over 2.5 cm. Through images of dielectric constant, we have observed

---

[*] Science, 1997, 276(27):2004

submicrometer-resolution profiles of periodic dielectric constant and evidence of a lattice-edge dislocation in a yttrium-doped LiNbO$_3$ superlattice crystal. The profiles of microwave energy loss yield nondestructive images of ferroelectric domain boundaries. By correlating the images of both dielectric constant and microwave energy loss, a growth-instability-induced defect in the periodic domain structure has been identified. These studies should contribute significantly to our understanding of the growth mechanism of the crystals. In addition, the imaging technique used here should prove useful in analyzing other ferroelectric and dielectric materials and thinfilm devices.

The design of our scanning-tip microwave near-field microscope involves a sharpened metal tip mounted on the center conductor of a high-quality factor (high-Q) coaxial resonator that extends beyond an aperture formed in the end wall of the resonator[6]. The spatial resolution of the microscope increases with the sharpness of the tip, without the exponential decrease in sensitivity that is a characteristic of hollow wave guides used in conventional near-field microscopes. Increases in microwave energy loss (absorption) and dielectric constant at the surface of the samples near the tip lead to a decrease in Q and resonant frequency f_0 of the resonator, respectively. By recording Q and f_0, we can image the surface profiles of dielectric constant and microwave energy loss in the sample.

Yttrium doping has proven to be an effective method of introducing a periodic ferroelectric domain structure (superlattice) in the as-grown LiNbO$_3$ single crystal[7]. For this study, an average of 0.5 weight ‰ yttrium-doped LiNbO$_3$ superlattice single crystal was grown along the a axis (x axis in Fig. 1) with the Czochralski technique at the NanJing University[7]. A 3-mm-thick wafer cut normal to the y axis and polished on both surfaces was used. The surface of the

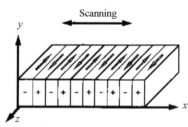

FIG.1. Schematic of a LiNbO$_3$ crystal containing periodic ferroelectric domains. Polarization is along the optical axis (z axis), and scanning is along the x axis.

crystal was examined with a profilometer to confirm the optical quality smoothness. In this configuration, the polarizations of the ferroelectric domains alternate along the c axis (z axis in Fig. 1). Two domain laminas with opposite polarization give rise to the same surface charge configuration (because there is no y-axis component of spontaneous polarization), simplifying the interpretation of our experimental results (charge-sensitive techniques fail to image the structures in this configuration). The microscope tip scans over the xz plane of the sample approximately along the x axis (with a typical scan speed of 1 to 5 μm/s in this study). The electrochemically etched tungsten tip (with typical radius of $\sim$0.1 μm) was kept in contact with the surface of the sample with a soft spring (with a spring constant of 4 N/m and a typical contact force of$<$20 μN). No surface damage was observed with this scanning format. Neither the scanning format nor the detected signal is sensitive to topography (that is, material features with identical dielectric constant and tangent loss but different height are not distinguishable in the image).

A periodic variation in the concentration of yttrium was generated by growing the $LiNbO_3$ crystal along the x axis in an asymmetric temperature field[1,8]. The periodic variation in dopant concentration induces (through the internal space-charge field) the formation of ferroelectric domains with alternating polarization when the crystal undergoes a para-to-ferroelectric phase transition. Although alternating polarization does not modulate its second-rank tensor dielectric constant ($LiNbO_3$ belongs to the $3m$ point group, for which the nonzero second-rank tensor ε_{11}, ε_{22}, and ε_{33} does not change its sign under the transformation of 180° rotation around the x axis), the periodic variation of dopant level should result in a periodic change in the dielectric constant. Therefore, in the Czochralski-grown $LiNbO_3$ superlattice crystals, two periodic structures (that is, dielectric constant and ferroelectric domain) should coexist, although the former has never been observed previously. We have observed both periodic structures through microwave imaging (Fig. 2). The image of resonant frequency f_0 reflects variations in the dielectric constant associated with changes in dopant levels, whereas the image of Q corresponds to losses in microwave energy, which are large at the ferroelectric domain boundaries (primarily as the result of movement of the domain walls under the influence of the microwave field). The change in resonant frequency is $\Delta f_0/f_0 = g\Delta\varepsilon'$[9], where $\varepsilon = \varepsilon' + i\varepsilon''$ is the complex dielectric constant and g was measured to be ~7×10^{-5} in this configuration. The total dielectric variation in Fig. 2 is estimated to be ~0.25 with a noise level of 0.03, which is the current sensitivity of the microscope. The change in cavity quality factor is $\Delta(1/Q) = g\Delta\varepsilon''$[9], and $\Delta\varepsilon''$ was calculated to be 7.1×10^{-2}. The total loss tangent variation in Fig. 2 is estimated to be about 1×10^{-2}. Detailed theoretical analysis and experimental calibration of the sensitivity and accuracy were

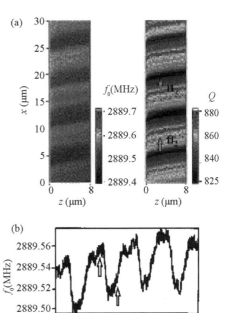

FIG.2. (a) Images of profiles of periodic dielectric constant and ferroelectric domain boundaries. The f_0 profile mainly indicates variations in dielectric constant when the change in loss is not large, and the Q profile indicates the microwave energy loss of the sample surface. (b) The f_0 and Q profiles of (a) along the x axis at $z = 4.0$ μm. The diffuse and sharp domain boundaries are marked B_1 and B_2, respectively.

performed[10].

The domain boundaries with relatively low loss[thin yellow stripe in Fig. 2(a) marked as B_2] are located at the dielectric constant minima[f_0 maxima in Fig. 2(b)], whereas the domain boundaries with high loss [wide dark-red stripe in Fig. 2(a) marked as B_1] are located at a critical value of the gradient in dopant concentration (where the polarization changes its direction) near (but not at) the dielectric constant maxima (ε_{max}). The full width at half maxima of B_1 (~2.5 μm) is greater than that of the B_2 (~1 μm). This result is consistent with the previous finding of sharp and diffuse boundaries by X-ray energy dispersive spectral analysis in these crystals[11] and the proposed scenario that diffuse domain boundaries do not occur at dopant maxima[12] if we assume that the doping of yttrium in the $LiNbO_3$ crystals monotonically increases the dielectric constant. Because diffuse domain boundaries are not always located at the dopant maxima (corresponding to ε_{max}), alternating polarization domains separated by domain boundaries (as observed in images of Q) may also have different widths. The width ratio of the two domains in one period of 7.0 μm (Fig. 2) is ~0.87.

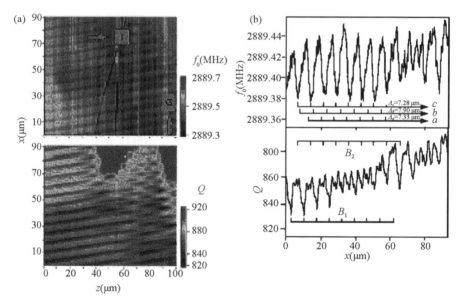

FIG.3. (a) The f_0 and Q profiles of a sample region, revealing a transition from periodic domain structure to an island-like domain caused by a growth instability. The island-like domain is marked as "I," and the growth direction, as "G." (b) The f_0 and Q profiles of (a) along the x axis at $z=$ 73 μm. In the f_0 profile, several modulation frequencies Λ (with periods of 7.33, 7.90, and 7.28 μm) are marked a, b, and c.

Because dopant (dielectric constant) and domain boundary profiles can be imaged simultaneously, it is possible to gain information about the growth mechanism of the crystal by correlating the detailed features of the profiles. For example, we observed a defect in the periodic domain structure (that is, an island-like domain caused by solid-liquid interface instability during growth) ["I" in Fig. 3(a)]. This type of defect is common in

Czochralski-grown LiNbO$_3$ crystals and is destructive to the periodic domain structure[12]. As is evident from the f_0 profile, the dopant modulation is more diffuse near the island, and the overall dopant level is higher in the island region. A corresponding transition from sharp (low loss) to diffuse (high loss) domain boundaries is clearly observable in the Q profile. These observations indicate that dopant aggregation, caused by temperature instabilities in the island region during the crystal growth, is the origin of the island-like domains. A gradual transition from sharp to diffuse domain boundaries is again evident in Q and f_0 profiles at $z=73$ μm [Fig. 3(b)], and the domain structure gradually loses its original periodicity approaching the island domain. For stable growth, the periodic modulation of dopant level should have a single frequency (which is the frequency of crystal rotation)[1,8], whereas disturbances in the system, such as melt or air convection and power fluctuations, may introduce additional modulations with different frequencies. When these additional frequencies are close to the crystal rotation frequency (with a period of ~7 μm in Fig. 3), interference eventually drives the solid-liquid interface out of stability, causing dopant aggregation and island domain formation during crystal growth [Fig. 3(b)]

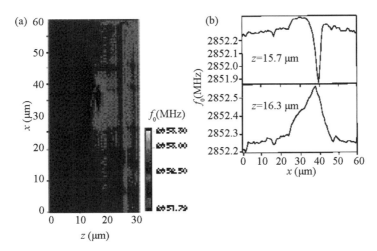

FIG.4. (a) The f_0 profiles of a sample region that indicate a lattice dislocation. (b) The f_0 profile of (a) along the z axis at $x=38.7$ μm.

A "butterfly" image of the dielectric constant is an indication of a lattice "edge dislocation" defect [Fig. 4(a)]. The large difference in dielectric constant adjacent to the dislocation is caused by compressive and tensile stresses in lattice structure induced by the dislocation. The stress-induced ε contour is qualitatively consistent with the theoretical prediction for a lattice "edge dislocation" defect[13]. The ε profile of a line scan along the z axis at $x=38.7$ μm [Fig. 4(b)], which cuts through regions with compressive and tensile stresses, shows large opposite changes in ε on either side of the edge defect. This profile also indicates that our spatial resolution in this configuration (mainly limited by the radius of the tip) is better than 1 μm, because between the two points (0.6 μm apart), ε changed from the minimum to near maximum. Lattice dislocations in LiNbO$_3$ crystals are not observable by optical microscopy with polarized light because of the crystal's large birefringence.

References and Notes

[1] D. Feng and N. B. Ming, *Appl. Phys. Lett.* **37**, 607 (1980).
[2] N. B. Ming *et al.*, *J. Mater. Sci.* **17**, 1663 (1982); G. A. Magel *et al.*, *Appl. Phys. Lett.* **56**, 108 (1990); Y. L. Lu and N. B. Ming, *ibid.* **69**, 1660 (1996).
[3] J. A. Armstrong and N. Bloembergen, *Phys. Rev.* **127**, 1918 (1962).
[4] See, for example, S. D. Cheng, Y. Y. Zhu, Y. L. Lu, N. B. Ming, *Appl. Phys. Lett.* **66**, 291 (1995), and references therein.
[5] T. Ozaki *et al.*, *J. Appl. Phys.* **80**, 1697 (1996); R. LeBihan and M. Maussion, *J. Phys.* **33**, C2-215 (1972); P. J. Lin *et al.*, *Philos. Mag. A* **48**, 251 (1983); F. Saurenbach and B. D. Terris, *Appl. Phys. Lett.* **56**, 1703 (1990).
[5] T. Wei, X. D. Xiang, P. G. Schultz, *Appl. Phys. Lett.* **68**, 3506 (1996).
[7] Y. L. Lu, Y. Q. Lu, N. B. Ming, *ibid.*, p. 2781.
[8] Y. L. Lu, L. Mao, N. B. Ming, *ibid.* **59**, 516 (1991).
[9] H. A. Bethe and J. Schwinger, *Perturbation Theory of Cavities* (National Defense Research Committee, Washington, DC, 1943), pp. D1-117.
[10] C. Gao, F. Duewer, X.-D. Xiang, in preparation.
[11] J. Chen, Q. Zhou, J. F. Hong, W. S. Wang, D. Feng, *J. Appl. Phys.* **66**, 336 (1989).
[12] Y. L. Lu, Y. Q. Lu, N. B. Ming, *Appl. Phys. Lett.* **68**, 2642 (1996), and references therein.
[13] J. P. Hirth and J. Lothe, *Theory of Dislocations* (Krieger, Malabar, FL, ed. 2, 1992).
[14] This work was supported by the director of Advanced Energy Projects Division, Office of Computational and Technology Research, U. S. Department of Energy, under contract DE-AC03-76SF00098.

第八章 光学超晶格的应用研究
Chapter 8 Engineered Quasi-phase-matching for Laser Techniques

8.1 Efficient Continuous Wave Blue Light Generation in Optical Superlattice $LiNbO_3$ by Direct Frequency Doubling a 978 nm InGaAs Diode Laser

8.2 Femtosecond Violet Light Generation by Quasi-phase-matched Frequency Doubling in Optical Superlattice $LiNbO_3$

8.3 Visible Dual-wavelength Light Generation in Optical Superlattice $Er:LiNbO_3$ through Upconversion and Quasi-phase-matched Frequency Doubling

8.4 Frequency Tuning of Optical Parametric Generator in Periodically Poled Optical Superlattice $LiNbO_3$ by Electro-optic Effect

8.5 Electro-optic Effect of Periodically Poled Optical Superlattice $LiNbO_3$ and Its Applications

8.6 High-power Red-green-blue Laser Light Source Based on Intermittent Oscillating Dual-wavelength $Nd:YAG$ Laser with a Cascaded $LiTaO_3$ Superlattice

8.7 Diode-pumped 1988-nm $Tm:YAP$ Laser Mode-locked by Intracavity Second-harmonic Generation in Periodically Poled $LiNbO_3$

8.8 Efficiency-enhanced Optical Parametric Down Conversion for Mid-infrared Generation on a Tandem Periodically Poled MgO-doped Stoichiometric Lithium Tantalate Chip

8.9 Polarization-free Second-order Nonlinear Frequency Conversion Using the Optical Superlattice

8.10 Polarization Independent Quasi-phase-matched Sum Frequency Generation for Single Photon Detection

8.11 DFB Semiconductor Lasers based on Reconstruction-equivalent-chirp Technology

8.12 High Channel Count and High Precision Channel Spacing Multi-wavelength Laser Array for Future PICs

第八章　光学超晶格的应用研究

陆延青

  应用是材料研究的终极目标。本章收集了介电体超晶格部分应用研究的文章。由于介电体超晶格在声学、量子光学领域的相关应用已经在第四、六章中分别有所涉及,因此本章中,我们着重介绍介电体超晶格在光学,特别是在非线性光学和全固态激光器件等方面的应用研究。

  上世纪八十年代末到九十年代初,光学存储技术得到迅猛的发展,所用的光源一般为近红外的半导体激光二极管。提高存储密度成为当时重要的技术需求。受衍射极限的限制,一个行之有效的解决方案就是改用短波长的激光。当时,蓝光激光二极管尚未问世,通过非线性光学获得倍频蓝光就成为一个自然的选择。波长减半带来的 4 倍存储密度显然是十分令人激动的愿景。为此,人们必须寻求高效率的非线性晶体。比起普通的双折射匹配非线性晶体,如 KDP、KTP 等,光学超晶格铌酸锂具有更高的非线性系数,更宽的工作波段。特别是由于使用准位相匹配带来的室温非临界匹配,使得具有周期极化的铌酸锂超晶格(PPLN)在半导体激光直接倍频方面具有显著的优势。这方面,我们的研究一直走在世界的前列。这些年来,我们的全固态激光相关工作向不同频段、不同脉宽、多个波长、电光调谐等方向拓展,展示出光学超晶格强大的技术潜力和应用价值。在本章中我们选择了 13 篇代表性的论文予以介绍。其中第一篇"Engineered Quasi-phase-matching for Laser Techniques"作为本章的英文序言。

  正如南京大学研究组首次全面验证了准位相匹配理论,我们也最早实现了毫瓦级的激光二极管直接蓝光倍频。代表性论文 8.1 报道了 1996 年利用提拉法,通过旋转型生长条纹的有效控制,制备了沿 x 轴生长的,周期为 5.2 μm,尺寸为 4 mm×4 mm×2 mm 铌酸锂光学超晶格样品。该样品的蓝光倍频是选用了一台 978 nm 输出波长的 InGaAs 半导体激光器作为基波光源,在 500 mW 输入时,得到了 1.27 mW 的直接倍频蓝光输出。虽然转换效率仅为 0.25%,但这是国际上第一个半导体激光器蓝光倍频的实验结果。该工作作为"八五"863 计划项目的重要成果通过了原国家教委组织的项目成果鉴定,被誉为"国际领先水平的工作"。

  另一项在国际上率先开展的是铌酸锂光学超晶格飞秒激光倍频的实验(8.2)。基于一台 90 fs,82 MHz 重复频率的 Ti:sapphire 激光,选用提拉法直接生长的样品,通过三阶准位相匹配,在 770 mW 基频光入射下,获得了 9.7 mW 的 390 nm 近紫外倍频光。由于飞秒激光往往具有较宽的线宽,这使得提拉法生长的样品十分适宜超短脉冲的准位相匹配频率转换。

  利用提拉法制备铌酸锂光学超晶格除了样品通光面大,制备过程简单直接外,另一个优点就是便于调节晶体组分和掺杂。自 1996 年我们率先在晶体中掺入 Nd 离子,以便能在铌酸锂光学超晶格晶体中产生 1064 nm 的激光激射，并同时实现对应激射波长的倍频，这就是所谓的自倍频。实际上，除了 Nd 离子，其他稀土或过渡金属离子也可方便地掺入晶体，实现一些独特的功能。8.3 即是其中的一篇有趣的代表性工作。我们在铌酸锂晶体生长过程中掺入 Er 离子，并且直接利用 Er 离子的周期性生长条纹所对应的晶体内电场，诱导出周期性的铁电畴结构。由于 Er 离子具有优异的上转换特性，这样，在近红外光泵浦下，利用超晶格结构，就可以同时在样品中得到紫色或蓝色的倍频光，以及明亮的上转换绿光。

除了非线性光学效应，在光学超晶格铌酸锂中电光效应也被结构调制。因为线性电光效应与二阶非线性效应同源，均由三阶物性张量描述。我们在国际上较早开展了该方面的系统研究。通过施加不同方向的电场，铌酸锂超晶格展示了不同的静态和动态的电光特性。因篇幅所限，我们仅选取了具有代表性的两篇工作予以重点介绍，即论文 8.4 与 8.5。在 8.4 报道了，在 Z 轴加场时，正负畴区的折射率椭球会发生相反的形变，但主轴方向并不改变。设计正负畴不等的铌酸锂超晶格，就可以通过调节外加电场改变匹配波长，实现光参量器件的快速电光调谐。

与 Z 向加场不同，Y 方向加场的铌酸锂光学超晶格会产生另外一种电光效应。此时，晶体的折射率椭球主轴绕 X 轴产生一个微小角度的旋转同时发生微弱形变（见 8.5）。这个微小的旋转角在不同畴区也刚好相反，形成类似于折叠 Solc 滤波器的结构。于是，合适设计晶体的畴周期，在线偏振光入射时，就可以使得出射光偏振旋转了一个与电场成正比的角度，实现快速的电光偏振控制。辅以偏振片，即可实现电调的快速滤波。此外，我们也曾基于铌酸锂超晶格提出过宽带高频（100 GHz 以上）的行波电光调制器，或者，选用其逆过程，产生 100 G 以上的相干微波或 THz 波。

继续回到铌酸锂光学超晶格在非线性光学方面的应用。我们结合多重准位相匹配原理和全固态激光技术，研制出了全固态红绿蓝三基色激光器原型，实现了"一台激光器、一块非线性晶体、三色激光输出"。相关工作引起了国内外的关注。（见 8.6）《Laser Focus World》杂志在"World News"栏目连续专文报导了我们的工作，并展望了光学超晶格多波长激光器在激光显示和激光诊疗等方面的应用前景。基于光学超晶格的准白光激光器，入选了 2008 年度中国光学重要成果。在上述工作基础上，我们进一步优化全固态激光技术和超晶格结构设计，将三基色激光的输出功率从几百毫瓦提高到数瓦量级。此外，我们还将光学超晶格从一维拓展至二维周期和二维准晶结构，研制出了可自动分光的非共线红绿蓝三色激光光源。

除多色激光器外，光学超晶格还可以用于激光锁模。我们利用光学超晶格可在非线性晶体透光范围内实现任意波长频率转换的优势，基于非线性镜和级联二阶非线性两种锁模技术，将光学超晶格锁模激光器的工作波段从 1.06 μm 首次延伸到 1.3 μm，并进一步拓展至中红外 2 μm 波段，实现了功率数瓦，重复频率几十 MHz，脉冲宽度几个皮秒的锁模激光输出。相关工作为研制中红外锁模皮秒激光器提供了新的技术方案。（8.7）

1.5～5 μm 波段的中红外激光器在新一代成像器件、分子和固态光谱、临床医疗诊断、近红外量子器件、功能晶体材料表征和大气探测、激光雷达、大气遥感、激光医疗和光电对抗等领域具有重要的应用。虽然中红外波段相干光源可以通过 Ho^{3+}、Er^{3+} 等稀土离子发射直接产生，但发光效率很低，目前波长被限制在 3 μm 以下。半导体量子级联激光器已经有产品，但是输出功率较低，调谐范围较窄，光束质量较差，有的还需在低温下工作，在远距离遥感探测和光电对抗等领域应用受到限制。鉴于明确的需求牵引和目前的技术现状,8.8 报道了基于光学超晶格的窄线宽、宽调谐大功率全固态 $1.5\sim5~\mu m$ 中红外激光器的研制。该激光器最高输出功率达到 30 W,稳定性、可靠性等指标满足实际应用需求,能够实现波长连续可调谐,具有连续、纳秒和皮秒三种脉冲输出方式,不论在前沿科学研究,还是在光电对抗、生物医疗和有毒/易爆气体探测等应用领域,该仪器都具有科学意义和重要价值。实际上,我们在远程有毒有害气体探测方面,已经获得了比较理想的测试结果。

  光学超晶格在激光技术方面的应用还远不限于此。由于其微结构的可控性以及丰富的内秉特性,一些有趣的应用也不断被提出。在 8.9 一文中利用一块超晶格中实现了正交偏振态频率转换过程相位的同时匹配,制备出应用急需的偏振无关变频器件。8.10 一文介绍一种分段结构的铌酸锂超晶格,利用超晶格独特的电光效应,通过另段施加电场,控制晶体中特定波长的光偏振,亦可实现偏振不敏感乃至偏振无关的倍频、和频和参量放大。这显然迥异于普通的非线性晶体,从而带来一些独特的应用。例如,量子光学研究中偏振纠缠光子对中的单个光子的偏振态在测量前是无法确定的。基于光学超晶格的偏振无关变频器件具有小型化、模块化、稳定性和效率高的优点,符合未来大规模量子网络和量子计算的要求。

  近年来,光学超晶格的学术思想不断深化,其应用研究也开始超出传统的铌酸锂等铁电晶体的范围。金属等离激元、半导体、光纤乃至液晶等领域纷纷引入光学超晶格、准位相匹配等理论和微结构设计思路,催生出一些新颖的功能与应用。这当中,基于所谓的"重构-等效啁啾"(REC)技术的半导体激光器阵列的设计和研制就是一个很典型的例子。

  8.11 一文介绍了如何将光学超晶格的相关的思想应用于半导体 DFB 激光器的研制。我们通过合适的设计,利用成熟的低精度光刻技术,制备相应的 REC 结构,实现原来需要电子束光刻等特殊手段才能实现的任意相移、啁啾、切趾等特殊光栅,进而实现应用功能多元化的高质量激光器芯片。该技术使得激光器芯片在制造难度大大降低的同时,波长精度得以显著提升。在此基础上已经设计出了单通道直调激光器模块、可调谐激光器模块、8 通道直调阵列模块等面向工业化应用的样品。进一步的理论和实验证明,REC 技术在半导体激光器芯片阵列研制上具有更加独特的优势,在非常关键的波长间隔控制上理论预言比传统技术好两个数量级。8.12 一文中报道了扩展到 60 个波长激光器阵列的实验研究,这是当前半导体激光器阵列信道数最多、波长间隔最准确的结果,展现出 REC 技术或者说光学超晶格的学术思想强大的生命力和可扩展性。

Chapter 8 Engineered Quasi-phase-matching for Laser Techniques*

—In Lieu of a Preface

X. P. Hu, P. Xu, and S. N. Zhu

National Laboratory of Solid State Microstructures and School of Physics, Nanjing University, Nanjing 210093, China

The quasi-phase-matching (QPM) technique has drawn increasing attention due to its promising applications in areas such as nonlinear frequency conversion for generating new laser light sources. In this paper, we will briefly review the main achievements in this field. We give a brief introduction of the invention of QPM theory, followed by the QPM-material fabrication techniques. When combing QPM with the solid-state laser techniques, various laser light sources, such as single-wavelength visible lasers and ultraviolet lasers, red-green-blue three-fundamental-color lasers, optical parametric oscillators in different temporal scales, and passive mode-locking lasers based on cascaded second-order nonlinearity, have been presented. The QPM technique has been extended to quantum optics recently, and prospects for the studies are bright.

1. Introduction to Quasi-Phase-Matching

The first laser action was demonstrated in ruby by Maiman in the early 1960s[1]. The high peak power and excellent coherent properties of the laser immediately opened up new research topics, including one of the most important ones, 'nonlinear optics'. In 1961, just one year after the invention of the ruby laser, Franken *et al.* observed second-harmonic generation(SHG) by projection of intense ruby laser light through crystalline quartz[2], which marked the birth of the field of nonlinear optics. By employing nonlinear optical frequency conversions, including sum-frequency generation(SFG), difference-frequency generation, and optical parametric oscillation and amplification, as well as the most well-known SHG, the wavelength of light can be converted to different wavelengths to obtain new laser light sources not readily available. In nonlinear optical interactions, because of the dispersive nature of crystal, the fundamental frequency and the generated frequency travel at different speeds in the material, so there is a phase mismatch between the interacting waves. This phase mismatch will cause the fundamental and the harmonic waves to go out of phase; thus the energy will flow back and forth between them, resulting in a low frequency conversion efficiency. The most common procedure for achieving phase-matching is to use the birefringence of some crystals, which may offset

---

* Photon. Res., 2013, 1(4):171

and obtain phase-matching in some wavelength region by choosing the dispersion orientation or the operating temperature of the crystal. This method is the so-called birefringence-phase-matching (BPM) technique, which was proposed by Kleinman in 1962[3]. In the same year, Armstrong et al.[4] suggested a scheme that uses a periodic modulation of the sign of the crystal's quadratic nonlinear coefficient $\chi^{(2)}$ to periodically offset the phase mismatch between the interacting waves by the reciprocal vector of the lattice structure. The proposed method is called the quasi-phase-matching (QPM) technique, and QPM has several advantages over BPM: first, QPM accesses the largest nonlinear coefficient d_{33} because the fundamental and harmonic waves can have the same polarization; second, QPM can avoid the spatial walk-off effect of BPM; third, QPM can be realized for any mixing interaction for which the material is transparent, even in crystals that have too weak birefringence for realizing BPM.

2. Material Fabrication

QPM can be realized by periodical inversion of the spontaneous polarization of ferroelectric nonlinear crystal(NLC) materials, and this kind of materials with modulated micrometer-scale domain structures is usually called optical superlattice. As mentioned above, QPM was proposed early in 1962, but could not be experimentally realized at that time because suitable fabrication techniques had not been developed. Early attempts to obtain periodical sign reversal of the second-order nonlinear coefficient were accomplished by utilizing stacked plates of a nonlinear material where adjacent layers are rotated by 180 deg[5—8]. An obvious disadvantage of this method is that multiple Fresnel reflections would introduce additional losses, hence lower the conversion efficiency. In addition, the application of this technique was limited in the far-infrared and terahertz spectral ranges because it is hard to fabricate such plates with thickness down to 100 μm.

In the 1980s, an important breakthrough came when Feng and Ming et al. of Nanjing University in China proposed the growth striation technique to obtain $LiNbO_3$ (LN) crystal with periodic ferroelectric domains in a Czochralski growth system. For this technique, the melt was doped with solute such as yttrium or indium. During crystal growth, a temperature fluctuation was introduced into the solid-liquid interface either by applying a modulated electric current or by an eccentric rotation. And this resulted in periodic solute concentration fluctuation in the crystal along the direction of growth, or the so-called growth striation. The periodic ferroelectric domain can be automatically induced by the growth striations when the growing crystal cools through the Curie temperature[9—10]. By exploiting the growth striation technique, Feng et al. prepared LN crystal with periodic laminar ferroelectric domains, and with this crystal, quasi-phase-matched SHG was experimentally verified[11]. Later, a similar procedure was applied to grow $LiTaO_3$ (LT) superlattices[12]. Besides Feng et al.'s work, in the 1980s, Fejer and co-workers prepared

LN single-crystal fiber using laser-heated pedestal growth[13,14], which was initiated in 1974 at Stanford University. Periodically poled LN (PPLN) single-crystal fibers with micrometer-scale domains were used for the generation of blue[15] and green light[16] by SHG.

Almost at the same time, efforts including chemical diffusion and substitution of impurities were made to control domain inversion at the surface of ferroelectric crystals. In 1989, the group led by Fejer used titanium diffusion followed proton exchange to fabricate an LN waveguide with periodic domain structure, and demonstrated the first quasi-phase-matched green[17] and blue generations[18] in waveguides. Similar results on the LN waveguide were obtained simultaneously by Webjörn et al. in Sweden[19] and in Japan[20], as well as the progress in KTP by Bierlein and co-workers[21,22].

Although direct growth of periodical structured NLCs might be adequate for device demonstration, this technique was not so flexible for mass production. In the 1990s, one new method, the lithographic-patterned electric-field poling technique, was developed for creating inverted domains with controlled period and duty cycle at a wafer scale. The significant break-through was achieved by Yamada et al. of Sony in LN crystal in 1993[23], following the work by Ito et al., which demonstrated fabrication of LN periodic domain structure by electron-beam poling[24]. Since then, much attention has been paid toward fabricating periodic patterns in thicker LN samples[25] as well as other ferroelectrics, such as LT[26], KTP[27], and SBN[28], using the electric-field poling technique. The processing recipe for electric-field poling is simple. As indicated in Fig. 1, a patterned electrode produced with a photolithographic process was made on the $+z$ surface of the single-domain crystal wafer, while a planar electrode was on the $-z$ surface. High voltage over the coercive field was applied to the plus($+$) electrode to reverse the polarization under the patterned electrode; thus the reversed domain was defined by the patterned electrode.

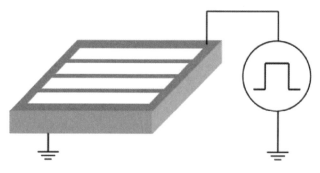

FIG.1. Schematic setpup of the electric-field poling technique.

With the aim of fabricating ferroelectric domain structures with high optical-quality, large-aperture, submicrometer periodicity over large crystal thickness, and with the development of new quasi-phase-matched nonlinear materials, up to now, the electric-field poling technique has been well-developed, and there are several improved methods for better domain control. Myers et al.[29] used an electrolyte as a liquid electrode to achieve field uniformity as well as avoid dielectric breakdown during poling LN crystals, and this method was adopted by many research groups. In 1995, 3 in. diameter LN wafers with modulated domain structures were achieved at Stanford University[30]. The extremely high coercive electric fields of congruent $LiNbO_3$(CLN), which are about 21 kV / mm, and the strong effect of domain widening limit the thickness of the poling crystals to 0.5 mm and the period to be in the range of 10 to 30 μm, which were suitable for infrared (IR) devices. To obtained smaller domain structures for visible light generation, Miller et al.[31] obtained PPLN with less than 10 μm periodicity, and later in 1999, a backswitching method was exploited by Batchko et al.[32] for fabrication of short-pitched PPLN with 4 μm period and was used for first-order continuous-wave (CW) single-pass 460 nm blue-light generation. An alternative approach for the fabrication of short-period domain in LN single crystals is the surface poling technique proposed by Busacca and co-workers[33,34] at the University of Southampton. This technique is based on conventional electric-field poling, but involves an intentional overpoling step that inverts all the material except a thin surface region directly below the patterned photoresist, and the produced domain period is~1 μm.

As is known, CLN crystals suffer from optical damage and the photorefraction effect thus limits the applications in highpower laser operations. Research revealed that when doping with 5.0 mol. % MgO, the optical damage threshold can be raised by 2 orders of magnitude[35] and the coercive of MgO-doped LN(MgO:LN) is reduced to be about 1 / 5 that of CLN[36]. One property of MgO:LN is that it behaves like a diode when the polarization inversion is formed[37]; thus local penetrations would degrade the uniformity of the electric field by the large leakage current of polarization-inverted regions. Sugita et al.[38] proposed a multipulse poling method, and this method can suppress current leakage in penetration of the inverted region by using a thick crystal and short-pulse application. Besides, Ishizuki et al.[39] reported that the coercive field of MgO:LN crystal was drastically reduced at elevated temperatures. Using the multipulse poling method at elevated temperatures, they succeeded in fabricating 3-mm-thick periodically poled MgO:LN (MgO:PPLN) of 30 μm period with smooth surfaces[40]. Ishizuki and Taira set the record of thickness to be 5 mm for MgO:PPLN in 2005[41], and 5 mm for MgO-doped periodically poled LT(MgO:PPLT) in 2010[42]. In 2012, a 10-mm-thick MgO:PPLN was successfully fabricated(see Fig. 2), and was used for construction of a high-energy optical parametric oscillator(OPO)[43]. Another advantage of the multipulse poling method is that it can suppress the side growth of the domain-inverted region, and thus is a promising approach for fabrication of short-period domain structures. By using the multipulse poling method, Sugita and co-workers[38,44] fabricated MgO:PPLN with the periods of 2.2 and 1.4 μm, respectively, and realized efficient first-order ultraviolet(UV) light generations by SHG.

FIG.2. 10-mm-thick MgO:PPLN with a poling period of 32.2 μm. Selected from Ref.[43].

There are also growing interests in light mediated ferroelectric domain engineering of LN single crystal, including light-assisted poling, UV laser-induced inhibition of poling, and all-optical poling methods, also overviewed by Ying et al.[45]. Briefly, laser irradiation, from deep-UV to the long-wavelength IR spectral range, can modify the coercive and nucleation field of LN single crystals, and can even induce domain inversion without applying any external electric field. And the light-mediated methods are suitable for fabrication of waveguide nonlinear optical devices, curved domain shapes for photonic applications, and thicker domain-engineered crystals for high peak power nonlinear applications due to the significant reduction of the coercive field.

In addition to LN and LN-family crystals, the poling techniques of KTP are progressing rapidly, and these were mainly accomplished by the Laurell group in Sweden. The demonstration of electric-field poling of hydrothermally grown KTP was realized in 1994[46]. Later, two improved methods, Rb-exchange-assisted electric-field poling[47] and a low-temperature poling[48], were developed to reduce the ion conductivity of the flux-grown KTP crystals. Due to the large anisotropy of KTP, the domain broadening during electric-field poling is reduced; thus it is favorable for submicrometer-periodic-domain inversion[49]. To further fabricate domain structure with submicrometer periodicity, Canalias et al.[50] utilized a chemical patterning method to avoid domain broadening due to the metal electrodes fringing fields. Using this technique, PPKTP crystal with a 720 nm period was fabricated for backward SHG[51], and one with 800 nm period was used to demonstrate experimentally the magic mirror-less optical parametric oscillation[52].

Besides the commonly used QPM materials, LN, LT, and KTP families, techniques for fabricating periodic structures in semiconductors such as GaAs, which is promising for mid-IR, and crystalline quartz, which is suitable for UV generation, are also of interest. For GaAs, epitaxial growth over an orientation-patterned GaAs (OP-GaAs) template was exploited[53]. Meanwhile, for crystalline quartz, the periodic structure can be realized by applying mechanical stress[54].

In the following, we will briefly review some of the solid-state lasers based on QPM materials, including single-wavelength lasers covering red to the UV, red-green-blue (RGB) three-color lasers based on the multiple-QPM technique, OPOs in different temporal scales, and passive mode-locking lasers using cascaded second-order nonlinearity.

3. Single-Wavelength Visible and UV Lasers Using Quasi-Phase-Matching

Nowadays, laser diodes(LDs) are used in many applications due to their advantages of compactness, high efficiency, good reliability, and low cost. Reliable laser light sources are available in the wavelength range from 630 to 2000 nm. However, many applications require wavelengths outside this wavelength range. For example, blue-green light sources are required in applications such as display, medicine, optical data storage, and color printing. The present blue-green light sources are based mostly on gas lasers or resonant frequency-converted solid-state lasers, which are complex, inefficient, or have a limited wavelength range. Frequency upconversions utilizing QPM optical superlattice offers a way to obtain compact, efficient solid-state laser light sources in the visible and down to the UV spectral range.

3.1 Red Lasers

Red light can be obtained through the following schemes. The first is direct frequency doubling of the IR lasers at around 1.3 or 1.5 μm. He et al.[55] obtained 840 mW red light at 671 nm by single-pass frequency doubling of a Nd:YVO$_4$ 1342 nm IR laser with a PPLT, the SHG efficiency being 60%. Hu et al.[56] used a 1.2-mm-thick periodically poled near stoichiometric LT(PPSLT) crystal for frequency doubling and obtained 1.4 W 671 nm red light, with a conversion efficiency of more than 50%. In stoichiometric crystals, the nonstoichiometric defects are significantly reduced, which leads to a reduction of the coercive field by one order; thus samples with thickness more than 1 mm could be easily fabricated with good penetrability; see Fig. 3. In addition, the laser damage threshold of SLT is increased by two or three orders compared with congruent ones. Due to the high optical quality of the PPSLT crystal, the NLC can handle more power, the output power of red was stable, and the output beam exhibited a circular shape with a Gaussian profile as shown in Fig. 4. Hu et al.[57] also scaled the output power to 2.4 W at 660 nm, a more bright red light, by using a high-power diode-side-pumped Nd:YAG laser operating at 1319 nm as the fundamental source, and the NLC was also a PPSLT but with different poling period. Figure 5 shows the model machine of the high-power 660 nm red laser made at Nanjing University. Thompson et al.[58] demonstrated generation of more than 900 mW tunable CW red light at 780 nm by single-pass frequency doubling a seeded fiber amplifier at 1560 nm with two cascaded PPLN crystals. The enhancement of SHG efficiency compared with using one single NLC is due to the presence of second-harmonic (SH) light from the first crystal acting as a seed for the second. Using the cascaded NLC scheme, Chiow et al.[59] obtained 43 W quasi-continuous 780 nm red light by single-pass frequency doubling two high-power coherently combined fiber amplifiers at 1560 nm with two cascading PPLN crystals. The total SHG efficiency is up to 66%.

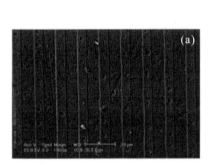

FIG.3. (a) Scanning electron microscopy micrograph of etched domain-inverted patterns on the $+C$ surface. (b) Cross-section view of Y face of a PPSLT sample. Selected from Ref. [56].

FIG. 4. Output red beam with a Gaussian profile selected from Ref.[56].

The second scheme is based on frequency mixing of 1.06 and 1.55 μm laser lights. Hart et al.[60] and Boullet et al.[61], respectively, obtained 630 nm red light by frequency mixing of an Er/Yb-codoped fiber laser operating at 1060 and 1550 nm with a PPLN. Champert et al.[62] presented the results on 1.4 W red generation by frequency mixing of seeded Yb and Er fiber amplifiers.

The third scheme for red light generation is by utilizing a direct OPO process in which the frequency of the generated red light is tunable, as indicated by Melkonian et al.[63]. They built the first CW oscillation of a single resonant OPO (SRO) operating from 619 to 640 nm with a MgO-doped periodically poled stoichiometric $LiTaO_3$ (MgO:PPSLT) crystal pumped by a 532 nm green laser, and 100 mW of single-frequency red light was obtained.

The fourth one is based on OPO cascaded SFG. Bosenberg et al.[64] reported a 2.5 W CW 629 nm red laser based on a two-step OPO and SFG process carried out in a single PPLN crystal having two periodic gratings in series.

FIG.5. High-power 660 nm red laser model machine with PPSLT.

3.2 Green Lasers

To obtain green light, frequency doubling is the commonly used way and most work is focused on high-power CW green light generation with high efficiency. In 1997, Miller et al.[65] first realized 2.7 W CW 532 nm green light generation by single-pass frequency doubling of a Nd:YAG laser with a PPLN crystal, and the power conversion efficiency is up to 42%. To construct a high-power room temperature green laser, MgO:LN is a good candidate. Pavel et al.[66] used MgO:PPLN to frequency double a Nd:GdVO$_4$ laser in a single-pass configuration; 1.18 W CW green light was obtained with the conversion efficiency being 16.8%. Yb-doped single-frequency fiber lasers were also used as the fundamental sources for efficient and high-power green light generation due to their high power, high beam quality, and narrow linewidth. Tovstonog et al.[67] reported 7 W CW 542 nm green generation at room temperature with 35.4% efficiency in a MgO:PPSLT crystal. Also, Kumar et al.[68] obtained 9.64 W CW 532 nm green light using a MgO:PPSLT and 6.2 W output with a PPKTP, both from a Yb-fiber laser at 1064 nm. To further improve the efficiency of green light generation, Ricciardi et al.[69] placed the PPLT crystal in a resonant enhanced cavity; 6.1 W green light was obtained with a conversion efficiency of 76%, which is the highest ever reported. In addition, Kumar et al.[70] used a novel cascaded multicrystal scheme[71] for efficient CW single-pass SHG, which provided more than 55% conversion efficiency. The multicrystal SHG scheme is shown in Fig. 6.

3.3 Blue Lasers

As for blue-light generation, there are several schemes. Frequency doubling or tripling of the near-IR light is the most commonly used approach. The challenge for first-order quasiphase-matched frequency doubling or tripling to obtain blue light is the rather small domain period, which is below 5 μm. Pruneri et al.[72] fabricated a PPLN crystal with 4.6

μm period by choosing optimized mark-to-space ratio during electric-field poling, and obtained 49 mW CW 473 nm blue light by frequency doubling with a 4.6% conversion efficiency. Batchko et al.[73] reported single-pass CW first-order QPM SHG of 60 mW blue light at 465 nm in a backswitched poled PPLN crystal with 4 μm periodicity. Due to the rather low coercive field, which is about 2 kV/mm, stoichiometric $LiTaO_3$ crystal is an alternative for small period (4.6 μm) sample fabrication, and can be used for frequency tripling to obtain blue light at 440 nm[74]. As for KTP, the low-temperature poling technique was exploited for creation of domain period down to 5 μm for efficient single-pass SHG of 460 nm blue light[75].

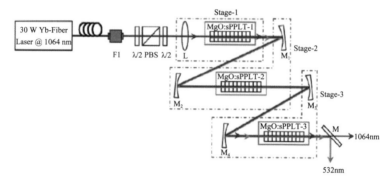

FIG.6. Schematic experimental setup for the multicrystal CW single-pass SHG. Selected from Ref.[70].

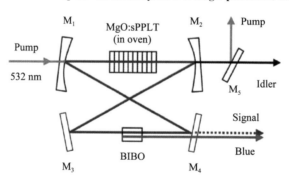

FIG.7. Schematic of an intracavity frequency-doubled MgO:PPSLT CW SRO for blue generation. Selected from Ref.[77].

Another approach is based on 532 nm green laser pumped OPO cascaded with a SHG or SFG process. This approach was first realized by Xu et al.[76] for blue-light generation with 20 nm tuning range in a quasi-periodic LT superlattice. Later, Samanta et al.[77] constructed a ring OPO cavity, as shown in Fig. 7, which comprised a MgO:PPSLT crystal placed between two concave mirrors and a BBO crystal placed between plane mirrors. Temperature tuning of MgO:PPSLT together with angular tuning of BBO produced a CW single-frequency blue laser source tunable in the wavelength range of 425—489 nm, and the output power of blue ranged from 45 to 448 mW. Lai et al.[78] made a special design, to realize 435 nm blue light through OPO and SHG in a single PPLT crystal. In addition, a tunable blue light can be obtained by intracavity frequency doubling in an external resonator geometry with spatial separation of the spectral components using a fanstructured PPLT crystal[79].

3.4 589 nm Yellow Lasers

Laser light sources at 589 nm are of important applications in medicine, communication, display technology, and sodium guide stars. There are mainly two ways to generate 589 nm yellow light: the first one is based on frequency mixing of the two emission lines of the Nd:YAG laser gain media, which are 1064 and 1319 nm, respectively, and the second one is by direct frequency doubling of a Raman fiber amplifier laser. For the first scheme, two IR lasers can be used with different NLCs such as LN[80], PPLN[81], PPSLT[82], and PPKTP[83], and a high average power of more than 16 W was obtained, though the laser system is complex. To make the system more compact, Zhao *et al.*[84] used a Nd:YAG dual-wavelength laser as the fundamental source and a PPLT crystal as the nonlinear frequency converter, as shown in Fig. 8. The three-mirror laser cavity, which was proposed by Chen and co-workers[85,86], can be used to adjust the power ratio between the two emission lines for efficient SFG and can improve the stability of the dual-wavelength laser system as well. By single-pass SFG, 500 mW 589 nm yellow with a 0.16 nm spectral width was obtained, as indicated in Fig. 9.

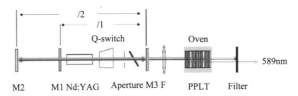

FIG. 8. Schematic setup for 589 nm yellow generation by frequency mixing a dual-wavelength IR laser. Selected from Ref. [84].

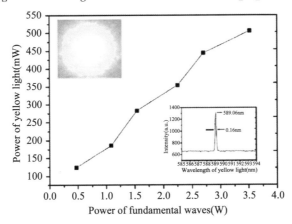

FIG. 9. Power dependence of the output yellow light on the total pumping power in [84]. The inset on the top left is a picture of the output yellow beam, and the insert on the bottom right is the spectrum of 589 nm yellow light.

Direct frequency doubling of a Raman fiber amplifier is an alternative approach for high-power narrow-linewidth CW 589 nm yellow light generation. Georgiev et al.[87] obtained 3 W CW 589 nm yellow light by frequency doubling a 1179 nm Raman laser using a MgO:PPLN crystal. Yuan et al.[88] also reported 4 W CW 589 nm laser from a Raman fiber amplifier with a MgO:PPSLT as the frequency-doubling crystal. Figure 10 shows the spectrum of the frequency-doubled 589 nm yellow light, the FWHM being about 0.2 pm, which is suitable for the sodium guide star application.

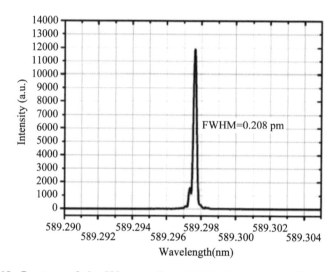

Fig.10. Spectrum of the 589 nm yellow light by frequency doubling a high-power Raman fiber amplifier. Selected from Ref.[88].

3.5 UV and Vacuum UV Lasers

Ferroelectric crystals such as LN[89,90], LT[91—94], and KTP[95—97] can be periodically poled for efficient UV generation. But the intrinsic band edge around 300 nm prevents the wavelength conversion to vacuum UV(VUV).

Crystal quartz, however, has a short UV band edge down to 150 nm, which is suitable for VUV generation. Recently, Kurimura et al.[54] created periodically twinned crystal quartz with modulated polarity by applying mechanical stress, and demonstrated QPM SHG emitting VUV light at 193 nm.

Besides crystal quartz, the ferroelectric fluoride $BaMgF_4$ shows a very wide transparency range extending from 125 nm to 13 μm, and by using periodical poling, QPM can be achieved in the whole transparent wavelength region[98]. To date, the shortest period obtained is 6.6 μm, which can be used for second-order SHG of 193 nm supposing a suitable duty cycle of the periodically poled crystal.

4. Red-Green-Blue Lasers Based on Multiple Quasi-Phase-Matching

Besides single-wavelength solid-state visible and UV lasers, multiwavelength solid-state lasers are also of great interest, such as RGB three-elementary-color lasers. As we all know, all the vivid beautiful colors of the visible world can be constituted through a weighted combination of only three fundamental colors: red, green, and blue. Laser-based projection display (LBPD) has some obvious advantages over the conventional display technique, such as high brightness, high spatial resolution, and great color gamut. One possible way is using three lasers, each emitting light with the wanted color. To date, many attempts for generation of RGB light have been made, and there are mainly two routes: the first one is based on frequency upconversion in series, such as SHG or/and SFG of the pump wavelengths or oscillating wavelengths, and the second approach is based on frequency downconversion cascaded with several frequency upconversion processes.

4.1 Multiple-QPM Technique

It is obvious that there always involves several nonlinear optical processes for RGB three-color generation, and in this subsection, we will introduce the multiple-QPM technique, which is favorable for compact and efficient RGB laser generation.

In the early years of QPM, research was focused on nonlinear materials with periodic modulated domain structure. As is known, the reciprocal vector of a one-dimensional(1D) periodic optical superlattice can be written as $G_m = m \cdot 2\pi / \Lambda$, where Λ is the grating period and m is the QPM order. And one can see that a periodic optical superlattice provides a series of reciprocal vectors, each of which is an integer times a primitive vector. Hence the reciprocal vectors are independent, and are usually used to efficiently realize one single nonlinear optical parametric process. In 1986, just two years after the invention of quasi-crystal[99], researchers at Nanjing University in China introduced quasi-periodic structure into optical superlattice design, and extended the study of QPM from periodic to quasi-periodic structure. In 1990, Feng et al.[100] established the multiple-QPM theory. Later on, various structures, such as aperiodic[101,102], dual-periodic[94,103], as well as two-dimensional(2D) optical superlattice[104-106], were proposed for multiple QPM.

Actually, a 1D quasi-periodic lattice can be created by the projection of 2D square lattice on a straight line[107]. The projection points on the straight line form two types of intervals, a and b. The arranged sequence of a and b depends on the projection angle θ. By analogy, a two-component quasi-periodic optical superlattice is constructed from two building blocks A and B with the widths D_A and D_B, respectively. Both A and B contain a pair of 180 deg antiparallel domains. Assume that all A and B blocks have the same width of positive domains, represented with l. According to the projection theory, the reciprocal vectors for this quasi-periodic structure are $G_{m,n}=2\pi(m+n\tau)/D$, where $D=\tau D_A+D_B$ is the average structure parameter, m and n are two integers, and $\tau=\tan\theta$. Since θ is an adjustable parameter, it offers the quasi-periodic structure additional design flexibility for QPM. Compared with a 1D periodic structure, a 1D quasi-periodic optical superlattice has a higher symmetry, and its reciprocal vectors are governed two integers m and n; thus some coupled optical parametric processes can be realized. For example, the QPM conditions for third-harmonic generation (THG) via cascaded quadratic nonlinearity are $\Delta k_1=k_2-2k_1-G_{m,n}=0$ for SHG and $\Delta k_2=k_3-k_2-k_1-G_{p,q}=0$ for SFG, where k_1, k_2, and k_3 are the wave vectors of the fundamental, the SH wave, and the third-harmonic wave, respectively; m, n, p, and q are integers representing the QPM order. In 1997, Zhu et al.[108] experimentally demonstrated direct THG in a quasi-periodic optical superlattice with the well-known Fibonacci sequence, of which the structure parameter τ is near the golden ratio, and the THG efficiency was up to 23%. And this pioneering work exhibited a possible important application of quasi-periodic superlattice in nonlinear optics.

4.1 RGB Laser Based On Frequency Upconversions

In this scheme, the fundamental light sources could be singlewavelength or a multiwavelength ones, and by SHG and/or SFG of the fundamental light, RGB lights can be obtained. Yamamoto et al.[109] utilized two LDs, respectively, operating at 1.3 and 0.86 μm as the fundamental sources; simultaneous generation of blue at 0.43 μm, green at 0.52 μm, and red at 0.65 μm by SHG and SFG in a proton-exchanged MgO:LN waveguide was demonstrated for the first time. In 1993, Laurell et al.[110] used two IR lasers as the pump light, and by combinations of SHG and SFG in a segmented KTP waveguide, simultaneous generation of UV, blue, and green light was obtained, and this may be the first published QPM multi-color generation. Not long after, Sundheimer et al.[111] reported RGB generation through simultaneous SHG, SFG, and fourth-harmonic generation of a single laser source at 1650 nm in a segmented KTP channel waveguide. Following the same idea, Baldi et al.[112] realized RGB three-color generation with a proton-exchanged PPLN channel waveguide. Rare-earth-doped LN crystal with modulated domain structure can also be used for simultaneous RGB generation with broad tunability. Cantelar et al.[113] fabricated a Zn-diffused Er^{3+}/Yb^{3+}-codoped aperiodically poled LN (aPPLN) channel waveguide. Red light ranging from 650 to 690 nm and green light from 520 to 575 nm arose from energy transfer upconversion processes between the two rare-earth ions, while the blue is produced by QPM SHG of the pump laser.

The above-mentioned RGB light sources were realized in waveguides; as for the bulk, Jaque et al.[114] experimentally demonstrated RGB laser generation from a $Nd:YAl_3(BO_3)_4$ employing two different pump wavelengths at 807 and 755 nm. Red at 669 nm was obtained by self-frequency-doubling of the laser oscillation at 1338 nm; the green at 505 nm and the blue at 481 nm were obtained by self-SFG of the fundamental and the two pump wavelength, respectively. And the need for a two-pump-source is the disadvantage of this scheme. Romero et al.[115] realized simultaneous CW RGB generation using a 0.88 μm pumped Nd:YVO$_4$ 1.34 μm laser with a SBN crystal as the intracavity frequency converter. Red and blue light were generated by SHG of the pump and the IR, respectively, while green light was produced by frequency mixing of the pump and IR laser radiation. Due to the broad distribution of ferroelectric domain sizes of SBN, no angle or thermal tuning is need for both SHG and SFG processes.

To realize simultaneous generation of RGB with compactness and efficiency, a promising approach combining the multiwavelength oscillating solid-state laser technique together with the multiple-QPM-theory-based optical superlattice has been developed recently. The Nd^{3+} ion is a very versatile laser active ion and has multiple allowed transitions leading to potential laser radiations around 1300, 1060, and 940 nm, respectively. If these three channels could be oscillated at the same time, simultaneous RGB generation could be implemented by the frequency-doubling technique. However, the line around 940 nm, which belongs to a quasi-three-energy-level system, is too difficult to oscillate due to its rather low gain. In addition, the existing gain competition among these three emission lines makes it almost impossible to oscillate three of them simultaneously. However, there is another possible way that employs frequency tripling of the line around 1300 nm to obtain blue light. With this method, only two fundamental waves around 1300 and 1060 nm are needed for simultaneous RGB generation. In general, there are three nonlinear processes during RGB generation; to make the system more compact, a special NLC is needed and optical superlattice based on multiple-QPM theory may be a good candidate. Capmany[116] realized 1084 and 1372 nm dual-wavelength operation in Nd^{3+}-doped aPPLN, and through frequency doubling of the two IR lights, 542 nm green and 686 nm red were obtained, while frequency mixing of the 744 nm pumping light with the 1084 nm IR produced 482 nm blue. In 2003, He et al.[117] and Liao et al.[118], both at Nanjing University in China, using a dual-wavelength Q-switched Nd:YVO$_4$ laser operating at 1064 and 1342 nm as the fundamental source, and aperiodically poled LT(aPPLT) crystals as the nonlinear frequency converter, experimentally realized red, yellow, and green "traffic signal lights," as well as RGB three-fundamental-color light. To combine RGB in a proper proportion for white-light generation, Li et al.[119] utilized a Nd:YVO$_4$ dual-wavelength laser and a dual-periodic LT optical superlattice for simultaneous RGB generation, and by changing the pump power and the operating temperature of the optical superlattice, the proportion of RGB was changed as well, and an average quasi-white-light laser power of 530 mW was achieved. In 2008, Hu et al.[120] realized 1 W quasi-white-light generation from an intermittent oscillating dual-wavelength Nd:YAG laser operating at 1064 and 1319 nm with a cascaded LT superlattice. The intermittent oscillating scheme composed of a Y-shaped laser cavity with two arms shared the same gain medium and the same output coupling mirror, but had two individual Q-switches with a delay between their opening time, thus avoiding the gain competition between the emission lines. (Please refer to Ref. [121] for a detailed description of the intermittent oscillation dual-wavelength laser technique) Another advantage of the intermittent oscillation dual-wavelength laser technique is that the power proportion of the two oscillating lines is adjustable, so that the power ratio of the RGB three fundamental colors is adjustable to obtain white light. Changing the crystal temperature can also adjust the power ratio of the output RGB beams. The temperature tuning curves of their scheme in Fig. 11 show that red, green, and blue reached the maximum output power at different temperatures, but the temperature tuning curves overlapped each other in a temperature region to ensure simultaneous RGB output. At 129.3℃, 780 mW red light, 146 mW green light, and 84 mW blue light were obtained, the power proportion of RGB being 9.3 : 1.7 : 1, which mixed up into 1.01 W cool white light according to the C.I.E. chromaticity diagram. The conversion efficiency from the power of the two IR beams to the power of the quasi-white light was 20%, and the output power fluctuation was measured to be ~6.5% within 1 h. Figure 12 shows a photo of the RGB beams separated by a prism from the setup.

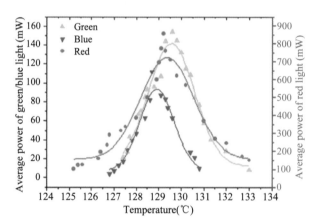

FIG.11. Temperature tuning curves of the red, green, and blue colors in Ref.[120].

4.3 RGB Laser Based On Frequency Downconversions

The dual-wavelength scheme described above for realizing RGB or quasi-white-light output involves a complex arrangement of optical components. They are thus relatively inefficient and are quite expensive to implement in commercial systems. Another scheme for realizing a monolithic RGB laser light source is based on frequency downconversion cascaded with some frequency upconversions.

Henrich et al. at University Kaiserslautern[122] developed a 19 W RGB solid-state laser source that is well suited for large frame laser projection displays, and they have brought it all the way to commercialization. The pumping source used is a high-power mode-locked Nd:YVO$_4$ 1064 nm laser. A KTA-OPO was synchronously pumped to generate a signal

FIG.12. Photo of the RGB beams separated by a prism from the setup. Selected from Ref.[120].

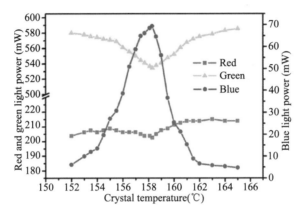

FIG.13. Temperature tuning curves of the RGB three colors selected from Ref.[124].

wave at 1535 nm and a mid-IR idler wave at 3470 nm. Frequency doubling of the IR laser generates 532 nm green. 629 nm red light results from single-pass sum-frequency mixing of the 1535 nm OPO signal wave and 1064 nm laser radiation in a KTA crystal. Another single-pass SFG process mixes the red light with residual OPO signal radiation to create a 446 nm blue beam. In 2001, Liu et al.[123] used a LT crystal with two periodic structure arranged in tandem pumped by a 532 nm green laser to produce red and blue light. Red light at 631 nm was generated as the signal wave by parametric downconversion of the incident green beam. Blue light at 460 nm was obtained by frequency mixing of the residual green light with the mid-IR idler of the former parametric process. Together with the residual green light, three fundamental colors were obtained. Following the same scheme, Gao et al.[124] created a monolithic RGB source with a tandem-structured SLT crystal wafer pumped by 1 W high-repetition-rate nanosecond 532 nm green light. Figure 13 gives the temperature tuning curves of RGB three colors at full pumping power of 1 W. At 158.3℃, 69.4 mW blue light was generated, which was the least of the three colors. By blocking the excess red and green powers, 304 mW of color-balanced white light, which included 140 mW red, 94.3 mW green, and 69.4 mW blue, was obtained. The optical-to-optical efficiency from the pump green to the white light was 30%. The generated white light separated with a prism is shown in Fig. 14. By raising the green pump power to 3.7 W, Xu *et al.*[125] scaled the output power of the white light to 1 W with a PPLT crystal. A unique choice in their work is that a cylindrical lens was used to shape the pump laser into an oblate spot so as to match the shape of the superlattice, as shown in Fig. 15. Besides the pulsed (picosecond and nanosecond time range) RGB laser light sources described above, Lin *et al.*[126] reported a CW watt-level RGB laser pumped by a Yb-fiber laser at 1064 nm. The red, green, and blue lights were obtained by frequency mixing of the pump and the signal wave, frequency doubling the pump IR light, and SFG of the red and the signal wave.

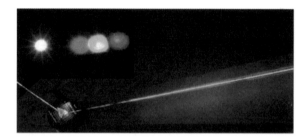

FIG. 14. Generated white light disperses into RGB three colors through a prism. Selected from Ref.[124].

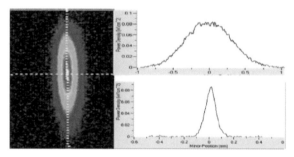

FIG.15. Elliptical spot at the focus plane of the pump. Selected from Ref.[125].

There are also several variants of the green light pumped parametric downconversion RGB source. Cudney *et al.*[127] presented a RGB source based on simultaneous QPM SHG and THG in a single LN crystal wafer with two periodic structure in tandem. In the first part, 1.43 μm signal was generated through optical parametric generation (OPG), and in the second section, SHG and THG of the signal wave produced the red and blue light. Green light was obtained by a nonphase-matched SFG between the signal and the pump wave.

Besides collinear RGB laser light sources from 1D optical superlattice, Xu *et al.*[128] and Zhao *et al.*[129] have demonstrated noncollinear RGB laser light sources from a 2D optical superlattice. The pump sources in both works are 532 nm green lasers; red and blue lights result from a green light pumped OPG process cascading two SHGs of the signal and idler waves in a single-pass setup. The NLCs are a hexagonally poled LT and a 2D nonlinear photonic quasi-crystal, respectively. Owing to the noncollinear reciprocal vectors provided by the 2D optical superlattice, the RGB light was autoseparated without using any optical separation elements, as indicated in Fig. 16.

5. Quasi-Phase-Matched Optical Parametric Oscillators

Nonlinear optical processes have different forms, and the most important ones include SHG, SFG, difference-frequency generation, and OPG. To extend the spectral range of the emitting light, OPG is of great interest. However, to realize a parametric process for practical use, the NLC is often placed into an optical resonator, and this is the well-known OPO. OPOs are versatile sources of coherent radiation; in particular, combined with QPM materials, they could provide optical radiation across an entire spectral range and over all the temporal scales.

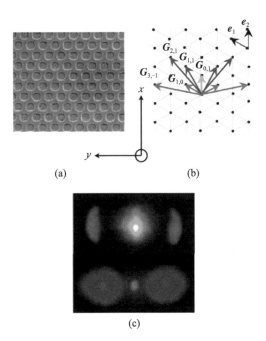

FIG.16. (a) Micrograph of the HexPLT sample, (b) reciprocal space of the sample, and (c) photograph of the noncollinear RGB colors from the HexPLT. Selected from Ref.[128].

In 1995, Myers et al.[130] demonstrated the first bulk QPM OPO, which was almost at the same time as the first QPM waveguide OPO realized by Bortz et al.[131]. The pumping source was a 1.064 μm Q-switched Nd:YAG laser, and the output was temperature tuned over the 1.66 – 2.95 μm spectral range. The NLC was fabricated using the electric-field poling technique with a thickness of 0.5 mm and an interaction length of 5.2 mm. The threshold was about 0.1 mJ, which was more than 1 order of magnitude below the crystal damage limit. For the nanosecond OPOs, the high energy of the pump pulse would demand on the material damage threshold. With the development of QPM materials and material fabrication techniques, optical superlattice with high damage threshold, large aperture, long interaction length, and large effective nonlinearity, together with the advances in pump-laser technology, could cause the output of nanosecond OPOs to produce more and more output. Ishizuki and Taira[42] reported a high-energy OPO of 118 mJ output in 10 ns pulse duration of 30 Hz repetition rate using a 5-mm-thick 39-mm-long MgO:PPLT with a poling period of 30 μm, the slope efficiency being about 70%. They also fabricated a 10-mm-thick Mg:PPLN for handling more energy. In the same 10 ns and 30 Hz pulse region, the output energy reached half-joule (540 mJ) with a total conversion efficiency of more than 76%[43]. Besides, Zukauskas et al.[132] demonstrated a high-energy OPO pumped at 1064 nm with 12 ns pulses at 100 Hz repetition rate, using a 5-mm-thick PPKTP with homogeneous and highquality domain gratings as the NLC. This OPO generated 60 mJ total energy with a 50% conversion efficiency at room temperature.

Ultrafast OPOs, using picosecond or femtosecond pulses as the pump, have also been developed since the 1990s. Pruneri et al.[133] reported a SRO in a PPLN crystal synchronously pumped with a 523.5 nm picosecond laser, producing 200 mW average power within the 10 μs envelope, and with a tuning range from 883 to 1285 nm. Almost at the same period, Galvanauskas et al.[134] demonstrated the first femtosecond QPM OPO in a PPLN crystal pumped at 777 nm. The output produced 300 fs pulses with a tuning range from 1 to 3 μm. In the ultrafast OPOs, a technique named "synchronous pumping" is necessary for amplification of the generated waves. In this technique, the interval between the arrival of adjacent pump pulses is arranged to match the round-trip time of the frequency downconverted pulse in the OPO cavity, so that the interacting pulses always meet in the NLC. Due to the extremely short pulse duration of the ultrafast pumping pulses, the input optical intensity is high for efficient frequency conversion. In addition, the energy content is rather small (typically from several tens of nanojoules to several hundreds of nanojoules); therefore, the material damage threshold is alleviated. For the QPM ultrafast synchronously pumped OPO, Nd-doped mode-locking lasers, Ti:sapphire lasers, and Yb-doped fiber lasers are commonly used as the pumping sources. Together with the QPM materials, such as PPLN and PPKTP, the ultrafast OPOs can provide pulses across the 1 to 5 μm spectral range from near-IR to mid-IR[135-140].

OPOs in the CW region are a big challenge because the low pump intensity and the subsequent low parametric gain prevent them from being practical. In CW OPOs, at least one of the generated waves should resonante in the cavity to overcome the round-trip loss. Among the different resonating configurations, the SRO is the simplest, with only one signal wave oscillating in the cavity. Compared with double resonant OPO (DRO), SRO does not suffer the stability problem, though the oscillation threshold may be 2 orders of magnitude that of DRO, hence requiring the QPM materials to have suitable interaction lengths and low losses. The first QPM CW SRO operation was reported by Myers et al.[141] in a 50-mm-long PPLN crystal with a threshold of 3 W. To get efficient and stable output, the pump power should be several times the oscillation threshold, and this can done by utilizing high-power single-frequency pumping sources[142], reducing the thresh-old by choosing a suitable interaction length[143], or using high-power broadband lasers as the pumps[144]. To extend the output spectral region to shorter wavelengths, especially in the visible, one can exploit frequency-doubled green laser as the pump to obtain red or yellow light, as well as cascading a SHG/SFG process to get blue light [77,145,146]. One of the challenges of the nonlinear materials used in CW OPO operation is the light absorption, and weak absorption losses will raise the pump threshold to several tens of watts. In addition, light absorption will cause refractive index change through thermal-optical or bulk photorefractive effect. To date, optical cleaning[147] or doping with MgO in LN-family crystals have turned out to be effective methods.

6. Passive Mode-Locking Lasers Using Optical Superlattice

In addition to high-power efficient single-wavelength lasers and RGB three-fundamental-color lasers through QPM, optical superlattice can be used to realize solid-state mode-locking picosecond lasers. There are two mechanisms to achieve passive mode-locking solid-state lasers in the picosecond regime. The first one is nonlinear mirror mode-locking (NLM), which was first proposed by Stankov and Jethwa in 1988[148]. The NLM scheme consists of a NLC for SHG, a dielectric dichroic mirror having partial reflection at the fundamental and total reflection at the SH wave, and a dispersive medium between the NLC and the dichroic mirror. The SH generated is completely reflected by the dichroic mirror, and if the SH wave experiences a proper phase shift with respect to the fundamental wave, it can be totally reconverted into a fundamental wave on its way back through the NLC. Hence, the reflectivity of the combination, a frequency-doubling crystal and a dielectric mirror, is intensity dependent. The NLM scheme is simple for picosecond pulse generation, but the longer duration of the output pulse in comparison with the pulse traveling inside the laser cavity is the main drawback. The second modelocking mechanism, which was first demonstrated by Cerullo et al.[149], is called cascaded second-order nonlinear mode-locking (CSM) or cascaded Kerr lens mode locking. The cavity configuration of CSM is essentially the one of NLM, but the dichroic mirror totally reflects both the fundamental wave and the SH wave. The frequency-doubling crystal is placed in off phase-matched condition; thus a nonlinear phase shift will be imprinted on the fundamental wave beam, resulting in nonlinear mode size variations, which can be converted into nonlinear loss modulations together with a suitably positioned intracavity hard or soft aperture. The modulation depth of this scheme will depend on the cavity design and it can be large even for small conversion efficiencies, so it does not exhibit the pulse duration limitation characteristic for NLM.

The nonlinear frequency-doubling crystal used for mode locking can be either BPM material, such as KTP, BBO, and LBO, or QPM material. There are several advantages when using QPM materials as the frequency doubler, as in the following: first, this approach is promising for scaling the power because the damage threshold of NLCs is an order of magnitude higher than that of SESAM and the low residual absorption at the fundamental wave enables operation at high average power; second, due to using QPM, mode locking can work at any wavelength in the NLC's transparent spectral region, so this technique is feasible in wide spectral regions, especially operating at wavelengths where no SESAMs exist. In 2002, Chen et al.[150] demonstrated, for the first time, 5.6 W and 20 ps nonlinear mirror mode-locked Nd:YVO$_4$ laser with a PPKTP; the long pulse duration was mainly due to the group velocity mismatch between the fundamental and the SH waves in the NLM scheme. The schematic experimental setup for diode-pumped passively mode-locking lasers is shown in Fig. 17. A four-mirror laser cavity configuration is usually used to ensure proper mode size in both the laser gain medium and the NLC. The nonlinear frequency-doubling crystals in the schematic are indicated as PPXX. As mentioned above, CSM scheme does not exhibit the pulse duration limitation characteristic compared with NLM. Holmgren et al.[151] demonstrated a Nd:GdVO$_4$ laser mode locked by a self-defocusing cascaded Kerr lens in PPKTP; the repetition rate was 200 MHz with an average power of 350 mW. A strong pulse shortening mechanism is produced by the interplay of group velocity mismatch and the cavity design, which leads to a pulse duration as short as 2.8 ps and with a bandwidth of 0.6 nm. Iliev et al.[152] utilized similar scheme, achieving stable and self-starting mode locking of a diode-pumped Nd:GdVO$_4$ by employing CSM in a MgO:PPSLT crystal; the maximum average output power reached 5 W and the shortest pulses were 3.2 ps. Besides the results relying on NLM or CSM alone, some hybrid mode-locking techniques have also been reported. In [153], Holmgren et al. reported on a Nd:YVO$_4$ laser mode locking with a hybrid active and passive modulator consisting of a single partially poled KTP crystal. The active phase modulation is done by the electro-optic effect of the unpoled region of the NLC, which initiates, enhances, and stabilizes the mode locking. The periodically poled part provides passive modulation by CSM, which is responsible for pulse shortening. The advantage of this hybrid scheme is that it has short pulses attributed to passive mode locking, but with the active modulation the requirements for self-starting the passive mode locking can be relaxed. By this means, the repetition rate of the laser was 94 MHz and the average output power was 350 mW, with the pulse lengths down to 6.9 ps. Moreover, Iliev et al.[154] demonstrated passive mode locking of a diode-pumped Nd:YVO$_4$ laser exploiting both NLM and CSM. The pulse repetition rate of the laser is 117 MHz with average power ranging from 0.5 to 3.1 W and pulse duration from 2.9 to 5.2 ps. The NLM ensures self-starting and self-sustaining mode locking, while the CSM process is the dominant pulse shortening mechanism that can generate pulse

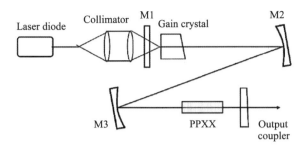

FIG. 17. Schematic experimental setup of passively mode-locking lasers with optical superlattice.

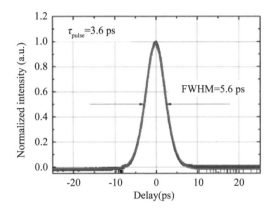

FIG. 18. Typical autocorrelation trace of passively mode-locked pulses with optical superlattice. Selected from Ref.[153].

duration below 10 ps. The advantages of this hybrid technique lie in both the high average output power and the short pulse-width down to several picoseconds. To date, most of the studies on intracavity SHG mode locking were focused on the 1.06 μm laser system, and more recently much attention has been paid to the development of 1.34 μm mode-locking lasers. Liu et al.[155] reported the first realization of an NLM mode-locked diode-pumped Nd:YVO$_4$ laser operating at 1342 nm with a PPLN crystal. The average output power is about ~1.5 W, and the pulsewidth and the repetition rate of the output laser are 9.5 ps and 101 MHZ, respectively. Almost at the same time, Iliev et al.[155] demonstrated CSM of a 1.34 μm Nd:YVO$_4$ laser in MgO:PPSLT crystal. A high average power of about 1 W and a short pulse duration of 3.6 ps were simultaneously obtained from this passively mode-locked 1.34 μm diode-pumped laser source. Figure 18 shows a typical autocorrelation trace of passively mode-locked pulses with optical superlattice, which is selected from [156].

7. Conclusion and Outlook

To summarize, from the 1980s, the QPM-material fabrication techniques have been well developed, and among them, the electric-field poling technique and its variations are applied to the commonly used nonlinear optical materials, such as LN-family, LT-family, and KTP crystals. Domain periods of a few micrometers to several tens of micrometers are fabricated for visible and IR applications. A challenge still remained in the fabrication of submicrometer or nanoscale ferroelectric domains, which have current and future applications such as efficient first-order QPM UV SHG, electro-optic Bragg gratings with an unrivalled capacity of integration, counterpropagation optical parametric interactions, and enhanced second-order optical processes in nonlinear photonic crystals[157,158].

With the development of QPM theory, nonlinear optical materials, materials fabrication techniques, all-solid-state laser pumping sources, and novel optical cavity geometries, new laser light sources based on parametric processes can extend the spectral range from the UV to the mid-IR, with temporal coverage in all time scales from CW to ultrafast femtosecond.

Nowadays, the QPM technique has entered a new regime. On the one hand, the QPM-based all-solid-state laser technique has brought out practical applications in modern laser industries, while, on the other hand, the QPM technique has paved a new way for fundamental research in quantum optics[159]. From the early study of multipartite continuous-variable entanglement[160] to the recent studies in the transformation of spatial and temporal properties of entangled photons[161,162], the domain-engineered crystal has taken a key role in the engineering of new types of photonic entanglement. In particular, for those QPM crystals capable of ensuring multiple nonlinear processes, several functions can be integrated into a single crystal wafer and on-chip steering of entangled photons can be achieved, by which the QPM crystal is proved to be a promising candidate for the future integrated quantum information processing.

References and Notes

[1] T. H. Maiman, "Stimulated optical radiation in ruby," Nature **187**, 493–494 (1960).
[2] P. A. Franken, A. E. Hill, C. W. Peters, and G. Weinreich, "Generation of optical harmonics," Phys. Rev. Lett. **7**, 118–119 (1961).
[3] D. A. Kleinman, "Theory of second harmonic generation of light," Phys. Rev. **128**, 1761–1775 (1962).
[4] J. A. Armstrong, N. Bloembergen, J. Ducuing, and P. S. Pershan, "Interactions between light waves in dielectric," Phys. Rev. **127**, 1918–1939 (1962).
[5] M. S. Piltch, C. D. Cantrell, and R. C. Sze, "Infrared second-harmonic generation in nonbirefringent cadmium telluride," J. Appl. Phys. **47**, 3514–3517 (1976).
[6] A. Szilagyi, A. Hordvik, and H. Schlossberg, "A quasi-phase-matching technique for efficient optical mixing and frequency doubling," J. Appl. Phys. **47**, 2025–2032 (1976).
[7] D. E. Thompson, J. D. McMullen, and D. B. Anderson, "Second-harmonic generation in GaAs" stack of plates "using high-power CO_2 laser radiation," Appl. Phys. Lett. **29**, 113–115 (1976).
[8] M. Okada, K. Takizawa, and S. Ieiri, "Second harmonic generation by periodic laminar structure of nonlinear optical crystal," Opt. Commun. **18**, 331–334 (1976).

[9] N. B. Ming, J. F. Hong, and D. Feng, "The growth striations and ferroelectric domain structures in Czochralski-grown LiNbO$_3$ single crystals," J. Mater. Sci. **17**, 1663–1670 (1982).

[10] Y. L. Lu, Y. Q. Lu, X. F. Chen, G. P. Luo, C. C. Xue, and N. B. Ming, "Formation mechanism for ferroelectric domain structures in a LiNbO$_3$ optical superlattice," Appl. Phys. Lett. **68**, 2642–2644 (1996).

[11] D. Feng, N. B. Ming, J. F. Hong, Y. S. Yang, J. S. Zhu, Z. Yang, and Y. N. Wang, "Enhancement of second-harmonic generation in LiNbO$_3$ crystals with periodic laminar ferroelectric domains," Appl. Phys. Lett. **37**, 607–609 (1980).

[12] W. S. Wang, Q. Zou, Z. H. Geng, and D. Feng, "Study of LiTaO$_3$ crystals grown with a modulated structure I. Second harmonic generation in LiTaO$_3$ crystals with periodic laminar ferroelectric domains," J. Cryst. Growth **79**, 706–709 (1986).

[13] M. M. Fejer, J. L. Nightingale, G. A. Magel, and R. L. Byer, "Laser-heated miniature pedestal growth apparatus for single-crystal optical fibers," Rev. Sci. Instrum. **55**, 1791–1796 (1984).

[14] Y. S. Luh, R. S. Feigelson, M. M. Fejer, and R. Byer, "Ferroelectric domain structures in LiNbO$_3$ single-crystal fibers," J. Cryst. Growth **78**, 135–143 (1986).

[15] G. A. Magel, M. M. Fejer, and R. L. Byer, "Quasi-phase-matched second-harmonic generation of blue light in periodically poled LiNbO$_3$," Appl. Phys. Lett. **56**, 108–110 (1990).

[16] D. H. Jundt, G. A. Magel, M. M. Fejer, and R. L. Byer, "Periodically poled LiNbO$_3$ for high-efficiency second-harmonic generation," Appl. Phys. Lett. **59**, 2657–2659 (1991).

[17] E. J. Lim, M. M. Fejer, and R. L. Byer, "Second-harmonic generation of green light in periodically-poled planar lithium nio-bate waveguide," Electron. Lett. **25**, 174–175 (1989).

[18] E. J. Lim, M. M. Fejer, R. L. Byer, and W. J. Kozlovsky, "Blue light generation by frequency doubling in periodically poled lithium niobate channel waveguide," Electron. Lett. **25**, 731–732 (1989).

[19] J. Webjörn, F. Laurell, and G. Arvidsson, "Blue light generated by frequency doubling of laser diode light in a lithium niobate channel waveguide," IEEE Photon. Technol. Lett. **1**, 316–318 (1989).

[20] S. Miyazawa, "Ferroelectric domain inversion in Ti-diffused LiNbO$_3$ optical waveguide," J. Appl. Phys. **50**, 4599–4603 (1979).

[21] J. D. Bierlein, D. B. Laubacher, J. B. Brown, and C. J. van der Poel, "Balanced phase matching in segmented KTiOPO$_4$ waveguides," Appl. Phys. Lett. **56**, 1725–1727 (1990).

[22] C. J. van der Poel, J. D. Bierlein, J. B. Brown Co., and S. Colak, "Efficient type I blue second-harmonic generation in periodically segmented KTiOPO$_4$ waveguides," Appl. Phys. Lett. **57**, 2074–2076 (1990).

[23] M. Yamada, N. Nada, M. Saitoh, and K. Watanabe, "First-order quasi-phase matched LiNbO$_3$ waveguide periodically poled by applying an external field for efficient blue second-harmonic generation," Appl. Phys. Lett. **62**, 435–436 (1993).

[24] H. Ito, C. Takyu, and H. Inaba, "Fabrication of periodic domain grating in LiNbO$_3$ by electron beam writing for application of nonlinear optical processes," Electron. Lett. **27**, 1221–1222 (1991).

[25] G. D. Miller, "Periodically poled lithium niobate: modeling, fabrication, and nonlinear-optical performance," Ph.D. thesis (Stanford University, 1998).

[26] S. N. Zhu, Y. Y. Zhu, Z. Y. Zhang, H. Shu, H. F. Wang, J. F. Hong, C. Z. Ge, and N. B. Ming, "LiTaO$_3$ crystal periodically poled by applying an external pulsed field," J. Appl. Phys. **77**, 5481–5483 (1995).

[27] W. P. Risk and S. D. Lau, "Periodic electric field poling of KTiOPO$_4$ using chemical patterning," Appl. Phys. Lett. **69**, 3999–4001(1996).

[28] Y. Y. Zhu, J. S. Fu, R. F. Xiao, and G. K. L. Wong, "Second harmonic generation in periodically domain-inverted Sr$_{0.6}$Ba$_{0.4}$Nb$_2$O$_6$ crystal plate," Appl. Phys. Lett. **70**, 1793–1795(1997).

[29] L. E. Myers, R. C. Eckardt, M. M. Fejer, R. L. Byer, W. R. Bosenberg, and J. W. Pierce, "Quasi-phase-matched opticalparametric oscillators in bulk periodically poled LiNbO$_3$," J. Opt. Soc. Am. B **12**, 2102–2116 (1995).

[30] L. E. Myers, "Quasi-phase-matched optical parametric oscillators in bulk periodically poled lithium niobate," Ph.D. thesis (Stanford University, 1995).

[31] G. D. Miller, R. G. Batchko, M. M. Fejer, and R. L. Byer, "Visible quasi-phase-matched harmonic generation by electric-fieldpoled lithium niobate," Proc. SPIE **2700**, 34–35(1996).

[32] R. G. Batchko, V. Y. Shur, M. M. Fejer, and R. L. Boyd, "Backswitch poling in lithium niobate for high-fidelity domain patterning and efficient blue light generation," Appl. Phys. Lett. **75**, 1673–1675 (1999).

[33] A. C. Busacca, C. L. Sones, V. Apostolopoulos, R. W. Eason, and S. Mailis, "Surface domain engineering in congruent lithium niobate single crystals: a route to submicron periodic poling," Appl. Phys. Lett. **81**, 4946–4948(2002).

[34] A. C. Busacca, C. L. Sones, R. W. Eason, and S. Mailis, "Firstorder quasi-phase-matched blue light generation in surfacepoled Ti: indiffused lithium niobate waveguides," Appl. Phys. Lett. **84**, 4430–4432(2004).

[35] G. Zhong, J. Jian, and Z. Wu, "Measurement of optically induced refractive-index change of lithium niobate doped with different concentration of MgO," in *Proceedings of the 11th International Quantum Electronics Conference* (1980), pp. 631–635.

[36] A. Kuroda, S. Kurimura, and Y. Uesu, "Domain inversion in ferroelectric MgO:LiNbO$_3$ by applying electric fields," Appl. Phys. Lett. **69**, 1565–1567(1996).

[37] S. Sonoda, I. Tsuruma, and M. Hatori, "Second harmonic generation in electric poled X-cut MgO-doped LiNbO$_3$ waveguides," Appl. Phys. Lett. **70**, 3078–3080(1997).

[38] T. Sugita, K. Mizuuchi, Y. Kitaoka, and K. Yamamoto, "Ultraviolet light generation in a periodically poledMgO:LiNbO$_3$ waveguide," Jpn. J. Appl. Phys. **40**, 1751–1753 (2001).

[39] H. Ishizuki, I. Shoji, and T. Taira, "Periodical poling characteristics of congruent MgO:LiNbO$_3$ crystals at elevated temperature," Appl. Phys. Lett. **82**, 4062–4064(2003).

[40] H. Ishizuki, T. Taira, S. Kurimura, J. H. Ro, and M. Cha, "Periodic poling in 3-mm-thick MgO:LiNbO$_3$ crystals," Jpn. J. Appl. Phys. **42**, L108–L110(2003).

[41] H. Ishizuki and T. Taira, "High-energy quasi-phase-matched optical parametric oscillation in a periodically poled MgO:LiNbO$_3$ device with a 5 mm × 5 mm aperture," Opt. Lett. **30**, 2918–2920 (2005).

[42] H. Ishizuki and T. Taira, "High energy quasi-phase matched optical parametric oscillation using Mg-doped congruent LiTaO$_3$ crystal," Opt. Express **18**, 253–258(2010).

[43] H. Ishizuki and T. Taira, "Half-joule output optical-parametric oscillation by using 10-mm-thick periodically poled Mg-doped congruent LiNbO$_3$," Opt. Express **20**, 20002–20010(2012).

[44] K. Mizuuchi, A. Morikawa, T. Sugita, and K. Yamamoto, "Efficient second-harmonic generation of 340-nm light in a 1.4-μm periodically poled bulk MgO:LiNbO$_3$," Jpn. J. Appl. Phys. **42**, L90–L91 (2003).

[45] C. Y. J. Ying, A. C. Muir, C. E. Valdivia, H. Steigerwald, C. L. Sones, R. W. Eason, E. Soergel, and S. Mailis, "Light-mediated ferroelectric domain engineering and micro-structuring of lithium niobate crystals," Laser Photon. Rev. **6**, 526–548(2012).

[46] Q. Chen and W. P. Risk,"Periodic poling of $KTiOPO_4$ using an applied electric field," Electron. Lett. **30**, 1516–1517 (1994).

[47] H. Karlsson and F. Laurell, "Electric field poling of flux grown $KTiOPO_4$," Appl. Phys. Lett. **71**, 3474–3476 (1997).

[48] G. Rosenman, A. Skliar, D. Eger, M. Oron, and M. Katz, "Low temperature periodic electrical poling of flux-grown $KTiOPO_4$ and isomorphic crystals," Appl. Phys. Lett. **73**, 3650–3652(1998).

[49] C. Canalias, V. Pasiskevicius, and F. Laurell, "Periodic poling of $KTiOPO_4$: from micrometer to sub-micrometer domain gratings," Ferroelectrics **340**, 27–47 (2006).

[50] C. Canalias, V. Pasiskevicius, R. Clemens, and F. Laurell, "Submicron periodically poled flux-grown $KTiOPO_4$," Appl. Phys. Lett. **82**, 4233–4235 (2003).

[51] C. Canalias, V. Pasiskevicius, M. Fokine, and F. Laurell, "Backward quasi-phase-matched second-harmonic generation in submicrometer periodically poled flux-grown $KTiOPO_4$," Appl. Phys. Lett. **86**, 181105 (2005).

[52] C. Canalias and V. Pasiskevicius, "Mirrorless optical parametric oscillator," Nat. Photonics **1**, 459–462 (2007).

[53] L. A. Eyres, P. J. Tourreau, T. J. Pinguet, C. B. Ebert, J. S. Harris, M. M. Fejer, L. Becouarn, B. Gerard, and E. Lallier, "All-epitaxial fabrication of thick, orientation-patterned GaAs films for nonlinear optical frequency conversion," Appl. Phys. Lett. **79**, 904–906(2001).

[54] S. Kurimura, M. Harada, K. Muramatsu, M. Ueda, M. Adachi, T. Yamada, and T. Ueno, "Quartz revisits nonlinear optics: twinned crystal for quasi-phase matching," Opt. Mater. Express **1**, 1367–1375 (2011).

[55] J. L. He, G. Z. Luo, H. T. Wang, S. N. Zhu, Y. Y. Zhu, Y. B. Chen, and N. B. Ming, "Generation of 840 mW of red light by frequency doubling a diode-pumped 1342 nm Nd:YVO_4 laser with periodically-poled $LiTaO_3$," Appl. Phys. B **74**, 537–539(2002).

[56] X. P. Hu, X. Wang, J. L. He, Y. X. Fan, S. N. Zhu, H. T. Wang, Y. Y. Zhu, and N. B. Ming, "Efficient generation of red light by frequency doubling in a periodically-poled nearly-stoichiometric $LiTaO_3$ crystal," Appl. Phys. Lett. **85**, 188–190(2004).

[57] X. P. Hu, X. Wang, Z. Yan, H. X. Li, J. L. He, and S. N. Zhu, "Generation of red light at 660 nm by frequency doubling a Nd:YAG laser with periodically-poled stoichiometric $LiTaO_3$," Appl. Phys. B **86**, 265–268 (2007).

[58] R. Thompson, M. Tu, D. Aveline, N. Lundblad, and L. Maleki, "High power single frequency 780 nm laser source generated from frequency doubling of a seeded fiber amplifier in a cascade of PPLN crystals," Opt. Express **11**, 1709–1713 (2003).

[59] S. Chiow, T. Kovachy, J. M. Hogan, and M. A. Kasevich, "Generation of 43 W of quasi-continuous 780 nm laser light via highefficiency, single-pass frequency doubling in periodically poled lithium niobate crystals," Opt. Lett. **37**, 3861–3863 (2012).

[60] D. L. Hart, L. Goldberg, and W. K. Burns, "Red light generation by sum frequency mixing of Er/Yb fibre amplifier output in QPM $LiNbO_3$," Electron. Lett. **35**, 52–53 (1999).

[61] J. Boullet, L. Lavoute, A. Desfarges Berthelemot, V. Kermène, P. Roy, V. Couderc, B. Dussardier, and A.-M. Jurdyc, "Tunable red-light source by frequency mixing from dual band Er/Yb codoped fiber

laser," Opt. Express **14**, 3936 – 3941 (2006).

[62] P. A. Champert, S. V. Popov, M. A. Solodyankin, and J. R. Taylor, "1.4 – W red generation by frequency mixing of seeded Yb and Er fiber amplifiers," IEEE Photon. Technol. Lett. **14**, 1680 – 1682 (2002).

[63] J. Melkonian, T. My, F. Bretenaker, and C. Drag, "High spectral purity and tunable operation of a continuous singly resonant optical parametric oscillator emitting in the red," Opt. Lett. **32**, 518 – 520 (2007).

[64] W. R. Bosenberg, J. I. Alexander, L. E. Myers, and R. W. Wallace, "2.5-W, continuous-wave, 629-nm solid-state laser source," Opt. Lett. **23**, 207 – 209 (1998).

[65] G. D. Miller, R. G. Batchko, W. M. Tulloch, D. R. Weise, M. M. Fejer, and R. L. Byer, "42%-efficient single-pass cw second-harmonic generation in periodically poled lithium niobate," Opt. Lett. **22**, 1834 – 1836 (1997).

[66] N. Pavel, I. Shoji, T. Taira, K. Mizuuchi, A. Morikawa, T. Sugita, and K. Yamamoto, "Room-temperature, continuous-wave 1-W green power by single-pass frequency doubling in a bulk periodically poled $MgO:LiNbO_3$ crystal," Opt. Lett. **29**, 830 – 832 (2004).

[67] S. V. Tovstonog, S. Kurimura, and K. Kitamura, "High power continuous-wave green light generation by quasiphase matching in Mg stoichiometric lithium tantalite," Appl. Phys. Lett. **90**, 051115 (2007).

[68] S. C. Kumar, G. K. Samanta, and M. Ebrahim-Zadeh, "High-power, single-frequency, continuous-wave second-harmonic-generation of ytterbium fiber laser in PPKTP and MgO:sPPLT," Opt. Express **17**, 13711 – 13726 (2009).

[69] I. Ricciardi, M. Rosa, A. Rocco, P. Ferraro, and P. Natale, "Cavity-enhanced generation of 6 W cw second-harmonic power at 532 nm in periodically-poled $MgO:LiTaO_3$," Opt. Express **18**, 10985 – 10994 (2010).

[70] S. C. Kumar, G. K. Samanta, K. Devi, and M. Ebrahim-Zadeh, "High-efficiency, multicrystal, single-pass, continuous-wave second harmonic generation," Opt. Express **19**, 11152 – 11169 (2011).

[71] G. C. Bhar, U. Chatterjee, and P. Datta, "Enhancement of second harmonic generation by double-pass configuration in barium borate," Appl. Phys. B **51**, 317 – 319 (1990).

[72] V. Pruneri, R. Koch, P. G. Kazansky, W. A. Clarkson, P. St, J. Russell, and D. C. Hanna, "49 mW of cw blue light generated by first-order quasi-phasematched frequency doubling of a diode-pumped 946-nm Nd:YAG laser," Opt. Lett. **20**, 2375 – 2377 (1995).

[73] R. G. Batchko, M. M. Fejer, R. L. Byer, D. Woll, R. Wallenstein, V. Y. Shur, and L. Erman, "Continuous-wave quasi-phase-matched generation of 60 mW at 465 nm by single-pass frequency doubling of a laser diode in backswitch-poled lithium niobate," Opt. Lett. **24**, 1293 – 1295 (1999).

[74] X. P. Hu, G. Zhao, C. Zhang, Z. D. Xie, J. L. He, and S. N. Zhu, "High-power, blue-light generation in a dual-structure, periodically poled, stoichiometric $LiTaO_3$ crystal," Appl. Phys. B **87**, 91 – 94 (2007).

[75] D. Woll, J. Schumacher, A. Robertson, M. A. Tremont, R. Wallenstein, M. Katz, D. Eger, and A. Englander, "250 mW of coherent blue 460-nm light generated by single-pass frequency doubling of the output of a mode-locked high-power diode laser in periodically poled KTP," Opt. Lett. **27**, 1055 – 1057 (2002).

[76] P. Xu, K. Li, G. Zhao, S. N. Zhu, Y. Du, S. H. Ji, Y. Y. Zhu, N. B. Ming, L. Luo, K. F. Li, and K. W. Cheah, "Quasi-phase-matched generation of tunable blue light in a quasi-periodic structure,"

Opt. Lett. **29**, 95 – 97 (2004).

[77] G. K. Samanta and M. Ebrahim-Zadeh, "Continuous-wave, single-frequency, solid-state blue source for the 425 – 489 nm spectral range," Opt. Lett. **33**, 1228 – 1230 (2008).

[78] C.-M. Lai, I.-N. Hu, Y.-Y. Lai, Z.-X. Huang, L.-H. Peng, A. Boudrioua, and A.-H. Kung, "Upconversion blue laser by intracavity frequency self-doubling of periodically poled lithium tantalate parametric oscillator," Opt. Lett. **35**, 160 – 162 (2010).

[79] J. Zimmermann, J. Struckmeier, M. R. Hofmann, and J. Meyn, "Tunable blue laser based on intracavity frequency doubling with a fan-structured periodically poled $LiTaO_3$ crystal," Opt. Lett. **27**, 604 – 606 (2002).

[80] J. D. Vance, C. Y. She, and H. Moosmüller, "Continuous-wave, all-solid-state, single-frequency 400-mW source at 589 nm based on doubly resonant sum-frequency mixing in a mono-lithic lithium niobate resonator," Appl. Opt. **37**, 4891 – 4896 (1998).

[81] J. Yue, C.-Y. She, B. P. Williams, J. D. Vance, P. E. Acott, and T. D. Kawahara, "Continuous-wave sodium D_2 resonance radiation generated in single-pass sum-frequency generation with periodically poled lithium niobate," Opt. Lett. **34**, 1093 – 1095 (2009).

[82] A. J. Tracy, C. Lopez, A. Hankla, D. J. Bamford, D. J. Cook, and S. J. Sharpe, "Generation of high-average-power visible light in periodically poled nearly stoichiometric lithium tantalite," Appl. Opt. **48**, 964 – 968 (2009).

[83] E. Mimoun, L. Sarlo, J. Zondy, J. Dalibard, and F. Gerbier, "Sum-frequency generation of 589 nm light with near-unit efficiency," Opt. Express **16**, 18684 – 18691 (2008).

[84] L. N. Zhao, J. Su, X. P. Hu, X. J. Lv, Z. D. Xie, G. Zhao, P. Xu, and S. N. Zhu, "Single-pass sum-frequency-generation of 589-nm yellow light based on dual-wavelength Nd: YAG laser with periodically-poled $LiTaO_3$ crystal," Opt. Express **18**, 13331 – 13336 (2010).

[85] Y. F. Chen, S. W. Tsai, S. C. Wang, Y. C. Huang, T. C. Lin, and B. C. Wong, "Efficient generation of continuous-wave yellow light by single-pass sum-frequency mixing of a diode-pumped Nd: YVO_4 dual-wavelength laser with periodically poled lithium niobate," Opt. Lett. **27**, 1809 – 1811 (2002).

[86] Y. F. Chen, "Cw dual-wavelength operation of a diode-end-pumped Nd: YVO_4 laser," Appl. Phys. B. **70**, 475 – 478 (2000).

[87] D. Georgiev, V. P. Gapontsev, A. G. Dronov, M. Y. Vyatkin, A. B. Rulkov, S. V. Popov, and J. R. Taylor, "Watts-level frequency doubling of a narrow line linearly polarized Raman fiber laser to 589 nm," Opt. Express **13**, 6772 – 6776 (2005).

[88] Y. Yuan, L. Zhang, Y. H. Liu, X. J. Lu, G. Zhao, Y. Feng, and S. N. Zhu, "Sodium guide star laser generation by single-pass frequency doubling in a periodically poled near-stoichiometric $LiTaO_3$ crystal," Sci. China Ser. B **56**, 125 – 128 (2013).

[89] R. T. White, I. T. McKinnie, S. D. Butterworth, G. W. Baxter, D. M. Warrington, P. G. R. Smith, G. W. Ross, and D. C. Hanna, "Tunable single-frequency ultraviolet generation from a continuous-wave Ti: sapphire laser with an intracavity PPLN frequency doubler," Appl. Phys. B **77**, 547 – 550 (2003).

[90] K. Mizuuchi, T. Sugita, and K. Yamamoto, "Generation of 360-nm ultraviolet light in first-order periodically poled bulk MgO: $LiNbO_3$," Opt. Lett. **28**, 935 – 937 (2003).

[91] J.-P. Meyn and M. M. Fejer, "Tunable ultraviolet radiation by second-harmonic generation in periodically poled lithium tantalite," Opt. Lett. **22**, 1214 – 1216 (1997).

[92] K. Mizuuchi and K. Yamamoto, "Generation of 340-nm light by frequency doubling of a laser diode in

bulk periodically poled LiTaO$_3$," Opt. Lett. **21**, 107–109 (1996).

[93] P. A. Champert, S. V. Popov, J. R. Taylor, and J. P. Meyn, "Efficient second-harmonic generation at 384 nm in periodically poled lithium tantalate by use of a visible Yb-Er-seeded fiber source," Opt. Lett. **25**, 1252–1254 (2000).

[94] Z. W. Liu, S. N. Zhu, Y. Y. Zhu, Y. Q. Qin, J. L. He, C. Zhang, H. T. Wang, N. B. Ming, X. Y. Liang, and Z. Y. Xu, "Quasi-Cw ultraviolet generation in a dual-periodic LiTaO$_3$ superlattice by frequency tripling," Jpn. J. Appl. Phys. **40**, 6841–6844 (2001).

[95] S. Wang, V. Pasiskevicius, J. Hellströa, F. Laurell, and H. Karlsson, "First-order type II quasi-phase-matched UV generation in periodically poled KTP," Opt. Lett. **24**, 978–980 (1999).

[96] S. Wang, V. Pasiskevicius, F. Laurell, and H. Karlsson, "Ultraviolet generation by first-order frequency doubling in periodically poled KTiOPO$_4$," Opt. Lett. **23**, 1883–1885 (1998).

[97] B. Zhang, Y. J. Ding, and I. B. Zotova, "Efficient ultrafast ultraviolet generation based on frequency doubling in short-period periodically-poled KTiOPO$_4$ crystal," Appl. Phys. B **99**, 629–632 (2010).

[98] E. G. Víllora, K. Shimamura, K. Sumiya, and H. Ishibashi, "Bire-fringent and quasi phase-matching with BaMgF4 for vacuum-UV/UV and mid-IR all solid-state lasers," Opt. Express **17**, 12362–12378 (2009).

[99] D. Shechtman, I. Blech, D. Gratias, and J. W. Cahn, "Metallic phase with long-range orientational order and no translational symmetry," Phys. Rev. Lett. **53**, 1951–1953 (1984).

[100] J. Feng, Y. Y. Zhu, and N. B. Ming, "Harmonic generation in an optical Fibonacci superlattice," Phys. Rev. B **41**, 5578–5582 (1990).

[101] B. Y. Gu, B. Z. Dong, Y. Zhang, and G. Z. Yang, "Enhanced harmonic generation in aperiodic optical superlattice," Appl. Phys. Lett. **75**, 2175–2177 (1999).

[102] H. Liu, S. N. Zhu, Y. Y. Zhu, N. B. Ming, X. C. Lin, W. J. Ling, A. Y. Yao, and Z. Y. Xu, "Multiple-wavelength second-harmonic generation in aperiodic optical superlattices," Appl. Phys. Lett. **81**, 3326–3328 (2002).

[103] M. H. Chou, K. R. Parameswaran, and M. M. Fejer, "Multiple-channel wavelength conversion by use of engineered quasi-phase-matching structure in LiNbO$_3$ waveguides," Opt. Lett. **24**, 1157–1159 (1999).

[104] V. Berger, "Nonlinear photonic crystals," Phys. Rev. Lett. **81**, 4136–4139 (1998).

[105] B. Q. Ma, T. Wang, Y. Sheng, P. G. Ni, Y. Q. Wang, B. Y. Cheng, and D. Z. Zhang, "Quasiphase matched harmonic generation in a two-dimensional octagonal photonics superlattice," Appl. Phys. Lett. **87**, 251103 (2005).

[106] L. H. Peng, C. C. Hsu, and Y. C. Shih, "Second-harmonic green generation from two-dimensional x(2) nonlinear photonic crystal with orthorhombic lattice structure," Appl. Phys. Lett. **83**, 3447–3449 (2003).

[107] R. K. P. Zia and W. J. Dallas, "A simple derivation of quasicrystalline spectra," J. Phys. A **18**, L341–L345 (1985).

[108] S. N. Zhu, Y. Y. Zhu, and N. B. Ming, "Quasi-phase-matched third-harmonic generation in a quasi-periodic optical superlattice," Science **278**, 843–846 (1997).

[109] K. Yamamoto, H. Yamamoto, and T. Taniuchi, "Simultaneous sum-frequency and second-harmonic generation from a proton-exchanged MgO-doped LiNbO$_3$ waveguide," Appl. Phys. Lett. **58**, 1227–1229 (1991).

[110] F. Laurell, J. B. Brown, and J. D. Bierlein, "Simultaneous generation of UV and visible light in

segmented KTP waveguides," Appl. Phys. Lett. **62**, 1872 – 1874(1993).

[111] M. L. Sundheimer, A. Villeneuve, G. I. Stegeman, and J. D. Bierlein, "Simultaneous generation of red, green and blue light in a segmented KTP waveguide using a single source," Electron. Lett. **30**, 975 – 976 (1994).

[112] P. Baldi, C. G. Trevino-Palacios, G. I. Stegeman, M. P. De Micheli, D. B. Ostrowsky, D. Delacourt, and M. Papuchon, "Simultaneous generation of red, green and blue light in room temperature periodically poled lithium niobate waveguides using single source," Electron. Lett. **31**, 1350 – 1351 (1995).

[113] E. Cantelar, G. A. Torchia, J. A. Sanz-García, P. L. Pernas, G. Lifante, and F. Cusso, "Red, green, and blue simultaneous generation in aperiodically poled Zn-diffused $LiNbO_3:Er^{3+}/Yb^{3+}$ nonlinear channel waveguides," Appl. Phys. Lett. **83**, 2991 – 2993(2003).

[114] D. Jaque, J. Capmany, and J. G. Sole, "Red, green, and blue laser light from a single $Nd:YAl_3(BO_3)_4$ crystal based on laser oscillation at 1.3 μm," Appl. Phys. Lett. **75**, 325 – 327 (1999).

[115] J. J. Romero, D. Jaque, J. F. Sole, and A. A. Kaminskii, "Simultaneous generation of coherent light in the three fundamental colors by quasicylindrical ferroelectric domains in $Sr_{0.6}Ba_{0.4}(NbO_3)_2$," Appl. Phys. Lett. **81**, 4106 – 4108 (2002).

[116] J. Capmany, "Simultaneous generation of red, green, and blue continuous-wave laser radiation in Nd^{3+}-doped aperiodically poled lithium niobate," Appl. Phys. Lett. **78**, 144 – 146(2001).

[117] J. L. He, J. Liao, H. Liu, J. Du, F. Xu, H. T. Wang, S. N. Zhu, Y. Y. Zhu, and N. B. Ming, "Simultaneous cw red, yellow, and green light generation, "traffic signal lights," by frequency doubling and sum-frequency mixing in an aperiodically poled $LiTaO_3$," Appl. Phys. Lett. **83**, 228 – 230(2003).

[118] J. Liao, J. L. He, H. Liu, H. T. Wang, S. N. Zhu, Y. Y. Zhu, and N. B. Ming, "Simultaneous generation of red, green, and blue quasi-continuous-wave coherent radiation based on multiple quasi-phase-matched interactions from a single, aperiodically-poled $LiTaO_3$," Appl. Phys. Lett. **82**, 3159 – 3161 (2003).

[119] H. X. Li, Y. X. Fan, P. Xu, S. N. Zhu, P. Lu, Z. D. Gao, H. T. Wang, Y. Y. Zhu, N. B. Ming, and J. L. He, "530-mW quasi-white-light generation using all-solid-state laser technique," J. Appl. Phys. **96**, 7756 – 7758 (2004).

[120] X. P. Hu, G. Zhao, Z. Yan, X. Wang, Z. D. Gao, H. Liu, J. L. He, and S. N. Zhu, "High-power red-green-blue laser light source based on intermittent oscillating dual-wavelength Nd:YAG laser with a cascaded $LiTaO_3$ superlattice," Opt. Lett. **33**, 408 – 410 (2008).

[121] X. W. Fan, J. L. He, H. T. Huang, and L. Xue, "An intermittent oscillation dual-wavelength diode-pumped Nd:YAG laser," IEEE J. Quantum Electron. **43**, 884 – 888 (2007).

[122] B. Henrich, T. Herrmann, J. Kleilbauer, R. Knappe, A. Nebel, and R. Wallenstein, "Concepts and technologies of advanced RGB sources," in *Advanced Solid-State Lasers Conference*(2002), pp. 179 – 181.

[123] Z. W. Liu, S. N. Zhu, Y. Y. Zhu, H. Liu, Y. Q. Lu, H. T. Wang, N. B. Ming, X. Y. Liang, and Z. Y. Xu, "A scheme to realize three-fundamental-colors laser based on quasi-phase matching," Solid State Commun. **119**, 363 – 366 (2001).

[124] Z. D. Gao, S. N. Zhu, S. Y. Tu, and A. H. Kung, "Monolithic red-green-blue laser light source based on cascaded wavelength conversion in periodically poled stoichiometric lithium tantalite," Appl. Phys. Lett. **89**, 181101 (2006).

[125] P. Xu, L. N. Zhao, X. J. Lv, J. Lu, Y. Yuan, G. Zhao, and S. N. Zhu, "Compact high-power red-green-blue laser light source generation from a single lithium tantalate with cascaded domain modulation," Opt. Express **17**, 9509–9514 (2009).

[126] S. T. Lin, Y. Y. Lin, R. Y. Tu, T. D. Wang, and Y. C. Huang, "Fiber-laser-pumped CW OPO for red, green, blue laser generation," Opt. Express **18**, 2361–2367 (2010).

[127] R. S. Cudney, M. Robles-Agudo, and L. A. Rois, "RGB source based on simultaneous quasi-phasematched second and third harmonic generation in periodically poled lithium niobate," Opt. Express **14**, 10663–10668(2006).

[128] P. Xu, Z. D. Xie, H. Y. Leng, J. S. Zhao, J. F. Wang, X. Q. Yu, Y. Q. Qin, and S. N. Zhu, "Frequency self-doubling optical parametric amplification: noncollinear red-green-blue light-source generation based on a hexagonally poled lithium tantalite," Opt. Lett. **33**, 2791–2793(2008).

[129] L. N. Zhao, Z. Qi, Y. Yuan, J. Lu, Y. H. Liu, C. D. Chen, X. J. Lv, Z. D. Xie, X. P. Hu, G. Zhao, P. Xu, and S. N. Zhu, "Integrated noncollinear red-green-blue laser light source using a twodimensional nonlinear photonic quasicrystal," J. Opt. Soc. Am. B **28**, 608–612(2011).

[130] L. E. Myers, G. D. Miller, R. C. Eckardt, M. M. Fejer, R. L. Byer, and W. R. Bosenberg, "Quasi-phase-matched 1.064-μm-pumped optical parametric oscillator in bulk periodically poled $LiNbO_3$," Opt. Lett. **20**, 52–54 (1995).

[131] M. L. Bortz, M. A. Arbore, and M. M. Fejer, "Quasi-phase-matched optical parametric amplification and oscillation in periodically poled $LiNbO_3$ waveguides," Opt. Lett. **20**, 49–51(1995).

[132] A. Zukauskas, N. Thilmann, V. Pasiskevicius, F. Laurell, and C. Canalias, "5 mm thick periodically poled Rb-doped KTP for high energy optical parametric frequency conversion," Opt. Mater. Express **1**, 201–206(2011).

[133] V. Pruneri, S. D. Butterworth, and D. C. Hanna, "Low-threshold picosecond optical parametric oscillation in quasi-phase-matched lithium niobate," Appl. Phys. Lett. **69**, 1029–1031(1996).

[134] A. Galvanauskas, M. A. Arbore, M. M. Fejer, M. E. Fermann, and D. Harter, "Fiber-laser-based femtosecond parametric generator in bulk periodically poled $LiNbO_3$," Opt. Lett. **22**, 105–107 (1997).

[135] S. D. Butterworth, P. G. R. Smith, and D. C. Hanna, "Picosecond Ti: sapphire-pumped optical parametric oscillator based on periodically poled $LiNbO_3$," Opt. Lett. **22**, 618–620 (1997).

[136] S. Chaitanya Kumar and M. Ebrahim-Zadeh, "High-power, fiber-laser-pumped, picosecond optical parametric oscillator based on MgO: sPPLT," Opt. Express **19**, 26660–26665 (2011).

[137] C. W. Hoyt, M. Sheik-Bahae, and M. Ebrahim-Zadeh, "High-power picosecond optical parametric oscillator based on periodically poled lithium niobate," Opt. Lett. **27**, 1543–1545(2002).

[138] M. V. O'Connor, M. A. Watson, D. P. hepherd, D. C. Hanna, J. H. V. Price, A. Malinowski, J. Nilsson, N. G. R. Broderick, D. J. Richardson, and L. Lefort, "Synchronously pumped optical parametric oscillator driven by a femtosecond mode-locked fiber laser," Opt. Lett. **27**, 1052–1054 (2002).

[139] T. Sudmeyer, J. Aus der Au, R. Paschotta, U. Keller, P. G. R. Smith, G. W. Ross, and D. C. Hanna, "Femtosecond fiber-feedback optical parametric osicalltor," Opt. Lett. **26**, 304–306 (2001).

[140] N. Coluccelli, H. Fonnum, M. Haakestad, A. Gambetta, D. Gatti, M. Marangoni, P. Laporta, and G. Galzerano, "25-MHz synchronously pumped optical parametric oscillator at 2.25-2.6 μm and 4.1-4.9 μm," Opt. Express **20**, 22042–22047 (2012).

[141] L. E. Myers, W. R. Bosenberg, J. I. Alexander, M. A. Arbore, M. M. Fejer, and R. L. Byer, "CW

singly resonant optical parametric oscillators based on 1.064-um pumped periodically poled LiNbO$_3$," in *Proceedings on Advanced Solid State Lasers*, S. A. Payne and C. R. Pollock, eds. (1996), p. 35–37.

[142] S. C. Kumar, R. Das, G. K. Samanta, and M. EbrahimZadeh, "Optimally-output-coupled, 17.5 W, fiber-laser-pumped continuous-wave optical parametric oscillator," Appl. Phys. B, **102**, 31–35 (2011).

[143] R. Sowade, I. Breunig, J. Kiessling, and K. Buse, "Influence of the pump threshold on the single-frequency output power of singly resonant optical parametric oscillators," Appl. Phys. B. **96**, 25–28 (2009).

[144] R. Das, S. C. Kumar, G. K. Samanta, and M. Ebrahim-Zadeh, "Broadband, high-power, continuous-wave, mid-infrared source using extended phase-matching bandwidth in MgO: PPLN," Opt. Lett. **34**, 3836–3838 (2009).

[145] U. Strossner, J. P. Meyn, R. Wallenstein, P. Urenski, A. Arie, G. Roseman, J. Mlynek, S. Schiller, and A. Peters, "Single-frequency continuous-wave optical parametric oscillator system with an ultrawide tuning range of 550 to 2830 nm," J. Opt. Soc. Am. B **19**, 1419–1424 (2002).

[146] T. Petelski, R. S. Conroy, K. Bencheikh, J. Mlynek, and S. Schiller, "All-soild-state, tunable, single-frequency source of yellow light for high-resolution spectroscopy," Opt. Lett. **26**, 1013–1015 (2001).

[147] J. R. Schwesyg, C. R. Phillips, K. Ioakeimidi, M. C. C. Kajiyama, M. Falk, D. H. Jundt, K. Buse, and M. M. Fejer, "Suppression of mid-infrared light absorption in undoped congruent lithium niobate crystals," Opt. Lett. **35**, 1070–1072 (2010).

[148] K. A. Stankov and J. Jethwa, "A new mode-locking technique using a nonlinear mirror," Opt. Commun. **66**, 41–46 (1988).

[149] G. Cerullo, S. De Silvestri, A. Monguzzi, D. Segaka, and V. Magni, "Self-starting mode-locking of a CW Nd:YAG laser using cascaded second-order nonlinearities," Opt. Lett. **20**, 746–748 (1995).

[150] Y. F. Chen, S. W. Tsai, and S. C. Wang, "High-power diode-pumped nonlinear mirror mode-locked Nd:YVO$_4$ laser with periocailly-poled KTP," Appl. Phys. B **72**, 395–397 (2001).

[151] S. J. Holmgren, V. Pasiskevicius, and F. Laurell, "Generation of 2.8 ps pulses by mode-locking a Nd:GdVO$_4$ laser with defocusing cascaded Kerr lensing in periodically poled KTP," Opt. Express **13**, 5270–5278 (2005).

[152] H. Iliev, I. Buchvarov, S. Kurimura, and V. Petrov, "High-power picosecond Nd:GdVO$_4$ laser mode locked by SHG in periodically poled stoichiometric lithium tantalite," Opt. Lett. **35**, 1016–1018 (2010).

[153] S. J. Holmgren, A. Fragemann, V. Pasiskevicius, and F. Laurell, "Active and passive hybrid mode-locking of a Nd:YVO$_4$ laser with a single partially poled KTP crystal," Opt. Express **14**, 6675–6680 (2006).

[154] H. Iliev, D. Chuchumishev, I. Buchvarov, and V. Petrov, "Passive mode-locking of a diode-pumped Nd:YVO$_4$ laser by intracavity SHG in PPKTP," Opt. Express **18**, 5754–5762 (2010).

[155] Y. H. Liu, Z. D. Xie, S. D. Pan, X. J. Lv, Y. Yuan, X. P. Hu, J. Lu, L. N. Zhao, C. D. Chen, G. Zhao, and S. N. Zhu, "Diode-pumped passively mode-locked Nd:YVO$_4$ laser at 1342 nm with periodically poled LiTaO$_3$," Opt. Lett. **36**, 698–700 (2011).

[156] H. Iliev, I. Buchvarov, S. Kurimura, and V. Petrov, "1.34-μm Nd:YVO$_4$ laser mode-locked by SHG-lens formation in periodically-poled stoichiometric lithium tantalite," Opt. Express **19**, 21754–

21759 (2011).

[157] V. Y. Shur, "Domain nanotechnology in ferroelectric single crystals: lithium niobate and lithium tantalate family," Ferroelectrics **443**, 71–82 (2013).

[158] J. J. Li, Z. Y. Li, and D. Z. Zhang, "Second harmonic generation in one-dimensional nonlinear photonic crystals solved by the transfer matrix method," Phys. Rev. E **75**, 056606 (2007).

[159] P. Xu and S. N. Zhu, "Quasi-phase-matching engineering of entangled photons," AIP Adv. **2**, 041401 (2012).

[160] Y. B. Yu, Z. D. Xie, X. Q. Yu, H. X. Li, P. Xu, H. M. Yao, and S. N. Zhu, "Generation of three-mode continuous-variable entanglement by cascaded nonlinear interactions in a quasiperiodic superlattice," Phys. Rev. A **74**, 042332 (2006).

[161] X. Q. Yu, P. Xu, Z. D. Xie, J. F. Wang, H. Y. Leng, J. S. Zhao, S. N. Zhu, and N. B. Ming, "Transforming spatial entanglement using a domain-engineering technique," Phys. Rev. Lett. **101**, 233601 (2008).

[162] H. Y. Leng, X. Q. Yu, Y. X. Gong, P. Xu, Z. D. Xie, H. Jin, C. Zhang, and S. N. Zhu, "On-chip steering of entangled photons in nonlinear photonic crystals," Nat. Commun. **2**, 429 (2011).

[163] P. Xu, H. Y. Leng, Z. H. Zhu, Y. F. Bai, H. Jin, Y. X. Gong, X. Q. Yu, Z. D. Xie, S. Y. Mu, and S. N. Zhu, "Lensless imaging by entangled photons from quadratic nonlinear photonic crystals," Phys. Rev. A **86**, 013805 (2012).

[164] H. Jin, P. Xu, J. S. Zhao, H. Y. Leng, M. L. Zhong, and S. N. Zhu, "Observation of quantum Talbot effect from a domain-engineered nonlinear photonic crystal," Appl. Phys. Lett. **101**, 211115 (2012).

[165] Y. F. Bai, P. Xu, Z. D. Xie, Y. X. Gong, and S. N. Zhu, "Mode-locked biphoton generation by concurrent quasi-phase-matching," Phys. Rev. A **85**, 053807 (2012).

[166] S. E. Harris, "Chirp and compress: toward single-cycle biphotons," Phys. Rev. Lett. **98**, 063602 (2007).

[167] This work was supported by the National Natural Science Foundation of China (Nos. 61205140, 11274165, and 11021403), the Jiangsu Science Foundation (BK2011545), the State Key Program for Basic Research of China (Nos. 2010CB630703 and 2011CBA00205), and the PAPD.

Efficient Continuous Wave Blue Light Generation in Optical Superlattice LiNbO₃ by Direct Frequency Doubling a 978 nm InGaAs Diode Laser*

Ya-lin Lu, Yan-qing Lu, Jian-jun Zheng, Cheng-cheng Xue, Xiang-fei Cheng, and Gui-peng Luo

National Laboratory of Solid State Microstructures, and Center for Advanced Studies in Science and Technology of Microstructures, Nanjing University, Nanjing 210093, People's Republic of China

Nai-ben Ming

CCAST (World Laboratory), P.O. Box 8370, Beijing 100080, People's Republic of China and National Laboratory of Solid State Microstructures, Nanjing University, Nanjing 210093, People's Republic of China

First-order quasiphase matched blue light generation in a LiNbO₃ optical superlattice was performed by direct frequency doubling of a continuous wave 978-nm diode laser. 1.27 mW output power of blue light was obtained with an incidence power of 500 mW. The frequency conversion efficiency is 0.25%.

Quasiphase matching (QPM)[1] techniques allow the light to be polarized such that the material's nonlinearity is maximized, and also permit use of material for which birefringent phase matching is not possible, e.g., the second-harmonic generation(SHG) of blue light in LiNbO₃ crystals using d_{33}, the maximum nonlinear coefficient. Periodic reversals in the sign of the nonlinear coefficient to compensate for dispersion is the main technique to achieve QPM and has recently been demonstrated in bulk[2] as well as in waveguide devices.[3,4] Alternating ferroelectric domains have been achieved in LiNbO₃, LiTaO₃, and KTP by modulating the dopant concentration during growth,[5] indiffusing dopants[6] applying electric fields,[7] or by techniques using electron beams[8] or SiO₂ masks.[9]

In LiNbO₃, QPM allows use of input radiation polarized along the z axis, for which the largest nonlinear coefficient $d_{eff}=2d_{33}/\pi=20.9$ pm/V can be used.[10] SHG of blue light can be achieved in LiNbO₃ crystal through QPM. Thus, it makes the material attractive for frequency doubling of diode lasers to construct an all solid-state blue laser. The Czochral-ski method for growing crystals is one of the most important techniques to obtain LiNbO₃ (Ref. [10]) or LiTaO₃ (Ref. [11]) crystals with periodic ferroelectric domain structures(i.e., optical superlattice). Feng et al.[5,10,12] produced Czochralski-grown LiNbO₃ crystals doped with 0.5 – 1 wt % yttrium with domain thickness of 3.4 μm to

---

* Appl.Phys.Lett.,1996,69(12):1660

frequency double the 1.064 μm Nd:YAG laser line. The observed conversion efficiency increased quadratically with the crystal length as expected for perfect domain periodicity up to crystal length of about 1.36 mm, corresponding to 400 domains. For longer crystals the increase was linear, revealing domain-boundary position errors on the order of the coherence length. These position errors are probably caused by variations in growth speed due to thermal fluctuations during the Czochralski growth. In practice, the creation of the required, finely spaced domain with sufficiently accurate periodicity is a challenging task. One of the authors of this paper has grown $LiNbO_3$ crystals with its modulating period ranging from 2.0 to over 15.0 μm.[13] The continuity and the periodicity of the domain structures are good for SHG experiments. Picosecond radiation at 430 and 490 nm(Ref. [14]) light was generated in a 0.78 mm long and a 1.56 mm long samples with frequency doubling efficiency of 4.2% and 24.0%, respectively. Picosecond light as 385 nm was generated in a 0.98 mm long sample with conversion efficiency of 1.6% through third-order QPM.[15] For direct frequency doubling of a 810 nm GaAlAs diode laser, 0.35 mW of blue light was obtained for an incidence power of 250 mW.[16]

In this letter, we report harmonic generation of 490 nm in $LiNbO_3$ optical superlattice crystals by direct frequency doubling of a continuous wave(cw), 980 nm InGaAs diode laser. 1.27 mW blue light was obtained with an incidence power of 500 mW.

$LiNbO_3$ optical superlattice crystals were grown along the a axis by the Czochralski method, as reported in previous papers.[5,10,12] Samples were cut parallel to the periodic domain wall that is not exactly in the $b-c$ plane. Two samples(A and B), which have dimensions of $4 \times 4 \times 2.2$ mm³ (here 2.2 mm is the crystal's length in the direction of propagation, which is nearly along the a axis) and $4 \times 4 \times 2.0$ mm³ (2.0 mm is also the propagation length), respectively, have been chosen for our experiments. No antireflection coating was used on the two sample's light transmitting surfaces. The observed average modulation periods of the two samples, which were observed by an optical microscope on crystal's acid-etched b face, are 5.22 and 5.20 μm, respectively. In the two samples, the thickness of the positive domain laminae is nearly equal to the t of the negative(i.e., the average duty cycle of the domain grating is about 1/2). The average period fluctuations in the two samples are below 5%.

For characterizing the periodicity of the two samples, the relationship between SHG efficiency and fundamental wavelength in the two samples have been measured by using a tunable optical parametric oscillator (OPO) as fundamental source. Using first-order QPM, sample A has maximal SHG efficiency at

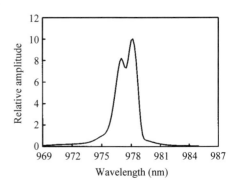

FIG. 1. The emission spectrum of the InGaAs diode laser.

fundamental wavelength of 975.8 nm when the fundamental light is at the normal incidence, and has an acceptance bandwidth of about 4.5 nm. Sample B has its maximal SHG efficiency at fundamental wavelength of 976 nm and has an acceptance bandwidth of about 4.0 nm. The large bandwidth of the two samples is primarily caused by the modulation period fluctuations along the growing direction (a axis). For frequency doubling the 978 nm light of our diode laser, the best QPM condition could be achieved by rotating the samples around their c axis(i.e., fundamental light is not at normal incidence).

The diode laser that we used was a SDL-6363-P1. The laser's output is continuous wave, multilongitudinal mode, and multitransverse mode. The diode laser has an emitting dimension of 100×1 μm^2 with beam divergence of θ_{+} 36°, $\theta_{\parallel} = 16°$, respectively. In its parallel far-field energy distribution, two peaks exist(i.e., the output of the laser shows two bright horizontal bars). The diode laser's emission spectrum is shown in Fig. 1. Two maxims at 977 and 978 nm are evident. The emission linewidth of the laser is about 3.0 nm, and fully falls within the acceptance bandwidth of the two samples. Figure 2 shows the experimental setup. The output of the diode laser is collected to be nearly a parallel beam by a lens with a focal length $f = 1$ cm, and is focused into the crystal by another lens with a focal length $f = 5$ cm. Three filers, which have high reflectivity at 978 nm($R > 99.6\%$), were used to block the transmitted infrared light. The generated blue light is focused by a lens into a detector. The full convergence angle of the beam onto the sample is about 20°. The spot of generated blue light is circular. The divergence angle of the generated blue light beam is about 8° and is much smaller than the beam divergence (about 20°) of the transmitted infrared light.

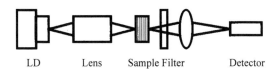

FIG.2. The experimental setup of the direct frequency doubling scheme.

The c axis of the crystal sample was parallel to the diode laser's polarization direction. The best QPM condition was achieved when the incidence angle of fundamental light is about 7°. Figure 3 shows the experimental results in samples A and B. We measured the relationship between the blue output power and the fundamental power(measured at the diode laser), in intervals of 50 mW, from 0 to 500 mW. The harmonic output increases nonlinearly with increasing fundamental power. When the incident fundamental power is 500 mW, the measured harmonic power generated in sample A and sample B reaches 1.27 and 1.0 mW, respectively, corresponding to a conversion efficiency of 0.25% and 0.20%, respectively. It is difficult to compare our observed efficiencies with the expected theoretical efficiency due to the difficulties in quantitative estimation of light beam quality. Our best obtained efficiency of 0.25% corresponds to about 2.27%/W cm and is lower than

the expected theoretical efficiency of 8%
W cm.[17] The expected efficiency is estimated
for the case of a continuous wave, optimally
focused, diffraction-limited, single-longitudinal-
mode EM_{00} beam. Better power at the blue
harmonic can be expected by improving the
quality of the fundamental radiation beam of the
diode laser through beam circularization and/or
astigmatism correction.

In conclusion, first-order, QPM, blue light
generation in a $LiNbO_3$ optical superlattice
crystal by direct frequency doubling of a cw
978 nm InGaAs diode laser has been
demonstrated. Second harmonic power 1.27 mW has been obtained at 489 nm.

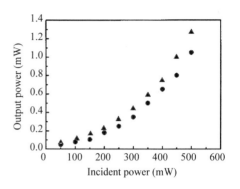

FIG. 3. The blue output power vs the fundamental power obtained both from sample A(symbol: ▲) and sample B(symbol: ●).

References and Notes

[1] J. A. Armstrong, N. Bloembergen, J. Ducuing, and P. S. Pershan, Phys. Rev. **127**, 1918(1962).
[2] D. H. Jundt, G. A. Magel, M. M. Fejer, and R. L. Byer, Appl. Phys. Lett. **59**, 2657(1991).
[3] K. Yamamoto, K. Mizuuchi, and T. Taniuchi, IEEE J. Quantum Electron. **28**, 1909(1992).
[4] C. J. van der Poel, J. D. Bierlein, J. B. Brown, and S. Colak, Appl. Phys. Lett. **57**, 2074(1990).
[5] N. B. Ming, J. F. Hong, and D. Feng, J. Mater. Sci. **17**, 1663(1982).
[6] E. J. Lim, M. M. Fejer, and R. L. Byer, Electron. Lett. **25**, 174(1989).
[7] A. Feisst and P. Koidl, Appl. Phys. Lett. **47**, 1125(1985).
[8] H. Ito, C. Takyu, and H. Inaba, Electron. Lett. **27**, 1221(1991).
[9] J. Webjorn, F. Laurell, and G. Arvidsson, IEEE Photonics Technol. Lett. **1**, 316(1989).
[10] D. Feng, N. B. Ming, J. F. Hong, and Y. N. Wang, Appl. Phys. Lett. **37**, 607(1980).
[11] W. S. Wang, Q. Zhou, and D. Feng, J. Cryst. Growth **79**, 706(1986).
[12] Y. H. Xue, N. B. Ming, and D. Feng, Acta Phys. Sin. **32**, 1515(1983).
[13] Y. L. Lu, L. Mao, and N. B. Ming, Appl. Phys. Lett. **59**, 516(1991).
[14] Y. L. Lu, Y. Q. Lu, and N. B. Ming, Appl. Phys. Lett. **68**, 2781(1996).
[15] Y. L. Lu, L. Mao, and N. B. Ming, Appl. Phys. Lett. **64**, 3029(1994).
[16] Y. L. Lu, L. Mao, and N. B. Ming, Opt. Lett. **19**, 1037(1994).
[17] D. Roberts, IEEE J. Quantum Electron. **28**, 2057(1992).
[18] This work was supported by a Ke-Li fellowship.

Femtosecond Violet Light Generation by Quasi-phase-matched Frequency Doubling in Optical Superlattice LiNbO₃ *

Yan-Qing Lu, Ya-Lin Lu, Chen-Chen Xue, Jian-Jun Zheng, Xiang-Fei Chen, and Gui-Peng Luo

National Laboratory of Solid State Microstructures, Nanjing University, Nanjing 210093, People's Republic of China

Nai-Ben Ming

CCAST(World Laboratory), P. O. 8730, Beijing 100080, and National Laboratory of Solid State Microstructures, Nanjing 210093, People's Republic of China

Bao-Hua Feng and Xiu-Lan Zhang

Optical Physics Laboratory, Institute of Physics, Chinese Academy of Sciences, Beijing 100080, People's Republic of China

We report the 390 and 385 nm violet light generation by frequency doubling of a tunable 90 fs, 82 MHz Ti: sapphire laser in two optical superlattice LiNbO₃ samples through third-order quasiphase matching (QPM). With the average incident infrared light power of 770 mW, 9.7 mW and 2.9 mW output second-harmonic lights were obtained without photorefractive damage. The QPM wavelength acceptance bandwidth measurement indicates that violet light generation with higher efficiency is possible.

With the development of the mode-locked femtosecond Ti: sapphire laser,[1] the generation of ultrashort pulses with high repetition rate has become possible in the near-infrared wavelength range.[2] However, for many applications, for example, experiments on molecules and wide-gap semiconductors, pulses at shorter wavelengths are required. Second-harmonic generation(SHG) in nonlinear optical crystals including LiNbO₃ and BBO represents a powerful technique to convert infrared pulses to the ultraviolet-blue spectral range. Up to date, a variety of frequency conversion schemes has been proposed and efficient blue lights generated. Among them, there are the directly frequency doubling,[3] intracavity,[4] and extracavity[5] SHG. However, all these schemes are based on the conventional phase-matching (PM) technique. To the best of our knowledge, the quasiphase-matching(QPM) technique[6] which is achieved in optical superlattice crystals (i.e., crystals with periodic nonlinear optical coefficient modulation) has not been applied in such an ultrashort pulse frequency conversion area. As the QPM technique allows

---

* Appl.Phys.Lett.,1996,69(21):3155

frequency doubling at room temperature over the crystal's entire transparency range using the largest nonlinear coefficient, the utility of a single material can be extended. For example, in LiNbO$_3$, due to the insufficiency of birefringence, PM condition cannot be satisfied when the fundamental wavelength is shorter than 1 μm, thus, the generation of blue or violet light is impossible; The maximum nonlinear coefficient d_{33}, which is about 7.5 times larger than the commonly used d_{31} cannot be used. Whereas in optical superlattice LiNbO$_3$ short wavelength light generation used d_{33} is possible, we may expect a QPM enhancement factor of $(d_{33}/d_{31})^2(2/\pi)^2 \approx 23$ in the harmonic generation process.[7] In our previous works, we have demonstrated the efficient blue or violet light generation by directly frequency doubling the CW diode laser or picosecond pulsed optical parametric oscillator.[8-11] For femtosecond short-wavelength light generation, the high peak power intensity of femtosecond infrared pulses must make the process of QPM SHG more efficient. On the other hand, the femtosecond pulses broadening arisen from the group-velocity mismatch (GVM) is a severe problem in the harmonic generation process, which requires a nonlinear crystal with high nonlinearity and shorter interaction length. Obviously, the QPM technique allows obtaining the same harmonic generation efficiency in a much thinner crystal, so the pulse broadening effect could be weakened. Thus, the optical superlattice crystal has great potentials in ultrashort harmonic generation. Furthermore, the possible higher photorefractive damage threshold caused by the periodic domain structure[12] is also beneficial for SHG of the high-frequency femtosecond Ti: sapphire amplifier.

In this letter we report the efficient femtosecond violet light SHG in two optical superlattice LiNbO$_3$ samples through third-order QPM by using a tunable femtosecond Ti: sapphire laser with high repetition rate. The QPM acceptance bandwidth was measured with a pulsed dye laser to analyze the experimental results.

An argon-pumped 90 fs, 82 MHz Ti:sapphire laser (Tsunami; Spectra-Physics) with a linewidth of about 8.0 nm was used as the fundamental light source. The diameter of the output infrared beam was 3 mm. After propagating through two convex lenses with their focusing lengths of 10 and 5 mm, the beam diameter was compressed to 1.5 mm and then normally injected into the sample. An infrared-cut filter was used to block the fundamental light and then let the passed harmonic light directly enter the detector.

Two optical superlattice LiNbO$_3$ samples grown by the Czochralski method as described earlier[13] were selected for the experiments. Their average modulation periods are 7.2 μm (sample 1) and 6.9 μm (sample 2) with the period fluctuation of 4.0% and 3.5%, respectively. Their thicknesses are 2.8 and 0.8 mm. Each sample has a 3×3 mm^2 ($b \times c$) light-transmitting face with the fundamental beam propagated in the uncoated a face. The polarization direction of the fundamental wave is along the c axis, just for satisfying the QPM condition.

After tuning the output wavelength of the Ti: sapphire laser, the optimum QPM wavelength of sample 1 was found to be 780 nm. Figure 1 shows the dependence of the output power of the 390 nm violet light on the input infrared light power with consideration of the violet light loss of the filter. Along with the increasing of the input power, the violet light power increases as a square relationship. When the infrared light power went up to 770 mW, 9.7 mW violet light generation was obtained without photorefractive damage in the crystal. The SHG conversion efficiency was 1.3%.

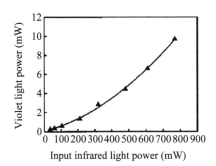

FIG. 1. The output power of the 390 nm violet light as a function of the fundamental light power in sample 1.

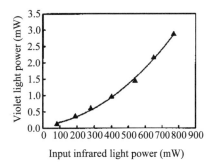

FIG. 2. The output power of the 385 nm violet light as a function of the input infrared light power of 770 nm infrared light in sample 2.

Figure 2 shows the experimental results in sample 2. The optimum fundamental light is at 770 nm and the 385 nm violet light was obtained. With the 770 mW incident infrared light, 2.9 mW violet light was generated, corresponding to the conversion efficiency of 0.4%. Although the thickness of this sample is only 0.8 mm and the QPM theory predicts that the conversion efficiency is proportional to the square of the sample thickness, from our experimental results,

$$(\eta/L^2)_{\text{sample 2}} : (\eta/L^2)_{\text{sample 1}} = 3.7 : 1 > 1,$$

where η is the conversion efficiency and L is the sample thickness. Furthermore, the thinner sample thickness of sample 2 also made the GVM smaller, so sample 2 is more suitable for the femtosecond light SHG. However, the reasons why the theory prediction has so large a deviation are complicated. We think that the better periodicity of sample 2 might be an influential factor, but the main reason must have something to do with the different wavelength acceptance bandwidth of the samples.

A tunable dye laser (model ND6000; Continnum Co.) with a pulsewidth of 8 ns and linewidth of 0.2 cm^{-1} was used as the fundamental light source for measuring the acceptance bandwidth. The experimental setup was similar to the femtosecond frequency doubling situation. Figure 3 shows the relationship between the energy of the generated violet pulse and the fundamental wavelength. The measured fundamental wavelengths with the highest conversion efficiency are also 780 nm (sample 1) and 770 nm (sample 2), which are the same as the results of the femtosecond SHG. From the figure, the wavelength acceptance bandwidth [full width at half-maximum (FWHM) of the wavelength tuning curve] of sample 1 is about 2.0 nm, which is greatly larger than the theoretical prediction of less than 1 nm according to Eq. (30) in Ref. [7]. As the equation is based on the ideal QPM condition (only consider the perfect periodic structure), the period fluctuation must be the major reason to the bandwidth expansion. Furthermore, the thickness of sample 1 is 3.5 times as much as that of sample 2 and their QPM wavelengths are nearly the same, the acceptance bandwidth of sample 2 should be also nearly 3.5 times as much as that of sample 1 according to the equation,[7] but the measured acceptance bandwidth of sample 2 was 6.4 nm, which is not in good accordance with the theoretical prediction. In fact, since the fundamental wavelength, the sample thickness, and the period fluctuation are all able to influence the wavelength acceptance bandwidth and even the detailed functional spatial form of the periodic fluctuation does something with the acceptance bandwidth, obtaining the acceptance bandwidth of a sample numerically is very difficult. However, whether in the case of sample 1 or sample 2, the measured acceptance bandwidth is smaller than the linewidth of the Ti: sapphire laser, thus, only a part of the fundamental light could be used for frequency doubling and a lot of fundamental light power was wasted. So the linewidth of the fundamental light and the acceptance bandwidth are two important factors for our consideration in the femtosecond frequency conversion process. Only if the linewidth is smaller than the sample's acceptance bandwidth can higher conversion efficiency be obtained. As we know, the femtosecond pulse usually has a wider linewidth, so it is beneficial for increasing the wavelength acceptance bandwidth of a sample. It is just the wider acceptance bandwidth of sample 2 that makes the $(\eta/L^2)_{\text{sample 2}}$ larger. To date, there have been some attempts to increase the QPM acceptance bandwidth, for example, using optimized domain grating[14] or the variable-spaced phase reversal technique.[15] Comparing to the unchangeable and the narrow acceptance bandwidth in the conventional PM SHG process, the width QPM bandwidth and its tunability seem very attractive. We hope that future acceptance bandwidth controlling techniques will make the QPM femtosecond SHG more efficient.

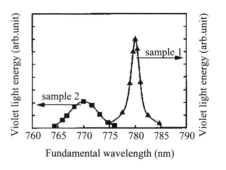

FIG.3. Second-harmonic light energy as a function of the fundamental wavelength in sample 1 and sample 2 measured by a pulsed dye laser.

In summary, we first demonstrated the femtosecond pulsed violet light SHG in optical superlattice $LiNbO_3$ crystals through third-order QPM. 9.7 mW 390 nm and 2.9 mW 385 nm lights were obtained. The conversion efficiencies were 1.3% and 0.4%, respectively. The acceptance bandwidth measurement indicated that SHG with higher conversion efficiency is possible.

References and Notes

[1] D. E. Spence, P. N. Kean, and W. Sibbett, Opt. Lett. **16**, 42 (1991).

[2] A. Stingl, M. Lenzner, Ch. Spielmann, F. Krausz, and R. Szipocs, Opt. Lett. **20**, 602 (1995).

[3] A. Nebel and R. Beigang, Opt. Lett. **16**, 1729 (1991).

[4] R. J. Ellingson and C. L. Tang, Opt. Lett. **17**, 343 (1992).

[5] S. Backus, M. T. Asaki, C. Shi, H. C. Kapteyn, and M. M. Murnane, Opt. Lett. **19**, 399 (1994).

[6] A. Amstrong, N. Bloembergen, J. Ducuing, and P. S. Pershan, Phys. Rev. **127**, 1918 (1962).

[7] M. M. Fejer, G. A. Magel, D. H. Jundt, R. L. Byer, IEEE J. Quantum Electron. **QE-28**, 2631 (1992).

[8] Y. L. Lu, L. Mao, S. D. Cheng, N. B. Ming, and Y. T. Lu, Appl. Phys. Lett. **59**, 516 (1991).

[9] Y. L. Lu, L. Mao, and N. B. Ming, Appl. Phys. Lett. **64**, 1092 (1994).

[10] Y. L. Lu, L. Mao, and N. B. Ming, Opt. Lett. **19**, 1037 (1994).

[11] Y. Q. Lu, Y. L. Lu, G. P. Luo, X. F. Cheng, C. C. Xue, and N. B. Ming, Electron. Lett. **32**, 336 (1996).

[12] G. A. Magel, M. M. Fejer, and R. L. Byer, Appl. Phys. Lett. **56**, 108 (1990).

[13] D. Feng, N. B. Ming, J. F. Hong, Y. Y. Shun, J. S. Zhu, Z. Yang, and Y. N. Wang, Appl. Phys. Lett. **37**, 607 (1980).

[14] J. Wu, C. Q. Xu, H. Okayama, M. Kauahara, T. Kondo, and R. Ito, Jpn. J. Appl. Phys. **33**, L1163 (1994).

[15] M. L. Bortz, M. Fujimura, and M. M. Fejer, Electron. Lett. **30**, 34 (1994).

[16] The authors would like to thank Dr. Jing-liang He and Dr. Yu-fei Kong of Institute of Physics, Chinese Academy of Sciences, for their help with the experiments.

Visible Dual-wavelength Light Generation in Optical Superlattice Er:LiNbO₃ through Upconversion and Quasi-phase-matched Frequency Doubling[*]

Jian-jun Zheng, Yan-qing Lu, Gui-peng Luo, Jing Ma, and Ya-lin Lu

National Laboratory of Solid State Microstructures, Nanjing University, Nanjing 210093, People's Republic of China

Nai-ben Ming

CCAST(World Laboratory), P.O. Box 8730, Beijing 100080, and National Laboratory of Solid State Microstructures, Nanjing University, Nanjing 210093, People's Republic of China

Jing-liang He and Zu-yan Xu

Institute of Physics, Chinese Academy of Science, Beijing 100080, People's Republic of China

Optical superlattice Er:LiNbO₃ was fabricated by inducing a periodic ferroelectric domain structure into the crystal during the growing process. Because of the combination of the nonlinear optical properties of LiNbO₃ and the spectral properties of Er^{3+}, the crystal can simultaneously emit the second harmonic light through quasi-phase matching and the green light through upconversion at room temperature. Pumped by infrared diode lasers, violet-and-green and blue-and-green light generation was demonstrated in two samples. The detailed absorption spectrum and emission spectrum of upconversion were measured. The possible physical mechanism was discussed.

Upconversion phenomena are being exploited for the development of short-wavelength solid state lasers, which have many technical applications including data storage, laser printing, underwater communications, and full color laser display. Erbium doped fibers were used to demonstrate the physical mechanism of two-photon as well as excited-state absorption under near-infrared diode laser excitation.[1-3] Room temperature green upconversion lasers were developed by pumping erbium-doped fibers with infrared lasers.[4-6] As for Er^{3+} doped bulk crystals, visible upconversion laser oscillation was also achieved at cryogenic temperatures.[7] On the other hand, the multi-wavelength laser has drawn much attention in recent years because of its applications in precise laser spectrum, laser radar, and nonlinear frequency conversion. Up to date, dual-wavelength laser oscillation has been demonstrated in several crystals including Nd:YAG[8-10] and Nd:YAP.[11-13] However, because of the difficulty of selecting the suitable laser crystal and designing the cavity mirrors so that the two spectral lines have the same oscillation

---

[*] Appl.Phys.Lett.,1998,72(15):1808

thresholds,[12] it is not easy to obtain the infrared dual-wavelength oscillation at the same time, let alone the visible dual-wavelength laser operation by frequency doubling.

In this letter, we first report on the fabrication of optical superlattice Er:LiNbO$_3$ (OSL ELN) which is able to simultaneously generate violet-and-green or blue-and-green light pumped by an infrared diode laser at room temperature by means of second-harmonic generation (SHG) and upconversion. The absorption and upconversion spectrums were measured and the phenomena of dual-wavelength light generation were demonstrated.

The optical superlattice (OSL) is a kind of man-made crystal, with its physical properties being modulated periodically in the span of several microns to several tens of microns, that is comparable with light wavelength. Because of the periodic structure, the OSL will exhibit some novel characteristics which cannot be found in ordinary materials.[14] For example, if the spontaneous polarization of LiNbO$_3$ (LN) is modulated periodically, the nonlinear optical coefficient will change its sign from the positive domains to the negative domains, so that the quasi-phase matching (QPM) technique can be achieved to compensate the phase velocity dispersion in frequency conversion applications.[15] A significant advantage of QPM is that any interaction within the transparency range can be noncritically phase matched at a specific temperature, even interactions for which birefringence phase matching is impossible. Another benefit is that the interacting waves can be chosen so that coupling occurs through the largest element of the $\chi^{(2)}$ tensor. In LN, QPM with all waves polarized parallel to the c axis yields a gain enhancement over the birefringence phase matched process of $(2d_{33}/\pi d_{31})^2 \approx 20$. In our laboratory, since the first demonstration of fabricating OSL LN by the Czochralski method in 1980,[16] we have grown various OSL LN crystals with different dopants and different modulation periods.[17,18] Up to date, CW, picosecond, and femtosecond visible light QPM SHG were obtained.[19—21] However, if we doped some erbium ions into the OSL LN, the crystal will combine the nonlinear optical properties of LN and the spectral properties of Er^{3+}. Pumped by adapt infrared light, the crystal is able to emit the upconversion light as well as the second harmonic light at the same time. Since the physical processes of SHG and upconversion are different and SHG has no pumping threshold, the complicated design of the cavity mirrors for obtaining ordinary dual-wavelength output becomes unnecessary. Thus it will be easier to obtain the visible dual-wavelength laser oscillation at the same time. Obviously, such a novel device will have many applications especially in the laser display area.

Our sample for SHG and spectral measurements is a Czochralski grown 4 mm×4 mm ×4 mm cubic OSL LN doped with 0.5 mol ‰ Er$_2$O$_3$. The average modulation period of the sample is 8.2 μm just corresponding to the 3rd-order QPM frequency doubling of 808 nm fundamental light. The two a faces of the sample were finely polished for light transmission.

Figure 1 shows the absorption spectrum of the OSL ELN sample measured at 300 K in the range of 300 – 1700 nm. The transmission rates of 808 nm light and 404 nm light are 88% and 90%, respectively. The spectrum appears in a sharp linelike structure, indicating that the spectral character of Er^{3+} ions in LN crystal is different from that of Er^{3+} doped glass fibers, in which only some peak envelopes can be observed.[21] In comparison with the spectra of Er^{3+} in YAG,[22,23] there are similarities in spectrum structure but dissimilarities in peak positions and peak intensities, indicating the effect of a different crystal field.

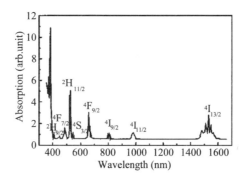

FIG. 1. Absorption spectrum of 0.5% Er^{3+} doped optical superlattice $LiNbO_3$ measured at 300 K. The corresponding energy levels of major absorption peaks were marked in the figure.

The diagram of the experimental setup for SHG and upconversion measurements is shown in Fig. 2. The output of an 808 nm GaAlAs diode laser was collimated and focused onto the a face of the sample. An infrared-cut filter was put behind the sample to block the transmitted fundamental light and pass the second harmonic light into the powermeter. A monochromator at the side of the sample was used to analyze the spectrum and a photomultiplier was used to receive the emitted light from the slit of the monochromator.

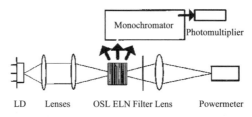

FIG. 2. Experimental setup for measuring the upconversion spectrum and second harmonic generation in optical superlattice $Er:LiNbO_3$.

In our experiments, when the 808 nm light injected into the sample, remarkable green fluorescence light and violet 404 nm violet second harmonic light were observed even with an incident power lower than 200 mW. With the increase of the output of the diode laser, the brightness of the two lights increased subsequently. When the input infrared power was 1.0 W, we got 0.8 mW stable violet SHG. The low conversion efficiency (0.08%) maybe due to several factors: for example the period fluctuation, the unpolarized incident infrared light, the 3rd-order QPM, the absorption of the infrared light, and the uncoated transmitted faces. Among them we believe that the period fluctuation is the most severe. In addition, the designed modulation period was calculated according to the refractive indices of the pure LN, not the doped Er:LN. Under the above condition, the fluorescence spectrum of upconversion was recorded as displayed in Fig. 3. It shows that the OSL ELN really possesses strong upconversion emission, at least for the green light of wavelength 500 – 570 nm. There are two twin-peak structures at this band, one at 547 nm due to the $^4S_{3/2}$–$^4I_{15/2}$ transition, the other at 523 nm due to the $^2H_{11/2}$–$^4I_{15/2}$ transition. The twin-peak structures of these two spectral lines are perhaps contributed by the perturbation of energy levels due to the small difference of crystal fields, since there are two different sites in LN crystal lattice, i.e., the Li site and the Nb site, which can be replaced by Er^{3+}. In addition to the green light, the OSL ELN also emitted blue and red fluorescences of upconversion excitation. The central peaks are at 460 and 661 nm which correspond to transitions between the states of $^4G_{9/2}$–$^4I_{13/2}$ and $^4F_{9/2}$–$^4I_{15/2}$, respectively. Among the four peaks, the 547 nm peak is the most intensive one. It is approximately six times larger than the 661 nm peak and 20 times larger than the 460 nm peak. That is why only green fluorescence can be observed with the naked eye.

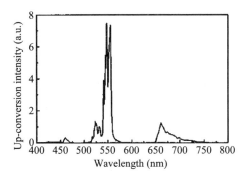

FIG. 3. Emission spectrum of upconversion pumped by an 808 nm GaAlAs diode laser.

For analyzing the physical process of upconversion, it is necessary to realize the relationship between the pumping power and the fluorescence intensity of upconversion. Figure 4 shows the variation of the 547 nm peak intensity as a function of the 808 nm laser power. Just as expected, the fluorescence intensity exhibits a square dependence on the pumping power indicating that the generation of green light is really a two-photon process. The possible detailed transition process may be as follows: Pumped by the 808 nm infrared light, the Er^{3+} is excited from the ground state to the $^4F_{9/2}$ state. Since the lifetime is sufficiently long, some of the ions in the $^4F_{9/2}$ level may absorb another 808 nm photon to populate the higher level $^2H_{9/2}$, which relaxes to the $^4S_{3/2}$ state, causing it to emit the 547 nm green light. Obviously, the above explanation is based on the mechanism of excited-

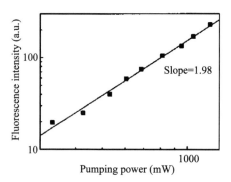

FIG. 4. **The intensity of emitted 547 nm green fluorescence of upconversion as a function of the pumped power.**

state absorption. However, the two-photon absorption process is also possible if there really exists some "virtual middle state" between the ground state and the high excited state or the Er^{3+} is able to absorb two photons simultaneously. Therefore investigating the mechanism of upconversion is needed.

Besides the violet-and-green dual-wavelength light generation, we also demonstrated the blue-and-green-dual-wavelength light emission in another OSL ELN sample with the modulation period of 5.3 μm for the first-order QPM of 980 nm light. By using a similar experimental setup and pumping the crystal with a 980 nm InGaAs diode laser, 490 nm second harmonic light and the green upconversion fluorescence were also achieved simultaneously.

In conclusion, we first demonstrated the visible dual-wavelength light generation in OSL ELN. Remarkable violet-and-green and blue-and-green lights were obtained through upconversion and QPM SHG at the same time. The detailed absorption and upconversion spectrums of upconversion were carefully measured. The possible physical mechanism was discussed. The phenomena demonstrated the possibility of constructing a novel kind of visible dual-wavelength coherent source that might be used in the laser display and other areas. Our future work is to increase the intensity of upconversion by increasing the concentration of Er^{3+} doping or by Yb^{3+} co-doping[24] and enhance the SHG by using the electric poling tenique[25] to fabricate the OSL ELN with better periodicity.

References and Notes

[1] S. Arahira, K. Watanabe, K. Shinozaki, and Y. Ogawa, Opt. Lett. **17**, 1679(1992).

[2] J. Thogersen, N. Bjerre, and J. Mark, Opt. Lett. **18**, 197 (1993).

[3] R. I. Laming, S. B. Poole, and E. J. Tarbox, Opt. Lett. **13**, 1084 (1988).

[4] J. F. Massicott, M. C. Brierley, R. Wyatt, S. T. Davey, and D. Szebesta, Electron. Lett. **29**, 2119 (1993).

[5] J. Y. Allain, M. Monerie, and H. Poignant, Electron. Lett. **28**, 111 (1992).

[6] T. J. Whitley, C. A. Millar, R. Wyatt, M. C. Brierly, and D. Szebesta, Electron. Lett. **27**, 1785

(1991).
[7] W. Lenth and R. M. Macfarlane, Opt. Photonics News **3**, 8 (1992).
[8] V. E. Nadtochev and O. E. Nanli, Sov. J. Quantum Electron. **19**, 444(1991).
[9] H. E. Tomaschke and G. A. Henderson, CLEO'90, 252 (1990).
[10] W. X. Lin, H. Y. Shen, R. R. Zeng, Y. P. Zhou, and G. F. Yu, Chin. J. Lasers **A21**, 334 (1994).
[11] F. Hanson and P. Poirier, J. Opt. Soc. Am. B **12**, 1311 (1995).
[12] H. Y. Shen, Chin. Phys. Lett. **7**, 174 (1990).
[13] H. Y. Shen, R. R. Zeng, Y. P. Zhou, G. F. Yu, C. H. Huang, Z. D. Zeng, W. J. Zhang, and Q. J. Ye, Appl. Phys. Lett. **56**, 1937 (1990).
[14] N. B. Ming, Y. Y. Zhu, and D. Feng, Ferroelectrics **106**, 935(1990).
[15] M. M. Fejer, G. A. Magel, D. H. Jundt, and R. L. Byer, IEEE J. Quantum Electron. **QE-28**, 2631 (1992).
[16] D. Feng, N. B. Ming, J. F. Hong, Y. S. Yang, J. S. Zhu, Z. Yang, and Y. N. Wang, Appl. Phys. Lett. **37**, 607 (1980).
[17] Y. L. Lu, Y. Q. Lu, C. C. Xue, and N. B. Ming, Appl. Phys. Lett. **68**, 1467 (1996).
[18] Y. L. Lu, Y. Q. Lu, X. F. Cheng, C. C. Xue, and N. B. Ming, Appl. Phys. Lett. **68**, 2781 (1996).
[19] Y. Q. Lu, Y. L. Lu, G. P. Luo, X. F. Cheng, C. C. Xue, and N. B. Ming, Electron. Lett. **32**, 336 (1996).
[20] Y. L. Lu, L. Mao, S. D. Cheng, N. B. Ming, and Y. T. Lu, Appl. Phys. Lett. **59**, 516 (1991).
[21] Y. Q. Lu, Y. L. Lu, C. C. Xue, J. J. Zheng, X. F. Chen, G. P. Luo, N. B. Ming, B. H. Feng, and X. L. Zhang, Appl. Phys. Lett. **69**, 3155(1996).
[22] E. Desurvire and J. R. Simpson, Opt. Lett. **15**, 547 (1990).
[23] H. Stange, K. Petermann, G. Huber, and E. W. Duczynski, Appl. Phys. B: Photophys. Laser Chem. **49**, 269 (1989).
[24] C. H. Huang, IEEE Photonics Technol. Lett. **9**, 599 (1997).
[25] S. N. Zhu, Y. Y. Zhu, Z. Y. Zhang, H. Shu, H. F. Wang, J. F. Hong, and C. Z. Ge, J. Appl. Phys. **77**, 5481 (1994).

Frequency Tuning of Optical Parametric Generator in Periodically Poled Optical Superlattice LiNbO₃ by Electro-optic Effect*

Yan-qing Lu, Jian-jun Zheng, Ya-lin Lu, and Nai-ben Ming

National Laboratory of Solid State Microstructures, Nanjing University,
Nanjing 210093, People's Republic of China

Zu-yan Xu

Institute of Physics, Chinese Academy of Science, Beijing 100080, People's Republic of China

Frequency tuning of optical parametric generators in periodically poled LiNbO₃ by applying a periodic electric field was demonstrated. Remarkable wavelength change was achieved. The dependence of the wavelength shift on the applied field shows a linear relationship. The tuning rates exceeding 3 nm/(kV/mm) were obtained. The phenomenon of dispersion in electro-optic tuning was predicted. Possible applications were discussed.

Recently, more and more research interests have been paid on a new artificial nonlinear material: the periodically poled optical superlattice LiNbO₃ (PPLN). Since the sign of the nonlinear optical coefficient of PPLN is modulated, the quasiphase-matching (QPM) technique can be used in frequency conversion applications instead of the conventional birefringent phase matching.[1] A significant advantage of the QPM technique is that any interaction within the transparency range of the material can be noncritically phase matched at a specific temperature by using the largest nonlinear coefficient.[2] In PPLN, the effective nonlinear coefficient reaches to 21.6 pm/V, which is much larger than many common nonlinear crystals including BBO and KTP.[2,3] Although there was concern that scattering or absorption induce by the periodic domain structure could contribute noticeable loss, careful measurement had shown that periodic poling adds no loss.[4] Since the intrinsic material loss of LiNbO₃ over most of its transparency range is negligible (absorption coefficient $\alpha \approx 0.002$ cm^{-1})[5,6] and even some research results showed that the PPLN has higher photorefractive damage threshold than single domain LiNbO₃,[7] thus the PPLN is really a good candidate for high performance optical parametric generators(OPG), which include optical parametric oscillators (OPO) and optical parametric amplifiers (OPA). In a 50-mm-long PPLN with the modulation period of 31 μm and with the pump intensity low to 1 GW/m² which can be achieved by compressing 10 W pump light to be a

---

* Appl.Phys. Lett., 1999,74(1):123

110-μm-diam beam), the calculated single pass parametric power amplification is

$$G(L) = \frac{|E_s(L)|^2}{|E_s(0)|^2} - 1 \approx \text{Sinh}^2\left[\sqrt{\frac{2\omega_s\omega_i d_Q^2 I_P}{n_s n_i n_p \varepsilon_0 c^3}} L^2\right] = 0.97, \qquad (1)$$

where d_Q is effective nonlinear coefficient of PPLN, I_p is the pump intensity, and L is the sample length. Comparing the results with that of other materials with the same sample length and pump intensity [e.g., for angle-tuned $LiNbO_3$ OPO at the nominal 47° phase-matching angle,[8] $G(L) \approx 0.044$; for type-II noncritically phase-matched KTP OPO,[9] $G(L) \approx 0.037$], the power gain of PPLN is so high that it makes the related devices more efficient. The pump thresh-old of an PPLN OPO can even be decreased to continuous wave (cw) level which is difficult to be achieved in ordinary crystals. In fact, in addition to various pulsed PPLN OPGs,[10—12] even the cw single resonant OPO pumped by a 1064 nm laser based on a 50-mm-long PPLN has already been developed by Bosenberg et al.[13] When the pump light was 13 W with the beam waist of 97 μm, the 1.25 W idler light at 3.25 μm and 0.36 W signal light at 1.57 μm were generated. (See Fig. 2 in Ref. [13]) However, almost all the devices above are tuned by changing the temperature, which is very slow and complicated. The reliability and stability are also not very good. These shortages limited the usage of the PPLN. It is very beneficial to find an alternative tuning method which is simple, rapid, and stable. Although the angle tuning has been widely used in conventional OPGs, it has not been applied in PPLN devices because of the beam deviation and Poynting vector walk off.[14] Electro-optical tuning of an OPG is also particularly attractive since it is nonmechanical and can be done very rapidly. This idea was firstly suggested in 1965[15] and demonstrated two years after.[16] A recent report by Ewbawk et al. showed that the electro-optic tuning range of up to ±44 nm can be achieved in $LiNbO_3$ with the applied voltage up to ±5 kV.[17] As for PPLN, to the best of our knowledge, the electro-optic tuning technique has not been proposed yet.

In this letter, OPG frequency tuning by applying a periodic electric field along the z direction of the PPLN sample was proposed. The detailed tuning properties were studied. Possible applications were discussed.

The PPLN sample fabricated by the electric poling method[4] usually has the thickness of about 0.3 - 0.5 mm along the z direction. The ferroelectric domain boundaries are at the y-z plane. In the typical QPM frequency conversion applications, all waves involved are the extraordinary waves. For simplifying the discussion, we assumed the pump wave normally injected into the crystal and propagated along the x axis, then produced the signal wave and idler wave along the same direction.

For a QPM OPG, the waves involved should satisfy the energy conservation condition and the QPM condition:

$$\Delta\omega = \omega_p - \omega_s - \omega_i = 0, \qquad (2)$$

$$\Delta k \cdot \Lambda = (k_p - k_s - k_i) \cdot (l_p + l_n) = 2m\pi (m=1,2,3...), \qquad (3)$$

where Δk is the wave vector mismatch, ω is the frequency, k is the wave vector, and subscripts p, s, and i represent the pump, signal, and idler wave, respectively. Λ is the modulation period which is the sum of the positive domain thickness l_p and the negative domain thickness l_n. If an electric field E is applied along the z axis, the new refractive index of the extraordinary light is

$$n'_e = n_e - \frac{1}{2} n_e^3 \gamma_{33} E, \qquad (4)$$

where $\gamma_{33} = 30.9$ pm/V[18] is the electro-optic coefficient of LiNbO$_3$. Since all the waves involved are extraordinary waves, the wave vectors of the pump, signal, and idler wave were changed subsequently

$$k'_{p,s,i} = \frac{1}{c}\omega_{p,s,i} n'_{p,s,i} = \frac{1}{c}\omega_{p,s,i}\left(n_{p,s,i} - \frac{1}{2} n_{p,s,i}^3 \gamma_{33} E\right). \qquad (5)$$

However, just as the nonlinear optical coefficient, the electro-optic coefficient also has the different signs in different domains. Thus the wave vector in positive domains is different from that in negative domains. In such a inhomogeneous media, the revised QPM condition[19] is

$$\Delta k' l_p + \Delta k'' l_n = 2m\pi \,(m = 1, 2, 3 \ldots), \qquad (6)$$

where $\Delta k'$ and $\Delta k''$ are the wave vector mismatch at the positive domain and at the negative domain, respectively, when applying the electric field. Assuming the domains have the same thickness $l_p = l_n = \Lambda/2$ and considering the different signs of γ_{33} in different domains, Eq. (6) becomes

$$(\Delta k' + \Delta k'')\frac{\Lambda}{2} = \Delta k \Lambda = 2m\pi \,(m = 1, 2, 3 \ldots), \qquad (7)$$

where Δk is the unperturbed wave vector mismatch. Equation (7) is just the same as the QPM condition without the electric field, which means the refractive indices of the signal wave and the idler wave do not need to change. Thus the electro-optic tuning cannot be accomplished.

A method that can solve the problem is applying a specific periodic electric field on the crystal. As we know, in the electric poling process, a periodically patterned electrode and a flat electrode were deposited on the $\pm z$ surface of the single domain wafer to make the spontaneous polarization be selectively reversed.[4] In the frequency conversion process, by using the same electrodes, we can apply the electric field only on one kind of domain, e.g., the positive domain. Under this circumstance, the QPM condition is

$$\Delta k' l_p + \Delta k l_n = \Delta k \Lambda - \frac{\Lambda}{4c}\gamma_{33} E(\omega_p n_p^3 - \omega_s n_s^3 - \omega_i n_i^3) = 2m\pi\,(m=1,2,3\ldots). \qquad (8)$$

To maintain the efficient OPG conversion, the energy conservation and the QPM condition should be satisfied which caused frequency shifts of the signal wave and the idler wave. Thus the electro-optic frequency tuning of a PPLN OPG is achieved. The detailed tuning curve can be obtained by solving Eqs. (2) and (8).

Chapter 8 Engineered Quasi-phase-matching for Laser Techniques

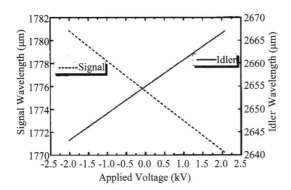

FIG. 1. Wavelength tuning for signal and idler waves at room temperature as a function of applied voltage in a 1064 nm pumped OPG based on a PPLN with the modulation period of 31 μm.

Figure 1 shows the calculated electro-optic wavelength tuning curve as a function of applied voltage on an OPG based on a typical 0.5-mm-thick PPLN with the modulation period of 31 μm, which is just the same as that in Ref. [10]. The wavelength of the pump light is 1064 nm. From the figure, the wavelength shifts of both the signal wave and the idler wave exhibit a near-linear dependence on the applied field. A tuning bandwidth of ± 10 nm was achieved in the idler wavelength for an applied voltage of ± 1.7 kV. The tuning rate (defined as the wavelength shift normalized by the applied electric field) is about 3 nm/(kV/mm). The above results are obtained at the room temperature $T=300$ K. When temperature changes (e.g., owing to optical absorption) the zero-field wavelength of the OPG changes subsequently. However, the electro-optic frequency tuning can also be achieved. The results in the same crystal are plotted in Fig. 2. Near-linear dependence of wavelength shift on applied electric field was observed at each temperature. However, the slopes of these lines differed slightly. The temperature dispersion of the electro-optic tuning rate is given in Fig. 3, where the data are expressed in the wave number so that the signal and the idler tuning rates are identical.

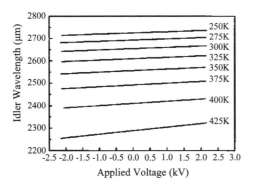

FIG. 2. Idler wavelength tuning as a function of applied voltage at different working temperatures.

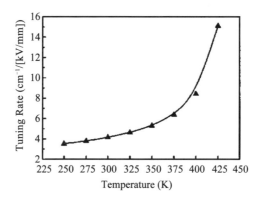

FIG.3. Dependence of signal(or idler) tuning rate on working temperature.

Besides OPG, the electro-optic tuning technique may be extended to other frequency conversion applications. For example, in an intracavity frequency doublor[20] or a self-frequency doublor[21] based on PPLN, the periodic electric field can influence both the second harmonic generation (SHG) power and the fundamental light output. The ratio of the fundamental light power and the SHG power is dependent on the intensity of the applied field so that a novel dual-wavelength source can be obtained. If applying this technique on a PPLN OPG or difference frequency generator[22] which has the output wavelength at 1.55 μm, the different intensity of the electric field is just corresponding to different output near the communication window, which means the multichannel wavelength conversion is possible. The most important usage of such a source is in the wavelength division multiplexing (WDM) optical networks.

In addition to above usage, another application is the possibility of constructing a novel ultrashort light source. As we know, the high peak power ultrashort pulses are obtained by compressing after amplifying the frequency-chirped laser pulses. The chirping is generated by complicated process, which makes the devices very expensive. However, in a PPLN OPA, if we applied a periodic electric field as we described when the pump pulse arrived in the crystal, the output signal and idler wave were modulated by the electric field. If the field was changed linearly, the wavelength output of the light formed a linear frequency sweep across the pulse, i.e., frequency chirping, which means we can even get tunable chirped pulses by using this technique. The obtained chirped pulse then might be compressed to generate ultrashort pulse directly. For example, from Fig. 1, applying a linearly electric pulse from -1.7 to 1.7 kV with the same pulse width of the pumping light (e.g., several nanoseconds) will result in a linear chirping with the span of 20 nm. The minimum pulse width that can be compressed is $\tau = 1/\Delta\omega = \lambda^2/(2\pi C\Delta\lambda) \approx 190$ fs.[23]

Although the electro-optic tuned PPLN devices have many applications, it will be more practical if the applied field can be decreased or the tuning range can be enlarged by choosing more adapt materials. Perhaps $Sr_{1-x}Ba_xNb_2O_6$ (SBN) is a good alternative. Its electro-optic coefficient is about 45 times larger than $LiNbO_3$ and their nonlinear coefficient were in the same order,[24] which means that the tuning rate may increase greatly, since the periodic domain structure can also be formed in SBN.[25] If this technique is applied in a periodically poled SBN, perhaps the performance of the devices will be better.

In conclusion, we proposed an electro-optic tuning technique that can tune the output wavelength of PPLN OPG quickly by applying a periodic electric field on the crystal. The dependence of the wavelength shift on the intensity of the applied electric field of an PPLN sample shows a nearlinear relationship. The tuning rates excess 3 nm/(kV/mm) are obtained. The phenomenon of dispersion in electro-optic tuning was predicted.

References and Notes

[1] D. Feng, N. B. Ming, J. F. Hong, Y. S. Yang, J. S. Zhu, Z. Yang, and Y. N. Wang, Appl. Phys. Lett. **37**, 607 (1980).

[2] M. M. Fejer, G. A. Magel, D. H. Jundt, and R. L. Byer, IEEE J. Quantum Electron. **28**, 2631 (1992).

[3] V. Druneri, S. D. Butterworth, and D. C. Hanna, Appl. Phys. Lett. **69**, 1029 (1996).

[4] L. E. Myers, R. C. Eckardt, M. M. Fejer, R. L. Byer, W. R. Bosenberg, and J. W. Pierce, J. Opt. Soc. Am. B **12**, 2102 (1995).

[5] G. D. Boyd, R. C. Miller, K. Nassau, W. L. Bond, and A. Savage, Appl. Phys. Lett. **5**, 234 (1964).

[6] D. J. Gettemy, W. C. Harker, G. Lindholm, and N. P. Barnes, IEEE J. **24**, 2231(1988).

[7] M. Taya, M. C. Bashaw, and M. M. Fejer, Opt. Lett. **21**, 857 (1996).

[8] S. J. Brosnan and R. L. Byer, IEEE J. Quantum Electron. **15**, 415 (1979).

[9] J. A. C. Terry, Y. Cui, Y. Yang, W. Sibbett, and M. H. Dunn, J. Opt. Soc. Am. B **11**, 758 (1994).

[10] L. E. Myers, G. D. Miller, R. C. Eckardt, M. M. Fejer, R. L. Byer, and W. R. Bosenberg, Opt. Lett. **20**, 52 (1995).

[11] M. L. Bortz, M. A. Arbore, and M. M. Fejer, Opt. Lett. **20**, 49 (1995).

[12] V. Pruneri, S. D. Butterworth, and D. C. Hanna, Appl. Phys. Lett. **69**, 1029 (1996).

[13] W. R. Bosenberg, A. Drobshoff, J. I. Alexander, L. E. Myers, and R. L. Byer, Opt. Lett. **21**, 713 (1996).

[14] L. E. Myers, R. C. Eckardt, M. M. Fejer, R. L. Byer, and W. R. Bosenberg, Opt. Lett. **21**, 591 (1996).

[15] J. A. Giordmaine and R. C. Miller, Phys. Rev. Lett. **14**, 973 (1965).

[16] L. B. Kreuzer, Appl. Phys. Lett. **10**, 336 (1967).

[17] M. D. Ewband, M. J. Rosker, and G. L. Bennett, J. Opt. Soc. Am. B **14**, 666 (1997).

[18] A. Yariv and P. Yeh, *Optical Waves in Crystals*(Wiley, New York, 1984), p. 232.

[19] C. Q. Xu, H. Okayama, and M. Kawahara, IEEE J. Quantum Electron. **31**, 981(1995).

[20] A. Harada, Y. Nihei, Y. Okazaki, and H. Hyuga, Opt. Lett. **22**, 805 (1997).

[21] Y. L. Lu, Y. Q. Lu, C. C. Xue, and N. B. Ming, Appl. Phys. Lett. **68**, 1467 (1996).

[22] C. Q. Xu, H. Okayama, and T. Kamijoh, Jpn. J. Appl. Phys., Part 2 **34**, L1543(1995).

[23] A. Yariv and P. Yeh, *Optical Waves in Crystals*(Wiley, New York, 1984), p. 308.

[24] A. Yariv and P. Yeh, *Optical Waves in Crystals*(Wiley, New York, 1984), p. 515.

[25] Y. Y. Zhu, J. S. Fu, R. F. Xiao, and G. K. L. Wong, Appl. Phys. Lett. **70**, 1793 (1997).

Electro-optic Effect of Periodically Poled Optical Superlattice LiNbO₃ and Its Applications[*]

Yan-Qing Lu, Zhi-Liang Wan, Quan Wang, Yuan-Xin Xi, and Nai-Ben Ming

National Laboratory of Solid State Microstructures, Nanjing University,
Nanjing 210093, People's Republic of China

The electro-optic effect of periodically poled optical superlattice LiNbO₃ (PPLN) was studied. Because of the periodic electro-optic (EO) coefficient, the reciprocal vector of the periodic structure can be used to compensate for the phase mismatch between the ordinary and extraordinary waves, which is similar to the nonlinear optical frequency conversion process. If the quasi-phase-matching condition is satisfied, polarization of a light propagated in PPLN can rotate linearly with the applied electric field, which shows that PPLN may be used as a precise spectral filter or an EO switch.

In recent years, more and more research attention has been paid to periodically poled optical superlattice LiNbO₃ (PPLN) because of its outstanding nonlinear optical properties.[1-3] The origin of the attractive characteristics is the periodic nonlinear coefficient caused by the periodic domain structure, thus the reciprocal vector may compensate the phase mismatch during the frequency conversion processes. This technique, called quasi-phase-matching (QPM), has many advantages compared to the conventional birefringence phase matching.[4-6] To date, various related devices have been demonstrated.[7-9] However, besides the nonlinear coefficient, other third-rank tensors are also modulated periodically due to the periodic ferroelectric domains in PPLN. Among them, there are the piezoelectric coefficient and electro-optic (EO) coefficient.[9-11] As a consequence, it is natural to ask the question: what will happen if the EO coefficient modulation is considered? Answering this question is important in the interest of fundamental physics and also has practical applications.

In this letter, we demonstrate that if an electric field is applied along the Y-axis of PPLN, coupling between the extraordinary wave and ordinary wave is established. Under the QPM condition, the polarization of light propagating in the crystal could be modulated by the applied electric field. Based on this effect, a novel EO filter may be developed.

LiNbO₃(LN) is a ferroelectric crystal with the symmetry of 3 m. In the negative domain, the crystal structure rotates 180° about the X axis, thus the electro-optic coefficients change subsequently under this operation. It is easy to demonstrate that all

---

[*] Appl.Phys.Lett.,2000,77(23):3719

elements of the electro-optic tensor have different signs in different domains.

Figure 1 shows the geometrical arrangement for studying the EO effect. In the presence of an external field along the Y axis, the index ellipsoid deforms to make the Y and Z axes rotate a small angle

$$\theta \approx \frac{\gamma_{51} E}{(1/n_e^2) - (1/n_0^2)}$$

about the X axis,[12] where E is the field intensity; γ_{51} is the EO coefficient; n_0 and n_e represent the refractive indices of the ordinary wave and extraordinary wave, respectively.

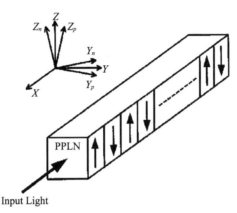

FIG.1. Experimental setup for studying the electro-optic effect of PPLN. X, Y, and Z represent the principal axes of the original index ellipsoid and $Y_{p,n}$, $Z_{p,n}$ are the perturbed principal axes of the positive domains and the negative domains, respectively. The arrows inside the PPLN indicate the spontaneous polarization directions.

With consideration of the periodic electro-optic coefficients, the dielectric constant of PPLN with an electric field along the Y axis thus can be written as

$$\varepsilon = \varepsilon(0) + \Delta\varepsilon f(x), \tag{1}$$

where

$$\varepsilon(0) = \varepsilon_0 \begin{bmatrix} n_0^2 & 0 & 0 \\ 0 & n_0^2 & 0 \\ 0 & 0 & n_e^2 \end{bmatrix}$$

is the original dielectric tensor,

$$\Delta\varepsilon = -\varepsilon_0 \gamma_{51} E n_0^2 n_e^2 \begin{bmatrix} 0 & 0 & 0 \\ 0 & 0 & 1 \\ 0 & 1 & 0 \end{bmatrix}$$

is the dielectric tensor change. In PPLN, a factor $f(x)$ should be included where

$$f(x) = \begin{cases} +1 & \text{if } x \text{ is in the positive domains} \\ -1 & \text{if } x \text{ is in the negative domains.} \end{cases} \tag{2}$$

The change of the dielectric tensor could be viewed as a disturbance, thus the coupled wave equations of the ordinary and extraordinary waves may be obtained as

$$\begin{cases} dA_2/dx = -iKA_3\exp(i\Delta\beta x) \\ dA_3/dx = -iK^*A_2\exp(-i\Delta\beta x) \end{cases}, \quad (3)$$

with

$$\Delta\beta = (\beta_2 - \beta_3) - G_m, \quad G_m = \frac{2\pi m}{\Lambda}$$

$$K = -\frac{\omega}{2c}\frac{n_0^2 n_e^2 \gamma_{51} E}{\sqrt{n_0 n_e}}\frac{i(1-\cos m\pi)}{m\pi}, (m=1,3,5...)$$

where A_2 and A_3 are the amplitudes of the ordinary wave and the extraordinary wave, respectively; β_2 and β_3 are the corresponding wave vectors. G_m is the mth reciprocal vector, Λ is the modulation period that is equal to twice domain thickness L if the duty cycle is 50%. Assuming the input light is an extraordinary wave by putting a vertical polarizer in front of the sample, the initial condition at $x=0$ is given by $A_3(0)=1$, $A_2(0)=0$. For studying the conversion from the extraordinary wave to ordinary wave, another horizontal polarizer should be put at the back of the sample as an analyzer. If the field is not intensive or the crystal length is short, the coupling between the extraordinary wave and the ordinary wave will be weak. In this case, the weak coupling approximation $A_3(x)=A_3(0)=1$ may be used and the power conversion efficiency from the extraordinary wave to the ordinary wave, i.e., the transmission for the extraordinary wave of this filter, is

$$T = \left|\frac{A_2(x)}{A_3}\right|^2 = K^2 x^2 \left(\frac{\sin(\Delta\beta \cdot x/2)}{\Delta\beta \cdot x/2}\right)^2. \quad (4)$$

This expression is similar to that of the QPM frequency conversion efficiency,[13] thus the research results for the QPM frequency conversion may be helpful in the study of the EO effect. From Eq. (4), the maximum conversion is achieved when $\Delta\beta = (\beta_2-\beta_3) - G_m = 0$, which means the reciprocal vector may also compensate for the wave vector mismatch. Similar to QPM frequency conversion, this condition could be called a QPM condition. Defining the coherence length $L_c = \lambda/2(n_0 - n_e)$, the QPM condition is satisfied for a given light of wavelength λ if each domain thickness is L_c or its multiple.

In the general case, for a given static field, the weak coupling approximation cannot be used, the transmission rate is thus obtained as

$$T = \left|\frac{A_2(x)}{A_3(0)}\right|^2 = |K|^2 \frac{\sin^2(Sx)}{S^2},$$

where

$$S^2 = |K|^2 + \left(\frac{\Delta\beta}{2}\right)^2. \quad (5)$$

Besides the QPM condition, the dynamical condition $|K|x = [(2u+1)\pi/2](u=0,1,2,...)$ should also be satisfied for 100% conversion to occur. Figures 2 and 3 show the influence of these two conditions on the transmission efficiency. For a sample with the domain thickness of 10.31 μm and the period of 500, the first-order QPM condition

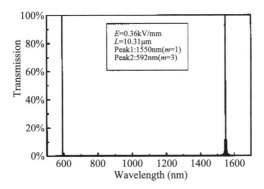

FIG.2. Calculated electro-optic transmission spectrum of a PPLN filter.

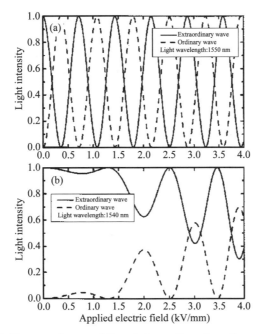

FIG.3. Power exchange relations between the extraordinary wave and the ordinary wave when (a) the QPM condition is satisfied and (b) the QPM condition is not satisfied. The solid line and the dashed line correspond to the intensity-varying curve of the two waves along the propagation direction, respectively.

corresponds to the wavelength of 1550 nm. From Fig. 2, if the dynamical condition is satisfied by applying a 0.36 kV/mm field, there is a transmission peak at 1550 nm that corresponds to 100% conversion from the extraordinary wave to the extraordinary wave. Besides this peak, there is another one at 592 nm that is due to the phase compensation of the third-order reciprocal vector. The corresponding concept in the frequency conversion process is the mth-order QPM.[13] The width of the 592 nm peak is narrower than that of the 1550 nm peak. In fact, the calculated transmission bandwidth for mth-order QPM is

$\Delta\lambda_{1/2} \approx 1.60(\lambda/2mN)$, where λ is the center wavelength and N is the period number. Since increasing m and N may decrease bandwidth, a precise spatial filter could be achieved. The influence of the dynamical condition may be obtained by studying the relation between the transmission and the applied electric field. Figure 3 shows the results. The power of the two waves is exchanged periodically in a sinusoidal fashion if the QPM condition is satisfied, as shown in Fig. 3(a). This effect may be useful for EO modulation or switching. On the other hand, the power relation of the two waves is complicated if the QPM condition is not satisfied, which is shown in Fig. 3(b).

From the theoretical results above, the QPM condition governs the passing frequency for a spectral filter based on PPLN. If the QPM condition is satisfied, the dynamic condition determines the transmission ratio of the passing light. Only when the dynamic condition is also satisfied, the polarization of the input extraordinary wave can rotate 90° and totally pass the analyzer. As for the polarization status changing of the light in the crystal along its propagation route, a simple theoretical analysis can give the details.

As we know, the principal axis X remains unchanged while the Y and Z axes rotate a small angle θ about the X axis after applying the field. The azimuth angle of the new Z axis thus rocks right and left from $+\theta$ to $-\theta$ successively due to the periodic EO coefficient, assuming each domain thickness is L_c to satisfy the QPM condition. In this case, each domain acts as a half-wave plate. For a 632.8 nm He-Ne laser, L_c is 3.74 μm which is achievable with current fabrication techniques. For an input extraordinary wave that is polarized along the Z axis, it will be polarized at $\psi = 2\theta$ after passing through the first domain. The second domain is oriented at angle $-\theta$, making an angle of 3θ with respect to the incoming polarization. At the output face of this domain, the polarization will be rotated by 6θ and oriented at azimuth angle 4θ. The final azimuth angle after N periods is $\psi = 4N\theta$, which produces a rotation of polarization. The solid line in Fig. 4 shows the calculated polarization rotation angle of the He-Ne laser as a function of the applied voltage in a sample with 300 domains. The domain thickness is 3.74 μm thus satisfying the QPM condition. From this figure, the polarization rotation angle is proportional to the intensity of the applied field. If an analyzer is employed, the transmission rate of this filter should exhibit a sinusoidal relationship with the applied field, which agree well with Fig. 3(a). The polarization of the input light can be rotated 90° with an applied electric field of 1.2 kV/mm. In this case, the input light can totally pass through the filter.

To verify these predictions, a PPLN crystal with an average domain thickness of 4.2 μm and the period fluctuation of less than 6% was used to do the experiment. The sample was fabricated using the Czochralski method.[14,15] The total domain number is 300. When a He-Ne laser was passed through the sample, the rotation angle increased with the applied field. A polarization rotation angle of 16.5° was obtained when the field was 1 kV/mm as shown in Fig. 4. However, for a single domain LN crystal with the same thickness, no remarkable polarization rotation was observed, which means that it is the periodic structure that is responsible for the rotation. However, perhaps because of the deviation of domain thickness from L_c and domain thickness fluctuation, the measured rotation angle was smaller than the theoretical prediction. One method that may improve the sample quality is using the electric-poling technique instead.[6] For a commercial available 5-cm-long PPLN, only a 27.6 V/mm field may make the polarization rotate by 90°, which is very attractive.

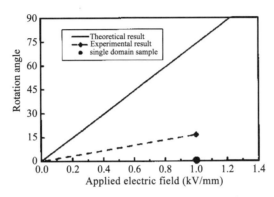

FIG. 4. Polarization rotation angle as a function of the applied electric field. The solid line and the dashed line correspond to the theoretical result and experimental result, respectively. The experimental result on a single domain LN sample is also displayed in the figure with a round dot.

References and Notes

[1] S. Bains, Laser Focus World **34**, 16 (1998).

[2] G. A. Magel, M. M. Fejer, and R. L. Byer, Appl. Phys. Lett. **56**, 108(1990).

[3] J. J. Zheng, Y. Q. Lu, G. P. Luo, J. Ma, Y. L. Lu, N. B. Ming, J. L. He, and Z. Y. Xu, Appl. Phys. Lett. **72**, 1808(1998).

[4] L. E. Myers, R. C. Eckardt, M. M. Fejer, R. L. Byer, W. R. Bosenberg, and J. W. Pierce, J. Opt. Soc. Am. B **12**, 2102(1995).

[5] V. Druneri, S. D. Butterworth, and D. C. Hanna, Appl. Phys. Lett. **69**, 1029 (1996).

[6] S. N. Zhu, Y. Y. Zhu, and N. B. Ming, Science **278**, 843 (1997).

[7] Y. Q. Lu, Y. L. Lu, C. C. Xue, J. J. Zheng, X. F. Chen, G. P. Luo, N. B. Ming, B. H. Feng, and X. L. Zhang, Appl. Phys. Lett. **69**, 3155(1996).

[8] M. Nakamura, T. Tsunekawa, H. Taniguchi, and K. Tadamoto, Jpn. J. Appl. Phys., Part 2 **38**, L1177 (1999).

[9] N. O'Brien, M. Missey, P. Powers, V. Dominic, and K. L. Schepler, Opt. Lett. **24**, 1750(1999).

[10] Y. Q. Lu, J. J. Zheng, Y. L. Lu, N. B. Ming, and Z. Y. Xu, Appl. Phys. Lett. **74**, 123 (1999).

[11] Y. Q. Lu, Y. Y. Zhu, Y. F. Chen, S. N. Zhu, N. B. Ming, and Y. Y. Feng, Science **284**, 1822(1999).

[12] J. F. Nye, *Physical Properties of Crystals* (Oxford, Oxford, 1985).

[13] M. M. Fejer, G. A. Magel, D. H. Jundt, and R. L. Byer, IEEE J. Quantum Electron. **28**, 2631(1992).

[14] Y. L. Lu, Y. Q. Lu, X. F. Chen, G. P. Luo, C. C. Xue, and N. B. Ming, Appl. Phys. Lett. **68**, 2642 (1996).

[15] V. Bermúdez, M. D. Serrano, P. S. Dutta, and E. Diéguez, J. Cryst. Growth **203**, 179(1999).

[16] This work is supported by the State Key Program for Basic Research of China, the National Natural Science Foundation Project of China (Contract No. 69708007), and the National Advanced Materials Committee of China.

High-power Red-green-blue Laser Light Source Based on Intermittent Oscillating Dual-wavelength Nd:YAG Laser with a Cascaded LiTaO$_3$ Superlattice[*]

X. P. Hu,[1] G. Zhao,[1] Z. Yan,[1] X. Wang,[1] Z. D. Gao,[1] H. Liu,[1] J. L. He,[2] and S. N. Zhu[1]

[1] *National Laboratory of Solid State Microstructures and Department of Physics, Nanjing University, Nanjing, 210093, China*

[2] *Institute of Crystal Materials, Shandong University, Jinan, 250100, China*

We demonstrate a high-power red-green-blue laser source based on the quasi-phase-matching and intermittent oscillating dual-wavelength laser technique. A cascaded LiTaO$_3$ superlattice was used to achieve the generation of red light at 660 nm, green light at 532 nm, and blue light at 440 nm to obtain the output of red-green-blue laser light from a diode-side-pumped Q-switched intermittent oscillating dual-wavelength Nd:YAG laser. The average output power of red-green-blue of 1.01 W was achieved under the total fundamental power of 5.1 W, which corresponds to the conversion efficiency of 20%.

Red, green, and blue(RGB) are the three elemental colors in the visible world, and most visual colors can be obtained by a weighted combination of these three colors. Hence RGB's three elemental colors can be used in projection displays and other high-tech applications. Diode-pumped infrared solid-state lasers based on an Nd^{3+} ion provide an excellent possibility to develop such devices when combined with nonlinear frequency conversion to the visible. An Nd^{3+} ion has multiple allowed transitions departing from the metastable level $^4F_{3/2}$ to the lower-lying-energy Stark sublevels $^4I_{13/2}$, $^4I_{11/2}$, and $^4I_{9/2}$, leading to potential laser radiations ~1.3, 1.06, and 0.94 μm, respectively. By frequency doubling the lines ~1.3 and 1.06 μm, and by frequency tripling the line ~1.3 μm, we can obtain simultaneous RGB generation, as several works have reported previously[1-3].

As is well known, for the Nd^{3+} ion, the two transitions, $^4F_{3/2}$–$^4F_{11/2}$ and $^4F_{3/2}$–$^4F_{13/2}$, share the same upper level $^4F_{3/2}$, and thus in dual-wavelength operation must compete for the energy stored by pumping to that level; this makes the dual-wavelength laser operation unstable. In this letter, we used an intermittent oscillating dual-wavelength Nd:YAG laser[4] operating at 1319 and 1064 nm as the fundamental source, which had a stable dual-wavelength output, and an optical superlattice in an LiTaO$_3$ crystal as the

---

[*] Opt.Lett.,2008,33(4):408

nonlinear frequency converter. With this scheme, we obtained 660 nm red light, 532 nm green light, and 440 nm blue light, i.e., RGB three-color light output; the total power reached 1 W, and the output is relatively stable.

Blue generation is a key factor in this letter, which can be realized by third-harmonic generation(THG) of the fundamental wave at 1319 nm in a quasi-periodic[5], aperiodic[6], or dual-periodic[7] structures. In this letter, we used a quasi-periodically optical superlattice in an LiTaO$_3$ crystal to obtain simultaneously, second-harmonic generation(SHG) at 660 nm(red) and THG at 440 nm(blue) of the fundamental wave at 1319 nm. The theory of a quasi-periodical superlattice for frequency conversion was described in [5]. Experimentally, the superlattice consists of five parallel channels and each of them has a similarly quasi-periodical domain reversion structure. These channels are 1 mm wide with a nominal phase-matching temperature at 140℃, for SHG at 660 nm, but with different phase-matching temperatures, 136℃, 138℃, 140℃, 142℃, and 144℃, for THG at 440 nm. As we all know, a small fabrication error of the domain widths may lead to a large deviation between the phase-matching temperature of SHG and THG; thus fewer red photons will be involved in the THG process, so we designed this multichannel structure providing an 8℃ temperature tuning range to let these two temperature tuning curves overlap more for efficient blue light generation. Also, we can optimize the proportion of red and blue by the selection of a proper channel. Obviously, the structure parameters for these five channels are different from one another. The detailed structure parameters of the five channels were given in Table 1. From Table 1 we can see that the parameter differs among five channels; however, they are very small due to very closely matching temperatures. The reciprocal vectors $G_{1,1}$ of these quasi-periodical channels were used for quasi-phase matching(QPM) frequency doubling of fundamental at 1319 nm to generate the red light at 660 nm, whereas $G_{3,4}$ was used to generate the blue light at 440 nm by QPM frequency adding fundamental and second harmonics. Theoretically, the effective nonlinear coefficients for the above two processes are $d_{eff(R)}=0.55d_{33}$ and $d_{eff(B)}=0.205d_{33}$, respectively, and they are almost the same for the five channels. The five-channel quasi-periodical superlattice is cascaded by a periodically optical superlattice with the period of 7.63 μm on the same LiTaO$_3$ wafer, which is used for the frequency doubling of another fundamental wave at 1064 nm to generate green light at 532 nm at 140℃. The corresponding nonlinear coefficient $d_{eff(G)}=0.64d_{33}$. The two segments, quasi-periodical and periodical, are arranged in a series and fabricated in the same LiTaO$_3$ wafer using the conventional electrical-poling technique[8,9]. The wafer is 0.5 mm in thickness, and the two segments are 30 and 10 mm, respectively, in length.

TABLE 1. Structure Paramters of the Five Quasi-Periodical Channels in First Segment[a]

Channel	T_{THG} (℃)	D_A (μm)	D_B (μm)	l (μm)	$\tan\theta$
1	136.0	17.850	13.744	6.870	0.054212
2	138.0	17.840	13.737	6.870	0.056165
3	140.0	17.830	13.730	6.860	0.058151
4	142.0	17.820	13.723	6.860	0.060144
5	144.0	17.820	13.715	6.860	0.062170

[a] D_A and D_B are the widths of blocks A and B, respectively; l is the width of the positive domain in both blocks A and B; and θ is the projection angle.

The experimental setup is shown in Fig. 1. The fundamental source was an intermittent oscillating dual-wavelength Nd:YAG laser, which was well described in [4]. The driving current of the Nd:YAG laser module (RD40-1C2, CEO Corp.) was adjustable within the range of 0–25 A, but when the driving current exceeded 19 A, multimode output would appear. Thus we set the operating current of the laser diode (LD) to be 19 A and the repetition rate of Q switch to be 3.2 kHz; then 5 W total average power and TEM_{00} profile were obtained. The output two IR radiations were both horizontal polarized, and the corresponding pulse durations of 1319 and 1064 nm are 260 and 90 ns, respectively. The lens F with the focus length of 100 mm was used to focus two polarized fundamental beams into the superlattice sample with a beam waist of 300 μm inside the crystal. A heater was used to heat the $LiTaO_3$ crystal to the phase-matching temperature with an accuracy of 0.1℃. The two ends of the sample were polished for optical measurement but not coated. A filter was put behind the sample to filter the IR at 1319 and 1064 nm for convenient measurement of the output power. The output RGB light was separated with a prism with transmittance of 100% and detected by a power meter, respectively.

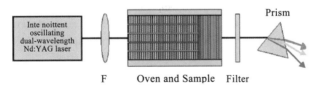

FIG.1. (color online) Experimental setup for RGB three colors generation.

The temperature tuning curves for RGB light are shown in Fig. 2. The curves were measured at a certain fundamental power, 3.9 W for 1319 nm and 1.2 W for 1064 nm; the corresponding LD driving current was 19 A and the delay time ΔT was 10 μs. From Fig. 2, one can see that the maximum output powers of RGB light are 856, 154, and 112 mW, and the corresponding phase-matching temperatures are 129.2℃, 129.6℃, and 128.7℃, respectively. The temperature tuning curves overlapped in a temperature region. At 129.3℃, we got 780 mW of red light, 146 mW of green light, and 84 mW of blue light from the second channel. The output laser operates in a high repetition manner (3.2 kHz), which is rather smaller than the time resolution of the human eyes, so we can see a quasi-white-light output from the instrument, which had the total power of 1.01 W. The power proportion of RGB at this temperature was 9.3 : 1.7 : 1, which was close to the cool-white point indicating in the Commission Internationale de l'Eclairage (CIE) chromaticity diagram[10]. The luminous flux corresponding to this cool white is 118 lines/m, and the corresponding color temperature is about 5000 K. At the same time, the blue light generated from other channels was also measured, and the corresponding average power was 4, 1, 0.25, and 0.05 mW, respectively, rather lower than that from the second channel due to larger phase mismatching at the measurement temperature. We changed the temperature from 128.0 ℃ to 130.0 ℃ in steps of 0.5 ℃, finding that outputs of RGB changed accordingly; however, the proportion of RGB outputs was still in the quasi-white-light region according to the CIE diagram. The wide temperature bandwidth indicates that the scheme is of practical use. Figure 3 is a photo of the RGB beams separated by a prism from the setup.

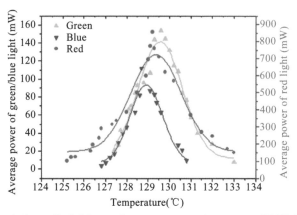

FIG.2. (color online) Measured temperature tuning curves of RGB lights. A simple Gaussian fit for three sets of data is to guide the eye. The Gaussian fit did not consider the wave coupling during the THG process, but the experimental data demonstrated the coupling effect.

FIG.3. (color online) Photo of the RGB beams separated by a prism from the setup.

The fluctuation of the total output power of RGB was measured to be ~6.5% within 1 h. The main reason leading to the fluctuation was the fluctuations of the fundamental wavelengths, 3.6% for 1319 nm and 2.6% for 1064 nm, respectively. Gain competition was observable in our experiment. Optimizing the parameters and structure of the laser cavity for raising the stability of output is still under consideration.

In summary, we used a cascaded $LiTaO_3$ optical superlattice to achieve the generation of red light at 660 nm, green light at 532 nm, and blue light at 440 nm to obtain RGB laser output from a diode-side-pumped, Q-switched, intermittent oscillating dual-wavelength Nd:YAG laser. A total average power of 1.01 W of the RGB laser was obtained at 129.3℃, and the corresponding fundamental powers were 3.9 W for 1319 nm and 1.2 W for 1064 nm, respectively. The result indicates that the scheme is an attractive way to construct a compact all-solid-state RGB laser, raising the fundamental power and using thicker periodically poled stoichiometric or MgO-doped stoichiometric $LiTaO_3/LiNbO_3$ crystals[11] that can easily scale the output to a higher power.

References and Notes

[1] H. X. Li, Y. X. Fan, P. Xu, S. N. Zhu, P. Lu, Z. D. Gao, H. T. Wang, Y. Y. Zhu, N. B. Ming, and J. L. He, J. Appl. Phys. **96**, 7756 (2004).

[2] J. Liao, J. L. He, H. Liu, H. T. Wang, S. N. Zhu, Y. Y. Zhu, and N. B. Ming, Appl. Phys. Lett. **82**, 3159 (2003).

[3] T. W. Ren, J. L. He, C. Zhang, S. N. Zhu, Y. Y. Zhu, and Y. Hang, J. Phys. Condens. Matter **16**, 3289 (2004).

[4] X. W. Fan, J. L. He, H. T. Huang, and L. Xue, IEEE J. Quantum Electron. **43**, 884 (2007).

[5] S. N. Zhu, Y. Y. Zhu, and N. B. Ming, Science **278**, 843 (1997).

[6] B. Y. Gu, Y. Zhang, and B. Z. Dong, J. Appl. Phys. **87**, 7629 (2000).

[7] Z. W. Liu, S. N. Zhu, Y. Y. Zhu, Y. Q. Qin, J. L. He, C. Zhang, H. T. Wang, N. B. Ming, X. Y. Liang, and Z. Y. Xu, Jpn. J. Appl. Phys., Part 1 **40**, 6841 (2001).

[8] S. N. Zhu, Y. Y. Zhu, Z. Y. Zhang, H. Shu, H. F. Wang, J. F. Hong, C. G. Ge, and N. B. Ming, J. Appl. Phys. **77**, 5481 (1995).

[9] V. Y. Shur, E. L. Rumyontsev, E. V. Nikolaeva, E. I. Shishkin, D. V. Fursov, R. C. Batchko, L. A. Eyres, M. M. Fejer, and R. L. Byer, Appl. Phys. Lett. **76**, 143 (2000).

[10] http://www.nd.edu/~sboker/ColorVision2/CIEColorSpace2.gif.

[11] S. V. Tovstonog, S. Kurimura, and K. Kitamura, Jpn. J. Appl. Phys., Part 1 **45**, L907 (2006).

[12] This work is supported by the National Natural Science Foundation of China (60578034, 10534020), by the National Key Projects for Basic Research of China (2006CB921804 and 2004CB619003), and by a grant from the National Advanced Materials Committee of China.

Diode-pumped 1988-nm Tm:YAP Laser Mode-locked by Intracavity Second-harmonic Generation in Periodically Poled LiNbO$_3$ *

H. Cheng, X. D. Jiang, X. P. Hu, M. L. Zhong, X. J. Lv, and S. N. Zhu

National Laboratory of Solid State Microstructures and School of Physics,
Nanjing University, Nanjing 210093, China

We report a diode-pumped intracavity second-harmonic generation mode-locked solid-state Tm:YAP laser operating at 1988 nm using a periodically poled congruent LiNbO$_3$ as the nonlinear crystal. The threshold of continuous wave mode locking is 11.6 W. The maximum output power is 1.67 W, while the shortest pulse obtained is 4.7 ps at a repetition rate of 97.09 MHz.

Solid-state ultrafast mode-locked lasers in the 2-μm wavelength region have received increasing interests in recent years due to their broad applications in photomedicine, remote sensing, optical communication, time-resolved molecular spectroscopy, minimally invasive surgery, as well as pumping sources for coherent X-ray generation and synchronously pumped optical parametric oscillators operating in the mid-infrared region around and above 5 μm[1-6]. In the past two decades, Tm^{3+}-doped, and Tm^{3+}, Ho^{3+}-codoped materials have been used as laser gain media to realize ultrafast 2-μm lasers through both active[7-9] and passive mode-locking approaches. Compared with an active mode-locking scheme, passively mode-locked 2-μm lasers not only have the advantages of reliability, simplicity and compactness, but also can generate much shorter pulses. In most passively mode-locked 2-μm lasers, ultrafast pulses are produced by using intracavity saturable absorbers (SAs). By using a semiconductor saturable absorber mirror (SESAM), pulses at around 2 μm with durations ranging from several tens of picoseconds[10] to a few picoseconds[11,12], even down to hundreds of femtoseconds[13,14], have been achieved. Besides, PbS quantum-dot-doped glass SAs are also promising candidates for 2-μm Q-switched mode locking[15,16]. Recently, single-walled carbon nanotubes[17-19], graphene oxide absorbers[20], and graphene SAs[21] showed great potential for 2-μm mode locking. The reported pulses durations range from a few picoseconds to hundreds of femtoseconds with different laser gain crystals.

Other than SA methods, intracavity frequency doubling is a promising passively

---

* Opt. Lett., 2014, 39(7):2187

mode-locking approach and there are mainly two mechanisms. The first one is a frequency-doubling nonlinear mirror (FDNLM)[22]. A FDNLM can provide a low self-starting threshold but has the disadvantage of longer pulse duration due to group velocity mismatch (GVM). The second mechanism is cascaded second-order nonlinear mode locking (CSM), which has a relatively high self-starting threshold, but it contributes substantially to pulse shortening, and can help to achieve continuous-wave mode locking (CWML)[23—26]. In both schemes, the frequency-doubling crystal can be either birefringence-phase-matching materials or quasi-phase-matching (QPM) ones. When QPM materials are used as intracavity frequency-doubling crystals, it is feasible to realize mode locking at any wavelength in the nonlinear crystal's transparent spectral range, especially for those lasers working at wavelengths where no SESAMs exist. Up to now, intracavity frequency-doubled mode locking using QPM materials has been reported at 1.064 μm[27] and 1.3 μm[28,29]; however, as far as we know, there are no reports on the realization of mode locking at around 2 μm using this scheme. In this letter, we demonstrated, for the first time to our knowledge, a diode-pumped Tm:YAP ultrafast laser operating at 1988 nm based on the CSM scheme. The nonlinear crystal used was a periodically poled $LiNbO_3$ (PPLN) and the average output power reached watt level.

The schematic experimental layout of the mode-locking laser at 2-μm spectral region is shown in Fig. 1. The laser gain media is an a-cut 3 at.% Tm:YAP crystal with the dimensions of 8 mm×4 mm×4 mm, and it is coated for antireflection ($R<1\%$) at 795 and 1988 nm on the two 4 mm×4 mm end faces. The laser crystal was mounted in a circulating-water-cooled copper holder and was end pumped through M1 by a 30-W fiber-coupled laser diode at 795 nm. The output beam from the fiber was imaged by an optical coupling system onto the laser crystal, generating a 200-μm beam diameter inside the laser gain crystal. M1 is a flat mirror placed very close to the Tm:YAP crystal and is coated for antireflection($T>98\%$)at 795 nm on both surfaces and high reflection ($R>99\%$) coated at 1988 nm on the left side. M2 and M3 are two concave mirrors coated for high reflection ($R>99\%$) at 1988 nm on their concave sides with radius of curvature being 500 and 200 mm, respectively. They were arranged with an incident angle below 5° to minimize astigmatism. M4, the output coupler, is a 5°-wedged dichroic mirror with high reflection ($R>99.5\%$)at 994 nm and partial transmittance ($T=3\%$) at 1988 nm on the right side.

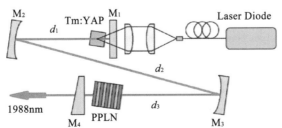

FIG.1. Scheme of intracavity SHG mode-locked 1988-nm laser with PPLN.

The 15 mm×3 mm×0.5 mm PPLN has a poling period of 28.75 μm, fabricated using the conventional electrical-field poling technique at room temperature[30]. The two end faces of PPLN are antireflection ($T>99\%$) coated at both 1988 and 994 nm. The PPLN crystal was put into a home-made oven with precise temperature control (± 0.1℃) and was placed very close to the output coupler M4. Both the PPLN and the laser crystal were tilted to avoid Fabry-Perot effect, which may interrupt mode locking. The distances d_1, d_2, d_3 (as indicated in Fig. 1) were designed to be 440, 950 and 132 mm, respectively, providing a mode size of 350 μm at the Tm: YAP crystal as well as an 148-μm one at PPLN. Considering the pump spot diameter, which is 200 μm, this set of cavity parameters can provide a soft aperture in the laser gain media.

In the experiment, the PPLN crystal was titled by 11.46°, which corresponded to an actual period of 28.88 μm, and this period can be used for frequency doubling of 1988 nm at 150℃. We measured the dependences of the output power on the incident pump power in both CW and ML cases, as shown in Fig. 2. At first, we set the temperature of PPLN to 100℃, which was far from the phase-matching temperature point, and studied the continuous-wave (CW) operation of the laser system. The system started to lase at 2.1-W pumping power, and a maximum output power of 2.74 W was obtained at a pumping power of 25.3 W, with the slope efficiency being ~12.9%. Then the crystal temperature was set to 162℃, and the temperature-induced phase mismatch was calculated to be+5.3 rad at this temperature using the Sellmeier equation of LN[31]. The +5.3 rad phase

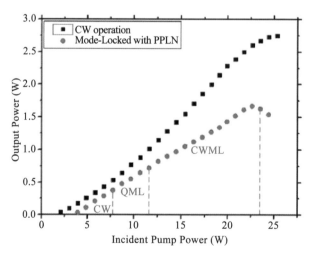

FIG.2. Output power versus incident pump power in CW and ML regimes.

mismatch corresponds to a negative nonlinear phase shifting, which represents a diverging cascaded $\chi^{(2)}$ lens. By careful alignment, oscillation started with a threshold of 3.9 W, but mode locking could not be activated at this pumping power level, the laser system was still working in CW mode. When the pumping power reached 7.7 W, Q-switched mode-locking (QML) pulses were observed with an output power of 380 mW. When the pumping power was further raised up to 11.6 W, stable 720-mW CWML was obtained. The output power increased with the pumping power, and the maximum average output power was 1.67 W under a pumping power of 22.6 W with a slope efficiency of 8.2%. As the pump power exceeded 23.5 W, the CWML output became unstable and the output power began to fall because of the thermal lens effect in the laser gain media.

The CWML waveform was detected with a high-speed InGaAs detector (ET-5000) and a digital oscilloscope (Tektronix DPO 7254). The oscilloscope traces are shown in Fig. 3. Here we used the DPX (Tektronix' digital phosphor technology) mode to better reflect the stability of the pulse train. As shown in Fig. 3(a), the mode-locked pulse train is clear and sharp, and the DC component is effectively reduced. The time interval between neighboring pulses is 10.3 ns, and this corresponds to a 97.09-MHz repetition rate, which agrees well with the 1522-mm long laser cavity. The CW characteristic of the mode-locked pulse train is evident in Fig. 3(b). The pulse-to-pulse amplitude fluctuation is estimated to be less than 5% in millisecond scale. The CWML output can be sustained for several hours with a power fluctuation less than 3%. Besides, the CWML operation can be repeated every time without any re-adjustment of the optical elements.

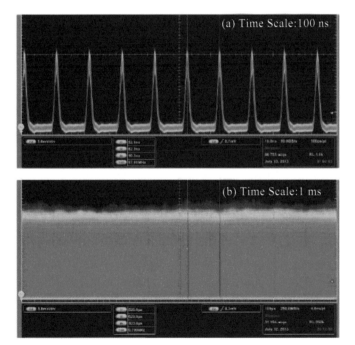

FIG.3. Oscilloscope traces of pulse train in different time scales: (a) 100 ns and (b) 1 ms.

Figure 4 shows the spectral properties in both CW and CWML operations. They were measured with an optical spectrum analyzer (YOKOGAWA AQ6375) with a resolution of 0.05 nm. The spectrum of the free-running CW Tm: YAP laser in Fig. 4(a) has several separated emitting lines at around 1988 nm, which was ensured by the coating properties of the mirrors as well as the cavity configuration used in the experiment. The maximum intensity was not located at 1988 nm, and this mainly attributed to the longitudinal mode competition in the free-running laser. The spectrum of the CWML laser has a Gaussian pulse shape with the peak at 1988 nm and a FWHM of 1.06 nm.

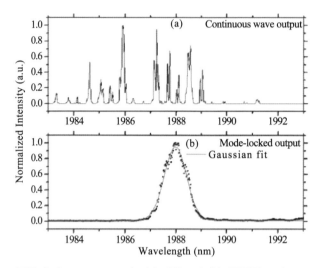

FIG. 4. Output spectra in (a) CW and (b) CWML regimes.

Theoretically, all longitudinal modes within the bandwidth of the laser gain crystal should be stimulated and oscillate in a mode-locked form. The fluorescence bandwidth of Tm:YAP is about 200 nm and it can support femtosecond mode-locked output according to the Fourier transform limit of the time-bandwidth product supposing no limiting factors. We calculate the dependences of the nonlinear phase shift and the double-pass second-harmonic generation (SHG) efficiency in the 15-mm-long PPLN on the phase mismatch, as shown in Fig. 5. The phase mismatch in the experiment is +5.3 rad, which lies in the envelope of the second peak of the SHG tuning curve. From Fig.5 we can see that the +5.3 rad phase mismatch corresponds to a proper nonlinear phase shift as well as a proper SHG efficiency. The "allowed" bandwidth acceptance within the envelope of the second peak is about 1.5 nm, which agrees well with the 1.06 nm pulse width obtained in the experiment.

The autocorrelation traces of CWML 1988-nm pulses were measured with an autocorrelator (FR-103XL), and the autocorrelator utilized a background-free (non-collinear) SHG for the measurement of ultrashort laser pulses. As shown in Fig. 6, the pulse duration is 4.7 ps at an output power of 1.05 W, while it is 6.5 ps at 1.67 W maximum output. The time-bandwidth products in both cases are 0.378 and 0.523, corresponding to 1.20 times and 1.66 times that of the Fourier transform limit with a $sech^2$ pulse shape, respectively. The asymmetrical shapes may result from the asymmetrical beam shapes of the incident beams as well as the GVM-induced pulse distortions. The pulse time-bandwidth product under the maximum average output power is larger than that under the lower power. This phenomenon is due to the inverse loss saturation of CSM[23], and a detailed analysis is as follows.

第八章 光学超晶格的应用研究 383
Chapter 8 Engineered Quasi-phase-matching for Laser Techniques

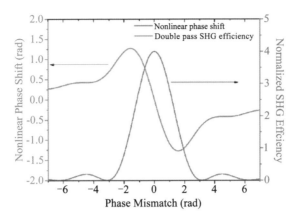

FIG. 5. Normalized double-pass SHG efficiency and nonlinear phase shift versus phase mismatch.

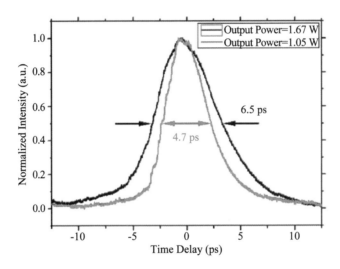

FIG. 6. Autocorrelation traces of mode-locked 1988-nm laser pulses at average output powers of 1.67 and 1.05 W, respectively.

Figure 7 shows how the mode size inside the Tm:YAP crystal varies with the focal length of the diverging cascaded $\chi^{(2)}$ lens. A critical point, where the mode size in the Tm:YAP crystal is equal to the 200-μm pump spot, is indicated as "F" in Fig. 7. The mode size decreases with the focal length of the diverging cascaded $\chi^{(2)}$ lens, and on the left side of the critical point "F," higher instantaneous intensity will lead to better overlapping between mode size and pump spot in the gain media and this positive feedback can help with the ML formation. When the pulse instantaneous intensity exceeds the critical point "F," negative feedback occurs and it can inhibit QML but may broaden the pulse width at the same time.

It should be mentioned that the double pass through the 15-mm-long PPLN would cause a 4.4 ps GVM between the fundamental wave and the second-harmonic wave, which

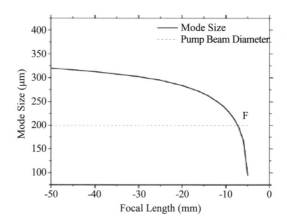

FIG.7. Mode size in Tm:YAP crystal versus focal length of the cascaded $\chi^{(2)}$ lens.

is comparable with the pulse duration obtained in the experiment. The 4.4-ps GVM would surely broaden the pulse duration to some extent. However, in the CSM scheme, GVM can produce a positive cascaded Kerr lens in the initial part of the pulse while it increases the negative Kerr lens in the trailing edge, thus will serve as an efficient pulse shortening mechanism[26]. Besides, the group velocity dispersion in the PPLN crystal, as well as in the laser gain crystal, was calculated to be less than 1 fs for a round trip in the laser cavity, which has little influence on pulse broadening. From the above analysis, we can see that GVM and the length of the nonlinear crystal are the two main limiting factors in the CSM approach. To further shorten the output pulses even down to femtoseconds, some optimum compensations should be implemented, and short PPLN crystals are preferred though requesting higher intracavity peak powers.

In summary, we have successfully constructed a picosecond solid-state laser in the 2-μm spectral region based on the CSM scheme. Near-transform-limit pulses with 6.5-ps and 4.7-ps durations were achieved at the output powers of 1.67 and 1.05 W, respectively, with the repetition rate being 97.09 MHz. Due to the flexible frequency-doubling characteristics of the optical superlattice, the mode-locking scheme based on intracavity SHG can easily extend ultrafast output from near-infrared to infrared so long as in the nonlinear crystal's transparent wavelength range.

References and Notes

[1] R. Targ, B. C. Steakley, J. G. Hawley, L. L. Ames, P. Forney, D. Swanson, R. Stone, R. G. Otto, V. Zarifis, P. Brockman, R. S. Calloway, S. H. Klein, and P. Robinson, Appl. Opt. **35**, 7117 (1996).

[2] P. A. Budni, L. A. Pomeranz, M. L. Lemons, C. A. Miller, J. R. Mosto, and E. P. Chicklis, J. Opt. Soc. Am. B **17**, 723 (2000).

[3] H. Xiong, H. Xu, Y. X. Fu, J. P. Yao, B. Zeng, W. Chu, Y. Cheng, Z. Z. Xu, E. J. Takahashi, K. Midorikawa, X. Liu, and J. Chen, Opt. Lett. **34**, 1747 (2009).

[4] T. Popmintchev, M. C. Chen, P. Arpin, M. M. Murnane, and H. C. Kapteyn, Nat. Photonics **4**, 822 (2010).
[5] B. M. Walsh, Laser Phys. **19**, 855 (2009).
[6] S. Amini-Nik, D. Kraemer, M. L. Cowan, K. Gunaratne, P. Nadesan, B. A. Alman, and R. J. D. Miller, PLoS One **5**, e13053 (2010).
[7] J. F. Pinto, L. Esterowitz, and G. H. Rosenblatt, Opt. Lett. **17**, 731 (1992).
[8] F. Heine, E. Heumann, G. Huber, and K. L. Schepler, Appl. Phys. Lett. **60**, 1161 (1992).
[9] D. Gatti, G. Galzerano, A. Toncelli, M. Tonelli, and P. Laporta, Appl. Phys. B **86**, 269 (2007).
[10] K. Yang, H. Bromberger, H. Ruf, H. Schäfer, J. Neuhaus, T. Dekorsy, C. V.-B. Grimm, M. Helm, K. Biermann, and H. Künzel, Opt. Express **18**, 6537 (2010).
[11] A. Lagatsky, F. Fusari, S. Calvez, J. A. Gupta, V. E. Kisel, N. V. Kuleshov, C. T. A. Brown, M. D. Dawson, and W. Sibbett, Opt. Lett. **34**, 2587 (2009).
[12] Q. Wang, J. Geng, T. Luo, and S. Jiang, Opt. Lett. **34**, 3616 (2009).
[13] R. C. Sharp, D. E. Spock, N. Pan, and J. Elliot, Opt. Lett. **21**, 881 (1996).
[14] S. Kivistö, T. Hakulinen, M. Guina, and O. G. Okhotnikov, IEEE Photon. Technol. Lett. **19**, 934 (2007).
[15] M. S. Gaponenko, V. E. Kisel, N. V. Kuleshov, A. M. Malyarevich, K. V. Yumashev, and A. A. Onushchenko, Laser Phys. Lett. **7**, 286 (2010).
[16] I. A. Denisov, N. A. Skoptsov, M. S. Gaponenko, A. M. Malyarevich, K. V. Yumashev, and A. A. Lipovskii, Opt. Lett. **34**, 3403 (2009).
[17] W. B. Cho, A. S. J. H. Yim, S. Y. Choi, S. Lee, F. Rotermund, U. Griebner, G. Steinmeyer, V. Petrov, X. Mateos, Ma. C. Pujol, J. J. Carvajal, M. Aguiló, and F. Díaz, Opt. Express **17**, 11007 (2009).
[18] M. A. Solodyankin, E. D. Obraztsova, A. S. Lobach, A. I. Chernov, A. V. Tausenev, V. I. Konov, and E. M. Dianov, Opt. Lett. **33**, 1336 (2008).
[19] K. Kieu and F. W. Wise, IEEE Photon. Technol. Lett. **21**, 128 (2009).
[20] J. Liu, Y. G. Wang, Z. S. Qu, L. H. Zheng, L. B. Su, and J. Xu, Laser Phys. Lett. **9**, 15 (2012).
[21] J. Ma, G. Q. Xie, P. Lv, W. L. Gao, P. Yuan, L. J. Qian, H. H. Yu, H. J. Zhang, J. Y. Wang, and D. Y. Tang, Opt. Lett. **37**, 2085 (2012).
[22] K. A. Stankov and J. Jethwa, Opt. Commun. **66**, 41 (1988).
[23] S. Mukhopadhyay, S. Mondal, S. P. Singh, A. Date, K. Hussain, and P. K. Datta, Opt. Express **21**, 454 (2013).
[24] H. Iliev, I. Buchvarov, S. Kurimura, and V. Petrov, Opt. Lett. **35**, 1016 (2010).
[25] H. Iliev, D. Chuchumishev, I. Buchvarov, and V. Petrov, Opt. Express **18**, 5754 (2010).
[26] S. J. Holmgren, V. Pasiskevicius, and F. Laurell, Opt. Express **13**, 5270 (2005).
[27] Y. F. Chen, S. W. Tsai, and S. C. Wang, Appl. Phys. B **72**, 395 (2001).
[28] Y. H. Liu, Z. D. Xie, S. D. Pan, X. J. Lv, Y. Yuan, X. P. Hu, J. Lu, L. N. Zhao, C. D. Chen, G. Zhao, and S. N. Zhu, Opt. Lett. **36**, 698 (2011).
[29] H. Iliev, I. Buchvarov, S. Kurimura, and V. Petrov, Opt. Express **19**, 21754 (2011).
[30] S. N. Zhu, Y. Y. Zhu, Z. Y. Zhang, H. Su, H. F. Wang, J. F. Hong, C. Z. Ge, and N. B. Ming, J. Appl. Phys. **77**, 5481 (1995).
[31] D. H. Jundt, Opt. Lett. **22**, 1553 (1997).
[32] This work was supported by the National Natural Science Foundation of China (61205140, 11021403,

and 91321312), the Jiangsu Science Foundation (BK2011545), the State Key Program for Basic Research of China (2010CB630703 and 2011CBA00205), and PAPD of Jiangsu Higher Education Institutions.

Efficiency-enhanced Optical Parametric Down Conversion for Mid-infrared Generation on a Tandem Periodically Poled MgO-doped Stoichiometric Lithium Tantalate Chip[*]

Y. H. Liu, Z. D. Xie, W. Ling, Y. Yuan, X. J. Lv, J. Lu, X. P. Hu, G. Zhao, and S. N. Zhu

National Laboratory of Solid State Microstructures, College of Physics,
Nanjing University, Nanjing 210093, China

We report an efficiency-enhanced mid-infrared generation via optical parametric down conversion. A tandem periodically-poled MgO-doped stoichiometric lithium tantalate crystal is used to realize on-chip generation and amplification of mid-infrared radiation inside an optical parametric oscillator cavity. We achieved 21.2% conversion efficiency (24% slope efficiency), which is among the highest efficiencies for the pump-to-mid-infrared conversion, with 1064 nm Nd class laser pump. The maximum average output power at 3.87 μm reached 635 mW with a 3.0 W pump.

1. Introduction

Mid-infrared (mid-IR) radiation around 3.8μm is of interest for a number of demanding applications, such as gas spectroscopy, environmental monitoring, atmospheric sensing, biomedical, and mid-IR countermeasures[1, 2]. One of the most widely-used scheme for mid-IR generation is by using the optical parametric down conversion, where a pump photon with higher frequency ω_p splits into two photons with lower frequencies ω_s and ω_i called signal and idler, respectively. However, when pumped by the most sophisticated Nd class solid state lasers at around 1064 nm, the power of idler light in the mid-IR range is limited by the low splitting ratio from the pump photon, i.e., the conversion efficiency is theoretically limited to about 1/4 even at 100% quantum efficiency[3, 4]. To increase the pump-to-mid-IR conversion efficiency, the pump lasers with lower frequencies may be used, but the performances of these lasers still need to be improved for high power applications. On the other hand, the efficiency can also be improved by extracting the residual energy from the signal light. For example, by using a cascaded optical parametric oscillator-optical parametric amplifier (OPO-OPA) process, it

---

[*] Opt.Express, 2011, 19(18): 17500

is possible to use the signal to amplify the idler. In the first section, the mid-IR light is generated in a similar way as the standard OPO; in the second section, the signal is used as pump to further amplify the existing mid-IR light. This method has been discussed as theoretical possibility[5—7] and experimentally demonstrated by using two independent nonlinear crystals with both synchronous[3, 6, 8] and Q-switched laser operating at 20 Hz repetition rate[9] pumping schemes. Later people found the OPO-OPA can be realized with better efficiency in a simpler setup by using a monolithic quasi-phase-matching material with complex domain structures, such as aperiodic and quasi-periodic structures[10, 11]. 14.58% pump-to-mid-IR conversion efficiency has been reported and 16.6% improvement has been achieved compared to a standard OPO.

In this letter, we present an intra-cavity OPO-OPA by a single tandem periodically-poled MgO-doped stoichiometric lithium tantalate (TPPMgSLT) crystal with two cascaded domain sections. These two sections are cascaded arranged and designed to compensate for the phase mismatching for the OPO and OPA processes, respectively. No complex domain structures are involved in this case, which simplifies the manufacturing processes. 21.2% pump-to-mid-IR conversion efficiency is obtained and it shows 77.6% efficiency enhancement compared to a single OPO. This value is among the highest efficiencies of mid-IR generation presented to date, using parametric down conversion approach with 1064 nm laser pump[2, 11, 12].

2. Method

To realize the phase matching for the OPO-OPA process, momentum conservation conditions for the OPO and OPA processes should be fulfilled simultaneously with the existences of the reciprocals of the two sections of the TPPMgSLT:

$$\Delta k_{OPO} = k_p - k_s - k_i - G_{OPO} = 0 \qquad (1)$$
$$\Delta k_{OPA} = k_s - k_{s2} - k_i - G_{OPA} = 0 \qquad (2)$$

where k_p, k_s, k_i are the wave vectors of the pump ω_p, signal ω_s, idler ω_i, and $G_{OPO} = \frac{2\pi}{\Lambda_{OPO}}$ is the reciprocal vector for the OPO; k_{s2} and $G_{OPA} = \frac{2\pi}{\Lambda_{OPA}}$ are the idler (ω_{s2}) wave vector and the reciprocal vector for the OPA, respectively. Λ_{OPO} and Λ_{OPA} are the poling periods of the OPO and OPA stages, respectively. According to the temperature-dependent Sellmeier equation of the 1.0 mol% MgO-doped stoichiometric lithium tantalate crystal (MgO:SLT)[13], we chose the poling periods $\Lambda_{OPO} = 28.83$ μm and $\Lambda_{OPA} = 31.59$ μm for the two stages in the following OPO-OPA process:

$$\text{OPO: } 1.064 \ \mu\text{m} \rightarrow 1.47 \ \mu\text{m} + 3.87 \ \mu\text{m} \qquad (3)$$
$$\text{OPA: } 1.47 \ \mu\text{m} \rightarrow 3.87 \ \mu\text{m} + 2.37 \ \mu\text{m} \qquad (4)$$

In these two stages, the phase matching curves of the OPO and OPA processes can be calculated from Eqs. (1) and (2). As shown in Fig. 1, the phase matching for the OPO-

OPA process occurs at the cross-over point for the two curves of around 145℃, which is measured to be 120℃ in the experiment. The difference between them is mainly due to the inaccuracies of the Sellmeier equation and the thermal expansion of the TPPMgSLT crystal. The angular separation of the temperature tuning curves for the OPO and OPA processes is small, and therefore a wide phase-matching bandwidth for the OPO-OPA can be expected, which is calculated to be about 100℃.

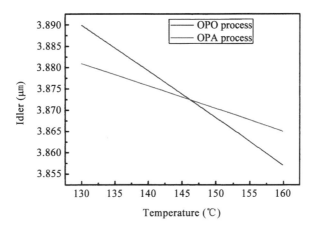

FIG. 1. The calculated phase-matching curves for the OPO and OPA processes.

Taking the plane-wave and the slowly varying envelope approximations, the parametric interaction between the amplitudes can be described by the coupling equations, under the perfect phase matching condition.

$$\frac{dA_p(z)}{dz} = i\kappa_{OPO} A_s(z) A_i(z)$$

$$\frac{dA_s(z)}{dz} = i\kappa_{OPO} A_i(z) A_i^*(z) + i\kappa_{OPA} A_i(z) A_{s2}(z)$$

$$\frac{dA_i(z)}{dz} = i\kappa_{OPO} A_p(z) A_s^*(z) + i\kappa_{OPA} A_s(z) A_{s2}^*(z)$$

$$\frac{dA_{s2}(z)}{dz} = i\kappa_{OPA} A_s(z) A_i^*(z) \tag{5}$$

in Eq.(5), $\kappa_{OPO} = \frac{d_{eff}}{c}\sqrt{\frac{\omega_p \omega_s \omega_i}{n_p n_s n_i}}$ and $\kappa_{OPA} = \frac{d_{eff}}{c}\sqrt{\frac{\omega_{s2} \omega_s \omega_i}{n_{s2} n_s n_i}}$ are the coupling coefficient, in which c is the vacuum light velocity. A_j, n_j are the amplitudes, and refractive indices of ω_j ($j = p, s, i, s2$), respectively. $d_{eff} = 0.637d$ is the effective second order nonlinear coefficient, where $d = \frac{\chi^{(2)}}{2}$.

3. Fabrication of the TPPMgSLT and Experimental Setup

The TPPMgSLT is fabricated using the electrical poling technique at room temperature from a 1 mm thick MgO:SLT wafer[14]. It is known that MgO:SLT, with large d_{eff}, shows high damage threshold, high resistance to the photorefractive effect, high thermal conductivity, and low IR absorption, especially compared to the commonly used lithium niobate. Furthermore the nonstoichometric defect is significantly reduced in MgO:SLT, and therefore it leads to a reduction of coercive field by~1 order of magnitude and facilitates the electrical poling of large aperture crystals with a good uniformity of domain structures. These characteristics collectively make MgO:SLT a promising material for the high efficiency frequency conversions at high power levels. Owing to the high intracavity power density inside a high-Q OPO cavity and the advantage of MgO:SLT, high efficiency can be expected. As shown in Fig. 2, the width of OPO section is 2 mm wider than the OPA section so that we can independently test the single OPO operation. The TPPMgSLT has the lengths of 24 mm and 19 mm for the OPO and OPA sections, respectively. And the total length of the crystal is about 44 mm. The two end faces are optically polished and coated for antireflection at 1064, 1400 – 1500, and 3800 – 4200 nm. The crystal is embedded in an oven whose temperature is controlled with accuracy of ±0.1℃ for stable operation.

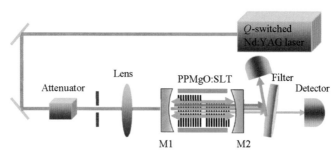

FIG.2. Schematic of the experiment setup.

The OPO-OPA apparatus is schematically shown in Fig. 2. We use a 100-mm-long linear OPO cavity, with concave cavity mirrors M1 and M2 of 100 mm radius. Both of them are high-reflection coated ($R>99\%$) for 1400 – 1500 nm to provide high-Q singly resonance for the signal, anti-reflection coated for 3800 – 4000 nm and anti-reflection coated for 1064 nm and 1650 nm to reduce unwanted resonances. The pump light is from a diode-pumped Nd:YAG laser working at 1064 nm, with repetition rate of 5 kHz and pulse duration of 40 ns. The pump beam is focus onto the crystal with a beam diameter of 200 μm, and it matches well with the waist of the cavity mode that is calculated to be 180 μm. A high-pass filter is used to split the pump and idler beams for the measurements.

4. Experimental Results and Discussion

We first characterized the OPO process while only the OPO section was illuminated by the pump beam. As shown in Fig. 3, a conversion efficiency of 12% and a slope efficiency of 14.5% could be calculated, that was 43.6% of quantum-limited performance. The oscillation started with a threshold below 500 mW and we obtained the maximum output power of 360 mW with 3 W pump. The thresholds for the pump power density of OPO and OPO-OPA processes could be calculated by solving the coupling Eq. (5), $I_{pt}=\frac{\varepsilon_0 n_p n_s n_i c^3}{4d^2 \omega_s \omega_i L^2} \cdot (1-R_s)$, where R_s and L were the equivalent reflectivity of the signal light and the length of the OPO section, respectively. The measured threshold power densities of the two processes were about 4 MW/cm², which were on the same order of calculated values of 1.5 MW/cm².

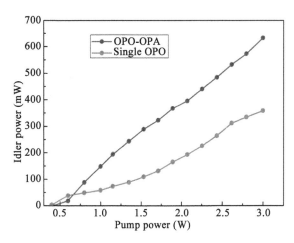

FIG.3. Measured mid-IR output power of the single OPO and the OPO-OPA versus pump power at 120℃.

Later, we moved the transverse position of TPPMgSLT so that the OPA section could take part in the nonlinear interaction. The temperature dependence of OPA was shown in Fig. 4, by measuring the output power of ω_{s2}. The measured FWHM bandwidth of OPO-OPA process was about 50℃, which was resulted from the broad phase-matching bandwidth as discussed in section 2.

At the phase matching temperature of 120℃, we measured the spectrum of the outputs using a spectrum analyzer (ANDO AQ-6315A). As shown in Fig. 5, the peaks at 1468 nm and 1187 nm were for ω_s and the second harmonic of ω_{s2}, respectively. From that, we could easily calculate the wavelengths of ω_{s2} and ω_i to be 2374 nm and 3847 nm respectively, which agreed well with the designed values in Eqs. (3) and (4). The output at other wavelength could not be recorded because of the limited response range of the spectrum analyzer. We also measured the dependence of mid-IR output power on pump power (Fig. 3). The OPO-OPA showed a similar threshold as the single OPO, but with much higher slope efficiency due to the existence of intracavity OPA process. When the pump power reached 3 W, the measured power of idler was 635 mW. The conversion and slope efficiencies for mid-IR light were 21.2% and 23.52%, respectively. Compared to the

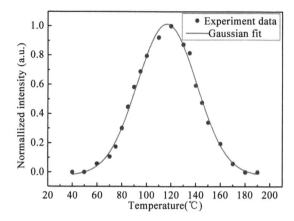

FIG.4. Normalized output power for ω_{s2} as a function of temperature.

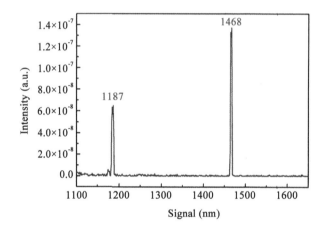

FIG.5. Output spectrum at 120℃ for the OPO-OPA device. The spectrum range is limited from 350 nm to 1750 nm by the spectrum analyzer.

single OPO, the OPO-OPA presented 76.4% improvement on the conversion efficiency and 65.6% improvement on the slope efficiency. The improvement was attributed to the extra gain in the OPA process. The high intensity signal light could be generated inside the high-Q OPO cavity, which served as pump of OPA process in the same cavity. According to the last formula of Eq. (5), the gain of OPA could be given by

$$g = \kappa_{OPA} |E_s| \qquad (6)$$

where E_s was the electric field of the signal light in the condition of steady state. The gain of OPA process could be calculated about $g = 0.6$ cm^{-1} from Eq. (6), which was measured to be $g = 0.5$ cm^{-1} in a well agreement.

5. Conclusion

In conclusion, we have demonstrated an efficiency enhanced mid-IR generation based on OPO-OPA using TPPMgSLT. A maximum conversion efficiency of 21.2% has been achieved, which corresponds to 77% of the quantum-limited (27.5%) performance for this pulsed OPO-OPA setup. The monolithic TPPMgSLT crystal has presented high efficiency and high intracavity power density capability and low loss for high-Q resonance in the OPO cavity. These features are also important for the future CW OPO-OPA devices and near quantum-limited performance could be expected. In this work, the pump laser has a maximum output power of about 3 W, corresponding to a power density of 45 MW/cm^2 in the experiment, which is much lower than the damage threshold of the MgO:SLT[12, 15]. Therefore, by replacing the pump source with higher power and TPPMgSLT with larger optical aperture, the OPO-OPA scheme can be readily scaled for higher power mid-IR output with high efficiency.

References and Notes

[1] R. Guoguang and H. Yunian, "Laser-based IRCM system defenses for military and commercial aircraft," Laser Infrared **36**, 1−6 (2006).

[2] Y. Peng, W. Wang, X. Wei, and D. Li, "High-efficiency mid-infrared optical parametric oscillator based on PPMgO:CLN," Opt. Lett. **34**(19), 2897−2899 (2009).

[3] M. E. Dearborn, K. Koch, G. T. Moore, and J. C. Diels, "Greater than 100% photon-conversion efficiency from an optical parametric oscillator with intracavity difference-frequency mixing," Opt. Lett. **23**(10), 759−761 (1998).

[4] J. M. Fraser and C. Ventalon, "Parametric cascade downconverter for intense ultrafast mid-infrared generation beyond the Manley-Rowe limit," Appl. Opt. **45**(17), 4109−4113 (2006).

[5] K. Koch, G. T. Moore, and E. C. Cheungy, "Optical parametric oscillation with intracavity difference-frequency mixing," J. Opt. Soc. Am. B **12**(11), 2268−2273 (1995).

[6] G. T. Moore and K. Koch, "Efficient High-Gain Two-Crystal Optical Parametric Oscillator," IEEE J. Quantum Electron. **31**(5), 761−768 (1995).

[7] A. Berrou, J.-M. Melkonian, M. Raybaut, A. Godard, E. Rosencher, and M. Lefebvre, "Specific architectures for optical parametric oscillators," C. R. Phys. **8**(10), 1162−1173 (2007).

[8] K. J. McEwan and J. A. C. Terry, "A tandem periodically-poled lithium niobate (PPLN) optical parametric oscillator (OPO)," Opt. Commun. **182**(4−6), 423−432 (2000).

[9] J. M. Fukumoto, H. Komine, W. H. Long, Jr., and E. A. Stappaerts, "Periodically Poled LiNbO$_3$ Optical Parametric Oscillator with Intracavity Difference Frequency Mixing," in: W. R. Bosenberg, M. M. Fejer (Eds.), Advanced Solid State Lasers, OSA Trends in Optics and Photonics Series, **19**, 245−248 (1998).

[10] H. C. Guo, Y. Q. Qin, Z. X. Shen, and S. H. Tang, "Mid-infrared radiation in an aperiodically poled LiNbO$_3$ superlattice induced by cascaded parametric processes," J. Phys. Condens. Matter **16**(47), 8465−8473 (2004).

[11] G. Porat, O. Gayer, and A. Arie, "Simultaneous parametric oscillation and signal-to-idler conversion for efficient downconversion," Opt. Lett. **35**(9), 1401−1403 (2010).

[12] N. E. Yu, S. Kurimura, Y. Nomura, M. Nakamura, K. Kitamura, J. Sakuma, Y. Otani, and A. Shiratori, "Periodically poled near-stoichiometric lithium tantalate for optical parametric oscillation,"

Appl. Phys. Lett. **84**(10), 1662–1664 (2004).

[13] W. L. Weng, Y. W. Liu, and X. Q. Zhang, "Temperature-Dependent Sellmeier Equation for 1.0 mol% Mg-Doped Stoichiometric Lithium Tantalate," Chin. Phys. Lett. **25**(12), 4303–4306 (2008).

[14] S. N. Zhu, Y. Y. Zhu, Z. Y. Zhang, H. Shu, H. F. Wang, J. F. Hong, C. G. Ge, and N. B. Ming, "$LiTaO_3$ crystal periodically poled by applying an external pulsed field," J. Appl. Phys. **77**(10), 5481–5483 (1995).

[15] X. P. Hu, X. Wang, J. L. He, Y. X. Fan, S. N. Zhu, H. T. Wang, Y. Y. Zhu, and N. B. Ming, "Efficient generation of red light by frequency doubling in a periodically-poled nearly-stoichiometric $LiTaO_3$ crystal," Appl. Phys. Lett. **85**(2), 188–190 (2004).

[16] This work was supported by the National Natural Science Foundations of China (Nos. 11021403, 10904066 and NSAF10776011), and the State Key Program for Basic Research of China (Nos. 2011CBA00205 and 2010CB630703).

Polarization-free Second-order Nonlinear Frequency Conversion Using the Optical Superlattice*

X. J. Lv, J. Lu, Z. D. Xie, J. Yang, G. Zhao, P. Xu, Y. Q. Qin, and S. N. Zhu

National Laboratory of Solid State Microstructure, Nanjing University, Nanjing 210093, China

We propose a universal and practical scheme to realize polarization-free second-order nonlinear frequency-conversion processes, including sum-frequency generation, difference-frequency generation, and optical parametric amplification. This scheme is based on the optical superlattice with a noncritical phase-matching condition, which is suitable for optical integration. A chirped dual-period structure is proposed to ensure the efficiency and stability of the conversions.

Usually second-order nonlinear frequency conversion (NLFC) is polarization sensitive owing to phase matching. However, polarization-free frequency conversion (PFFC) is required in a number of practical cases, where the state of polarization (SOP) is cross polarized or varying with time. For example, in fiber-related systems including fiber lasers, laser fiber delivery systems, and optical fiber communications, the SOP may be varying or unpolarized; in quantum communication and computation using polarization-entangled photon pairs, the SOP is unpredictable. Additionally, optical integration of fiber, waveguide, and nonlinear convertors has attracted a great deal of attention. Efficient, compact, and integratable polarization-free frequency converters will play important roles in these applications.

PFFC schemes have been demonstrated on semiconductor material[1], quasi-phase-matched (QPM) materials[2], and in different types of optical fibers[3]. But the above-mentioned methods usually need a complicated setup to split and recombine the beams. Recently, multi phase matching of two NLFC processes with a different SOP was demonstrated for a polarization switch[4] and internal interference[5]. By theoretical analysis and numerical calculation, we find that multi phase matching of the polarization is a good way to realize an efficient, single-pass, polarization-free frequency converter. Combined with the waveguide technique, it may be integrated in an optical circuit to realize generation and conversion of a randomly polarized laser, polarization-entangled photon pairs, or polarization-modulated photons.

Second-order nonlinear conversion is usually expressed as a three-wave coupling process that has three angular frequencies, $\omega_3 = \omega_2 + \omega_1$, where $\omega_3 > \omega_2 \geqslant \omega_1 > 0$. For

---

* Opt. Lett., 2011, 36(1):7

crystals belonging to the 3 m point group of symmetry and a beam propagating along the x axis, there are four kinds of noncritical phase-matching methods ($\sigma_1 \sim \sigma_4$)—described in Table 1. Here we assume that E_1, E_2, and E_3 identify waves with angular frequencies ω_1, ω_2, and ω_3 respectively, and subscripts y and z denote the polarization directions. Because of crystallographic symmetry, we have $d_{32} \equiv d_{24}$. For example, for the most popular QPM material, periodically poled $LiNbO_3$, d_{33} equals 25.2 pm/V at 1064 nm and $d_{32} \equiv d_{24}$ equals 4.6 pm/V at 1064 nm[6]. From left to right, columns 2, 3, and 4 of Table 1 are the polarization directions of frequencies ω_3, ω_2, and ω_1, respectively. Columns 5 and 6 are nonlinear coefficients and phase-matching types, respectively. By multi phase matching any two of the four processes, a PFFC process will be generated, if the efficiency of the two processes is equal. At the same time, the conversion gives a one-to-one mapping between the input and output polarization. The following part of this letter will show that cascaded and dual-periodic superlattices will meet the above conditions, and a practical scheme for polarization-free sum frequency generation (SFG) is demonstrated as an example.

TABLE 1. Four Kinds of Noncritical Phase-Matching Processes

	E_3	E_2	E_1	d_σ	Type
σ_1	z	z	z	d_{33}	0
σ_2	z	y	y	d_{32}	I
σ_3	y	z	y	d_{24}	II
σ_4	y	y	z	d_{24}	II

For simplicity, we define a symbol system such as $E_{n\sigma}$, where $n \in \{1, 2, 3\}$ labels the frequency and $\sigma \in \{\sigma_1, \sigma_2, \sigma_3, \sigma_4\}$ labels the QPM process. The specific polarization direction is decided by Table 1 according to n and σ. For example, if $\sigma = \sigma_4$, according to row 5 of Table 1, we have $E_{3\sigma} = E_{3y}$, $E_{2\sigma} = E_{2y}$, and $E_{1\sigma} = E_{1z}$. We designate the following equation as $F(\alpha, \sigma)$:

$$\frac{dE_{\alpha\sigma}}{dx} = -i \frac{2\omega_\alpha d_\sigma f_\sigma}{n_{\alpha\sigma} c} E_{\beta}^{\star} E_{\gamma}^{\star}, \tag{1}$$

α, β and $\gamma \in \{1,2,3\}, \alpha \neq \beta \neq \gamma$,

$$E_{\beta}^{\star} = \begin{cases} E_{\beta\sigma}^{\star}, \alpha \neq 3 \text{ and } \beta \neq 3 \\ E_{\beta\sigma}, \text{else} \end{cases},$$

$$E_{\gamma}^{\star} = \begin{cases} E_{\gamma\sigma}^{\star}, \alpha \neq 3 \text{ and } \beta \neq 3 \\ E_{\gamma\sigma}, \text{else} \end{cases}. \tag{2}$$

ω_α is the angular frequency, d_σ is the nonlinear coefficient of process σ, c is the light velocity in vacuum, $n_{\alpha\sigma}$ is the refractive index of specified frequency and polarization according to Table 1, and f_σ denotes the pattern of the superlattice in real space. Using the plane wave and slowly varying amplitude approximation, the coupled three-wave nonlinear equations can be expressed as $F(1,\sigma)$, $F(2,\sigma)$, and $F(3,\sigma)$, where F is declared in Eq. (1). Now we extend the property of $(E, \alpha, \beta, \gamma)$ in Eq. (2) to $(A, \alpha_1, \beta_1, \gamma_1)$ and $(A, \alpha_2, \beta_2, \gamma_2)$, where $A=(n/\omega)^{1/2} E$ is the normalized complex amplitude of light. Under a small-signal condition, we assume that the power and SOP of light $E_{a_3\sigma}$ with frequency ω_{a_3} is invariable with x and time, $E_{a_2\sigma}(x=0)$ is random polarized and $E_{a_1\sigma}(x=0)=0$. To describe the second-order NLFC processes in a unified presentation, we do not specify the frequencies of these waves. For example, $\{\alpha_3=3\}$ describes the optical parametric amplification (OPA) or difference-frequency generation (DFG) processes, and $\{\alpha_3=1, \alpha_2=2\}$ describes an SFG process. Considering the reciprocal vector intensity is $g_\sigma=\frac{1}{L}F \times f_\sigma(x)$, we have the following equations when it is phase matched:

$$\frac{dA_{a_1\sigma}}{dx}=-iG_\sigma A^\star_{\beta_1\sigma}A^\star_{\gamma_1\sigma}, \frac{dA_{a_2\sigma}}{dx}=-iG_\sigma A^\star_{\beta_2\sigma}A^\star_{\gamma_2\sigma}, \quad (3)$$

where

$$G_\sigma=\frac{2d_\sigma g_\sigma}{c}\sqrt{\frac{\omega_1\omega_2\omega_3}{n_{1\sigma}n_{2\sigma}n_{3\sigma}}}.$$

Equation (3) can be readily solved as

$$I_{a_1\sigma}(x)=(\omega_{a_1}/\omega_{a_2})C_{a_1\sigma}I_{a_2\sigma}(0), I_{a_2\sigma}(x)=C_{a_2\sigma}I_{a_2\sigma}(0), \quad (4)$$

where $I=\frac{1}{2}c\varepsilon_0\omega|A|^2$, and

$$C_{a_1\sigma}=\begin{cases}\sinh^2(|G_\sigma A_{a_3\sigma}|x) & \alpha_3=3 \\ \sin^2(|G_\sigma A_{a_3\sigma}|x) & \alpha_3\neq 3\end{cases},$$

$$C_{a_2\sigma}=\begin{cases}\cosh^2(|G_\sigma A_{a_3\sigma}|x) & \alpha_3=3 \\ \cos^2(|G_\sigma A_{a_3\sigma}|x) & \alpha_3\neq 3\end{cases}.$$

From Table 1, we select two processes σ_a and σ_b with E_{a_3} holding the same polarization. As mentioned above, the gain of the two processes should be equal: $|G_{\sigma_a}|=|G_{\sigma_b}|$. So we design the reciprocal vectors with the following ratio:

$$\left|\frac{g_{\sigma_a}}{g_{\sigma_b}}\right|=\left|\frac{d_{\sigma_b}}{d_{\sigma_a}}\right|\sqrt{\frac{n_{1\sigma_a}n_{2\sigma_a}n_{3\sigma_a}}{n_{1\sigma_b}n_{2\sigma_b}n_{3\sigma_b}}}. \quad (5)$$

And noticing that $A_{a_3\sigma_a}\equiv A_{a_3\sigma_b}$, we obtain $C_{a_1\sigma_a}=C_{a_1\sigma_b}$, $C_{a_2\sigma_a}=C_{a_2\sigma_b}$, and the output power will be independent with the SOP of $I_{a_2}(0)$.

The multi-phase-matching technique includes cascaded, dual-periodic[7,8], and quasi-periodic[9,10] structures. Variably spaced phase-reversed[11] and chirped[12] QPM gratings have been used to increase the bandwidth of the second harmonic generation. Here we introduce a chirped dual-periodic structure, which combines multi phase matching and bandwidth control techniques. Designing two functions $f_1(x)$ and $f_2(x)$ with periods of Λ_1 and Λ_2, respectively, the corresponding reciprocal vectors are $G_1=2\pi/\Lambda_1$ and $G_2=2\pi/\Lambda_2$. The dual-periodic structure is given as $F(x)=f_1(x)f_2(x)$, with reciprocal

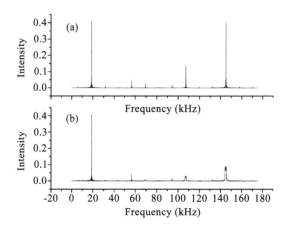

FIG. 1. Reciprocal vectors of chirped dual-periodic structure, $L = 2$ cm. (a) $\phi_1 = \phi_2 = 0$. (b) $\phi_1 = 0, \phi_2 = \pi/8$.

vectors $G_{1,1} = G_1 + G_2$ and $G_{1,-1} = G_1 - G_2$. The intensity of the vectors is $g_{1,1} = g_{1,-1} = g_1 g_2$ [Fig. 1(a)]. Now we extend this idea to a chirped dual-periodic structure, which can modulate the width and intensity of the reciprocal vectors. Assume $\Lambda(x)$ is chirped slightly around $\Lambda(0)$, $x \in (-L/2, L/2)$. We have $G(x) = 2\pi/\Lambda(x)$. If we expand the $\Delta G_{1,-1} L$ to $\pm \phi_1$ and $\Delta G_{1,1} L$ to $\pm \phi_2$, that is,

$$\begin{cases} [G_{1,-1}(x) - G_{1,-1}(0)]L = 2\phi_1 x/L \\ [G_{1,1}(x) - G_{1,1}(0)]L = 2\phi_2 x/L \end{cases}, \tag{6}$$

the two periods can be solved as

$$\Lambda_1(x) = \frac{2\pi}{(\phi_1 + \phi_2)x/L^2 + G_1},$$

$$\Lambda_2(x) = \frac{2\pi}{(\phi_1 - \phi_2)x/L^2 + G_2}. \tag{7}$$

Figure 1(b) shows the reciprocal vectors of the chirped dual-periodic structure, with $\phi_1 = 0$ and $\phi_2 = \pi/8$, $g_{1,-1}/g_{1,1} \approx d_{33}/d_{32} \approx 5.5$, which is quite fit for the PFFC schemes

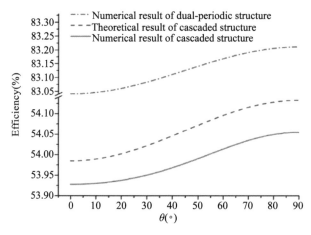

FIG. 2. (color online) Polarization-free SFG scheme.

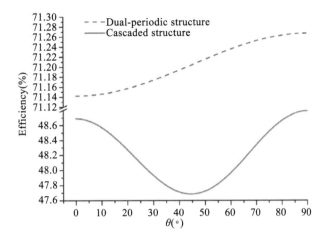

FIG.3. (color online) SFG efficiency $[I_{631nm}(L)/I_{1550nm}(0)]$ when $I_{1550nm}(0)/I_{1064nm}(0)=0.01$.

using LiNbO$_3$ and process σ_1. Fortunately, for LiNbO$_3$, there are three processes with identical d_σ; we can always find two workable processes and use a dual-periodic structure. However, the chirped dual-periodic structure offers a solution to balance the two processes with extremely high gain (e.g., OPA), where a slight difference between the two process (caused by n_σ) may induce a great discrepancy between the output powers.

The SFG process of 1064 nm+1550 nm → 631 nm is shown as an example (Fig. 2). We assume that the 1550 nm wave is randomly polarized and the 1064 nm wave is linearly polarized at the y axis with an input intensity of 10 MW/cm^2. We choose σ_2 and σ_4 as the two processes to be phase matched by a cascaded or dual-periodic superlattice with a total length of $L=2$ cm. The cascaded sections are equal to each other, and the periods can be calculated as 52.98 and 6.90 μm for LiNbO$_3$ using the Sellmeier equation given by [13]. The reciprocal vectors of the dual-periodic structure are shown in Fig. 1(a). We describe the 1550 nm wave as an elliptically polarized light. The angle between the major axis and the z axis is θ. We numerically calculate the coupled three-wave equations using Eq. (4), in which the phase mismatch superlattice pattern in real space and pump depletion are considered. The calculation method is elaborated in our previous works[14]. We compare the theoretical result given by Eq. (4) with the numerical result (Fig. 3). When pump depletion is very small, the numerical result is in good agreement with the theoretical result. Because of the difference between refractive indices, the G_{σ_2} is slightly greater than G_{σ_4}, so the efficiency at $\theta=90°$ is larger than that at $\theta=0°$. The instability is about 0.10% using the cascaded and dual-periodic superlattice. But the conversion efficiency of the dual-periodic superlattice is higher than the cascaded one, because the intensity of the reciprocal vectors in the dual-periodic structure is larger. When the pump depletion is nonignorable, the conversion efficiency is decreased due to the strong depletion of the pump, but the stability is kept good in the dual-periodic case. The instability of the cascaded structure is ±1.0% and the dual-periodic one is ±0.09% (Fig. 4). This is caused by the synchronous pump depletion of the two processes in the dual-periodic structure, while in the cascaded structure it is asynchronous.

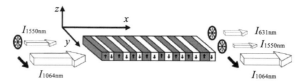

FIG. 4. SFG efficiency$[I_{631nm}(L)/I_{1550nm}(0)]$ when $I_{1550nm}(0)/I_{1064nm}(0)=1$.

In conclusion, a scheme based on an optical superlattice is demonstrated to realize polarization-free SFG, DFG, and OPA with high efficiency and a compact configuration. A chirped dual-periodic structure is proposed for the first time, to modify the two reciprocal vectors independently. The polarization-free conditions are given by theoretical deduction under the small-signal condition; numerical simulation shows that even under strong depletion of the pump, this method is still feasible. The highly efficient, single-pass, and integratable features make it promising in fiber-related systems, quantum optics, and optical circuits.

References and Notes

[1] J. Tang, P. Spencer, and K. Shore, Appl. Phys. Lett. **75**, 2710 (1999).

[2] I. Brener, M. H. Chou, E. Chaban, K. R. Parameswaran, M. M. Fejer, S. Kosinski, and D. L. Pruitt, Electron. Lett. **36**, 66 (2000).

[3] K. Wong, M. Marhic, K. Uesaka, and L. Kazovsky, IEEE Photon. Technol. Lett. **14**, 1506 (2002).

[4] A. Ganany-Padowicz, I. Juwiler, O. Gayer, A. Bahabad, and A. Arie, Appl. Phys. Lett. **94**, 091108 (2009).

[5] B. Johnston, P. Dekker, M. Withford, S. Saltiel, and Y. Kivshar, Opt. Express **14**, 11756 (2006).

[6] I. Shoji, T. Kondo, A. Kitamoto, M. Shirane, and R. Ito, J. Opt. Soc. Am. B **14**, 2268 (1997).

[7] R. Lifshitz, A. Arie, and A. Bahabad, Phys. Rev. Lett. **95**, 133901 (2005).

[8] Z. Liu, Y. Du, J. Liao, S. Zhu, Y. Zhu, Y. Qin, H. Wang, J. He, C. Zhang, and N. Ming, J. Opt. Soc. Am. B **19**, 1676 (2002).

[9] S. Zhu, Y. Zhu, and N. Ming, Science **278**, 843 (1997).

[10] K. Fradkin-Kashi, A. Arie, P. Urenski, and G. Rosenman, Phys. Rev. Lett. **88**, 023903 (2001).

[11] M. L. Bortz, M. Fujimura, and M. M. Fejer, Electron. Lett. **30**, 34 (1994).

[12] M. A. Arbore, O. Marco, and M. M. Fejer, Opt. Lett. **22**, 865 (1997).

[13] G. J. Edwards and M. Lawrence, Opt. Quantum Electron. **16**, 373 (1984).

[14] X. Lv, Z. Sui, Z. Gao, M. Li, Q. Deng, and S. Zhu, Opt. Laser Technol. **40**, 21 (2008).

[15] We thank Z. Sui of the China Academy of Engineering Physics for fruitful discussions. This work is supported by the National Natural Science Foundation of China (NSFC) under grants 10534020 and NSAF 10776011).

Polarization Independent Quasi-phase-matched Sum Frequency Generation for Single Photon Detection*

Xiao-shi Song, Zi-yan Yu, Qin Wang, Fei Xu, and Yan-qing Lu

College of Engineering and Applied Sciences and National Laboratory of Solid State Microstructures, Nanjing University, Nanjing 210093, China

Polarization independent sum frequency generation (SFG) is proposed in an electro-optic (EO) tunable periodically poled Lithium Niobate (PPLN). The PPLN consists of four sections. External electric field could be selectively applied to them to induce polarization rotation between the ordinary and extraordinary waves. If the domain structure is well designed, the signal wave with an arbitrary polarization state could realize efficient frequency up-conversion as long as a z-polarized pump wave is selected. The applications in single photon detection and optical communications are discussed.

1. Introduction

Recently, more and more attention has been paid to single photon detection at telecom wavelength making use of Sum-frequency generation (SFG)[1—3]. The infrared single photons are converted to the visible regime for high detection efficiency, which is very important in quantum communication. However, the nonlinear optical up conversion schemes are basically polarization sensitive[1,2]. Although some techniques could be adopted to solve this problem, such as employing a 90° rotated second nonlinear crystals[3] or inserting a wave plate between a pair of nonlinear materials, they are not monolithic thus may have stability and reliability concerns in practical applications. As a consequence, it is desired to develop a kind of compact and robust configuration to realize the polarization independent SFG[4]. On the other hand, quasi-phase-matched (QPM) SFG or cascaded SFG/difference frequency generation (DFG) in ferroelectric waveguides has been proposed for various photonic applications. For example, all-optical wavelength converters[5,6] and logic gates[6—8], which are two key functional elements in future photonic networks.

In this letter, a four-section periodically poled LiNbO$_3$ (PPLN) is proposed to realize the polarization independent SFG. The signal wave with an arbitrary state of polarization could be converted to the sum-frequency wave (SFW) in this special PPLN. The z-

---

* Opt.Express,2011,19(1):380

polarized pump wave is injected into the sample then external DC electric fields along the y-axis are selectively applied to manipulate the lights' polarization states. The polarization dependence of SFG thus could be greatly suppressed with a suitable PPLN structure. The interaction among these waves is investigated through coupling wave equations. Related mechanism and future applications are also discussed.

2. Theory and Simulation

Figure 1 shows the schematic diagram of a polarization independent PPLN containing four sections. The first and the fourth sections are identical whose period is designed just for QPM SFG. However, the periods in the second and third sections are different; they are designed only for polarization rotation of the signal wave and the SFW, respectively[9]. External DC electrical fields are applied at the y-surfaces of the second and third sections. With a suitable electric field, the second and third sections could just act as 90° polarization rotators for signal wave and SFW, respectively. In order to utilize the largest nonlinear coefficient d_{33} of $LiNbO_3$, the lights should propagate along the crystal's x-axis and the escorting pump wave should be z-polarized. Because the wavelength bandwidths for polarization rotation in the second and third sections are very sharp[9], the polarization state of the pump wave is not affected inside the sample.

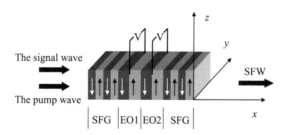

FIG.1. Schematic diagram of a four-section PPLN. External electric fields are applied along the y-axis at the second and the third sections, representing with EO1 and EO2 in the figure.

When a z-polarized signal wave is propagated through the sample, the signal wave will be converted to the z-polarized SFW basically in the first section and then the generated z-polarized SFW is further totally converted to a y-polarized SFW in the third section. At last the SFW passes through the last section without nonlinear frequency conversion. On the other hand, when a y-polarized signal wave is injected, it passes through the first section without SFG. Then it is converted to a z-polarized signal wave in the second section. Because the third section has no impact to it, the generated z-polarized signal wave is further frequency up-converted to a z-polarized SFW after it enters the fourth section. In a word, a y-polarized signal wave generates a z-polarized SFW while a z-polarized signal wave induces a y-polarized SFW. As long as the first and the last sections have the same length, the PPLN may have the same frequency up-conversion capability for y- and z- polarized signal waves. Because all normally-incident lights could be divided into y- and z- polarized components, polarization independent frequency up-conversion thus could be expected.

To analysis the above processes in detail and obtain the numerical results, the corresponding coupling equations can be deduced under the plane-wave approximation with consideration of both SFG and electro-optic (EO) interactions[10,11].

$$\begin{cases} \dfrac{dE_{sy}}{dx} = -i\dfrac{\omega_s}{n_{sy}c}\varepsilon_{23}^{(s)}(x)E_{sz}e^{i\Delta k'_2 x}, \\ \dfrac{dE_{sz}}{dx} = -i\dfrac{\omega_s}{n_{sz}c}\left[\varepsilon_{23}^{(s)}(x)E_{sy}e^{-i\Delta k'_2 x} + d_{33}(x)E_{pz}^*E_{oz}e^{-i\Delta k'_1 x}\right], \\ \dfrac{dE_{pz}}{dx} = -i\dfrac{\omega_p}{n_{pz}c}d_{33}(x)E_{sz}^*E_{oz}e^{-i\Delta k'_1 x}, \\ \dfrac{dE_{oy}}{dx} = -i\dfrac{\omega_o}{n_{oy}c}\varepsilon_{23}^{(o)}(x)E_{oz}e^{i\Delta k'_3 x}, \\ \dfrac{dE_{oz}}{dx} = -i\dfrac{\omega_o}{n_{oz}c}\left[\varepsilon_{23}^{(o)}(x)E_{oy}e^{-i\Delta k'_3 x} + d_{33}(x)E_{sz}E_{pz}e^{i\Delta k'_1 x}\right]. \end{cases} \quad (1)$$

Where $E_{j\xi}$, $\omega_{j\xi}$, $k_{j\xi}$ and $n_{j\xi}$ (the subscript $j=s$, p, o refer to the signal wave, the pump wave and the SFW respectively, and $\xi=y,z$ represent the polarizations) are the external field amplitudes, the angular frequencies, the wave-vectors and the refractive indices, respectively. c is the speed of light in vacuum. The converted wave is generated at frequency $\omega_o = \omega_s + \omega_p$. $d_{33}(x) = d_{33}f(x)$ is the modulated nonlinear coefficient, $\varepsilon_{23}^{(s)}(x) = -n_{sy}^2 n_{sz}^2 \gamma_{51}(x)E$ for the signal wave and $\varepsilon_{23}^{(o)}(x) = -n_{oy}^2 n_{oz}^2 \gamma_{51}(x)E$ for the SFW, where $\gamma_{51}(x) = \gamma_{51}f(x)$ is the modulated EO coefficient in PPLN and E refers to the external DC electric field. As no electric field is applied to the first and the fourth section, the corresponding E is 0. $\Delta k'_1 = k_{oz} - k_{sz} - k_{pz}$, $\Delta k'_2 = k_{sy} - k_{sz}$, $\Delta k'_3 = k_{oy} - k_{oz}$ are the wave vector mismatch for SFG and polarization rotation of the signal wave and the SFW, respectively. The asterisk denotes complex conjugation.

In a PPLN, the structure function $f(x)$ changes its sign form $+1$ to -1 periodically in different domains. It can be expanded as Fourier series, $f(x) = \sum_m g_m \exp(-iG_m x)$ where G_m are the reciprocal vectors and g_m are the amplitudes of the reciprocal vectors. Without loss of generality, the reciprocal vector G_1 (the subscript 1 is ignored hereinafter) is adopted to compensate the wave vector mismatch for efficient conversion. For example, in the first and fourth section, $G_{1,4}$ (here the second subscript denotes which section the reciprocal vector is in) is adopted as $\Delta k'_1 = k_{oz} - k_{sz} - k_{pz} = G_{1,4}$ to compensate the nonlinear phase mismatch for SFG and the wave vector mismatch becomes $\Delta k_1 = k_{oz} - k_{sz} - k_{pz} - G_{1,4} = 0$. In the same way, G_2 in the second section is adopted to compensate the phase mismatch for the signal wave's polarization rotation as $\Delta k'_2 = k_{sy} - k_{sz} = G_2$ (i.e., $\Delta k_2 = k_{sy} - k_{sz} - G_2 = 0$). And G_3 in the third section is adopted to compensate the phase mismatch for

SFW's polarization rotation as $\Delta k'_3 = k_{oy} - k_{oz} = G_3$ ($\Delta k_3 = k_{oy} - k_{oz} - G_3 = 0$). In this case, the coupling Eqs. (1) can be simplified as

$$\begin{cases} \dfrac{dA_{sy}}{dx} = iK_2 A_{sz} e^{i\Delta k_2 x}, \\ \dfrac{dA_{sz}}{dx} = iK_2^* A_{sy} e^{-i\Delta k_2 x} - iK_1 A_{pz}^* A_{oz} e^{-i\Delta k_1 x}, \\ \dfrac{dA_{pz}}{dx} = -iK_1 A_{sz}^* A_{oz} e^{-i\Delta k_1 x}, \\ \dfrac{dA_{oy}}{dx} = iK_3 A_{oz} e^{i\Delta k_3 x}, \\ \dfrac{dA_{oz}}{dx} = iK_3^* A_{oy} e^{-i\Delta k_3 x} - iK_1^* A_{sz} A_{pz} e^{i\Delta k_1 x}. \end{cases} \quad (2)$$

with $A_j = \sqrt{\dfrac{n_j}{\omega_j}} E_j$, $K_1 = \dfrac{d_{33} g_1}{c} \sqrt{\dfrac{\omega_{sz} \omega_{pz} \omega_{oz}}{n_{sz} n_{pz} n_{oz}}}$, $K_2 = \dfrac{i\gamma_{51} g_1 \omega_s E}{2c} \sqrt{n_{sy}^3 n_{sz}^3}$, $K_3 = \dfrac{i\gamma_{51} g_1 \omega_o E}{2c} \times \sqrt{n_{oy}^3 n_{oz}^3}$. $K_i (i = 1, 2, 3)$ represents the coupling coefficients for SFG and EO effect.

Numerical solutions to Eqs. (2) have been done to verify our design. A 1064 nm pump wave is employed and the injected signal wave has the wavelength of 1550 nm. In order to convert the signal wave to the SFW completely, we set the intensity of the signal wave at 0.2 MW/cm^2, which is far weaker than the intensity of the pump wave at 10 MW/cm^2. The PPLN period in the first and fourth section is designed at $\Lambda = 11.62$ μm to satisfy the SFG QPM condition at room temperature 25℃. The corresponding periods of these two sections are 500 for complete up-conversion. $\Lambda = 20.48$ μm in the second section and $\Lambda = 7.27$ μm in the third section are selected for polarization rotation of the signal wave and the SFW, respectively. $d_{33} = 25.2$ pm/V and $\gamma = 32.6$ pm/V are used for simulation. To realize the complete conversion between the ordinary and extraordinary waves, we set the electric field at 710 V/mm and the periods of the second and the third sections at 250 and 260 respectively. The sample length is about 18.6 mm.

Figures 2 and 3 show the numerical simulation results. Figure 2 describes the light intensity evolution of the signal wave and the SFW inside our devised PPLN. It can be seen that no matter with what kind of polarization states, the injected signal wave is almost totally converted to SFW. The light intensity of the SFW remains about 0.5 MW/cm^2 for the y-polarized [Fig. 2(a)] and z-polarized [Fig. 2(b)] signal waves. Figure 3 shows the spectral response. Figures 3(a) and 3(b) correspond to the results when a y-polarized signal wave is injected. Figures 3(c) and 3(d) correspond to the results when a z-polarized signal wave is injected. For example, when a y-polarized signal wave is injected, the light intensity of the generated z-polarized SFW reaches the highest and the light intensity of the y-and z-polarized signal wave vanishes if the phase matching condition is satisfied in accordance with Fig. 2. In addition, the light intensity of the z-polarized signal wave shows double peaks at the wavelength which has a slight offset from the phase matching point.

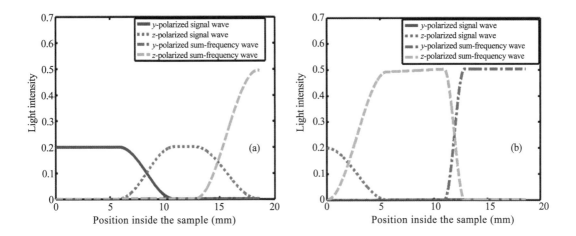

FIG.2. The light intensity at different positions inside the four-section PPLN when the signal wave is y-polarized (a) or z-polarized (b). Solid, dotted, dash-dotted and dashed curves represent y-polarized signal wave, z-polarized signal wave, y-polarized SFW and z-polarized SFW, respectively. The unit of the intensity is MW/cm^2.

The results are consistent when a z-polarized signal wave is injected but the light intensity of the z-polarized signal wave has small fluctuations near the phase matching point. On the other hand, the y-polarized SFW is negligible at side wavelengths when a y-polarized signal wave is injected. This is because when a y-polarized signal wave near the phase matching point is injected, the z-polarized SFW will not be generated in the first section so that there is no y-polarized SFW being converted in the third section. As a result, the y-polarized sum-frequency wave will not be observed at the side wavelengths. However, when a z-polarized signal wave at side wavelengths is injected, the signal wave will not be fully converted into the z-polarized SFW in the first section and the left signal wave will not be totally converted into the y-polarized signal wave. The remaining z-polarized signal wave is partly frequency up-converted into SFW in the forth section so we observe a non-negligible generation of z-polarized SFW at side wavelengths. Because the total energy of involved lights should be conserved and there are some z-polarized SFWs at side wavelengths, the light intensity of the z-polarized signal wave thus shows small side fluctuations (e.g., double peaks) at corresponding wavelengths.

Our previous results only show the SFG behaviors of pure y- and z-polarized signal waves, which is not enough. To well testify the polarization independency, an arbitrarily polarized signal should be considered. We calculated the SFW intensifies when the signal waves have different intensity ratio and phase difference between the y- and z-polarized components. Figure 4 shows the simulation results. The signal wave with arbitrary polarization state could be converted to the SFW with basically the same intensity at

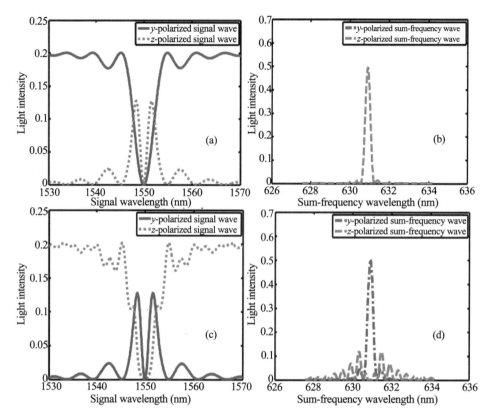

FIG.3. The light intensity versus the wavelength when the signal wave is y-polarized ((a) and (b) correspond to the signal wave and SFW respectively) or z-polarized ((c) and (d) correspond to the signal wave and SFW respectively). The unit of the intensity is MW/cm^2.

$0.5 MW/cm^2$. The intensity fluctuation is less than 1%. We can conclude that our multi-section PPLN really may realize the polarization independent SFG perfectly. Considering the different photon energies of the signal and SFW, it is found that the signal photons almost have been totally converted into SFW photons in visible regime. Although the calculated quantum efficiency is not exactly at 100%, it might be due to the intrinsic slowly varying approximation of the coupled wave approach. As a consequence, a "fully" polarization independent SFG conversion is still expectable if no propagation loss is considered. This would be very attractive in the single photon detection and other applications. In comparison with other techniques, our approach has many advantages. For example, it's monolithic and easily applied in future photonic systems. In addition to the quantum detection, we believe our technique is also applicable for classical optical communications, because it may open a new window in eliminate the polarization dependent issues in a fiber-optic link with wavelength conversion.

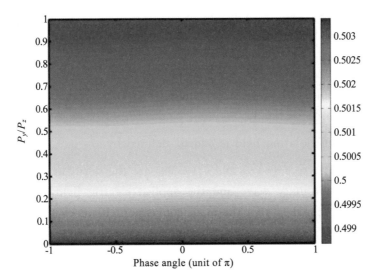

FIG.4. The light intensity of SFW at arbitrary polarization states. The unit of the intensity is MW/cm².

3. Discussion

In addition to bulk PPLN, Lithium niobate waveguide also has been utilized in many optical devices and applications. In comparison with bulk crystal, the waveguide has non-uniform mode distribution while with high field confinement. Efficient fiber to waveguide coupling has been realized for years so that a PPLN waveguide could be well compatible with most optical communication devices. Although our simulation above is based on the plane-wave approximation, the key coupling equations, $i.e.$, Eqs. (2), still can be used in the waveguide case with minor revision about the coupling coefficients and wave vector mismatches. The overlap integral of the mode fields also should be added in the coupling coefficients. The propagation constants and effective refractive indices take place of the wave vectors and bulk wave's refractive indices, respectively. We take a typical Ti-diffused PPLN waveguide as an example.

In the first and fourth sections for frequency up-conversion, the coupling coefficient K_1 in Eqs. (2) should be modified by an overlap factor $\sim \iint e_{sf}(y,z) e_s^*(y,z) e_p^*(y,z) \mathrm{d}y \mathrm{d}z$[12], where $e_{sf}(y,z)$, $e_s^*(y,z)$ and $e_p^*(y,z)$ are the normalized field profiles of the SFW, signal wave and pump wave, respectively. This modification may result in a decrease of the coupling coefficient, so the power density of the input light should be elevated accordingly to compensate the coupling coefficient reduction. As a result, the conversion efficiency in these two sections still may be kept in a desired level. On the other hand, the TE/TM mode conversion realizes in the second and the third sections, which correspond to the polarization rotation in a bulk PPLN. In this case, the coupling coefficients K_2 and K_3 should be multiplied by the overlap factors of the involved mode fields and the applied electric field[13]. We take K_2 for the second section as an example. It changes to $K'_2 = \frac{i\gamma_{51}g_1\omega_s E}{2c}\sqrt{n^3_{sy,\text{eff}}n^3_{sz,\text{eff}}}\nu$. Different from the original coefficients in Eqs. (2), the refractive indices of the ordinary and extraordinary signal waves are replaced by the effective counterparts of the TE and TM modes respectively. An overlap factor of the mode fields and the modulating field is added, defined by $\nu = \iint e_o^*(y,z)e_y(y,z)e_e(y,z)dydz$, where $e_o(y,z)$, $e_e(y,z)$ and $e_y(y,z)$ are the normalized field profiles of the ordinary wave, extraordinary wave and the applied electric field, respectively. As the overlap factor is less than 1 so that the mode conversion efficiency will be affected. Unlike the SFG in the first and fourth sections, the TE/TM mode conversion is a linear process so it is not influenced by the light intensity. However, a 100% mode conversion is still easily achievable by extending the length of the corresponding second section or just simply increasing the applied field. In the third section, the situation is similar for the mode conversion of the SFW. A word, same results could be obtained in all four PPLN sections, thus the simulation curves shown in Fig. 2–4 should also be valid in PPLN waveguides.

Unlike the bulk PPLN with negligible inherent loss, the waveguide normally has some propagation loss that has to be taken into account. For a typical Ti-diffused PPLN waveguide, the propagation losses of SFW and signal waves are ~0.2 dB/cm and ~0.1 dB/cm, respectively[14,15]. Even if these losses are considered, the SFG photon to photon quantum efficiency still could reaches ~94%. Because the commercial detectors in telecom wavelength only has 1/3–1/4 single photon detection probability than that in visible band, our up-conversion scheme are still very favorable for single photon detection operating at infrared wavelengths[1].

4. Conclusion

In summary, we proposed a polarization independent PPLN containing four well-designed sections. External DC electric fields are applied to the given sections to induce the polarization rotation between the ordinary and extraordinary waves. A five-wave-coupling approach is proposed to study the optical power transferring among these waves. The SFG dependency on the signal wave's polarization states are investigated, which exhibits great polarization-insensitivity. The light intensity of the converted SFW keeps at the same level with arbitrary signal wave's polarization sates. The related mechanism and future applications in the single photon detection and optical communications were also discussed.

References and Notes

[1] R. V. Roussev, C. Langrock, J. R. Kurz, and M. M. Fejer, "Periodically poled lithium niobate

waveguide sumfrequency generator for efficient single-photon detection at communication wavelengths," Opt. Left. **29**(13), 1518 – 1520 (2004).

[2] X.R. Gu, K. Huang, Y. Li, H. F. Pan, E. Wu, and H. P. Zeng, "Temporal and spectral control of single-photon frequency upconversion for pulsed radiation," Appl. Phys. Lett. **96**(13), 131111 (2010).

[3] A. P. Vandevender, and P. G. Kwiat, "High efficiency single photon detection via frequency up-conversion," J. Mod. Opt. **51**, 1433 (2004).

[4] M.A. Albota, F. N. C. Wong, and J. H. Shapiro, "Polarization-independent frequency conversion for quantum optical communication," J. Opt. Soc. Am. B **23**(5), 918 (2006).

[5] S. Yu, and W. Y. Gu, "Wavelength conversions in quasi-phase-matched $LiNbO_3$ waveguide based on doublepass cascaded $\chi^{(2)}$ SFG+DFG interactions," IEEE J. Quantum Electron. **40**(12), 1744 (2004).

[6] J. Wang, J. Q. Sun, and Q. Z. Sun, "Experimental observation of a 1.5 μm band wavelength conversion and logic NOT gate at 40 Gbits/s based on sum-frequency generation," Opt. Lett. **31**(11), 1711 – 1713 (2006).

[7] T. Suhara, and H. Ishizuki, "Integrated QPM sum-frequency generation interferometer device for ultrafast optical switching," IEEE Photon. Technol. Lett. **13**(11), 1203 – 1205 (2001).

[8] Y.L. Lee, B. A. Yu, T. J. Eom, W. Shin, C. Jung, Y. C. Noh, J. Lee, D. K. Ko, and K. Oh, "All-optical AND and NAND gates based on cascaded second-order nonlinear processes in a Ti-diffused periodically poled $LiNbO_{(3)}$ waveguide," Opt. Express **14**(7), 2776 – 2782(2006).

[9] Y.Q. Lu, Z. L. Wan, Q. Wang, Y. X. Xi, and N. B. Ming, "Electro-optic effect of periodically poled optical superlattice $LiNbO_3$ and its applications," Appl. Phys. Lett. **77**(23), 3719 (2000).

[10] A. Yariv, and P. Yeh, *Optical Waves in Crystals* (John Wiley and Sons, New York, 1984), Chap. 12.

[11] Y. Kong, X. F. Chen, and Y. Xia, "Competition of frequency conversion and polarization coupling in periodically poled lithium niobate," Appl. Phys. B **91**(3 – 4), 479 – 482(2008).

[12] P.F. Hu, T. C. Chong, L. P. Shi, and W. X. Hou, "Theoretical analysis of optimal quasi-phase matched second harmonic generation waveguide structure in $LiTaO_3$ substrates," Opt. Quantum Electron. **31**(4), 337 – 349 (1999).

[13] C. Y. Huang, C. H. Lin, Y. H. Chen, and Y. C. Huang, "Electro-optic Ti:PPLN waveguide as efficient optical wavelength filter and polarization mode converter," Opt. Express **15**(5), 2548 – 2554 (2007).

[14] L. G. Sheu, C. T. Lee, and H. C. Lee, "Nondestructive measurement of loss performance in channel waveguide devices with phase modulator," Opt. Rev. **3**(3), 192 – 196 (1996).

[15] E. Pomarico, B. Sanguinetti, N. Gisin, R. Thew, H. Zbinden, G. Schreiber, A. Thomas, and W. Sohler, "Waveguide-based OPO source of entangled photon pairs," N. J. Phys. **11**(11), 113042 (2009).

[16] This work is supported by National 973 program under contract No. 2011CBA00205 and 2010CB327803, and the National Science Foundation of China (NSFC) program No. 60977039 and 10874080. The authors also acknowledge the support from New Century Excellent Talents Program and the Specialized Research Fund for the Doctoral Program of Higher Education.

DFB Semiconductor Lasers based on Reconstruction-equivalent-chirp Technology[*]

Yitang Dai and Xiangfei Chen

Microwave-Photonics Technology Laboratory, National Key Lab of Microstructure,
Nanjing University, Nanjing 210093, China

A distributed feedback (DFB) semiconductor laser with equivalent phase shifts and chirps is proposed for the first time to our knowledge and is investigated numerically. As an example, it is shown that the desired $\lambda/4$ phase shift in a phase-shifted laser can be obtained equivalently by a specially designed sampling structure instead of an actual phase shift, while the external characteristics are unchanged. This novel DFB structure is advantageous in that it can be fabricated by standard holographic technology. Hence, the proposed scheme is expected to provide a low-cost method for fabricating a high-performance DFB semiconductor laser with complex structures.

1. Introduction

A distributed feedback (DFB) semiconductor laser is one of the most widely used laser sources in fiber-optic communications for its excellent performance and compact size. Because of mode degeneracy in the uniform grating structure quarter-wave ($\lambda/4$) shift[1], multiple phase shifts or chirps are usually used to achieve stable single-longitude-mode (SLM) operation. A number of methods can be applied to form the above-mentioned phase shifts and chirps. Electron-beam (E-beam) lithography is mature and reliable for the production of exact phase shifts or complex chirps in the DFB structure. However, it is still expensive for mass production. From a commercial point of view, holographic exposure may be the cheapest way to fabricate DFB structures directly on the diode chip. Some improved holographic methods, including varying the stripe width[2] or using a special photoresist[3], were introduced. Compared with E-beam technology, these methods are still difficult and complicated for producing both complex phase shifts and chirps during real fabrication. A technology wherein the fabrication is simple, as is E-beam technology, and where the cost is cheap, as is standard holographic technology, is required and necessary for high-end DFB laser diodes with very low cost.

In this study we propose a novel DFB semiconductor laser where the equivalent phase shift and chirp can be fabricated by the conventional holographic technique. The proposed

---

[*] Opt.Express,2007,15(5):2348

DFB structure is based on a uniform grating that is sampled with unequal spacing, and a SLM lasing can be obtained at the wavelength corresponding to the −1st-order channel of the sampled structure. As an example, it is shown numerically that a λ/4 shift can be obtained equivalently by a special unequally spaced sampling, which makes the proposed DFB laser act the same as one with an actual π-phase shift, including the threshold, the P-I characteristics, the light intensity distribution, and the side-mode suppression ratio (SMSR). Because there is no actual phase shift in the novel DFB laser, its sampling structure can be fabricated by conventional holographic technology.

Sampling is usually used to achieve multiwavelength lasing in a DFB structure. However, it acts in a completely different way in this study. The proposed special unequally spaced sampling allows an effective control of the main lasing peak position relative to the Bragg wavelength, i.e., it acts the same as actual phase shifts. This equivalent technique comes from reconstruction-equivalent-chirp (REC) technology and has been used successfully in fiber Bragg gratings (FBGs)[4]. Many relevant devices, such as the fiber DFB laser (which contains a single equivalent π-phase shift)[5] and the phase en/de-coder for optical code division multiple access system (which contains numerous equivalent π-phase shifts)[6], have been experimentally demonstrated based on REC technology. REC technology has greatly simplified the fabrication of phase shifts and chirps by realizing precise optical phase control using only μm-level precision. On the other hand, nm-level precision is required for achieving precise optical phase control during conventional fabrication. In this study, REC technology is applied, for the first time to our knowledge, to the design of DFB semiconductor lasers and is expected to mitigate the fabrication difficulties of high-quality DFB semiconductor lasers.

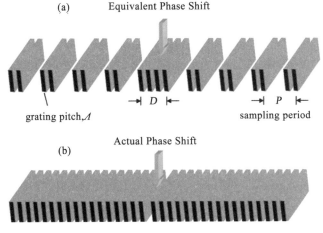

FIG.1. Schematic of the DFB structure with (a) equivalent phase shift and (b) actual phase shift.

2. Principle

A schematic of the proposed DFB structure is shown in Fig. 1(a) compared with the conventional phase-shift structure shown in Fig. 1 (b). The novel structure is a uniform Bragg grating sampled with period P (6 μm in this study). The length of each sample (containing tens of grating pitches) equals $P/2$, corresponding to a duty cycle of 0.5. However, the sample length at the center is extended to D.

There are many channels for the sampled structure illustrated in Fig. 1 (a). Based on REC technology, an equivalent phase shift θ will be introduced into the $-$1st-order channel (hereinafter referred to as the resonant channel, for convenience) by D as follows[7]:

$$\theta = 2\pi \left(\frac{D}{P} - \frac{1}{2} \right), \frac{P}{2} \leqslant D < \frac{3P}{2} \tag{1}$$

Then, by changing D, one can control the position of the low-threshold mode (the so-called gap mode) within the stop-band of the resonant channel. For example, if $D=P$, an equivalent $\lambda/4$ (π-phase) shift is obtained, and it is possible to achieve SLM lasing at the center of the resonant channel.

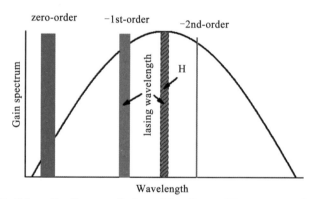

FIG. 2. Schematic diagram of channel locations with respect to the gain curve: H: with decreased sampling period and unchanged grating period, the lasing wavelength will be shifted rightward.

However, the possibility of lasing in channels of other orders should be considered carefully. Usually the 3dB bandwidth of the semiconductor laser gain spectrum is about 50 nm. To obtain good dynamic performance, the lasing wavelength is selected at 15 – 20 nm shorter than the wavelength that corresponds to peak gain. In the DFB laser proposed here, the lasing wavelength falls within the resonant channel. For possible lasing at other wavelengths, only a zero-order channel and a $-$2nd-order channel should be considered, as shown schematically in Fig. 2. Other nonzero-order channels have much less gain or effective index modulation (i.e., κL) than the resonant channel and can be completely neglected. When the duty cycle is 0.5, the index modulation of the $-$2nd-order channel is, from Fourier analysis, almost zero. Thus the $-$2nd-order channel is much weaker than the resonant channel, as shown in Fig. 2 (the thicker it is, the stronger it is in index modulation). As a result, only the possibility of lasing in the zero-order channel(less gain but larger κL) should be considered. Because the channel spacing is large (about 50 nm in this study, corresponding to $P=6$ μm), the optical gain decreases greatly in the zero-order channel. Therefore, it is expected that the threshold in the zero-order channel is much higher than that in the resonant channel.

In this study, lasing operation in both the resonant and the zero-order channels is analyzed by simulation. We will see that: first, the threshold margin (difference in lasing threshold between resonant and zero-order channels) is large enough to maintain SLM operation even at high output; and second, external characteristics of the proposed structure are the same as those of the conventional phase-shifted DFB semiconductor lasers.

3. Simulation

A spectral domain model developed in Refs.[8, 9] is used in our simulation. Using this model, static characteristics of the proposed DFB with equivalent $\lambda/4$ shifts are studied, while the conventional λ/shifted DFB laser is also analyzed for comparison. Parameters used for the simulation are listed in Table 1 except for the index modulation, which is 50 cm^{-1} and 150 cm^{-1} for the lasers with actual and equivalent $\lambda/4$ shift, respectively. (as the Fourier coefficient in the$-$2nd-order is 1/3, in order to get the same characteristics, the index modulation required by the DFB structure using REC technique is usually three times the conventional one)

TABLE 1. Parameters used in the simulation.

Effective refractive index(n_{eff})	3.283
Slope of gain-carrier density relationship(α)	3×10^{-16} cm^2
Transparency carrier density(N_t)	1.5×10^{18} cm^{-3}
Coupling of mode to active layer(Γ)	0.35
Line width enhancement factor(β_C)	1.5
Linear carrier lifetime(τ)	4 ns
Waveguide absorption and scattering loss(α_L)	25 cm^{-1}
Active layer thickness(d)	0.18 μm
Current stripe width(W)	3.5 μm
Bimolecular carrier recombination coefficient(B)	1×10^{-10} cm^{3}s^{-1}
Effective optical width perpendicular to junction(d_{eff})	0.47 μm
Far-field pattern FWHM perpendicular to junction(θ_d)	50°
Effective optical width parallel to junction(W_{eff})	3.5 μm
Far-field pattern FWHM parallel to junction(θ_w)	20°
Width of spontaneous emission(Δf_{sp})	80 nm
Cavity length(L)	400 μm

The left part of Fig. 3(a) shows the calculated *P-I* curve of a DFB semiconductor laser with an actual $\lambda/4$ shift (dotted line) or an equivalent $\lambda/4$ shift (solid line). The corresponding thresholds are 15.7 mA and 15.8 mA, respectively, while their lasing wavelengths are both around 1550 nm. When the injection current = 17.5 mA (corresponding to 8 mW output power for both lasers), light intensity distribution along the lasers are plotted in Fig. 3(b). Unsmooth intensity distribution in the equivalent $\lambda/4$-shifted DFB laser results from the sampled structure.

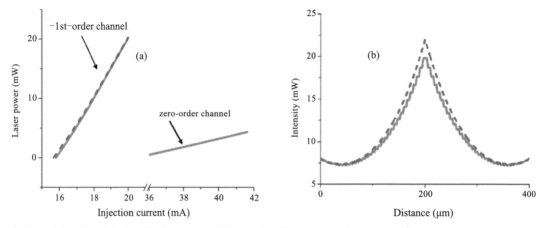

FIG.3. (a) Left: Calculated *P-I* curve of DFB semiconductor laser with a real $\lambda/4$ shift (dotted line) or an equivalent $\lambda/4$ shift (solid line). Right: *P-I* curve of one of the gap-modes in the zero-order channel of the DFB laser with equivalent $\lambda/4$ shift. (b) Light intensity distribution in the DFB laser with a real $\lambda/4$ shift (dotted line) or an equivalent $\lambda/4$ shift (solid line). The output power is 8 mW for both lasers.

The spectrum of the DFB semiconductor laser can be obtained based on the model in Ref.[9], as shown in Fig. 4. It can be seen that both lasers have the same excellent SMSR (about 75 dB). Obviously, Figs. 3 and 4 show that the DFB laser with an equivalent $\lambda/4$ shift acts almost the same as the laser with an actual $\lambda/4$ shift.

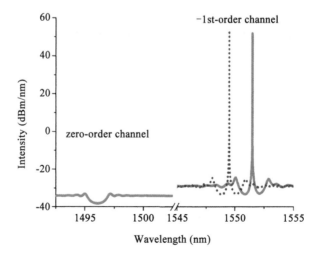

FIG.4. The simulated lasing spectra of the DFB lasers with actual (dotted line) and equivalent (solid line) π-phase shift. Power output for both lasers is about 8 mW.

Since the wavelength difference between the zero-order and the resonant channels is as large as ~50 nm, gain in the former case is much less than that in the latter case, and it is assumed to be 1 : 6 in this study. The spectrum corresponding to the zero-order channel is calculated and plotted in Fig. 4, where it shows no increase in the SMSR (the side mode induced by the zero-order channel has lower intensity than that of the resonant channel). It can be seen that the spectrum generated by the spontaneous emission in the zero-order channel is similar to that of a DFB laser with uniform grating[10]. This is because there is no equivalent phase shift or chirp in the zero-order channel, and the two symmetrical gap-modes in the zero-order channel are out of the stop-band, which significantly restrains the possible lasing in the zero-order channel. If the injection current is large enough, lasing will occur in the zero-order channel. The calculated threshold of one of the gap-modes in the zero-order channel is about 36 mA, which is much higher than that in the resonant channel (15.8 mA) and shows a wide-enough threshold margin. The calculated P-I curve of the gap-mode is plotted in the right half of Fig. 3(a).

Based on the above analysis, one can conclude that the sampled structure, which generates multi-wavelength operation traditionally, shows no impact on the characteristics of the DFB laser with an equivalent $\lambda/4$ shift, especially on the SMSR.

4. Discussion

Based on the REC technique, various equivalent phase shifts or chirps (which may be more complicated than the DFB laser mentioned above with only one phase shift) can be obtained simply by adjusting the sampling period of the sampled structure. This is quite useful for high-quality DFB semiconductor lasers. For example, the DFB structure with only one phase shift has some disadvantages such as heavy spatial hole-burning, and more complex structures like multi-phase-shifts[11] or even chirped DFB structures have been proposed to solve this problem. With a specially-designed sampled structure, almost all of the semiconductor lasers with a complex DFB structure can be obtained without any actual phase shifts or chirps. Figure 5 shows an example with two equivalent phase shifts: 0.5 π occurs at $z=0.33 L$ and $z=0.66 L$, respectively. The threshold is 16.5 mA. Figure 5(a) shows the master mode P-I curve of this DFB laser diode, while Fig. 5(b) indicates the corresponding intensity distributions along the cavity under an output power of 9.03 mW. Compared with Fig. 3, the maximum intensity decreases a lot under similar output power. Additionally, since the DFB structure in the zero-order channel is always a uniform one, a large threshold margin will then confirm the equivalence between the characteristics of the DFB laser with actual and equivalent phase shifts or chirps.

In our design, the magnitude of the sampling period is 6 μm and the minimum line width is 3 μm, so it is possible to use the conventional holographic technology to fabricate DFB semiconductor lasers with equivalent phase shifts and chirps. For example, a specially

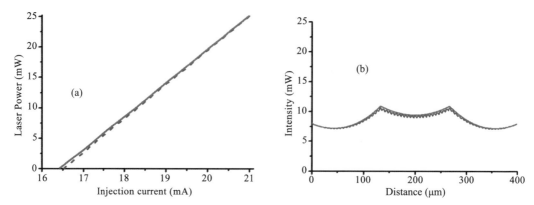

FIG.5. (a) Calculated *P-I* curve of a DFB semiconductor laser with multiple real phase shifts(dotted line) or multiple equivalent phase shifts (solid line). (b) Light intensity distribution in the DFB laser with multiple real phase shifts (dotted line) or multiple equivalent phase shifts(solid line). The output power is ~9.03 mW for both lasers.

designed sampling structure corresponding to a certain phase shift and chirp profile can be first encoded into a photomask, and then the DFB structure is fabricated by holographic technology based on this photomask. In fact, a minimum line-width of 1 μm and a sampling period of 2 μm are very common for photomasks. In terms of the technology we proposed, the fundamental grating (zero-order channel) provides only a basic feedback, and the lasing is actually determined by the resonant channel. Because the characteristics of the resonant channel can be varied by adjusting the sampling period (2 - 6 μm) and duty cycle in each sample, the lasing wavelength and some lasing performance can be controlled easily. With a decreasing sampling period, the lasing wavelength will increase in terms of the Fourier theory. Hence, precise control of the lasing wavelength can be realized by carefully adjusting the sampling period. For example, as shown in the dashed plot H of Fig. 2, we can decrease the sampling period and the duty cycle so that (1) lasing occurs at the gain peak, and (2) the resonant channel can be weakened a little. That is to say, when the corresponding pattern is encoded in the photomask, tens of thousands of laser diodes on a laser chip can have different lasing wavelengths and performances, even between two neighboring laser diodes. It should be mentioned again that standard holographic technology can provide such great flexibilities. Thus the proposed technology may provide a lower-cost and highly flexible method for fabricating high-end DFB semiconductor lasers with complicated structures.

5. Conclusion

A novel DFB semiconductor laser with equivalent phase shifts and chirps based on REC technology is proposed and numerically studied in this research. The equivalent phase shifts or chirps can be obtained by a specially designed sampling structure so that no actual phase shifts or chirps are required. Simulation results show that DFB lasers with equivalent phase shifts and chirps may have the same external characteristics as traditional DFB lasers. However, the novel structure shows great advantage in fabrication easiness, requiring only conventional holographic techniques and photomasks, even for very complex structures.

References and Notes

[1] S. Akiba, M. Usami, and K. Utaka, "1.5-μm λ/4-shifted InGaAsP/InP DFB lasers," IEEE J. Lightwave Technol. **5**, 1564–1573 (1987).

[2] J. Hong, W. P. Huang, T. Makino, and G. Pakulski, "Static and dynamic characteristics of MQW DFB lasers with varying ridge width," IEE Proc. Optoelectron. **141**, 303–310 (1994).

[3] W. K. Chan, J. Chung, and R. J. Contolini, "Phase-shifted quarter micron holographic gratings by selective image reversal of photoresist," Appl. Opt. **127**, 1377–1380 (1988).

[4] Y. Dai, X. Chen, L. Xia, Y. Zhang, and S. Xie, "Sampled Bragg grating with desired response in one channel by use of a reconstruction algorithm and equivalent chirp," Opt. Lett. **29**, 1333–1335 (2004).

[5] D. Jiang, X. Chen, Y. Dai, H. Liu, and S. Xie, "A novel distributed feedback fiber laser based on equivalent phase shift," IEEE Photon. Technol. Lett. **16**, 2598–2600 (2004).

[6] Y. Dai, X. Chen, J. Sun, Y. Yao, and S. Xie, "High-performance, high-chip-count optical code division multiple access encoders-decoders based on a reconstruction equivalent-chirp technique," Opt. Lett. **31**, 1618–1620 (2006).

[7] Y. Dai, X. Chen, D. Jiang, S. Xie, and C. Fan, "Equivalent phase shift in a fiber Bragg grating achieved by changing the sampling period," IEEE Photon. Technol. Lett. **16**, 2284–2286 (2004).

[8] G. P. Agrawal and A. H. Bobeck, "Modeling of distributed feedback semiconductor lasers with axiallyvarying parameters," IEEE J. Quantum Electron. **24**, 2407–2414 (1988).

[9] J. E. A. Whiteaway, G. H. B. Thompson, A. J. Collar, and C. J. Armistead, "The design and assessment of λ/4 phase-shifted DFB laser structures," IEEE J. Quantum Electron. **25**, 1261–1279 (1989).

[10] H. Soda and H. Imai, "Analysis of the spectrum behavior below the threshold in DFB lasers," IEEE J. Quantum Electron. **22**, 637–641 (1986).

[11] S. Nilsson, T. Kjellberg, T. Klinga, R. Z. Schatz, J. Wallin, and K. Streubel, "Improved spectral characteristics of MQW-DFB lasers by incorporation of multiple phase-shifts," IEEE J. Lightwave Technol. **13**, 434–441 (1995).

High Channel Count and High Precision Channel Spacing Multi-wavelength Laser Array for Future PICs[*]

Yuechun Shi[1], Simin Li[1,2], Xiangfei Chen[1], Lianyan Li[1,3], Jingsi Li[4], Tingting Zhang[1], Jilin Zheng[1], Yunshan Zhang[1], Song Tang[1], Lianping Hou[2], John H. Marsh[2] and Bocang Qiu[5]

[1] National Laboratory of Solid State Microstructures, College of Engineering and Applied Sciences, Microwave-Photonics Technology Laboratory, Nanjing University, Nanjing, 210093, China,

[2] School of Engineering, University of Glasgow, Glasgow G12 8QQ, U.K,

[3] Photonics Research Group, Department of Information Technology (INTEC), Ghent University, Sint-Pietersnieuwstraat 41, 9000 Ghent, Belgium,

[4] Microelectronic Research Center, Department of Electrical Engineering, The University of Texas at Austin, Austin, Texas 78758, USA,

[5] Suzhou Institute of Nano-Tech and Nano-Bionics, Chinese Academy of Sciences, Suzhou 215123, China.

Multi-wavelength semiconductor laser arrays (MLAs) have wide applications in wavelength multiplexing division (WDM) networks. In spite of their tremendous potential, adoption of the MLA has been hampered by a number of issues, particularly wavelength precision and fabrication cost. In this paper, we report high channel count MLAs in which the wavelengths of each channel can be determined precisely through low-cost standard μm-level photolithography/holographic lithography and the reconstruction-equivalent-chirp (REC) technique. 60-wavelength MLAs with good wavelength spacing uniformity have been demonstrated experimentally, in which nearly 83% lasers are within a wavelength deviation of ±0.20 nm, corresponding to a tolerance of ±0.032 nm in the period pitch. As a result of employing the equivalent phase shift technique, the single longitudinal mode (SLM) yield is nearly 100%, while the theoretical yield of standard DFB lasers is only around 33.3%.

Since the proposal of the concept of photonic integrated circuits (PICs), tremendous progress has been made. In 2005, Infinera Corp. rolled out the first commercial PICs, in which hundreds of optical functions were integrated onto a small form factor chip for wavelength multiplexing division (WDM) systems[1], and a monolithically integrated 5 × 100 Gb/s WDM chip has now been demonstrated. Despite the advances made in recent years, there are still some general challenges associated with PICs, such as materials[2,3], integration of the isolators[4], and ultra-low-cost fabrication[5]. Of the issues indicated above, the critical issue to be addressed is how to increase the integration density at a very

---

[*] Sci.Rep.,2014,4:7377

low cost. Multi-wavelength laser arrays (MLAs) with a high channel count are considered to be an engine of PICs, but high-volume production of MLAs with accurate wavelength control and low manufacturing cost remains a huge challenge. Currently, the distributed feedback (DFB) lasers used in MLAs are fabricated using electron beam lithography (EBL), which offers high resolution fabrication but low throughput because of the long writing time[6]. It is also well-known that EBL suffers from drawbacks such as blanking or deflection errors and shaping errors. Very few references have discussed the non-uniformity of the wavelength spacing of the devices fabricated using EBL. In Ref. [6], it is shown that using EBL, only 35% lasers have a wavelength variation of less than ±0.2 nm. To the best of our knowledge, this is the most recent paper reporting a statistically significant data set concerning wavelength accuracy. Ref. [7] shows that the error associated with the EBL process may be as large as 3 nm. No further reports with detailed information are available. Such issues greatly decrease the yield of monolithically integrated WDM PICs and significantly increase their manufacturing cost.

The yield and cost of DFB laser arrays are considerably different from those of individual lasers. At present, the manufacturing cost of an individual laser is very low. However, when the yield is 80% for an individual component, the yield of a 60-laser array is only around 0.0015%, and the cost will be several tens of thousand times more than that of the individual lasers, which makes it impossible to manufacture high-channel-count, monolithically integrated WDM chips. Furthermore the fine wavelength tuning required for each channel can only be accomplished using a sophisticated chip structure and complex auxiliary systems, leading to extra power consumption and degraded laser performances[8].

The yield issue is one of the major obstacles in manufacturing ultra-large scale PICs. In this paper, 60-channel WDM laser arrays based on the reconstruction-equivalent-chirp (REC) technique are reported, in which nearly 83% of the lasers are within a wavelength deviation of ±0.20 nm. Furthermore, the lasers are fabricated using only standard commercial semiconductor processes and μm-level photolithography. The good experimental results imply that the two great obstacles to manufacture MLAs for very-large-scale PICs, namely poor wavelength accuracy and the low yield, which have impeded progress for nearly three decades, have been essentially overcome.

Results

Principle of the REC technique. Although the basic principle of REC has been well illustrated[9,10], here the REC technique is further explained by a new view of the well-known wave-vector conservation principle. The grating-vector conversion relation in a sampled grating with a uniform seed (basic) grating and arbitrary sampling pattern can be expressed as,

$$K_g(z) = K_0 + K_s(z) \text{ or } \frac{1}{\Lambda_m(z)} = \frac{1}{\Lambda_0(z)} + \frac{mf'(z)}{P} \tag{1}$$

Here K_g is the wave-vector of the mth order sub-grating of the sampled grating, K_s is the wave-vector of the mth order Fourier component of the sampling pattern, and K_0 is the wave-vector of the seed grating. $f'(z)$ describes the arbitrary profile of the sampling pattern. P is the reference sampling period, which is of μm scale. From equation (1), it can be seen that an additional component $\frac{mf'(z)}{P}$ with large scale of P is introduced to manipulate the grating structure.

As illustrated in Fig. 1, an additional wave-vector $K_s(z)$, resulting from the large scale sampling pattern, is introduced artificially. Hence, the wave-vector of the sampled grating $K_g(z)$ can be manipulated by altering the value of $K_s(z)$. If the phase match condition between the light and $K_g(z)$ is satisfied, the interaction takes place. As a consequence, the optical properties of the grating can be equivalently realized by designing a μm-scale sampling pattern, and the wavelength precision can be improved by a factor of $mf'(z)/(P/\Lambda_0 + mf'(z))$ [2,11,12]. Therefore, the fabrication tolerance can be relaxed and the fabrication cost can be dramatically reduced. This principle is very similar to quasi-phase matching (QPM) in nonlinear materials for high efficiency light wavelength conversion[13,14], where an artificial periodic structure with larger period is introduced to produce a desired phase matching condition. So the basic principle of the REC technique can be regarded as microstructure based QPM.

FIG.1. Schematic illustration of the wave-vector conversion of REC technique. The sampling pattern provides an additional wave-vector $K_s(z)$ to form a new wave-vector of $K_g(z)$ to manipulate the light behavior in a waveguide with uniform seed (basic) grating wave-vector of K_0. Through this vector conversion, complex nanometer grating structures can be equivalently realized by μm-level structures.

DFB laser array based on REC technique. 60-wavelength DFB laser arrays with a π-equivalent phase shift (π-EPS) were designed and fabricated. The Bragg wavelength of the 0th order sub-grating is about 1640 nm (seed grating period is about 258.7 nm), a wavelength where the gain is small enough to avoid unwanted lasing. The designed wavelength spacing of the arrays is 0.2 nm, 0.4 nm and 0.8 nm, respectively. The sampling periods for 0.8 nm wavelength spacing are from about 2.82 μm to 4.58 μm. The cavity length is 600 μm and the lateral pitch of lasers is 125 μm. A 2 μm ridge waveguide is used to guide light and <1.0% Anti-reflection facet coatings(AR/AR) are applied to avoid the influence of random phase reflections from the facets. The normalized coupling coefficient of the ±1st sub-grating is around 2.5. Fig. 2 shows a schematic illustration of the DFB laser array with π-EPS.

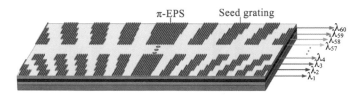

FIG.2. Schematic of the DFB laser array with π-EPS. The seed grating is uniform with pre-designed sampling pattern with μm-scale for equivalently realizing the nano-fine grating structures. The wavelength can be tailored by sampling period and the single-longitudinal-mode can be guaranteed by π-EPS.

Wavelength accuracy performances of the arrays. DFB MLAs were fabricated on two wafers with different seed grating periods. Seven 60-wavelength MLAs from one wafer were randomly selected for a detailed analysis of the lasing wavelength precision. All the lasers were measured at the same bias current of 80 mA and at an ambient temperature of 23℃. The second wafer, with a different seed grating period, was used to verify the effectiveness of the REC technique.

The relative wavelength accuracy was determined as follows. The measured wavelengths of the lasers in arrays were measured and linearly fitted. The wavelength residual, which indicates wavelength deviations (δ) from the fitted line, could then be obtained. The slope of the fitted line denotes the wavelength spacing of the fabricated MLA. The detailed wavelength residual frequency counts of the seven measured arrays are shown in Fig. 3. For all the laser arrays, the mean lasing wavelength residuals of 83.3% of the lasers are within ±0.20 nm, and 93.5% are within ±0.30 nm. Lasers with a wavelength deviation of >0.50 nm are less than 1.0% of the total laser count. The single-longitudinal-mode (SLM) yield of lasers from the seven arrays are 98.3%, 100%, 100%, 100%, 98.3%, 93.3% and 100% respectively, with the average SLM yield being 98.6%.

During cleaving and measurement, some laser bars were damaged, especially for the MLAs with a large channel number. In order to further investigate the lasing wavelength precision, we analyzed the statistics from 871 randomly selected lasers with the seed grating Bragg wavelength of 1,640 nm, among which some of the laser bars were broken but residual lasers within the arrays could be operated. A further 781 lasers from the second wafer with a seed grating Bragg wavelength of 1,660 nm were also randomly selected for comparison. A Gaussian distribution was fitted to the frequency count of the residual wavelengths.

As shown in Fig. 4, the standard deviations are 0.159 nm and 0.147 nm for the two different seed grating periods, which means 68.26% of the laser wavelengths are lied within ±0.159 nm and ±0.147 nm, respectively. The close values of the standard

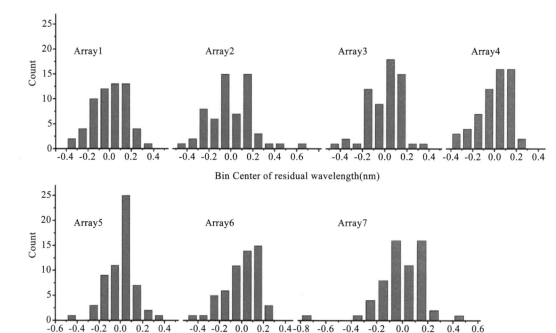

FIG.3. The statistical characteristics of the measured 7 arrays. The ratios of the wavelength residuals within ±0.20 nm are 81.4%, 71.7%, 90.0%, 85.0% 88.1% 82.1% and 85.0% respectively and the mean value is 83.3%. The single-longitudinal-mode yield of the arrays are 98.3%, 100%, 100%, 100%, 98.3%, 93.3% and 100% respectively.

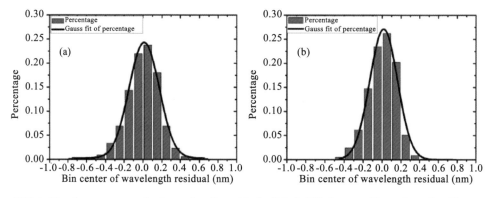

FIG.4. The frequency count of wavelength residual of (a) 871 lasers with seed grating Bragg wavelength of 1,640 nm and (b) 781 lasers with seed grating Bragg wavelength of 1,660 nm. The standard deviations of the two groups of lasers are 0.159 nm and 0.147 nm respectively.

deviations (~0.01 nm difference) shows that the REC method is effective for different seed grating periods, and confirms the flexibility of the fabrication technique.

Absolute wavelength accuracy is another key parameter that must be evaluated. We randomly selected 6 laser array bars with 15 wavelengths, for which the same wavelength spacing of 0.8 nm was obtained, as shown in Fig. 5(a). Fig. 5(b) illustrates the maximum wavelength differences in each channel for the 6 laser arrays, which vary from 0.36 nm to 1.17 nm with a mean value of 0.788 nm (±0.394 nm) and standard deviation of 0.222 nm. It should be noted that the wafer with epitaxy (epi-wafer) is a commercially available product and is briefly described in following part of Methods and the relative distance between two measured arrays may be very large, which would magnify the effects of non-uniformity of the wafer, non-uniformity in the holographic lithography and imperfection in fabrication. As a result, the absolute accuracy is worse than the relative wavelength accuracy. However, the deviation of 0.394 nm in the mean absolute wavelength is sufficiently small that it can be easily compensated by adjusting the operating temperature of the chip, as the thermally induced wavelength shift is about 0.09 nm/℃.

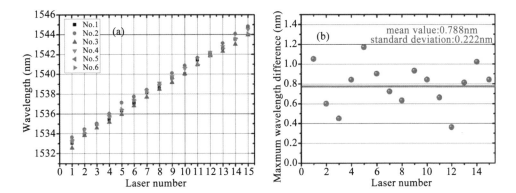

FIG. 5. (a) The lasing wavelengths of 6 laser arrays with wavelength spacing of 0.8 nm. (b) The maximum wavelength differences for the 15 channels of the 6 randomly selected laser arrays. The mean value is 0.788 nm (±0.394 nm) and standard deviation is 0.222 nm.

An example of 60-wavelength DFB laser array. The detailed lasing spectrum of one array (Array No.5 in Fig. 3) was randomly selected. Its wavelength spacing is 0.8 nm, as shown in Fig. 6(a). One laser out of the 60 elements shows dual mode, so the SLM ratio of the array is as high as 98.3%. Fig. 6(b) shows the lasing wavelengths as well as the linear fitting curve. In order to further analyze the deviation of the measured data from the fitted curve, the wavelength residual values of 59 lasers (the laser with dual mode was ignored) after fitting is given in Fig. 6(c), which shows that 88% of the laser wavelengths lie within a deviation of ±0.20 nm [Fig. 3 Array No.5 shows detailed information]. The threshold currents are between 25 mA and 35 mA. The high threshold mainly results from the long cavity length (600 μm), and could be reduced by optimizing parameters such as the cavity length, coupling coefficient of the grating, by using a buried-heterostructure waveguide[15].

Discussion

Other than the grating error, some other factors may also influence the wavelength accuracy. These include non-uniformity of the epi-wafer, waveguide inhomogeneity (e.g.

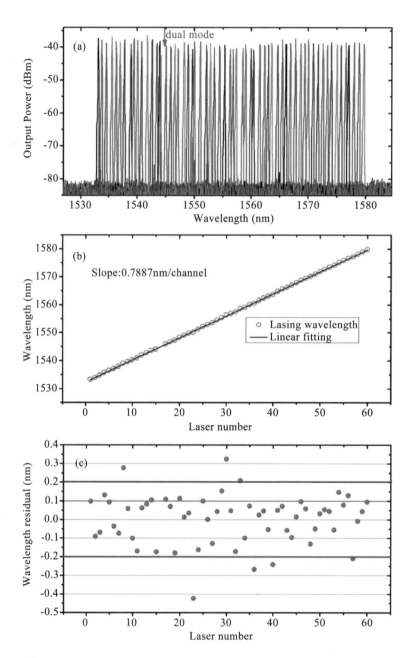

FIG.6. (a) The measured lasing spectra of one 60-wavelength array which is corresponding to Array No.5 in Fig. 3. One laser is dual mode. (b) The lasing wavelengths and the linear fitting curve with the slope of 0.7887 nm/channel (the deigned value is 0.80 nm/channel). (c) The wavelength residuals after linear fitting which is also plotted in Fig. 3 Array No.5 for detailed statistical data.

width variations) and imperfect facet coatings. The epi-wafer may have spatial variations arising during material growth. Imperfect fabrication of the waveguide can lead to index variations along its length. In both cases, the lasing wavelength may be slightly shifted. In addition, imperfect facet coatings can lead to residual facet reflections with random phase which can also shift the wavelength. Therefore, in order to further improve the wavelength accuracy, the other processes including material growth, fabrication and post-processing should be improved in the future.

To eliminate the remaining wavelength errors, the absolute wavelength can be readily compensated by changing the temperature of the entire chip by only a few degrees, as already analyzed above. The residual errors in the wavelength spacing can be reduced by improving the fabrication processes or can be compensated directly by fine tuning the injection currents, with a measured current-wavelength slope of about 0.012 nm/mA, to exactly meet the DWDM standard.

DFB laser arrays with grating structures with special properties can also be fabricated using this approach, for example the equivalent of a DFB laser array with 3 phase shifts, which is usually designed for long-cavity narrow-linewidth lasing[12]. It is possible for all the lasers to share the same seed grating, and then various lasers or arrays can be realized simultaneously when defining the sampling patterns. In addition to active components, the REC technique can also be applied to passive filters no matter what material is used[16], where, similarly to the laser array, the Bragg waveguide can also be well controlled[17]. Therefore, both active and passive photonic components can be integrated on the same chip using the REC technique, and the wavelength precision of the active and passive components can be strictly guaranteed simultaneously.

In conclusion, the statistical properties of high channel count DFB laser arrays fabricated using the REC technique have been experimentally studied. Excellent lasing wavelength precision has been achieved. An example of a 60-wavelength DFB laser array has also been demonstrated. All the lasers were fabricated using conventional holographic lithography and photolithography processes; as a result of using the REC technique, high reproducibility is obtained at a remarkably low fabrication cost. Moreover, this technique can also be applied to realize other passive photonic devices based on Bragg grating structures. This report shows that low-cost fabrication of MLAs with high-precision wavelength spacing has been solved. We believe this will provide a platform for volume manufacture of large-scale PICs using a low cost fabrication process based on defining μm-scale features.

References and Notes

[1] Nagarajan, R. *et al*. Large-scale photonic integrated circuits. *IEEE J. Sel. Top. Quan. Electron.* **11**,

50 - 65(2005).

[2] Polman, A. Photonic materials: Teaching silicon new tricks. *Nat. Mater.* **1**, 10 - 12(2002).

[3] Welch, D. F. et al. The realization of large-scale photonic integrated circuits and the associated impact on fiber-optic communication systems. *J. Lightwave Technol.* **24**, 4674 - 4683 (2006).

[4] Oliver, G. Silicon photonics: Integrated isolators. *Nat. Photonics.* **5**, 571 - 571(2011).

[5] Koch, T. L. & Koren, U. Semiconductor photonic integrated circuits. *IEEE J. Quantum Electron.* **27**, 641 - 653(1991).

[6] Lee, T.-P. et al. Multiwavelength DFB laser array transmitters for ONTC reconfigurable optical network testbed. *J. Lightwave Technol.* **14**, 967 - 976 (1996).

[7] Zanola, M., Strain, M. J., Giuliani, G. & Sorel, M. Post-growth fabrication of multiple wavelength DFB laser arrays with precise wavelengthspacing. *IEEE Photon. Technol. Lett.* **24**, 1063 - 1065(2012).

[8] Felipe, D. et al. Hybrid InP/Polymer optical line terminals for 40-Channel 100-GHz spectrum-sliced WDM-PON. *39th European Conference and Exhibition on Optical Communication(ECOC 2013)*, 237 - 239; DOI: 10.1049/cp.2013.1352(2013).

[9] Dai, Y. & Chen, X. DFB semiconductor lasers based on reconstruction-equivalent-chirp technology. *Opt. Expr.* **15**, 2348 - 2353 (2007).

[10] Li, J. et al. Experimental demonstration of distributed feedback semiconductor lasers based on reconstruction-equivalent-chirp technology. *Opt. Expr.* **17**, 5240 - 5245(2009).

[11] Shi, Y. et al. Experimental demonstration of eight-wavelength distributed feedback semiconductor laser array using equivalent phase shift. *Opt. Lett.* **37**, 3315 - 3317 (2012).

[12] Shi, Y. et al. Study of the multiwavelength DFB semiconductor laser array based on the reconstruction-equivalent-chirp technique. *J. Lightwave Technol.* **31**, 3243 - 3250 (2013).

[13] Armstrong, J. A., Bloembergen, N., Ducuing, J. & Pershan, P. S. Interaction between light waves in a nonlinear dielectric. *Phys. Rev.* **127**, 1918 - 1939 (1962).

[14] Zhu, S., Zhu, Y. & Ming, N. Quasi-phase-matched third-harmonic generation in a quasi-periodic optical superlattice. *Science* **278**, 843 - 846 (1997).

[15] Utaka, K., Akiba, S., Sakai, K. & Matsushima, Y. Room-temperature CW operation of distributed-feedback buried-eterostructure InGaAsP/InP lasers emitting at 1.57 μm. *Electron. Lett.* **17**, 961 - 963 (1981).

[16] Sun, J., Holzwarth, C. W. & Smith, H. I. Phase-shift bragg grating in silicon using equivalent phase-shift method. *IEEE Photon. Technol. Lett.* **24**, 25 - 27 (2012).

[17] Sun, J. et al. Uniformly spaced $\lambda/4$-shifted Bragg grating array with wafer-scale CMOS-compatible process. *Opt. Lett.* **38**, 4002 - 4004 (2013).

[18] The authors would like to acknowledge the National Nature Science Foundation of China under Grant 61090392, National "863" project under Grand 2011AA010300 and LuxNet Corp. for post-processes of MLAs.

[19] Y.S. and X.C. conceived the idea, conducted theoretical analysis. Y.S., X.C. S.L., L.Li, J.L., J.Z. and Y.Z. performed the experiments and analyzed the data. T.Z. and S.T. carried out the measurements. L. H., J.M., B.Q. discussed the experiment results and revised the manuscript. All authors contributed to writing the paper.

[20] Device fabrication. The laser epi-wafer is fabricated by conventional two-stage low-pressure metal-organic vapor phase epitaxy (MOVPE) with a multiple-quantum-well (MQW) structure as the active region. On an S-doped n-type (100)-oriented InP substrate, an InP buffer layer, a lower AlGaInAs separate-confinement-heterostructure (SCH) layer, a multiple-quantum-well (MQW) structure, an upper AlGaInAs SCH layer, an InP etch stop layer and a 1.25Q InGaAsP grating layer are successively grown. The MQW structure contains five undoped 6nm-thick AlGaInAs quantum well layers with $\pm 1.2\%$ compressive strain and six 9nm-thick AlGaInAs barrier layers with -0.45% tensile strain. The photoluminescence (PL) peak of the MQW is around 1540 nm at room temperature. The two SCH layers are asymmetric structure with gradual variations. Then the sampled grating is defined by conventional holographic lithography combined with photolithography. A p-type InP cladding layer and a p-InGaAs contact layer are re-grown. After regrowth, a 2 μm ridge waveguide is formed by wet etching. Anti-reflection coatings are deposited on both facets with a reflectivity $<1.0\%$.

第九章　总结与展望
Chapter 9　Summary and Outlook

第九章 总结与展望

朱永元 祝世宁

介电体超晶格是闵乃本于二十世纪八十年代中提出的微结构材料体系,经过三十多年的努力,它已经从最初的概念演变成当今的现实,从微结构材料研究领域的弄潮儿发展成为功能材料滚滚洪流中重要方面军。经过多年的探索,介电体超晶格内涵不断扩展,从验证准相位匹配,实现激光高效倍频发展到今天实际上包含三种不同功能的介电微结构晶体:光学超晶格、声学超晶格和离子型声子晶体。本书由南京大学有关研究组历年发表的介电体超晶格研究论文中精选出的 90 篇代表性论文编辑而成。全书根据内容分成九个章节,除了第一章绪论和第九章总结和展望外,其余七章每一章都自成体系,对应着介电体超晶格发展的一个主要方向。读者可以从每一章的阅读中体会出每篇文章作者的研究思路和这一方向发展的时间脉络。材料是时代的标志,是经济和科学发展程度的象征。要使材料研究经久不衰,首先就必须使其建立在坚实的科学基础之上。介电体超晶格既是一种新功能材料,也是一个完善的科学体系。它的发展经过提出基本概念,建立基础理论,实验验证基本效应,直到现今的领域拓展和应用研究。可以看出介电体超晶格的研究有其特定的内涵,但它又不是简单地从定义出发,它的发展遵循开放模式。界域的开拓与不同知识的融入,使其内涵日臻丰富,外延不断拓展,功能不断发掘,最后达到材料研究的最高境界——被应用所青睐,由需求所牵引。

微结构功能材料的研究源于半导体超晶格。1970 年 Esaki 和 Tsu 等提出[IBM J. Research and Development 14,61(1970)]用两种具有不同禁带宽度半导体异质结构形成的量子阱超结构来剪裁材料原有的能带,改变材料的光子发射特性,在光电子领域获得了重要应用。从那时起通过人工微结构来优化、重塑甚至创造材料的新性能逐步成为材料研究的一种"范式"。介电体超晶格的研究也遵循了这一范式。虽然介电体超晶格的概念是类比于半导体超晶格提出的,但它的最重要的理论基础之一——准相位匹配原理早在 1962 年就已经由 Bloembergen 等人提出了[Phy. Rev. 127(6),1918(1962)]。准相位匹配的实验验证一直受到材料研究的困扰,直至二十世纪七十年代末、八十年代初,发现了铁电畴的自发极化矢量取向决定于晶体生长层中交变的溶质浓度梯度,于是发展了聚片多畴的生长层技术,制备成周期性聚片多畴铌酸锂晶体,在其中完成了倍频实验,给出了准相位匹配最为直接的实验验证[Beijing/Shanghai Proceeding of an International Conference on Lasers,May,1980,China Academic Publishers & John Wiley & Sons,1983,283]。到了二十世纪八十年代中后期,由于在周期铁电畴调制的铌酸锂(LN)、钽酸锂(LT)晶体中更多的物理效应被揭示,介电体超晶格概念被及时提了出来。特别是将准晶结构引入超晶格体系,通过准周期光学超晶格,将 Bloembergen 等提出的准相位匹配理论拓展到多重准相位匹配,对非线性光学研究产生了重要影响。介电体超晶格的众多预言最终在畴工程的关键环节——室温电场极化技术1994年取得突破后相继被实验证实。此后介电体超晶格研究受到越来越广泛的关注和重视，最为重要的事件是1997年研制出Fibonacci准周期光学超晶格，实验证实了多重准相位匹配理论，实现了高效的多波长激光的倍频和激光直接三倍频。介电体超晶格研究从周期、准周期扩展到多周期、非周期，从一维扩展到二维、三维，相位匹配方式也从单一准相位匹配、多重准相位匹配，发展出局域准相位匹配（非线性惠更斯原理）直到提出非线性菲涅耳全息。畴工程对铁电畴及其分布实现了精准设计和控制，给研究带来更大的想象空间，极大地推动了介电体超晶格的理论与实验研究。由于铁电晶体中受畴调控的除了有二阶非线性光学系数外还有热电、压电、电光等所有奇数阶张量物性参数，介电体超晶格的研究内容也从最初的准相位匹配激光倍频、三倍频等发展出利用畴工程实现电光调制、电声调制、声光调制、高频体声波激发以及通过光波与声波耦合产生极化激元等新效应，最终形成了光学超晶格、声学超晶格和离子型声子晶体三类不同功能的介电体超晶格晶体，发展出相关理论，研制出一批有应用价值的材料和器件。

1987年E.Yablonovitch[Phys.Rev.Lett.58,2059(1987)]和S.John[Phys. Rev.Lett.58,2486(1987)]分别提出了另外一种介电微结构材料——光子晶体。光子晶体中调制的物性参数是介电常数或折射率。对于光子晶体，光沿着三维方向都受到强烈的布拉格散射，以至于在三维布里渊空间发生光子带隙的重叠，形成光子全带隙。人们能够通过光子晶体中光子能带的设计控制光的方向、偏振、位相、波速等自由度，甚至通过点缺陷和线缺陷的引入，使电磁场的态密度局域在晶体中某一位置，全带隙光子晶体提供了对原子自发辐射等光学过程的控制新途径。基于同样原理，光子晶体概念很容易推广到声的领域，声子晶体也随之产生。声子晶体中也存在声子能带和带隙，这起源于弹性系数空间调制对声波的散射和声波之间的干涉效应。声子晶体开辟了微结构材料在声学领域研究和应用的新天地。光学超晶格与声学超晶格概念的提出与研究和光子晶体与声子晶体基本同步，虽然材料构成和功能不尽相同，但互有渗透与交流，这在本书的第三章和第四章中有所反映。例如利用电光与光折变效应，可以在铌酸锂一类非线性晶体中写入空间光栅，这类折射率周期性弱调制的结构也能与光产生相互作用，并对光的传播特性产生有效的调制。可见在光学超晶格的早期研究中就已经考虑到折射率调制的情况并运用了光子带隙的概念。有趣的是，由于两者平行发展，概念之间也相互借用与参考，越来越多的研究人员将二价非线性系数（对应于三阶张量）周期或有序调制的光学超晶格简称之为"非线性光子晶体"，以求与折射率强调制"光子晶体"相对应，尽管这种所谓"非线性光子晶体"并不存在线性折射率的调制而带来的光子带隙，更何况具有折射率调制的光子晶体本身也可能具有光学非线性。在由光折变效应构成折射率调制的二维光子晶体结构中就观察到了非线性响应所导致光学双稳、光学失稳和混沌等一系列在具有光学反馈的非线性系统中所特有的光学效应，这些效应遵循着四波非线性动力学理论。在我们关于光子带隙中非线性问题的研究开展之后，引起国际同行的关注，一批光子晶体中非线性光学新效应陆续被揭示。

离子型声子晶体是在九十年代末在声学超晶格研究基础上提出的一个新概念，在压电微结构材料中电磁波可以与超晶格声波耦合构成一种新的元激发，这十分类似于离子晶体中的光波与光学支声子之间的耦合形成的"极化激元"。从物理上理解，这一物理过程与本世纪初Pendry等提出的metamaterials（超构材料）[Science 312,1780(2006)]概念也是相通的，两者的差别只是受激的基本共振单元不一样。在Pendry最初提出的metamaterials

中基础共振与发射单元是由开口金属环构成的谐振器,它能被入射电磁波共振激发,继而发射电磁波,并与原电磁波耦合导致介质在谐振频率附近色散关系的改变,影响波的传输。我们利用的是铁电畴的压电效应产生的共振,超晶格的每一个结构单元都可以看作压电谐振子,能接收和发射电磁波或声波,因此它就是一种压电型的超构材料。在结构单元的共振区,介电体超晶格会出现由共振导致反常色散,介电常数和折射率都呈负值,具有左手材料的某种性质。极化激元也有带隙,该带隙起源于电磁波与超晶格振动的耦合,而在光子晶体中其带隙来源于周期单元的布拉格反射。在离子晶体中由极化激元导致的介电异常发生在远红外波段,而离子型声子晶体的介电异常发生在微波波段。其波长与超晶格的压电共振频率有关,因而是可设计的。这预示介电体超晶格也可用在微波电磁器件和电声器件。由此可以看出在超构材料概念形成过程中,我们 1999 年就开始的离子型声子晶体研究应该有其独特的贡献。

有关介电体超晶格的应用研究一直受到广泛关注并被不断开发。在声学领域,声学超晶格超高频体波谐振器、换能器和滤波器已经在无线通讯网站找到实际应用。在激光领域,光学超晶格的电光、声光调控可用于研制结构更加紧凑的全固态脉冲激光器。运用高的非线性增益和相移,光学超晶格可代替半导体锁模器件应用于激光锁模和高重复频率超短脉冲激光产生。半导体锁模材料和染料有特定的工作波长,而光学超晶格却不受限制,可以设计工作在超晶格透明窗口的任何波长,这为研制中红外锁模激光器和超快激光器提供了技术方案。光学超晶格在激光技术方面最大的应用可能就在中红外领域,这是因为铌酸锂光学超晶格的红外吸收边可达 5 微米,这填补了 3~5 微米缺乏可适用的红外非线性晶体和激光晶体的空白。已经研制出的可调谐中红外激光器能覆盖从 1.5 微米至 5 微米近中红外谱段,能在从连续到纳秒、皮秒不同工作模式下工作。采用光学超晶格与远红外非线性晶体级联有可能将全固态激光输出拓展到 10 微米左右的远红外波段。光学超晶格中红外激光技术的发展也带动了相关应用研究,研制中的远程大气监测雷达就采用光学超晶格中红外激光器技术。光学超晶格在高性能激光器研究方面仍有很大的空间,包括高精度光学频率梳,激光脉冲压缩和阿秒技术等。在光信息处理方面,光学超晶格周期分布反馈结构已被用在了半导体激光器阵列的设计中,作为激光光源集成到硅基光电芯片上,用于光电信息的高速转换与处理,或被刻写进光纤中用于提高探测灵敏度和信号的收集与处理。

光学超晶格用于量子光学是光学超晶格研究重要的新拓展,虽然一些初步的结果是预料中的,但是随着研究的深入它所展示的前景却让人始料未及。光学超晶格正在逐步取代 LBO、BBO 和 KTP 等常规非线性晶体,作为首选应用于各种量子光学实验和量子通讯、量子信息处理。相比于常规非线性晶体,光学超晶格有两大优势:一是光子产率高,二是易于集成。其中第一大优势是因为光学超晶格有更大的有效非线性系,能产生更高亮度的单光子和纠缠光子,这对量子信息处理和量子光学实验非常重要,意味更快的速度和更大的容量。第二大优势易于集成有两层含义,一层含义是指超晶格在高效产生不同类型纠缠光子的同时还能通过结构设计来调整光子的波前和位相,实现光子的分束、汇聚等功能,这相当于把一些分立光学元件功能集成到光学超晶格中,这能简化后续的实验光路,增加实验可靠性;另一层含义是指实践证明畴工程与光波导技术是工艺兼容的,可以在 LN、LT、KTP 晶片上通过畴工程完成超晶格图型写入和光波导制作,将纠缠光子源、电光调制器、波导分束器、波导干涉器等分立光学元件集成到一块芯片上,完成特定的量子信息处理功能。这样的功能芯片具有可扩展性。上述两层意义上的集成技术上都已基本成熟，我们已研制出了有集成功能的光学超晶格晶体和芯片，在芯片上演示特定光量子逻辑操作也已无困难。尽管这样，目前的进展离在光子芯片上完成通用量子计算的目标仍然相距甚远，还有许多重要技术需要突破，其中最为关键的是单光子可控存储。虽然现在已经有多种技术能演示单光子的存储，但要将这些存储单元集成到 LN 芯片上并实现可控操作仍是一个很大的难题。除此以外半导体激光器、单光子探测器等元件也需要考虑到在芯片上去集成的问题，最终目标的实现仍有很大的挑战。不过这种挑战对其他材料体系和技术方案也同样存在，竞争在所难免。除了解决通用量子计算的长远目标外，光量子芯片还有很大的发展空间。目前阶段可考虑先易后难地研制一些具有特定量子信息处理功能的芯片，如光量子随机行走芯片，这些芯片能演示量子搜索、量子模拟功能，也能解决某类具体的数学问题或者模拟真实物理系统难以演示的物理实验，就像一台小型专用量子计算机，正如费曼在二十世纪八十年代所预言的那样。除此以外，光学超晶格光子芯片一个持续热点将是用它来产生各类可调控的单光子、纠缠光子、高维纠缠甚至超纠缠光子。对光子来说 LN 晶片无疑是一个条件优异的舞台，通过集成技术，它可优雅地演示出原先需要非常复杂技术和光路组合才能获得的不同性能。可以预言高阶 W 态单光子源、可预知的单光子源、EPR 态多光子源等会借助光学超晶格 LN 晶片在不久的将来被陆续研制出来，用于各类量子光学实验。有关量子力学基础物理实验如延迟选择、波粒二相性、隐形传态等都可以从自由空间移植到光子芯片上，在更小的物理空间对引人关注的思想实验进行检验，在这其中光学超晶格能发挥的作用毋庸置疑。

"凡是过去，皆为序章"（摘自莎士比亚《暴风雨》），未知世界的探索永无止境。在探索过程中，有苦也有乐，有成功也会有失败。介电体超晶格的研究成果凝结着所有从事过该研究的研究人员心血，希望该书的出版能引起从事微结构材料的研究人员的兴趣，特别是年轻人能够从中有所借鉴。

Chapter 9 Summary and Outlook

Yongyuan Zhu   Shining Zhu

The concept of dielectric superlattice was firstly proposed by Naiben Ming in the eighties of the last century. After laser was invented, scientists began to consider light instead of electrons as the information carrier, because light has many advantages over electrons for information processing and information transmission. Dielectric materials can transmit light like metal or semiconductor transmitting electrons. However, light propagation in a homogeneous dielectric crystal is the same as in a continuous medium because its wavelength is much larger than the size of lattice unit. Ming and his coauthors considered constructing microstructure in a dielectric crystal to control light, in analogy to control electron in semiconductor or semiconductor superlattice. After 30 years of efforts, the idea for dielectric superlattice has evolved from the original concept into today's reality. Three kinds of dielectric microstructured crystal, optical superlattice, acoustic superlattice and ionic-type phononic crystal, have been developed. They exhibit many excellent functions to manipulate light and photon, sound and phonon. We make effort in collecting the 90 representative papers in these fields for this book. These papers are classified and compiled into eight chapters according to their contents. Each chapter is relatively independent and introduces one aspect about the researches. We hope that the audience of this book can understand the research thoughts of authors by reading each article, and can grasp the development sequence of each major aspect from reading each chapter. Material is the symbol of the times, and stands for the developing level of economy and science. In order to make new materials research durable, it is necessary to lay it on a solid scientific basis. Dielectric superlattice not only is one type of new functional material, but also is a well-developed scientific system, from the basic concepts, to the fundamental theory, and to the experimental validation. It contains a specific essence, but it does not restrict itself by a simple definition. Its development has followed an open mode. Along with more new knowledge merging into it, its connotation will be further extended and more applications will be found.

In 1970, Esaki and Tsu proposed the concept of semiconductor superlattice [IBM J. Research and Development 14, 61 (1970)], which is composed of two kinds of semiconductor materials with different band gaps. The original band gap of material is tailored by the superlattice structure, and therefore, new photon emission properties appear. Since then, it has gradually become a 'paradigm' of new material research, that is, optimizing, promoting, even creating material's new functions by constructing artificial microstructures. The study of the dielectric superlattice follows this paradigm. Although the concept of the dielectric superlattice is analogous to that of a semiconductor superlattice, its most important physical basis, the quasi-phase-matching(QPM) principle, was initiated by Bloembergen et al. in 1962 [Phy. Rev. 127(6), 1918(1962)]. This was earlier than the artificial band gap theory of semiconductor superlattice. The experimental verification of QPM has been plagued due to the difficulty on material fabrication, until the late seventies of the last century. Researchers in Nanjing University showed the first clear QPM experiment result—the enhanced second-harmonic-generation in $LiNbO_3$ (LN) crystals with periodic laminar ferroelectric domains [Beijing/Shanghai Proceeding of an International Conference on Lasers, May, China Academic Publishers & John Wiley & Sons, 1983, 283]. Through to the middle eighties, due to the more and more novel physical effects were discovered from the crystals with the periodic domains, the concept of dielectric superlattice is formally proposed as a new category of micro-structured material. Meantime, the quasi-crystal structure was introduced into nonlinear crystal, and the one-dimensional(1D) quasi-periodic superlattice was successfully developed. In a quasi-periodic superlattice, two or three optical parametric processes can realize QPM, simultaneously. This is different from the parametric process in a periodic superlattice in which only a single QPM may be satisfied. Thus the theory of QPM was extended from single parametric process to multiple parametric processes and the concept of QPM was extended to the multiple quasi-phase-matching (MQPM) accordingly. A new door was opened for nonlinear optics. Since then, more and more attentions have been paid to the research of dielectric superlattice. An important event occurred in 1997, that is, the first quasi-periodic optical superlattice with Fibonacci sequence was successfully prepared. As expected, a multi-wavelength frequency doubling and a direct third harmonic generation were demonstrated with high efficiency from this quasi-periodic superlattice. The idea of MQPM was experimentally proved eventually. Following this succees, the study of dielectric superlattice was further extended from periodic, quasi-periodic, to aperodic, from one-dimensional (1D) to 2D, even to 3D micro-structured materials. The scheme of QPM was extended from single QPM, to MQPM, to local QPM(nonlinear Huygens principle), and recently to the nonlinear Fresnel hologram. Nowadays different domain configurations can be designed and controlled with very high accuracy and precision in ferroelectric crystals, which brings more imaginable space for the study of superlattice, and greatly promotes the merging of theoretical and experiment researches.

In addition to the second order nonlinear optical coefficient, in ferroelectric crystals, there are other physical properties that can be modulated by domain order, including thermoelectric coefficient, piezoelectric coefficient, electro-optic coefficient etc. These coefficients are all odd order tensors, therefore can change their signs from "+" to "−"

with the polarization direction of the domain. In this case, according to the same principle, the related physical properties can be redesigned or optimized in the dielectric superlattices by domain engineering. Even different physical processes may couple each other at certain condition, like generating the polariton by coupling of photon and phonon. Various novel devices, such as electro-optic, electro-acoustic, high frequency transducer and microwave devices etc., have been developed to serve different applications and engineering demands. According to the introduction above, the dielectric superlattice actually contains three kinds of microstructure crystals: optical superlattice, acoustic superlattice and ionic-type acoustic crystal, which is classified based on their functions and physical properties.

In 1987 E. Yablonovitch[Phys. Rev. Lett. 58, 2059 (1987)] and S. John[Phys. Rev. Lett. 58, 2486 (1987)] respectively proposed another microstructured dielectric material, so-called photonic crystal. The physical parameter modulated in photonic crystal is dielectric coefficient or refractive index. The light propagating in photonic crystal would be strongly scattered in all three dimensions so as to result in a full photonic band gap (PBG). By photonic band gap engineering, the wave vector, polarization, phase and even velocity of light could be controlled. Through the introduction of point defect or line defect, the light wave could even be localized. The existence of full PBG provides a new method to control the spontaneous radiation of atoms. Based on the same principle, the concept of photonic crystal can be easily extended to phonons. Likewise the phononic crystal has band structure and band gap for phonons which originates from the interference of the scattered acoustic waves due to the modulation of elastic coefficients. The study on the optical superlattice and acoustic superlattice begins almost at the same time with photonic crystal and phononic crystal. Although the material constituents and functions are not the same, they all deal with the interaction of waves with materials, as presented in this book. Interestingly, due to the parallelism of study on the optical superlattice and acoustic superlattice, photonic crystal and phononic crystal, some concepts are used for reference by each other. In fact as early as in 1988, we studied the dynamic evolution of light propagating in refractive index modulated structures with the concept of photonic band gap. By using the electro-optic and photorefractive effect, the gratings could be written into some nonlinear optical crystals such as LN, forming a superlattice. Such refractive-weakly-modulated structure could also control the propagation of light. Nonlinear optical responses such as optical bistability, instability and chaos, which exist usually in nonlinear systems with optical feedback, were observed in such a photonic crystal (2D optical superlattice). These phenomena lead to the development of the nonlinear four-wave dynamical theory. Later on, more and more researchers called optical superlattice with the modulation of the second-order nonlinear optical coefficient "the nonlinear photonic crystal", though no photonic band gap exists inside it. Of course, it is only an analogy with photonic crystal in which the refractive index of medium is modulated.

At the end of 1990s, the idea of ionic-type phononic crystal was proposed by our team

(Chapter 5). We believed that electromagnetic wave propagates in a domain-engineered superlattice could couple with superlattice acoustic vibration excited by piezoelectric effect, which results in a new-type polariton. This is highly similar to the polariton generating in an ion crystal due to the coupling of light with transverse optical phonons. Interestingly, because of the anisotropic of piezoelectric effect, the electromagnetic wave can be coupled not only with transverse superlattice vibration but also in some cases with longitudinal superlattice vibration. The above polariton concept can be easily understood from the point of view of metamaterials proposed by Pendry et al. [Science 312, 1780(2006)] at the beginning of this century as well. The only difference between the ionic-type phononic crystal and metamaterial is that the basic resonant unit stimulated is different. For Pendry's model, the basic resonant unit in metamaterial is the metallic split-ring excited by electromagnetic wave, including light wave, and then reemit it, thus the dispersion characteristic of medium changes. Whereas in our model, each domain in ionic-type phononic crystal is a piezoelectric resonant unit excited by and reemit either the electromagnetic or acoustic wave. Therefore it can be taken as a piezoelectric metamaterial. The superlattice can show abnormal dispersion nearby the piezoelectric resonance, even the dielectric coefficient and refractive index appears in negative value, which is similar to the characteristic of left-handed materials. For ionic crystals, the abnormal dispersion occurs at the far-infrared region, whereas for ionic-type phononic crystal the abnormal dispersion occurs at the microwave region. The resonance and the abnormal dispersion are related to the structure of the superlattice and thus can be designed and controlled artificially. Furthermore, the ionic-type phononic crystal possesses a band gap different from that of photonic crystal. It originates from the coupling of electromagnetic wave with the superlattice vibration rather than from the Bragg reflection. The above phenomena show that the dielectric superlattice can be used for some microwave and electro-acoustic devices.

Research on the applications of dielectric superlattice has been widely concerned and has been developed incessantly. In the acoustic field, the acoustic resonator, the transducer and the filter with ultra-high frequency have already found practical applications in wireless communication base stations. In the laser field, the most possible applications of optical superlattice may be in mid-infrared field as an effective nonlinear crystal. LN crystal has a wide optical transparence range from 0.4 to 5 microns, which can be used to develop mid-infrared laser by QPM optical-parametric technique. As a result the tunable mid-infrared lasers have been developed with the output covering the spectrum range from 1.5 to 5 microns and with the working modes at continuous and pulses with pulse width from nanosecond to picosecond. This is very valuable, because there is still a lack of alternative infrared nonlinear crystals and laser crystals at this moment. A scheme to expand the laser output to 10 microns or so by using the optical superlattice cascaded by a far infrared nonlinear crystal has been proposed. Infrared laser has many important applications, such as molecule spectrum and fire-alarm prediction. The development of infrared laser has also brought along the progress in related technologies, such as gas remote monitoring radar for environment protection. The electro-optic and acousto-optic modulator made from the optical superlattice can be used to develop a compact all-solid-state pulse laser. With high nonlinear gain and large phase shift, the optical superlattice has been used to the novel mode-locked laser with high repetition rate. In general, mode-locked laser takes dye or semiconductor as mode-locked medium and works in an appointed wavelength. The optical superlattice can be designed to work in any wavelength, which provides a technical solution for the development of mode-locked laser and ultrafast laser in mid-infrared range. There is large room in the research of high performance laser with optical superlattice as nonlinear crystal, such as high precision optical frequency comb, laser pulse compression for attosecond technique and so on. In optically integrated chip, the periodically distributed feedback structure has been used in the design of the semiconductor laser array. Such laser array has been integrated onto silicon chips for high speed switching and processing of optical information. And periodic structure has even been written into a fiber used to improve the detection sensitivity for signal collection and processing.

It is a great expansion for the optical superlattice to be used to quantum optics and quantum information. Although some preliminary results can be expected, the rapid progress and the attractive prospect are still going beyond the previous estimation. Optical superlattice is gradually replacing conventional nonlinear crystals, such as LBO, BBO and KTP, as the first choice for various single photonic and entangled photonic sources for quantum optical experiments, quantum communication and quantum information processing. Compared with the conventional nonlinear crystals, the optical superlattice has two advantages: one is the high photon yield and the other is the ease to integrate. The optical superlattice has larger effective nonlinear coefficients than the conventional nonlinear crystals, which leads to higher yield, for either single photons or entangled photon-pairs. This is very important, because the higher yield means the faster speed and the more capacity in quantum information processing and quantum optics experiments. The advantages of being easily integrated are two-fold. The first one refers to the fact that the superlattice not only generates single photons and entangled photons but also adjusts at the same time the features of photons such as phase, path and wave-front etc., realizing beam splitting, reflecting, convergence and other functions. This is equivalent to integrate a number of discrete optical components into an optical superlattice chip. The functional integration may simplify the experiment system, and increase its reliability. The second one means that both domain engineering and optical waveguide is compatible with each other in technology. One can prepare superlattices and waveguides onto the same crystal chips. Therefore, more discrete components, such as photon sources, electro-optic and acousto-optic modulators, photon splitters and interferometers can be integrated into the same chip in order to complete more complex quantum information processing. Such a featured chip has high scalability and high stability, hence, becomes a developing direction of quantum technology. The integrated techniques mentioned above for the photonic chip have basically matured, and the related progresses are introduced in this book. There is not any big difficulty to demonstrate the quantum logic operation on such a chip at present. Even so, however, there is a long way to go to realize the universal quantum computing on the photonic chip. There are many bottlenecks that have to be broken through. One of the most critical is the single photon controllable memory. Although there are already a variety of techniques to demonstrate the single photon memory, it is still a big challenge to integrate these memory cells into the LN chip and realize the controllable operation. In addition, how to integrate the semiconductor laser, low noise single photon detector and other components onto the same LN chip also need to be taken into account. The ultimate goal remains a great challenge. However, the same challenge exists for other competitive material systems and technology. Competition in the field is inevitable. In addition to the long-term goal of realizing universal quantum computing, our strategy at present is to develop some chips, from easy to difficult, with specific functions, such as quantum random walk and quantum simulation. Such chips can demonstrate quantum searching, solve a specific class of mathematical problems or perform some physical simulation that is difficult to realize in the present world. Such an individualized chip looks like a small special quantum computer to solve individually scientific problem as Feynman prophesied in the eighties of the last century. In addition, a continuous hot topic in optical superlattice is to consider it as a good candidate for the single and entangled photon sources, because it can be designed to produce various kinds of adjustable photonic states. Many results have proved that the LN chip is an excellent platform for manipulating photons from the view of integration photonics. The high-order single photon W state, the heralded single photon state, multi-photon EPR state can be predicted to realize soon on LN chips by using domain engineering and integration photonics technology. Currently, they are still very complex and time-consuming tasks, realized only on the optical table. Those famous thought experiments, which people pay much attention to, such as delay choice, wave particle duality, quantum teleportation etc. could be transplanted from free space into the photonic chip. The validity of quantum mechanics and relativity would be further checked in a smaller physical space.

"What's past is prologue" (《Tempest》, Shakespeare). Exploration of the unknown world is endless. During the exploration process, happiness and pains, success and failure coexist. The achievements on dielectric superlattice embody the painstaking efforts of all the researchers who have worked in this field. We hope that readers can benefit from reading this book, especially those young researchers.

附录 最新的重要成果收录
Appendix　The Collection of the Latest Important Papers

Surface Phononic Graphene[*]

Si-Yuan Yu[1], Xiao-Chen Sun[1], Xu Ni[1], Qing Wang[1], Xue-Jun Yan[1], Cheng He[1], Xiao-Ping Liu[1,2],
Liang Feng[3], Ming-Hui Lu[1,2] and Yan-Feng Chen[1,2]

[1] *National Laboratory of Solid-State Microstructures and Department of Materials Science and Engineering, College of Engineering and Applied Sciences, Nanjing University, Nanjing 210093, China*

[2] *Collaborative Innovation Center of Advanced Microstructures, Nanjing University, Nanjing 210093, China*

[3] *Department of Electrical Engineering, The State University of New York at Buffalo, Buffalo, New York 14260, USA*

Strategic manipulation of wave and particle transport in various media is the key driving force for modern information processing and communication. In a strongly scattering medium, waves and particles exhibit versatile transport characteristics such as localization[1,2], tunnelling with exponential decay[3], ballistic[4], and diffusion behaviours[5] due to dynamical multiple scattering from strong scatters or impurities. Recent investigations of graphene[6] have offered a unique approach, from a quantum point of view, to design the dispersion of electrons on demand, enabling relativistic massless Dirac quasiparticles, and thus inducing low-loss transport either ballistically or diffusively. Here, we report an experimental demonstration of an artificial phononic graphene tailored for surface phonons on a $LiNbO_3$ integrated platform. The system exhibits Dirac quasiparticle-like transport, that is, pseudo-diffusion at the Dirac point, which gives rise to a thickness-independent temporal beating for transmitted pulses, an analogue of Zitterbewegung effects[7—9]. The demonstrated fully integrated artificial phononic graphene platform here constitutes a step towards on-chip quantum simulators of graphene and unique monolithic electro-acoustic integrated circuits.

The Dirac equation[10] constitutes the foundation of relativistic quantum mechanics and facilitates the development of quantum field theory. While a variety of intriguing effects on free relativistic quantum particles, such as the Klein paradox[11], were theoretically predicted decades ago, the corresponding validations were not initialized due to the lack of possible experimental platforms. The recent discovery of two-dimensional graphene has inspired research worldwide towards Dirac physics, because graphene's symmetry enables a Dirac conical singularity at the corner of the Brillouin zone. Prior studies near this singularity point have led to several important scientific discoveries about the electron transport behaviour[12—14]. However, possible structural defects/impurities and coherence

---

[*] Nat. Mater., 2016, 4743

degradation of electron waves due to electron-electron and electron-phonon scattering have hindered the further exploration of Dirac physics in graphene. For instance, 'Zitterbewegung'[7—9,15,16] (ZB), a unique fundamental outcome of Dirac relativity manifested as an oscillatory motion as a result of the superposition of a relativistic electron and positron, has never been observed in graphene or other solid-state electronic systems, not only because the dephasing issue, as a result of coherence degradation, reduces the interface contrast in ZB, but also because the large oscillation frequency and small oscillation amplitude of the ZB, if it still exists, are beyond the measurement capabilities of today's technology. Therefore, designing artificial graphene[17—26] beyond the restrictions of graphene, which can then enable high-fidelity phase-sensitive measurements and accommodate different space and time measurement scales, becomes crucial both in investigating some of the long-standing theoretical speculations related to Dirac relativity and in expanding the fundamental research and application scope of Dirac physics.

During the past decades, several artificial graphene systems have been developed to facilitate the exploration of Dirac relativity and its related physical phenomena. These include nanopatterned two-dimensional electron gases in semiconductor heterostructures[18], carbon monoxide molecular assemblies on a copper surface[19], ultracold atoms trapped in an optical lattice[20], photonic crystals[15,21—24], phononic crystals[9,25,26], and so on. Among them, the condensed-matter-based systems provide the most elegant laboratory tools for studying Dirac physics. Unfortunately, the high capital and maintenance expense of such tools restricts research activities on Dirac physics to only a small number of research groups. Besides, all these platforms suffer from coherence degradation over time/space and/or (ambient and quantum) noise, which may prevent them from investigating high-fidelity Dirac-dispersion-related transport, especially phase-sensitive phenomena. Hence, it is highly desirable to explore an alternative platform that could, in principle, overcome such limitations.

Surface acoustic waves (SAWs), as a mechanical perturbation travelling on the surface of solids, have received much attention in industrial applications, for example, microwave communication (cellular, Wi-Fi, GPS, and so on). The use of SAW systems to construct artificial graphene has many potential advantages over previous systems. This includes, but is not limited, to the four following key facts. First, with a careful choice of substrate, the SAW propagation can be extremely low loss[27], which is of great advantage for phase-sensitive measurements. Second, the tensor nature of surface phonons adds extra design freedom, enabling various physical phenomena that are otherwise impossible (see Supplementary Information). Third, the Dirac dispersion of the system can be surface engineered, and gauge potentials can be created, for example, by curving the two-dimensional lattice or by time-dependent modulation. These two factors combined can be further harnessed to create systems to explore topological Dirac physics and relevant edge states. Last, the monolithically integrated surface phononic graphene system is very resistant to electromagnetic radiation and ambient acoustic noise. It thus can act as a potential candidate to interface with superconducting qubits[28] or single-electron transistors (SETs)[29], which may enable strong phononic interactions and greatly expand the research scope to the quantum phononics region, an emerging research field. All these merits make the surface phononic platform very promising for implementing and studying Dirac relativity, for example, 'Zitterbewegung', which in turn may help to expand the radiofrequency (RF) signal processing capabilities of the SAW platform by replacing the conventional defect/interface-sensitive SAW Fabry-Perot and whispering gallery mode resonance with a robust oscillation scheme.

Here we explore, based on this SAW platform, a fully integrated artificial graphene solution and investigate the role of Dirac conical dispersion in the transportation of surface phonons, the quanta of SAW. In particular, we show that the SAW transmission through our on-chip artificial graphene features a pseudo-diffusion characteristic manifested by an almost constant product of the transmittance and the sample thickness. Additionally, we demonstrate that the linear dispersion near the Dirac point induces temporal oscillation for the transmitted SAW, leading to a successful realization of the surface phononic 'Zitterbewegung' effect originating from the interference of two Bloch modes in the upper and lower branches of the Dirac cone. This fully integrated chip-scale surface phononic graphene establishes a platform to facilitate the study of the Dirac Hamiltonian and enable energy-efficient and anti-jamming signal transfer with an ultrahigh signal-to-noise ratio.

Our artificial surface phononic graphene consists of a honeycomb-latticed surface phononic crystal on a piezoelectric $LiNbO_3$ substrate, as shown in Fig. 1(a). The metallic micro-pillar array in our phononic crystal introduces periodic potential variation to the SAWs. Figure 1(b) shows a scanning electron microscopy (SEM) picture of the fabricated surface phononic graphene, where nickel (Ni) micro-pillars were electrochemically grown onto the $LiNbO_3$ substrate (see Methods). Although pillars with perfect vertical sidewalls would be preferred, the actual fabricated pillars, as shown in Fig. 1(c), turned out to be a coneshape because of fabrication limitations. Nevertheless, these 'cone'-shaped pillars can still introduce perturbation to form an artificial surface phononic graphene. To study the SAW transport near the Dirac point, a ribbon sample was prepared and oriented such that the SAW propagates along the $\Gamma - K$ direction. Two broadband planar electro-acoustic transducers, one acting as a SAW emitter and the other as a SAW receiver, were then patterned on the same substrate to create our ambient-noise-resistant fully integrated chip-scale test platform (see Supplementary Information).

Dirac dispersion can be realized accidentally in a regular two-dimensional (2D) lattice[30,31] or deterministically in a 2D triangular/honeycomb lattice at the corner of Brillouin zone. For example, our phononic crystal shown in Fig. 1 is expected to have a deterministic Dirac singularity cone with linear dispersion at K and its equivalent points. A full-wave finite element method (see Methods and Ref. [32]) is used to numerically study the energy dispersion of our surface phononic graphene. The band structure shown in Fig. 2 (a) for the Γ - K direction possesses a clear band crossing at the K or K' point for two bands denoted by Curve Ⅰ and Curve Ⅱ. The Poynting vector of the Bloch modes for these two bands has two components, one along the crystallographic y-axis and the other along the crystallographic z-axis with an exponentially decay into the substrate, confirming the evanescent nature of SAWs. As shown in the inset of Fig. 2(a), the band denoted by Curve Ⅰ contains Bloch modes with a symmetrical field distribution with respect to the propagating direction k_x, implying an in-phase motion of the two 'atoms' in a unit cell, while the band denoted by Curve Ⅱ corresponds to anti-symmetrical Bloch modes with out-of-phase motion (see also Supplementary Fig. 4). A careful examination of the calculated band crossing reveals that these two bands form a Dirac cone with a pair of linear dispersion bands(see the zoom-in-band structure in Supplementary Fig. 1). Our theoretical analysis also reveals the same results via the derived Dirac Hamiltonian near the K or K' point (see Supplementary Information):

$$H(\boldsymbol{k})\psi = \begin{bmatrix} 0 & \hbar v_\kappa e^{-i\phi} \\ \hbar v_\kappa e^{i\phi} & 0 \end{bmatrix} \begin{pmatrix} \psi_1 \\ \psi_2 \end{pmatrix} = E(\boldsymbol{k}) \begin{pmatrix} \psi_1 \\ \psi_2 \end{pmatrix} \tag{1}$$

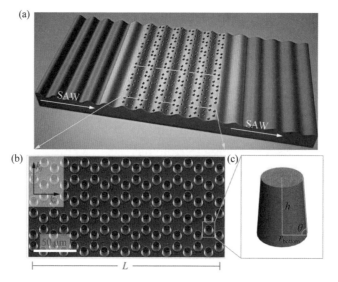

FIG. 1. Configuration of an integrated surface phononic graphene. (a), Surface phononic graphene ribbons are precisely fabricated on a 50 nm Cr-Au seed layer on top of a crystallographic y-cut lithium niobate(LiNbO$_3$) substrate. It interacts with surface acoustic waves propagating along crystallographic z-direction with a frequency range from 60 MHz to 80 MHz. (b), SEM picture of one representative phononic graphene. Honeycomb-latticed nickel micro-pillars are electroplated on the LiNbO$_3$ substrate with a lattice constant $a = 24$ μm. (c), Configuration of the cone-shaped pillar: pillar height h, bottom radius r_{bottom} and side wall angle θ are 8.90 μm, 5.35 μm and 85°, respectively.

where $\boldsymbol{k}$ is the reciprocal lattice vector, $\psi = \begin{pmatrix} \psi_1 \\ \psi_2 \end{pmatrix}$ represents the amplitudes of the two Bloch modes at one of the corners of the hexagonal first Brillouin zone, v is the Dirac velocity, (κ, ϕ) represents a wavevector originating from K point, and E is the eigenvalue.

The measured SAW transmission spectrum through a surface phononic graphene ribbon with 12 repeat units along the Γ-K direction is shown in Fig. 2(b). This spectrum differs significantly from that obtained for a reference sample without any surface structural features. Specifically, at frequencies above 74.5 MHz, the transmission through the phononic graphene ribbon is strongly suppressed. This is because the band denoted by Curve II is effectively a deaf band[33], since the anti-asymmetric Bloch modes in this band cannot be excited by a normally incident planar SAW. In addition, a transmission dip can be easily distinguished at the spectral location from 70.7 MHz to 72.9 MHz with a centre frequency at 71.8 MHz. Such a unique transmission feature can be attributed to the pseudo-diffusion effect of energy fluxes at the Dirac point[34—36], and the centre frequency of the measured transmission dip corresponds exactly to the Dirac frequency of our fabricated surface phononic graphene. Note that this Dirac frequency derived from experiments matches well (within an error of 0.4%) with numerical simulations (see Supplementary Information). The physics behind this pseudo-diffusion effect can be revealed by examining the equi-frequency contours near the Dirac frequency. As shown in Fig. 2(c), the linear dispersion near the Dirac frequency in the Brillouin zone gives rise to a series of concentric equi-frequency ring contours. However, in a honeycomb lattice, deterministic band crossing occurs, such that the equi-frequency contour at the Dirac frequency is degenerated from a ring to a single singularity point, where the phononic dispersion is non-differentiable. Hence, the energy flux, of which the direction is determined by the corresponding group velocity $v_g = (\partial \boldsymbol{k}/\partial \omega)^{-1}$, becomes omni-directional, leading to diffusion-like transport and a dip in the transmission spectrum. The right panel of Fig. 2(d) illustrates the representative field distribution obtained from simulations for two different SAW excitation cases, one at the frequency below the Dirac frequency (top) and the other at the Dirac frequency (bottom). In the first case, due to the directional pseudo-gap along the Γ-M direction and the supercollimation effect[37], the SAW transmission exhibits diffraction-less ballistic transport with an almost invariant beam width through the ribbon sample. In the second case, the field, however, clearly 'diffuses' out almost immediately inside the ribbon. In experiments, due to the lack of acoustic imaging systems with high spatial resolution at this frequency regime, a different characterization method is used to distinguish these two transport effects, which involves measuring the SAW transmittance through a series of surface phononic ribbon samples with different thicknesses (see Supplementary Information). This method is similar to that routinely used in studying wave transport in disordered media[38], where a constant product of the transmittance and the sample thickness (denoted as TL product) determines a diffusion

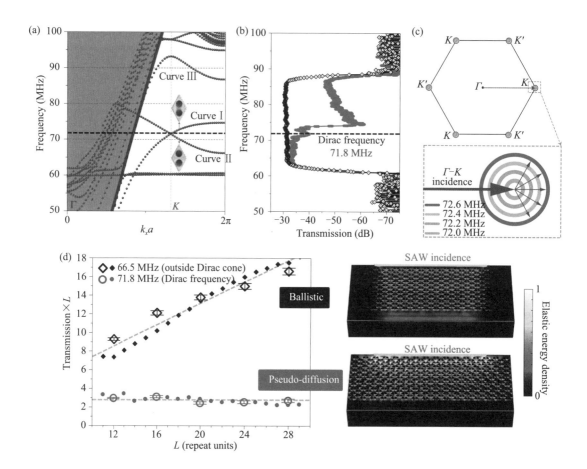

FIG.2. SAW transport behaviours in different regimes.(a), Calculated dispersions of the phononic graphene along the Γ-K(k_x) direction. Surface phononic dispersions are under the sound line(red solid line)(see Methods). The band denoted by Curve Ⅰ, containing symmetrical Bloch modes, and the band denoted by Curve Ⅱ, containing anti-symmetrical Bloch modes, become degenerate at 71.8 MHz, forming the conical Dirac cone. (b), Experimental transmission spectra. Black curve corresponds to the SAW transmission through only a plain substrate as a reference; red curve corresponds to the SAW transmission through a ribbon sample with 12 repeat units along the Γ-K direction. A distinct transmission dip, caused by the pseudo-diffusion effect, appears at the Dirac frequency(71.8 MHz). For frequencies above 74.5 MHz, the SAW transmission is forbidden. (c), Equi-frequency contours(EFCs) near the Dirac point. At the K/K' points, the SAW energy flux can exist omni-directionally in the two-dimensional space. (d), Product of the SAW transmission coefficient T and the ribbon sample thickness L(TL) versus thickness L. Here, thickness L is measured in the number of repeat units along the SAW injection direction(Γ-K(k_x)), which is chosen to be 12,16,20,24 and 28, respectively, in experiments and from 11 to 29 in steps of 1 in simulations. Red open circles correspond to an incident SAW frequency of 71.8 MHz, which represent an almost constant TL product, implying a pseudo-diffusion regime. Black open diamonds correspond to an incident SAW frequency of 66.5 MHz. In this case, the TL product grows almost linearly with L, indicating a nearly constant transmission and thus a diffraction-less ballistic transport regime. The red filled circles and black filled diamonds are the simulated TL products at the corresponding SAW frequencies. The right insets show the simulated representative elastic energy densities for these two transport regimes.

transport behaviour while a linear relation between this product and the sample thickness indicates a ballistic transport behaviour. Figure 2(d) shows our experimentally derived TL parameter. Clearly, at the Dirac frequency this key parameter is almost constant within our experimental error limit, implying the occurrence of diffusion-like transport or simply a pseudo-diffusion effect of the SAW. However, this kind of effect vanishes when the SAW frequency deviates from the Dirac frequency due to the broken degeneracy/singularity. In this scenario, as also illustrated in Fig. 2(d), the TL product grows linearly with the sample thickness, indicating ballistic transport for the SAW.

Different from the diffusion effect in disordered media, our Dirac-dispersion-related pseudo-diffusion behaviour has further significant impacts on the temporal dynamics of different SAW pulses. For SAW waves with a temporal Gaussian shape, whose spectral components are away from the Dirac frequency, for example, a centre frequency of $f_c = 66.5$ MHz and 3dB bandwidth bw of 0.045 (bw $= \Delta f_{3dB}/f_c$) as shown in Fig. 3(a), the dispersion effect dominates their temporal dynamics, which only slightly alters the temporal width of the transmitted pulse. In stark contrast to this case, the pulse dynamics changes dramatically for the SAW pulses with a centre frequency at the Dirac frequency $f_D = 71.8$ MHz. The experimentally recorded dynamic, shown in Fig. 3(b)-(d), exhibits strong temporal oscillations or beating. In general, the beating patterns are very similar for SAW pulses with different bandwidths and transmitting through different thickness ribbons. They are all characterized with multiple beating tones and an exponentially decaying beating strength. The beating period or frequency is found to be almost independent of the ribbon thickness, but scales with the bandwidth of the input SAW, as shown in Fig. 3(e). A strictly linearly proportional relation can be identified for smaller bandwidths, for example, $<0.35 f_D$, which is a direct consequence of the linear dispersion near the Dirac point. However, as the bandwidth increases, the spectral components start to experience a quadratic dispersion outside of the linear Dirac dispersion region and, as a result, the proportional relation deviates from linearity and evolves into an asymptotic relation.

Fourier analysis of the temporal dynamics shown in Fig. 3(b)-(d) suggests the existence of two major Gaussian spectral components with their peak frequencies located on either side of the Dirac frequency (see Supplementary Information). Hence, transporting these two major spectral components in our surface phononic graphene is via two separated Bloch modes, one in the upper branch and the other in the lower branch of the Dirac cone. The observed beating effect can then be attributed to the interference of these two Bloch modes, which is a direct and fundamental emulation of the interference of a positron and electron in the 'Zitterbewegung' phenomenon[7] initially proposed in relativistic electron systems. Apart from temporal beating, another unique feature for 'Zitterbewegung' is the oscillatory time-dependent velocity of the wavepacket, $\bar{v}(t)$. Because of practical diffculties in measuring $\bar{v}(t)$, we resort to a different but measurable physical parameter, that is, the

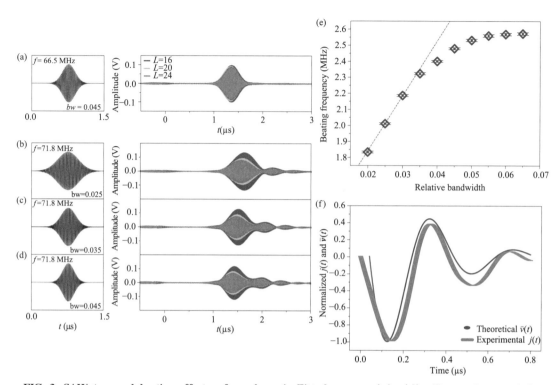

FIG. 3. SAW temporal beating effect-surface phononic Zitterbewegung. (a)-(d), Temporal transmission spectra for input SAW pulses with a Gaussian temporal shape. Left panels show SAW packets with different centre frequencies f_c (66.5 MHz and 71.8 MHz) and relative bandwidths bw (0.025, 0.035 and 0.045); right panels plot the transient dynamics of the transmitted SAW Gaussian pulses, with blue, green and red curves representing different ribbon thicknesses (16, 20 and 24, respectively) measured along the SAW injection direction ($\Gamma - K(k_x)$). The oscillatory beating is evident in all these cases as long as the centre frequency of the SAW pulses is set at the Dirac frequency. (e), Relationship between the bandwidth of the incident Gaussian pulses and the beating frequency obtained via the Fourier transform of the temporal dynamics for a ribbon sample with thickness $L=16$ (repeat units along the SAW injection direction). The beating frequency increases linearly with the bandwidth when the excitation frequency of the SAW pulses is with in the linear dispersion region of the Dirac cone, as indicated by the linear grey dotted line for a clear illustration. The error bars originate from signal fluctuations caused by electronic Johnson noise in time-domain measurements. (f), Temporal dynamics of the theoretically calculated SAW pulse average velocity $\bar{v}(t)$ and the rate of change $j(t)$ of the experimentally measured SAW pulse energy flux for the same ribbon sample. Here the amplitude illustrated is normalized and the zero-time corresponds to the time when the second beating tone just starts to form.

changing rate of the energy flux $j(t) = dI(t)/dt$, which is proportional to $\bar{v}(t)$. A quantitative comparison in Fig. 3(f) shows a very good agreement between the measured $j(t)$ and the theoretically calculated $\bar{v}(t)$ along the Γ-K direction of our surface phononic graphene (see Supplementary Information). The slight discrepancy between them stems from well-known issues, that is, the finite size effect and the interface scattering in experiments. Nevertheless, the dynamics for $j(t)$ and $\bar{v}(t)$ illustrates clear oscillation patterns as suggested in the 'Zitterbewegung' theory. This further confirms that the observed SAW beating effect in our surface phononic graphene is a direct analogue of the 'Zitterbewegung' of relativistic electrons.

In summary, we explore artificial phononic graphene and its associated Dirac physics on a fully integrated surface phononic platform. We show intriguing SAW transport behaviours in this unique artificial phononic graphene, which include ballistic transport, but more importantly Dirac-dispersion-related pseudo-diffusion transport and 'Zitterbewegung' effects. Our study highlights the potential of the commercially available surface phononic platform for both fundamental research and practical application. Our surface phononic graphene integrated on this platform can serve as a promising platform for researching Dirac relativistic quantum mechanics and its practical implementation, which are otherwise impossible with graphene. For instance, the shear elasticity unique to our surface photonic system may enable pseudo-angular-momentum/spin[39] (see Supplementary Information), and with appropriate pseudo spin-orbital coupling mechanisms introduced into the Dirac model of our system, a bandgap and in certain scenarios a band inversion effect can be realized, which greatly facilitates the investigation of topological edge states and the vortex effect in graphene-like lattices. In addition, the absence of interaction between surface phonons and substrate bulky phonons, coupled with the low-loss propagation of surface phonons, allows for high-fidelity phase-sensitive measurements, which opens the door to exploring a wide range of interface-based subjects related to the Dirac Hamiltonian in experiments—for example, Bloch-Zener oscillation and Dirac oscillators—in addition to the ZB oscillation studied here.

References and Notes

[1] Schwartz, T., Bartal, G., Fishman, S. & Segev, M. Transport and Anderson localization in disordered two-dimensional photonic lattices. *Nature* **446**, 52–55 (2007).

[2] Hu, H., Strybulevych, A., Page, J. H., Skipetrov, S. E. & van Tiggelen, B. A. Localization of ultrasound in a three-dimensional elastic network. *Nat. Phys.* **4**, 945–948 (2008).

[3] Beenakker, C. W. J. Colloquium: Andreev reflection and Klein tunneling in graphene. *Rev. Mod. Phys.* **80**, 1337–1354 (2008).

[4] Du, X., Skachko, I., Barker, A. & Andrei, E. Y. Approaching ballistic transport in suspended graphene. *Nat. Nanotech.* **3**, 491–495 (2008).

[5] Labeyrie, G. *et al.* Slow diifusion of light in a cold atomic cloud. *Phys. Rev. Lett.* **91**, 223904 (2003).

[6] Novoselov, K. S. *et al.* A roadmap for graphene. *Nature* **490**, 192–200 (2012).

[7] Schrödinger, E. Über die kräftefreie Bewegung in der relativistischen Quantenmechanik. *Sitz. Preuss. Akad. Wiss. Phys.* **24**, 418–428 (1930).

[8] Gerritsma, R. *et al.* Quantum simulation of the Dirac equation. *Nature* **463**, 68–71(2010).

[9] Zhang, X. & Liu, Z. Extremal transmission and beating effect of acoustic waves in two-dimensional sonic crystals. *Phys. Rev. Lett.* **101**, 264303 (2008).

[10] Thaller, B. *The Dirac Equation* (Springer, 1992).

[11] Katsnelson, M. I., Novoselov, K. S. & Geim, A. K. Chiral tunnelling and the Klein paradox in

[12] Novoselov, K. S. A. *et al*. Two-dimensional gas of massless Dirac fermions in graphene. *Nature* **438**, 197–200 (2005).

[13] Ponomarenko, L. A. *et al*. Chaotic Dirac billiard in graphene quantum dots. *Science* **320**, 356–358 (2008).

[14] Hsieh, D. *et al*. A tunable topological insulator in the spin helical Dirac transport regime. *Nature* **460**, 1101–1105 (2009).

[15] Zhang, X. Observing Zitterbewegung for photons near the Dirac point of a two-dimensional photonic crystal. *Phys. Rev. Lett.* **100**, 113903 (2008).

[16] Vaishnav, J. Y. & Clark, C. W. Observing Zitterbewegung with ultracold atoms. *Phys. Rev. Lett.* **100**, 153002 (2008).

[17] Polini, M., Guinea, F., Lewenstein, M., Manoharan, H. C. & Pellegrini, V. Artificial honeycomb lattices for electrons, atoms and photons. *Nat. Nanotech.* **8**, 625–633 (2013).

[18] Singha, A. *et al*. Two-dimensional Mott-Hubbard electrons in an artificial honeycomb lattice. *Science* **332**, 1176–1179 (2011).

[19] Gomes, K. K., Mar, W., Ko, W., Guinea, F. & Manoharan, H. C. Designer Dirac fermions and topological phases in molecular graphene. *Nature* **483**, 306–310 (2012).

[20] Tarruell, L., Greif, D., Uehlinger, T., Jotzu, G. & Esslinger, T. Creating, moving and merging Dirac points with a Fermi gas in a tunable honeycomb lattice. *Nature* **483**, 302–305 (2012).

[21] Zandbergen, S. R. & de Dood, M. J. Experimental observation of strong edge effects on the pseudodiffusive transport of light in photonic graphene. *Phys. Rev. Lett.* **104**, 043903 (2010).

[22] Bittner, S. *et al*. Observation of a Dirac point in microwave experiments with a photonic crystal modeling graphene. *Phys. Rev. B* **82**, 014301 (2010).

[23] Bellec, M., Kuhl, U., Montambaux, G. & Mortessagne, F. Topological transition of Dirac points in a microwave experiment. *Phys. Rev. Lett.* **110**, 033902 (2013).

[24] Plotnik, Y. *et al*. Observation of unconventional edge states in 'photonic graphene'. *Nat. Mater.* **13**, 57–62 (2014).

[25] Torrent, D. & Sánchez-Dehesa, J. Acoustic analogue of graphene: observation of Dirac cones in acoustic surface waves. *Phys. Rev. Lett.* **108**, 174301 (2012).

[26] Lu, J. *et al*. Dirac cones in two-dimensional artificial crystals for classical waves. *Phys. Rev. B* **89**, 134302 (2014).

[27] Szabo, T. L. & Slobodnik, A. J. The effect of diffraction on the design of acoustic surface wave devices. *IEEE Trans. Sonics Ultrason.* **20**, 240–251 (1973).

[28] Gustafsson, M. V., Santos, P. V., Johansson, G. & Delsing, P. Local probing of propagating acoustic waves in a gigahertz echo chamber. *Nat. Phys.* **8**, 338–343 (2012).

[29] Gustafsson, M. V. *et al*. Propagating phonons coupled to an artificial atom. *Science* **346**, 207–211 (2014).

[30] Huang, X., Lai, Y., Hang, Z. H., Zheng, H. & Chan, C. T. Dirac cones induced by accidental degeneracy in photonic crystals and zero-refractive-index materials. *Nat. Mater.* **10**, 582–586 (2011).

[31] He, C. *et al*. Acoustic topological insulator and robust one-way sound transport. *Nat. Phys.* http://dx.doi.org/10.1038/nphys3867 (in the press).

[32] Khelif, A., Achaoui, Y., Benchabane, S., Laude, V. & Aoubiza, B. Locally resonant surface acoustic wave band gaps in a two-dimensional phononic crystal of pillars on a surface. *Phys. Rev. B* **81**, 214303

(2010).

[33] Sánchez-Pérez, J. V. et al. Sound attenuation by a two-dimensional array of rigid cylinders. *Phys. Rev. Lett.* **80**, 5325 – 5328 (1998).

[34] Katsnelson, M. I. Zitterbewegung, chirality, and minimal conductivity in graphene. *Eur. Phys. J. B* **51**, 157 – 160 (2006).

[35] Tworzyd.o, J., Trauzettel, B. & Titov, M. et al. Sub-Poissonian shot noise in graphene. *Phys. Rev. Lett.* **96**, 246802 (2006).

[36] Sepkhanov, R. A., Bazaliy, Y. B. & Beenakker, C. W. J. Extremal transmission at the Dirac point of a photonic band structure. *Phys. Rev. A* **75**, 063813 (2007).

[37] Rakich, P. T., Dahlem, M. S. & Tandon, S. et al. Achieving centimetre-scale supercollimation in a large-area two-dimensional photonic crystal. *Nat. Mater.* **5**, 93 – 96 (2006).

[38] Bouchaud, J. P. & Georges, A. Anomalous diifusion in disordered media: statistical mechanisms, models and physical applications. *Phys. Rep.* **195**, 127 – 293 (1990).

[39] Zhang, L. & Niu, Q. Chiral phonons at high-symmetry points in monolayer hexagonal lattices. *Phys. Rev. Lett.* **115**, 115502 (2015).

[40] We thank Z. Shi and X. Zhang for helpful discussions. This work was jointly supported by the National Basic Research Program of China (Grant No. 2012CB921503, No. 2013CB632904 and No. 2013CB632702), the National Nature Science Foundation of China (Grant No. 11134006, No. 11474158, and No. 11404164), and the Natural Science Foundation of Jiangsu Province (BK20140019). We also acknowledge the project funded by the Priority Academic Program Development of Jiangsu Higher Education (PAPD) and China Postdoctoral Science Foundation (Grant No. 2012M511249 and No. 2013T60521). L. F. acknowledges support from the National Science Foundation (DMR-1506884 and ECCS-1507312).

[41] M.-H.L. and X.-P.L. conceived the idea. M.-H.L., X.-P.L. and Y.-F.C. coordinated and guided the project. S.-Y.Y. designed the devices, fabricated the samples and carried out the measurements. X.-C. S. performed the theoretical analysis. All the authors contributed to discussion of the project. S.-Y.Y., X.-P.L., L.F. and M.-H.L. prepared the manuscript with revisions from other authors.

[42] Supplementary information is available in the online version of the paper.

[43] Sample fabrication. Fabrication of the phononic graphene sample involves the LIGA technique. The acronym, LIGA, stands for the German words for the three important processes: Lithographie, Galvanoformung and Abformung. The fabrication procedure consists of the following steps: a 10nm Cr layer deposition followed by a 40 nm Au seed-layer deposition on a 500 μm thickness $LiNbO_3$ substrate; spin-coating of thick negative resist MicroChem SU-8, followed by a contact lithographic and development process to define the hollow honeycomb pattern in the resist; electrochemical deposition of Ni (nickel) on the exposed Au seed layer in an electrolyte environment; resist removal in a reactive-ion etching (RIE) chamber.

[44] Calculation of SAW dispersion. A three-dimensional finite element method is used to calculate the dispersion of the surface phononic graphene. A triangular lattice, consisting of two Ni pillars on the $LiNbO_3$ substructure, is defined as the unit cell of the honeycomb structure. The height of the unit cell is chosen carefully (larger than ten times of the wavelength of the SAW) to minimize the disturbance of the bulk modes on the calculation of the surface modes. Periodic boundary conditions are used for the interfaces between the nearest unit cells according to Bloch-Floquet theorem. The elastic parameters of the Ni pillars used in the calculations are density $\rho_{Ni} = 8,906$ kg $\cdot$ m^{-3}, Young's modulus $E_{Ni} = 110 \times$

10^9 Pa and Poisson's ratio 0.30. Note that these values are specific to our growth technique. Detailed methods for determining the elastic parameters are provided in the Supplementary Information. The phononic band structure is acquired by varying the wavevector in the first Brillouin zone and solving the eigenvalue problem. On the band diagram, the surface modes can be distinguished under the slowest bulk mode dispersion line, which approximately corresponds to the dispersion of the Rayleigh wave on a pure $LiNbO_3$ substrate.

[45] Methods and any associated references are available in the online version of the paper.
[46] Si-Yuan Yu and Xiao-Chen Sun contributed equally to this work.

Experimental Realization of Bloch Oscillations in a Parity-time Synthetic Silicon Photonic Lattice[*]

Ye-Long Xu[1], William S. Fegadolli[2], Lin Gan[3], Ming-Hui Lu[1,4],
Xiao-Ping Liu[1,4], Zhi-Yuan Li[3], Axel Scherer[2] & Yan-Feng Chen[1,4]

[1] *National Laboratory of Solid State Microstructures, Nanjing University, Nanjing, Jiangsu 210093, China*

[2] *Department of Physics and Kavli Nanoscience Institute, California Institute of Technology, Pasadena, California 91125, USA*

[3] *Laboratory of Optical Physics, Institute of Physics, Chinese Academy of Sciences, Beijing 100190, China*

[4] *Collaborative Innovation Center of Advanced Microstructures, Nanjing University, Nanjing 210093, China*

As an important electron transportation phenomenon, Bloch oscillations have been extensively studied in condensed matter. Due to the similarity in wave properties between electrons and other quantum particles, Bloch oscillations have been observed in atom lattices, photonic lattices, and so on. One of the many distinct advantages for choosing these systems over the regular electronic systems is the versatility in engineering artificial potentials. Here by utilizing dissipative elements in a CMOS-compatible photonic platform to create a periodic complex potential and by exploiting the emerging concept of parity-time synthetic photonics, we experimentally realize spatial Bloch oscillations in a non-Hermitian photonic system on a chip level. Our demonstration may have significant impact in the field of quantum simulation by following the recent trend of moving complicated table-top quantum optics experiments onto the fully integrated CMOS-compatible silicon platform.

Bloch oscillations (BO) effect is a fundamental electron transport phenomenon in condensed matter, which is characterized as the coherent oscillatory motion of electron in a periodic potential driven by an external DC electric field[1]. The electron is accelerated by the electric field, then reflected when it reaches a momentum satisfying the Bragg reflection condition in the periodic potential and, subsequently, decelerated to the initial state to form a BO cycle. However, because of dephasing effects, it is difficult to observe BO in natural crystals. Owing to the development of semiconductor technology, BO have been experimentally observed in semiconductor superlattices with periodically arranged layers of different semiconductor materials[2]. Due to the equivalence between the Schrödinger equation in quantum mechanics and the classical wave equation, BO can be also observed for matter waves[3-6], optical waves[7-13], acoustic waves[14] and surface

[*] Nat.Commun.,2016,7:11319

plasmon polariton waves[15]. In fact, studying BO and related wave transportation phenomena in some of these waveforms can be very advantageous compared to electrons. For example, coherent optical sources with much longer coherent length are readily available, so the dephasing effect would not become a hurdle. Moreover, advanced nanofabrication technologies allow for the creation of very complicated optical potentials, which has already become a major driving force behind the rapid development of the latest nano-photonic technologies. It is worth pointing out that almost all the experimental studies involving BO are primarily focused on the wave dynamics in Hermitian systems with only real-valued potentials. Recently, a particular class of non-Hermitian photonic systems with parity-time (PT) symmetry has drawn great research interest[16—25]. It is shown that such systems can possess a spontaneous symmetry-breaking phase transition with an eigen energy transition from real spectra into complex spectra. The phase transition point shows all the characteristics of an exceptional point, that is, having non-Hermitian degeneracy, where the real and imaginary parts of eigenvalues are identical. Different from conventional Hermitian degeneracy, not only the eigenvalues but also the eigenvectors coalesce at an exceptional point, giving rise to abundant intriguing physical phenomena. These include unidirectional BO[26], unidirectional invisibility[27—30] and unidirectional transmission[31]. In addition, the exceptional point singularity also has broadened the scope of metamaterials[32—34].

Here we report an experimental study of photonic BO and related wave dynamics in Hermitian and non-Hermitian systems realized with complementary metal-oxide-semiconductor(CMOS)-compatible fabrication processes on a silicon-on-insulator (SOI) platform. A real-valued periodic potential in a Hermitian photonic lattice in conjunction with a linear gradient potential perturbation acting as a constant acceleration force is created by bending an array of identical SOI waveguides, while a complex PT synthetic potential in a non-Hermitian photonic lattice is realized with the same lattice but with an additional optical loss modulation layer, that is, a chrome (Cr) layer on top of every other bent silicon waveguide. Compared with all other platforms for studying BO and its related classic or quantum wave dynamics[2—15], our integrated platform demonstrated here shows great advantages in emerging applications requiring ultimate scalability and stability, for instance, quantum optics on a chip[35—40]. In addition, by using scanning near-field optical microscope (SNOM), we are able to directly visualize the wave dynamics responsible for the BO continuously for both the Hermitian and non-Hermitian photonic lattice in spatial domain, which is another key advantage over the large-scale discretized PT-symmetric optical networks[30], where the wave dynamics is discrete in temporal domain and all waves are needed to be multiplexed with actively controlled optical switching components to construct the BO in temporal domain. Our spatial wave dynamics recorded by the SNOM reveal a prominent secondary emission around the BO recovery point for the non-Hermitian system in contrast to the classical picture of BO in a Hermitian system. This feature in such a PT synthetic photonic lattice is related to the spontaneous PT symmetry breaking.

Results

BO in a Hermitian photonic lattice. As depicted in Fig. 1(a), our Hermitian photonic lattice consists of an array of straight equally spaced identical SOI stripe waveguides, which support only a fundamental transverse electric mode at the wavelength of 1,550 nm. In the tight-binding approximation, the propagation of optical wave in this Hermitian photonic lattice is described by coupled-mode equations

$$\frac{d}{dz}A_n = i\kappa(A_{n-1} + A_{n+1}) \tag{1}$$

where A_n is the propagating amplitude in the nth waveguide, and κ is the coupling coefficient between adjacent waveguides. The corresponding dispersion relation is determined (see Methods) to be

$$\beta_H = \pm 2\kappa \cos(q_H d) \tag{2}$$

where d is the centre distance between adjacent waveguides and β_H and q_H are the longitudinal propagation constant and the transverse Bloch momentum of the Hermitian photonic lattice, respectively. A plot of this dispersion relation or band diagram is shown in Fig. 1(b), where the blue and red curves represent the real and imaginary parts of the band, respectively. As expected, in this Hermitian photonic lattice the imaginary part of the band is always equal to zero regardless of the Bloch momentum. To obtain Hermitian BO, the Hermitian photonic lattice is bent to modify its spatial distribution of optical potential, as sketched in Fig. 2(a). The curvature is seen as inertial force acting on the optical wave, and the linear position dependence of curvature in the radial direction thus can result in a linear potential gradient, which ultimately gives rise to an equivalent DC field required for BO. The corresponding dynamic evolution of this Hermitian bent photonic lattice becomes

$$\frac{d}{dz}A_n = i\kappa(A_{n-1} + A_{n+1}) + iFnA_n \tag{3}$$

where A_n is the mode envelop for the nth waveguide, and $F = \frac{2\pi n_{eff}d}{R\lambda}$ is the equivalent DC field corresponding to the equivalent index gradient in this particular kind of photonic lattice systems (Supplementary Note 1). In the presence of this DC field, the eigenmode can be represented as $A_n^m(z) = \Psi_m(n)e^{i\beta_{WSL}^m z}$, where $\beta_{WSL}^m = mF$, $m = 0, \pm 1, \pm 2, ...$ is Wannier-Stark ladder and $\Psi_{m(n)}$ are the corresponding Wannier-Stark eigenstates in the Wannier representation (Supplementary Note 2). We can see that the equivalent DC field F acts as a constant Wannier-Stark ladder spacing, which induces field recovery with Bloch period $z_B = \frac{2\pi}{F}$, and Wannier-Stark eigenstate is a localized optical mode caused by total internal

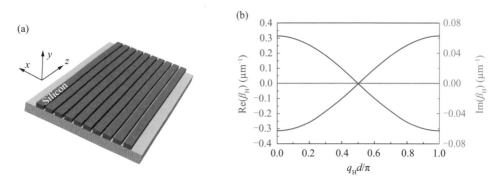

FIG.1. Structure and band diagram of the Hermitian straight photonic lattice. (a) Schematic of the photonic lattice consisting of an array of straight SOI waveguides. All SOI waveguides are stripe waveguides 400 nm wide and 220 nm thick and the array period is 500 nm. The individual SOI waveguide has a propagation constant $k = n_{\text{eff}} k_0 = 2.114\ k_0$, while the adjacent SOI waveguides have a coupling coefficient $\kappa = 0.157\ \mu m^{-1}$ in this lattice configuration. (b) The band diagram for this photonic lattice. Blue and red curves represent the real and imaginary parts of the longitudinal propagation constants, respectively.

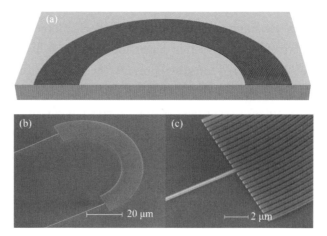

FIG.2. Structure of the Hermitian bent photonic lattice. (a) Schematic of the Hermitian bent photonic lattice. The bending radius of the silicon waveguides ranges from $R = 22.7$ to $34.7\ \mu m$. (b) SEM picture of the fabricated device where the input/output waveguide is located at the centre of the photonic lattice and has a bending radius $R = 28.7\ \mu m$. (c) Zoom-in SEM view of the device. (b), (c) Scale bar, 20 μm and 2 μm, respectively.

reflection on the low-index side and Bragg reflection on the opposite high-index side.

To explore the wave dynamics of this photonic lattice, a single waveguide is selectively excited at the input plane. The corresponding evolution of the mode in this waveguide resembles a breathing mode[9,10]. To verify this characteristic, we simulate this photonic lattice by means of 3D finite-difference time-domain(FDTD) simulations where the electric field distribution is shown in Fig. 3(a). The optical wave initially couples from the centre waveguide into the parallel waveguides towards the transverse and forward direction, and then after propagating pass a certain curvature point (approximately half of Bloch Period) it starts to converge back to a recovering point, which forms a Bloch period $z_B = \frac{2\pi}{F} = \frac{R\lambda}{n_{eff}d}$, thus finishing a breathing cycle or a BO cycle. The Bloch period can also be conveniently expressed as an in-plane polar angle $\theta_B = \frac{z_B}{R} = \frac{\lambda}{n_{eff}d}$. As seen from Fig. 3(a), the analytical calculated value $\theta_B \approx 84°$ agrees well with the simulation. In experiment, this Hermitian bent photonic lattice is fabricated using CMOS-compatible nanofabrication processes (see Methods). A scanning electron micrograph of the fabricated device is shown in Fig. 2(b), (c). The input waveguide used as the single waveguide excitation is located at the centre of the bent photonic lattice. The electric field distribution of the optical wave recorded by a SNOM (see Methods) in the area of 30 μm 15 μm within the vicinity of the first BO recovery point is shown in Fig. 3(c). As would be expected for a Hermitian BO system, the experimental result indicates that the dynamics of the optical wave has a breathing motion and possesses a very distinct recovery point. This experimental observation in fact agrees very well with the simulated electric field distribution shown in Fig. 3(b).

BO in a non-Hermitian photonic lattice. PT synthetic photonic structures exhibit intriguing and unusual wave dynamics due to the presence of symmetry breaking, resulting in phase transition from a Hermitian system to a non-Hermitian system. To obtain strict photonic PT symmetry, it requires precise engineering of the spatial distribution of complex optical potential by using optical gain and loss media. However, gain media is not straightforward to be incorporated into the CMOS-compatible fabrication process; therefore, instead of gain, a class of dissipative PT symmetry systems[19,29] with an average background loss have been widely adopted in the literature to study PT-related physics. Here the dissipative element consists of an array of lossy metal stripes deposited on top of every other silicon waveguide of the photonic lattice, as shown in Fig. 4(a). By referencing to the background loss, the optical wave in silicon waveguides without (with) metal strips experiences an equivalent gain (loss). The dynamic evolution of the straight photonic lattice, initially Hermitian, is drastically altered by the presence of the synthetic dissipative PT symmetric potential. This change is reflected in the coupled-mode equations governing the dynamics of the mode amplitude $A_{n'}$:

$$\frac{d}{dz}A_{n'} = i\kappa(A_{n-1'} + A_{n+1'}) - \frac{\gamma}{2}(1 + (-1)^n)A_{n'} \tag{4}$$

where there exists an alternating mode attenuation $(0 \leftrightarrow \gamma/2)$ in adjacent waveguides in addition to the mode coupling κ between them. The dispersion relation governing the wave dynamics of this PT synthetic photonic lattice is determined (see Methods) as

$$\beta_{NH} = \frac{i\gamma}{2} \pm \sqrt{4\kappa^2 \cos^2(q_{NH}d) - \frac{\gamma^2}{4}} \tag{5}$$

which is plotted in Fig. 4(b). This class of dissipative PT symmetric systems possesses two exceptional point singularities at $q_{NH}d = \arccos\left(\pm\dfrac{\gamma}{4\kappa}\right)$ as highlighted by the black points, where both the real and imaginary parts of the band diagram are degenerate. Between these two exceptional point singularities, the state of the system falls into a symmetry-broken phase with a coalesced real part of the band but bifurcated imaginary part. Since for any $\gamma > 0$ there will also be complex eigenvalues near the Brillouin Zone edge ($q_{NH} = \pi/(2d)$), the PT symmetry breaking is spontaneous or in other words threshold-less. However, it is worth noting that the threshold-less PT symmetry breaking in the straight photonic lattice discussed here is local, where PT symmetry breaking only happens for particular k-vectors along the transverse direction[20], which is strongly related to optical excitation conditions.

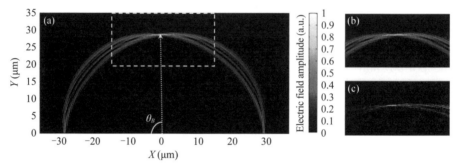

FIG.3. Simulation and experimental observation of the Hermitian BO. (a) Simulated electric field amplitude distribution when a single waveguide, that is, the input waveguide is excited. (b) Zoom-in view of the simulated electric field amplitude distribution in an area of 30 μm × 19 μm in the vicinity of the first Bloch recovery point. (c) Experimentally recorded electric field distribution for the same area using a scanning near-field optical microscope.

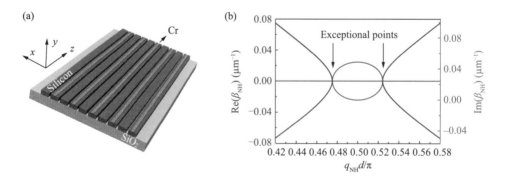

FIG.4. Structure and band structure of the non-Hermitian straight photonic lattice. (a) Schematic of the non-Hermitian photonic lattice consisting of a dissipative PT silicon waveguide array. This non-Hermitian photonic lattice has the exact same geometric design as the Hermitian photonic lattice, except for the additional 100-nm-wide and 4-nm-thick chrome (Cr) layer on top of every other silicon waveguide to introduce periodic optical loss modulation with a filed attenuation coefffcient $\gamma = 0.00486$ μm^{-1}. (b) The band diagram of such non-Hermitian photonic lattice. Blue and red curves represent the real and imaginary parts of the longitudinal propagation constants, respectively. The exceptional singularity points are indicated and highlighted as the black points.

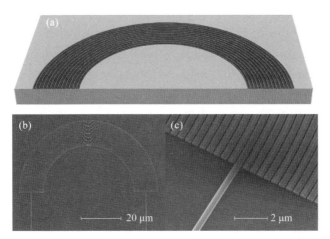

FIG. 5. Structure of the non-Hermitian bent photonic lattice. (a) Structure of the non-Hermitian bent photonic lattice. This non-Hermitian lattice has the exact same geometrical design as the Hermitian counterpart. (b) SEM picture of the fabricated device where the input/output waveguide is located at the centre of the photonic lattice. (c) Zoom-in SEM view of the device. The 100-nm-wide and 4-nm-thick Cr layer can be seen on top of every other silicon waveguide. (b),(c) Scale bar, 20 and 2 μm, respectively.

Similarly to the realization of BO in the Hermitian photonic lattice, a constant DC field is also mimicked by a circular bending of the lattice, as sketched in Fig. 5(a). In the presence of this DC field the spectra have two interleaved Wannier-Stark ladders $\beta'^m_\pm = 2mF + \frac{i\gamma}{2} \pm i\varepsilon\left(\frac{\kappa}{F}, \frac{\gamma}{F}\right)$, $m = 0, \pm 1, \pm 2, \ldots$, where $\varepsilon\left(\frac{\kappa}{F}, \frac{\gamma}{F}\right)$ is a complex functiion of $\frac{\kappa}{F}$ and $\frac{\gamma}{F}$ (Supplementary Fig. 1 and Supplementary Note 3). In other words, the eigen energy of a classical BO is split into two in a complex energy plane to reflect the emergence of two localized modes, one with pure optical loss and the other with equivalent optical gain biased by the background loss. The splitting of eigen energy level is a direct outcome of the PT symmetry breaking near the Brillouin boundary. Note that the PT symmetry breaking phenomenon for a particular Bloch mode of the previously described non-Hermitian straight photonic lattice occurs when $\gamma > 4\kappa\cos(q_{NH}d)$ as shown in Fig. 4(b). In this bent configuration the PT symmetry breaking can always be experienced by the optical wave under any excitation condition, because the transverse wave vector in the initial excitation will change under the acceleration force caused by the bending and it will eventually approache the edge of the Brillouin zone, where the threshold-less PT symmetry breaking takes place. To elucidate this, the optical wave dynamics in the non-Hermitian bent photonic lattice is simulated and shown in Fig. 6(a). The dynamics of BO in this case still possesses a periodic feature, but with decreasing intensity due to the dissipative background of this system. The BO here, however, is accompanied by a noticeable secondary emission at the BO recovery point as indicated in the zoom-in view of Fig. 6(b). In experiment, the designed non-Hermitian bent photonic lattice is fabricated with an additional Cr metal layer on top of the Hermitian bent photonic lattice(see Methods). An example of a fabricated device is shown in Fig. 5(b),(c), where the waveguide bending radius and the input/output waveguide follow the same design as its Hermitian counterpart. The near-field electric field distribution of the optical wave in this on-Hermitian system is recorded by a SNOM and is shown in Fig. 6(c). The overall electric field distribution obtained experimentally near the BO recovery point(indicated as red dots)

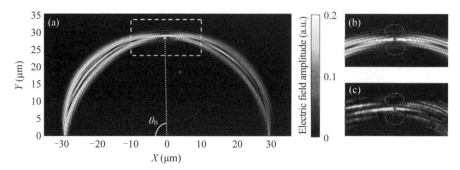

FIG.6. Simulation and experimental observation of the non-Hermitian BO. (a) Simulated electric field amplitude distribution when a single waveguide, that is, the input waveguide, is excited. (b) Zoom-in view of the simulated electric field amplitude distribution in an area of 20 μm×10 μm in the vicinity of the first Bloch recovery point. (c) Experimentally recorded electric field distribution for the same area using a scanning near-field optical microscope. The white dotted circles mark the secondary emission region, while red dots mark the recovery point in (b),(c).

shows good agreement with the theoretical results in Fig. 6(b). Although some of the finest features are not well resolved due to the instrumental limitation, the secondary emission (highlighted in while dotted circles) is clearly visible. Such a emission is originated from the presence of two Wannier-Stark ladders as a result of the spontaneous PT symmetry breaking experienced by the optical wave as it propagates down the photonic lattice with its transverse wave vector evolving to touch the Brillouin zone edge ($q_{NH} = \pi/(2d)$) under the gradient force F as shown in Fig. 4(b).

Discussion

By engineering the spatial distribution of the complex potential for photonic lattices on a SOI photonic platform, we have experimentally demonstrated photonic BO in both a Hermitian and a non-Hermitian system. Apart from the classic oscillatory evolution of the optical wave in Hermitian BO, the dynamics of non-Hermitian BO exhibits an intriguing feature characterized by a secondary emission formed near the BO recovery point. This unique feature is closely related to the spontaneous/threshold-less PT symmetry-breaking phase transition for the BO in our dissipative PT synthetic photonic lattice. The fully integrated non-Hermitian photonic lattice provides a major step forward in expanding the application scope of on-chip photonics in quantum simulation[35—42]. In fact, the threshold-less PT symmetry-breaking characteristic in our passive system makes it a perfect candidate for studying the quantum version of non-Hermitian BO as long as the loss modulation is kept low enough to allow for a meaningful high-fidelity quantum measurement, which, however, is difficult to conduct in a gain-loss balanced non-Hermitian system because of the degradation of the initial quantum input state by the inherent quantum noise inside the gain medium. Moreover, thanks to state-of-the-art nanofabrication technology, which allows for the deliberately designed complex potential distribution to be created, this CMOS-compatible photonic platform with great flexibility opens the door towards the exploration of unprecedented non-Hermitian transport phenomena on a chip[43—48].

References and Notes

[1] Bloch, F. Quantum mechanics of electrons in crystal lattices. *Z. Phys.* **52**, 555 – 600 (1928).

[2] Waschke, C. *et al.* Coherent submillimeter-wave emission from Bloch oscillations in a semiconductor superlattice. *Phys. Rev. Lett.* **70**, 3319 – 3322 (1993).

[3] Ben Dahan, M., Peik, E., Reichel, J., Castin, Y. & Salomon, C. Bloch oscillations of atoms in an optical potential. *Phys. Rev. Lett.* **76**, 4508 – 4511 (1996).

[4] Wilkinson, S. R., Bharucha, C. F., Madison, K. W., Niu, Q. & Raizen, M. G. Observation of atomic Wannier-Stark ladders in an accelerating optical potential. *Phys. Rev. Lett.* **76**, 4512 – 4515 (1996).

[5] Anderson, B. P. Macroscopic quantum interference from atomic tunnel arrays. *Science* **282**, 1686 – 1689 (1998).

[6] Morsch, O., Müller, J., Cristiani, M., Ciampini, D. & Arimondo, E. Bloch oscillations and meanfield effects of Bose-Einstein condensates in 1D optical lattices. *Phys. Rev. Lett.* **87**, 140402 (2001).

[7] Sapienza, R. *et al.* Optical analogue of electronic Bloch oscillations. *Phys. Rev. Lett.* **91**, 263902 (2003).

[8] Morandotti, R., Peschel, U., Aitchison, J. S., Eisenberg, H. S. & Silberberg, K. Experimental observation of linear and nonlinear optical Bloch oscillations. *Phys. Rev. Lett.* **83**, 4756 – 4759 (1999).

[9] Lenz, G., Talanina, I. & de Sterke, C. M. Bloch oscillations in an array of curved optical waveguides. *Phys. Rev. Lett.* **83**, 963 – 966 (1999).

[10] Pertsch, T., Dannberg, P., Elflein, W., Brauer, A. & Lederer, F. Optical Bloch oscillations in temperature tuned waveguide arrays. *Phys. Rev. Lett.* **83**, 4752 – 4755 (1999).

[11] Chiodo, N. *et al.* Imaging of Bloch oscillations in erbium-doped curved waveguide arrays. *Opt. Lett.* **31**, 1651 – 1653 (2006).

[12] Breid, B. M., Witthaut, D. & Korsch, H. J. Bloch-Zener oscillations. *New J. Phys.* **8**, 110 (2006).

[13] Dreisow, F. *et al.* Bloch-Zener oscillations in binary superlattices. *Phys. Rev. Lett.* **102**, 076802 (2009).

[14] Sanchis-Alepuz, H., Kosevich, Y. & Sánchez-Dehesa, J. Acoustic analogue of electronic Bloch oscillations and resonant Zener tunneling in ultrasonic superlattices. *Phys. Rev. Lett.* **98**, 134306 (2007).

[15] Block, A. *et al.* Bloch oscillations in plasmonic waveguide arrays. *Nat. Commun.* **5**, 3483 (2014).

[16] Bender, C. M. & Boettcher, S. Real spectra in non-Hermitian Hamiltonians having PT symmetry. *Phys. Rev. Lett.* **80**, 5243 – 5246 (1998).

[17] Bender, C. M. Making sense of non-Hermitian Hamiltonians. *Rep. Prog. Phys.* **70**, 947 – 1018

(2007).

[18] Rüter, C. E. et al. Observation of parity-time symmetry in optics. *Nat. Phys.* **6**, 192-195 (2010).

[19] Guo, A. et al. Observation of PT-symmetry breaking in complex optical potentials. *Phys. Rev. Lett.* **103**, 093902 (2009).

[20] Makris, K. G., El-Ganainy, R., Christodoulides, D. N. & Musslimani, Z. H. Beam dynamics in PT symmetric optical lattices. *Phys. Rev. Lett.* **100**, 103904 (2008).

[21] Chong, Y. D., Ge, L. & Stone, A. D. PT-symmetry breaking and laser-absorber modes in optical scattering systems. *Phys. Rev. Lett.* **106**, 093902 (2011).

[22] Longhi, S. PT-symmetric laser absorber. *Phy. Rev. A* **82**, 031801 (2010).

[23] Schindler, J. et al. PT-symmetric electronics. *J. Phys. Math. Theor.* **45**, 444029 (2012).

[24] Feng, L. et al. Single-mode laser by parity-time symmetry breaking. *Science* **346**, 972-975 (2014).

[25] Hodaei, H. et al. Parity-time-symmetric microring lasers. *Science* **346**, 975-978 (2014).

[26] Longhi, S. Bloch oscillations in complex crystals with PT symmetry. *Phys. Rev. Lett.* **103**, 123601 (2009).

[27] Lin, Z. et al. Unidirectional invisibility induced by PT-symmetric periodic structures. *Phys. Rev. Lett.* **106**, 213901 (2011).

[28] Longhi, S. Invisibility in PT-symmetric complex crystals. *J. Phys. Math. Theor.* **44**, 485302 (2011).

[29] Feng, L. et al. Experimental demonstration of a unidirectional reffectionless parity-time metamaterial at optical frequencies. *Nat. Mater.* **12**, 108-113 (2013).

[30] Regensburger, A. et al. Parity-time synthetic photonic lattices. *Nature* **488**, 167-171 (2012).

[31] Peng, B. et al. Parity-time-symmetric whispering-gallery microcavities. *Nat. Phys.* **10**, 394-398 (2014).

[32] Sun, Y., Tan, W., Li, H. Q., Li, J. & Chen, H. Experimental demonstration of a coherent perfect absorber with PT phase transition. *Phys. Rev. Lett.* **112**, 143903 (2014).

[33] Fleury, R., Sounas, D. L. & Alu, A. Negative refraction and planar focusing based on parity-time symmetric metasurfaces. *Phys. Rev. Lett.* **113**, 023903 (2014).

[34] Lawrence, M. et al. Manifestation of PT symmetry breaking in polarization space with terahertz metasurfaces. *Phys. Rev. Lett.* **113**, 093901 (2014).

[35] Lebugle, M. et al. Experimental observation of N00N state Bloch oscillations. *Nat. Commun.* **6**, 8273 (2015).

[36] Broome, M. A. et al. Photonic boson sampling in a tunable circuit. *Science* **339**, 794-798 (2013).

[37] Spring, J. B. et al. Boson sampling on a photonic chip. *Science* **339**, 798-801 (2013).

[38] Tillman, M. et al. Experimental boson sampling. *Nat. Photonics* **7**, 540-544 (2013).

[39] Crespi, A. et al. Integrated multimode interferometers with arbitrary designs for photonic boson sampling. *Nat. Photonics* **7**, 545-549 (2013).

[40] Gräfe, M. et al. On-chip generation of high-order single-photon W-states. *Nat. Photonics* **8**, 791-795 (2014).

[41] Corrielli, G., Crespi, A., Della Valle, G., Longhi, S. & Osellame, R. Fractional Bloch oscillations in photonic lattices. *Nat. Commun.* **4**, 1555 (2013).

[42] Longhi, S. Quantum-optical analogies using photonic structures. *Laser Photon. Rev.* **3**, 243-261 (2009).

[43] Longhi, S. Bloch oscillations in non-Hermitian lattices with trajectories in the complex plane in non-Hermitian physics. *Phys. Rev. A* **92**, 042116 (2015).

[44] Della Valle, G. & Longhi, S. Spectral and transport properties of time-periodic PT-symmetric tight-binding lattices. *Phys. Rev. A* **87**, 022119 (2013).

[45] Hatano, N. & Nelson, D. R. Localization transitions in non-Hermitian quantum mechanics. *Phys. Rev. Lett.* **77**, 570–573 (1996).

[46] Longhi, S., Gatti, D. & Della Valle, G. Robust light transport in non-hermitian photonic lattices. *Sci. Rep.* **5**, 13376 (2015).

[47] Zeuner, J. M. *et al*. Observation of a topological transition in the bulk of a non-hermitian system. *Phys. Rev. Lett.* **115**, 040402 (2015).

[48] Poli, C., Bellec, M., Kuhl, U., Mortessagne, F. & Schomerus, H. Selective enhancement of topologically induced interface states in a dielectric resonator chain. *Nat. Commun.* **6**, 6710 (2015).

[49] This work was jointly supported by the National Basic Research Program of China (Grants 2012CB921503, 2013CB632700 and 2015CB659400) and the National Nature Science Foundation of China (Grants 11134006, 11474158 and 61378009). M.-H.L. also acknowledges the support of Natural Science Foundation of Jiangsu Province (BK20140019) and the support from Academic Program Development of Jiangsu Higher Education (PAPD). W.S.F. and A.S. acknowledge Boeing for their support under their SRDMA program and also thank the NSF CIAN ERC (Grant EEC-0812072).

[50] Y.-L.X., W.S.F., L.G., these authors contributed equally to this work. Y.-L.X., W.S.F., M.-H.L. and X.-P.L. conceived the idea and designed the devices. Y.-L.X. performed the FDTD simulations, W.S.F. designed the chip layout and fabricated the devices, and L.G. carried out the SNOM measurements. All the authors contributed to discussion of the project. A.S., Z.-Y.L. and Y.-F.C. guided the project. Y.-L.X., X.-P.L. and M.-H.L. wrote the manuscript with revisions from other authors.

[51] Supplementary Information accompanies this paper at http://www.nature.com/naturecommunications

[52] Methods. Sample fabrication. The Hermitian photonic lattice was fabricated by means of a single step of electron-beam (e-beam) lithography to define the silicon waveguide layer, which was followed by resist development and inductively coupled plasma etching using a mixture of SF_6 and C_4F_8. The non-Hermitian photonic lattice was fabricated using two steps of aligned e-beam lithography. First, the Cr layer was defined by using a positive e-beam resist and developed, followed by 4 nm of Cr deposition by means of e-beam deposition and liftoff. The silicon waveguides were defined by means of a second step of aligned e-beam lithography using a negative resist, followed by development and finally inductively coupled plasma dry etching using a mixture of etchant aforementioned to define the waveguides. Both photonic lattices went through additional final processing steps before optical testing. These include (1) a photolithographic process to fabricate SU-8 polymer waveguide-based mode converters surrounding the silicon input/output waveguides and (2) dicing and facet polishing.

Dispersion calculation. Coupled mode equations describing the wave dynamics in both the Hermitian and non-Hermitian photonic lattices were solved to obtain the dispersion spectrum or band diagram. This was done by inserting a plane wave ansatz $A_n \propto \exp(i\beta z - inqd)$ for the nth waveguide's mode amplitude into the coupled mode equation, that is, equations (1) and (4), and by enforcing the non-trivial solutions for A_n. The corresponding transverse Bloch momentum q-dependent dispersion β in a complex wave-vector domain represents the underlying dispersion spectrum or band diagram.

Measurement. In measurements, a near-infrared optical wave at the wavelength of 1550 nm from a tunable external cavity laser (Photonetics, 1500–1640 nm) was launched into the SU-8 mode converter through a lensed tapered single-mode fibre. The testing chip and the lensed fibre are bonded onto a glass slide so that the relative position of the fibre to the waveguide is fixing when the testing chip is translated on a moving stage under a SNOM (SNOM-100 Nanonics). The SNOM is equipped with a metal-coated fibre tip with a 50 nm radius to map the intensity distribution. A schematic of our measurement setup is shown in Supplementary Fig. 2 and details of this setup are included in Supplementary Note 4.

Photonic Topological Insulator with Broken Time-reversal Symmetry*

Cheng He[1], Xiao-Chen Sun[1], Xiao-Ping Liu[1,2], Ming-Hui Lu[1,2],
Yulin Chen[3], Liang Feng[4], and Yan-Feng Chen[1,2]

[1] *National Laboratory of Solid State Microstructures & Department of Materials Science and Engineering,*
Nanjing University, Nanjing 210093, China
[2] *Collaborative Innovation Center of Advanced Microstructures, Nanjing University, Nanjing, 210093, China*
[3] *Clarendon Laboratory, Department of Physics, University of Oxford, Oxford OX1 3PU, United Kingdom*
[4] *Department of Electrical Engineering, University at Buffalo,*
The State University of New York, Buffalo, NY 14260

A topological insulator is a material with an insulating interior but time-reversal symmetry-protected conducting edge states. Since its prediction and discovery almost a decade ago, such a symmetry-protected topological phase has been explored beyond electronic systems in the realm of photonics. Electrons are spin-1/2 particles, whereas photons are spin-1 particles. The distinct spin difference between these two kinds of particles means that their corresponding symmetry is fundamentally different. It is well understood that an electronic topological insulator is protected by the electron's spin-1/2 (fermionic) time-reversal symmetry $T_f^2=-1$. However, the same protection does not exist under normal circumstances for a photonic topological insulator, due to photon's spin-1 (bosonic) time-reversal symmetry $T_b^2=1$. In this work, we report a design of photonic topological insulator using the Tellegen magnetoelectric coupling as the photonic pseudospin orbit interaction for left and right circularly polarized helical spin states. The Tellegen magnetoelectric coupling breaks bosonic time-reversal symmetry but instead gives rise to a conserved artificial fermionic-like-pseudo time-reversal symmetry, T_p ($T_p^2=-1$), due to the electromagnetic duality. Surprisingly, we find that, in this system, the helical edge states are, in fact, protected by this fermionic-like pseudo time-reversal symmetry T_p rather than by the bosonic time-reversal symmetry T_b. This remarkable finding is expected to pave a new path to understanding the symmetry protection mechanism for topological phases of other fundamental particles and to searching for novel implementations for topological insulators.

Topological description of electronic phase has now become a new paradigm in the classification of condensed matters[1—10]. Electronic topological insulators (TIs) are time-reversal symmetry (TRS) protected topological phase, exhibiting gapless edge/surface states in their bulk bandgap due to strong spin-orbit coupling[11,12]. This intriguing classification of the topologically protected phase has been applied to study other systems

---
* Proc. Natl. Acad. Sci. USA, 2016, 113(18): 4924

as well, for example, photonic systems. The photonic analog of the integer quantum Hall effect has been proposed and extensively studied with a single transverse electric (TE) polarization state[13,14] or magnetic field (TM) polarization state[15,16] in gyrotropic photonic crystals with broken TRS in the presence of an external magnetic bias; the fractional quantum Hall effect has also been investigated in a correlated photonic system[17]. To date, to design photonic TIs (PTIs) with TRS, several models have been reported[18—26] using a pair of degenerate photonic states, e.g., hybrid TE + TM/TE − TM states[21,22,24] or pseudospins represented by clockwise/counterclockwise helical energy flow states[19,20,23], where the degenerate states "see" opposite effective magnetic gauge fields. Additionally, the photonic Floquet TIs have also been explored with a helix photonic structure under a broken spatial inversion symmetry[18]. However, the robustness of these topological phases has not been fully investigated and validated against a comprehensive set of TRS invariant impurities[18—27]. According to the topological theory developed for condensed matters, having a Kramers doublet is the key to constructing a TI. It is well known that the spin quantum number of photons is different from that of electrons. Hence, they belong to two different classes of particles, boson and fermion, and have different TRS operators: $T_b = \tau_z K$ for photons ($T_b^2 = 1$) and $T_f = i\tau_y K$ for electrons ($T_f^2 = -1$). Here, $\tau_z = diag\{1, -1\}$ and $\tau_y = [0, -i; i, 0]$ are the z and y components, respectively, of a Pauli matrix, and K is the complex conjugation. Kramers degeneracy theorem states that there cannot exist any Kramers doublet for photons under T_b. Therefore, the bosonic TRS T_b is highly unlikely to be the TRS requirement for the PTIs, and thus the underlying symmetry protection mechanism requires further investigation. Consequently, the robustness of those previously reported photonic topological states may be questionable and could be readily jeopardized by certain impurities[26,28].

In this work, we explore this mysterious symmetry protection issue in a Tellegen photonic crystal medium[29], where the Tellegen magnetoelectric coupling can be realized in piezoelectric (PE) and piezomagnetic (PM) superlattice constituents[30,31]. The presence of Tellegen magnetoelectric coupling enables a photonic pseudospin-orbital coupling effect, which breaks the bosonic TRS but, at the same time, creates a fermionic-like pseudo TRS due to the electromagnetic duality. Consequently, two pseudospin states that are helical states represented by left and right circular polarizations (or any two desired modes/polarization states) can be carefully designed and matched to form Kramers degeneracy under the fermionic-like pseudo TRS. In addition, we show that a pair of degenerate gapless edge states for the Kramers doublet, i.e., left and right circular polarizations, exists in the bulk bandgap of the Tellegen photonic crystal. Such edge states exhibit pseudospin-dependent transportation, which characterizes a type of PTI with broken bosonic TRS in the presence of inherent magnetoelectric coupling[32]. Our further analysis shows that this PTI is protected by the fermionic-like pseudo TRS T_p ($T_p^2 = -1$), and thus the gapless edge states are robust against T_p - invariant impurities rather than T_b -

invariant impurities.

> **Significance**
>
> Topological insulators are first discovered in electronic systems. A key factor is the Kramers doublet for the spin-1/2 electrons under fermionic time-reversal symmetry $T_f^2 = -1$. Unlike electrons, photons are massless bosons with spin-1. Therefore, the Kramers degeneracy theorem cannot readily apply to photons under the bosonic time-reversal symmetry. So far, there has been no coherent physical explanation for the symmetry protection mechanism behind the photonic topological insulator. Here, we design a photonic topological insulator that violates the bosonic time-reversal symmetry but complies with a fermionic-like pseudo time-reversal symmetry. The analyses and results, through comprehensive investigations on the properties of edge states, validate that the topological edge states are, in fact, protected by the fermionic-like pseudo time-reversal symmetry $T_p (T_p^2 = -1)$.

Models

Previous studies on polarization degeneracy-based PTIs mainly focus on a pair of linearly polarized states, e.g., TE/TM states [22] and TE+TM/TE−TM states[21,24] as shown in a Poincaré sphere [Fig. 1(a), Left]. There also exists another important pair of polarization states, left circular polarization (LCP: positive helix)/right circular polarization (RCP: negative helix). These two states are commonly used to mimic the electronic spin states [Fig. 1(a), Right], due to their opposite helicity-orbital couplings. Herein, they are referred as ψ_{LCP} and ψ_{RCP}.[33]

$$\begin{bmatrix} \psi_{LCP} \\ \psi_{RCP} \end{bmatrix} = \begin{bmatrix} 1 & i \\ i & 1 \end{bmatrix} \begin{bmatrix} E_z \\ H_z \end{bmatrix}. \tag{1}$$

These photonic polarization states do not generally form a Kramers doublet under T_b, because $T_b \psi_{LCP} = \psi_{LCP}^*$ and $T_b \psi_{RCP} = -\psi_{RCP}^*$. It is therefore necessary to identify a different symmetry operator, under which Kramers degeneracy required for constructing TIs can be fulfilled. From the electromagnetic duality, it is not difficult to image that an interchange operator $\tau_x = [0, 1; 1, 0]$ (the x component of a Pauli matrix) should also be taken into account during the time-reversal operation because of the magnetoelectric interaction $E_z \leftrightarrow H_z$ in our model. A new pseudo TRS operator thus becomes $T_p = T_b \tau_x = \tau_x K \tau_x = i\tau_y K$. Surprisingly, this symmetry operator T_p has the same form as the fermionic TRS operator T_f, under which ψ_{LCP} and ψ_{RCP} become a photonic Kramers doublet, opening the door to building a PTI.

Clearly, the key to realizing a photonic Kramers doublet in our system is to implement the required magnetoelectric coupling. Here we construct a quasi-2D superlattice photonic crystal (SLPC) consisting of arrayed meta-atoms[34,35]: superlattices with alternating PE/

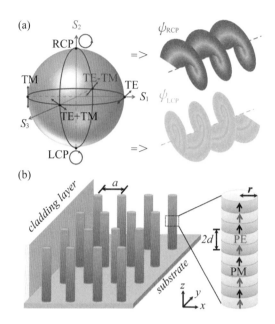

FIG.1. Schematic of SLPC. (a)(Left) The polarization on the Poincaré sphere, where the three Stokes parameters are $S_1 = |E_y|^2 - |E_x|^2$, $S_2 = 2|E_x||E_y|\sin\delta$, and $S_3 = 2|E_x||E_y|\cos\delta$, and δ is the phase difference between the x and y axis field components. (Right) LCP and RCP photonic states (helicity-orbital coupling). (b) A 2D square lattice SLPC consisting of stacked PE and PM superlattice constituents in vacuum background. The positive domain and stacking direction of PE (red arrows) and PM (black arrows) media are along the z axis. Lattice constants of photonic crystal and PE/PM superlattice are a and $2d$, respectively. The radius of cylinder is $r = 0.11a$. LCP and RCP edge states propagate in x-y plane at the boundary (with a separation distance of $0.5a$) between SLPC and cladding layer.

PM layers with the periodicity at a deep subwavelength scale along the z direction [Fig. 1(b)]. To realize the pseudospin degenerate states, the point groups of PE and PM materials are chosen to be 622 and 6 mm(or 422 and 4 mm) with nonzero PE coefficient e_{14} and PM coefficient m_{15}, respectively. For simplicity, the impedance matching condition is assumed. The values for the density, elastic coefficient, and PM coefficient used in our study are based on the $BaTiO_3$ - $CoFe_2O_4$ superlattice[30,31] (*SI Appendix*, *Part A*). The effective bianiso-tropic constitutive relation for these deep-subwavelength superlattices can be described as

$$\bm{D} = \overleftrightarrow{\varepsilon}\bm{E} + \overleftrightarrow{\xi}\bm{H}, \bm{B} = \overleftrightarrow{\zeta}\bm{E} + \overleftrightarrow{\mu}\bm{H} \quad (2)$$

where $\overleftrightarrow{\varepsilon} = diag\{\varepsilon_{xx},\varepsilon_{xx},\varepsilon_{zz}\}$, $\overleftrightarrow{\mu} = diag\{\mu_{xx},\mu_{xx},\mu_{zz}\}$, $\overleftrightarrow{\xi} = \overleftrightarrow{\zeta}^T$, $\xi_{xy} = -\xi_{yx} = -A_1 e_{14} m_{15}$, $A_1 = 2/[d^2\rho(\omega^2 - \omega_T^2 + i\gamma_T\omega)]$, and $\omega_T = \pi/d \times \sqrt{C_{44}/\rho}$ for the transverse vibration frequency (details in *SI Appendix*, *Part A*). The elastic coefficient is C_{IJ} with the damping coefficient γ_T, and ρ is the material density. In a lossless case ($\gamma_T = 0$), $\overleftrightarrow{\xi} = \overleftrightarrow{\zeta}^\dagger$ (superscript dagger denotes the complex conjugate transpose), and it only has real tensor elements, forming the so-called Tellegen media[29], in which the magnetoelectric coupling effect (off-diagonal terms) breaks the TRS and reciprocity ($\vec{\vec{\xi}} \neq -\vec{\vec{\zeta}}^T$)[32,36]. It is also worth noting that the broken TRS is originated from the polariton induced by the coupling between the elastic vibration of the superlattice and incident electromagnetic waves. The coupling strength in this case is much larger than that in multiferroic materials[37]. This approach for generating magnetoelectric coupling is fundamentally different from the previously proposed T_b-invariant chiral medium with purely imaginary magnetoelectric parameters[21]. Therefore, the Hamiltonian of our designed SLPC for LCP and RCP states can be derived as

$$\mathcal{H}\begin{bmatrix} \psi_{LCP} \\ \psi_{RCP} \end{bmatrix} = \begin{bmatrix} \mathcal{H}_0 - i\mathcal{H}_1 & 0 \\ 0 & \mathcal{H}_0 + i\mathcal{H}_1 \end{bmatrix} \begin{bmatrix} \psi_{LCP} \\ \psi_{RCP} \end{bmatrix} = 0 \qquad (3)$$

where $\mathcal{H}_0 = k_0^2 \varepsilon_{zz} + \partial_x \frac{1}{\mu_\parallel} \partial_x + \partial_y \frac{1}{\mu_\parallel} \partial_y$, $\mathcal{H}_1 = \partial_x \kappa \partial_y - \partial_y \kappa \partial_x$, $\mu_\parallel = (\mu_{xx}\varepsilon_{xx} - \xi_{xy}^2)/\varepsilon_{xx}$, $\kappa = -\xi_{xy}/(\mu_{xx}\varepsilon_{xx} - \xi_{xy}^2)$ is a pseudospin-orbital coupling parameter. This coupling parameter exhibits a polariton dispersion relation near the resonant frequency ω_T [Fig. 2(a) and (b)]. The $\pm i\mathcal{H}_1$ operator indicates the pseudospin-orbital coupling and the resulting opposite effective gauge fields for LCP and RCP states, which can further lead to the nonreciprocal transmission of the LCP/RCP polarization states, forming an isolator that is free of external magnetics, as shown in Fig. 2(c). In contrast to conventional isolators, such an isolator exhibits a clear pseudospin-dependent transmission manifested by the opposite transmission direction of the LCP and RCP states. This unique one-way transmission property can be further leveraged to construct PTIs.

Results and Discussion

Gapless Edge State. In this work, the thickness of each PE and PM layer is set to be $d = 500$ nm; the transverse vibration frequency is $\omega_T = 18.05$ GHz. The lattice constant of the SLPC is set to be $a = 9$ mm. The material property PE/PM superlattice inside the SLPC at frequency of $1.076\omega_T$ (19.42 GHz) are $\kappa = 0.3$ obtained from Fig. 2, $\mu_\parallel = \varepsilon_\parallel = 2.5$, and $\mu_{zz} = \varepsilon_{zz} = 14.2$. Without considering pseudospin-orbital coupling, i.e., $\kappa = 0$, the band structure of the SLPC [Fig. 3(a)] clearly shows band crossing at M point of the Brillouin zone for the second and third bulk bands. However, after introducing pseudospin-orbital coupling, e.g., $\kappa = 0.3$, the degeneracy at M point is lifted, creating a bandgap for the bulk states with their normalized frequency ranging from 0.55 to 0.61 (corresponding to wavelengths from $1.82a$ to $1.64a$, i.e., 16.38 mm to 14.76 mm). In the bandgap, there exist gapless helical edge states for LCP and RCP states [Fig. 3(b)], which exhibit pseudo-spin-dependent edge state transportation. Because LCP and RCP states experience opposite effective gauge fields, their power flux transportation behavior is completely reversed, i.e., counter-clockwise for LCP and clockwise for RCP [Fig. 3 (c)-(f)].

Robustness of Pseudospin-Dependent Transportation. The robustness for the edge states

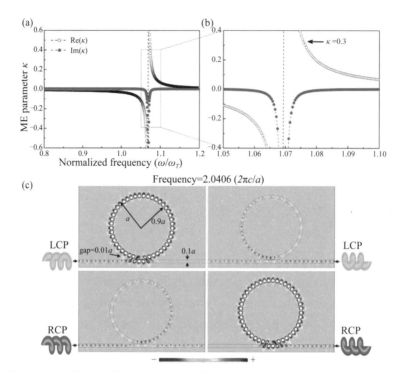

FIG. 2. Pseudospin-orbit coupling parameter κ. (a) Dispersion of κ, where open and solid circles represent the real and imaginary parts, respectively. (b) Zoom-in view of the rectangular frequency area around frequency of interest. (c) Pseudospin-dependent nonreciprocal transmission of LCP/RCP state in an airborne ring resonator constructed with Tellegen media with material parameters $\varepsilon=2.5$, $\mu=2.5$, and $\kappa=0$ for the straight bus waveguide and $\varepsilon=2.5$, $\mu=2.5$, and $\kappa=0.3$ for circular waveguide. Colorbar represents the field strength.

transportation is studied in various SLPC configurations (Fig. 4). Without any defects, the pseudospin-dependent transportation is obvious [Fig. 4(a)], because the LCP (RCP) source only excites one-way clockwise (counterclockwise) light transportation (*SI Appendix*, Part B). This one-way robust transportation is then verified and confirmed against three different types of geometric defects, an L-shaped slab obstacle, a cavity obstacle, and a strongly disordered domain inside the cavity as shown in Fig. 4(b)-(d). Note that the impact of propagation loss can be safely omitted in the above analysis (*SI Appendix*, Part C), because the frequency of operation is far away from the resonance condition of the superlattice [as shown in Fig. 2(a)].

Next, the robustness of such photonic pseudospin-dependent transportation is further checked against a comprehensive set of impurities with optical properties satisfying all possible combination of T_b and T_p symmetries, i.e., (i) a uniaxial dielectric impurity that is T_b-invariant but not T_p-invariant; (ii) a Tellegen impurity that is neither T_b-invariant nor T_p-invariant; (iii) a chiral impurity that is T_b-invariant but not T_p-invariant; and (iv) a chiral impurity that is both T_b- and T_p-invariant. In cases i and ii, the LCP state can be backscattered into the RCP state by impurities [Fig. 5(a) and (b), Left]. Such findings imply that the pseudospin-dependent transportation is not robust if impurities' T_p symmetry is broken, regardless of their T_b status. This conclusion is also consistent with the projected band structure [Fig. 5(a) and (b), Right], which indicates the emergence of a bandgap from an initially gapless band structure. In case iii, the eigen equation of the chiral medium can be treated as a superposition of two orthogonal LCP and RCP eigen states with different Bloch wave vectors. Although the impurity cannot backscatter the RCP state [Fig. 5(c), Left], the LCP and RCP edge states decouple, and both states exhibit an independent integer photonic quantum Hall effect[Fig. 5(c), Right]. In case iv, the RCP state can be excited at the impurity site as shown in Fig. 5(d), Left. However, the excited RCP state is localized in the vicinity of the impurity site and no RCP state is backscattered because T_p-invariant impurity is a polarization degenerate medium for TE+TM and TE−TM hybrid states. With LCP state incidence, this state decouples into these two degenerate hybrid states inside the impurity, and it reemerges as LCP light after transmitting through it. The destructive interference between two backscattering paths for the RCP light eliminates backscattering, leading to robust one-way transmission. This unique transportation characteristic can be further confirmed by examining the corresponding projected band structure. As shown in Fig. 5(d), Right, the edge states remain gapless, and such a system still supports pseudospin-dependent transportation. The above analysis strongly indicates that LCP and RCP eigen states can only be considered as a pair of pseudospin states under the fermionic-like pseudo TRS T_p operator, i.e., $T_p\psi_{LCP}=\psi_{RCP}^*$ and $T_p\psi_{RCP}=-\psi_{LCP}^*$. The Hamiltonian of our SLPC shown in Eq. 3 commutes with T_p: $T_p \mathcal{H} T_p^{-1}=\mathcal{H}$ but does not commute with T_b: $T_b \mathcal{H} T_b^{-1} \neq \mathcal{H}$, confirming our proposed Tellegen system is a bosonic TRS broken system. It is thus evident that our proposed PTI is protected by the fermionic-like pseudo TRS T_p rather than by the bosonic TRS T_b.

Note that, in our proposed PTI, a unity normalized impedance ($\eta=\sqrt{\mu/\varepsilon}=1$) is assumed. Therefore, strictly speaking, back-scattering immune propagation is only applicable to impurity with an equally valued relative permeability and permittivity (e.g., vacuum). A slight mismatch between the relative permeability and permittivity may be overcome by using sufficiently large pseudospin-orbital(helicity-orbital) coupling κ (refer to Eq. 3 for details). For a large mismatch, two elliptical polarized eigen states $\varphi_{LCP}=E_z+i\eta H_z$, $\varphi_{RCP}=iE_z+\eta H_z$, shall be chosen instead of LCP and RCP to construct a T_p protected PTI.

Topological Values. It should be noticed that each energy band in Fig. 3(a) is doubly degenerate for LCP and RCP states experiencing opposite gauge fields. Thus, although the total Chern number is zero in our broken TRS system, the spin Chern number for LCP and RCP is $+1$ and -1, respectively[38]. Therefore, in such a system, the photonic quantum spin Hall effect (or PTI) rather than the photonic quantum Hall effect is expected. On the other hand, because the pseudo TRS T_p extracted from our system has the same

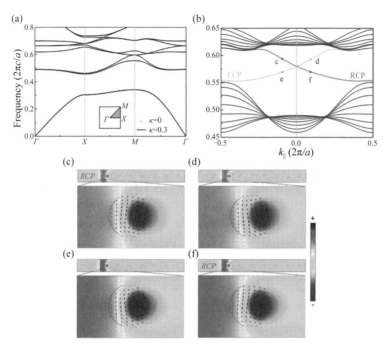

FIG.3. Topological edge states in SLPC. (a) Bandstructures without pseudospin coupling effect (dotted lines, $\kappa = 0$) and with pseudospin coupling effect (solid lines, $\kappa = 0.3$). (b) The projected bandstructures with $\kappa = 0.3$, where bulk states are denoted by blue lines and gapless LCP and RCP edge states are denoted by green and red lines, respectively. (c-f) The Bloch field distributions of supercell configuration corresponding to points c(C), d(D), e(E), and f(F) marked in b. Zoom-in view shows the opposite rotating power flux near the boundary represented by black arrows. Colorbar represents the field strength.

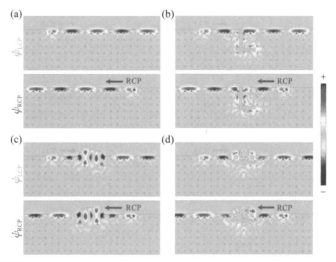

FIG.4. Field distribution of backscattering immune pseudospin-dependent transportation. (a) LCP (*Top*) and RCP (*Bottom*) light propagation without structural defects. (b-d) Robustness for LCP (*Top*) and RCP (*Bottom*) light against different types of defects: (b) an L-shape slab obstacle (index of refraction, $-i$) with thickness $0.5a$, (c) a cavity obstacle, and (d) a strongly disordered domain in the cavity. Operating frequency is 0.6 $(2\pi c/a)$. LCP and RCP point excitation sources are indicated by green and red stars, respectively. Colorbar represents the field strength.

mathematic form as the fermionic TRS T_f, the Z_2 invariant of our PTI can be characterized under the constructed pseudo TRS T_p. Based on the inversion symmetry of our model[39], the Z_2 invariant can be determined by the quantity equation, $\delta_i = \prod_{m=1}^{N} \xi_{2m}(\Gamma_i)$, where $\xi_{2m}(\Gamma_i)$ is the parity eigenvalue of the $2m$th occupied energy band at four T_p-invariant momenta Γ_i in the Brillouin zone. The Z_2 invariant $\nu = 0, 1$, which distinguishes different quantum spin Hall phases, is governed by the product of all of the δ_i: $(-1)^\nu = \prod_i \delta_i$. The quantity equation can be further simplified as $\delta_i = \xi(\Gamma_i)$, in which $\xi(\Gamma_i)$ is the parity eigenvalue of lower energy band at Γ_i. Because our system is of T_p invariance, the eigen states in reciprocal space are degenerate at Γ_i. It is thus straightforward to demonstrate that the Bloch field distribution at (0, 0) in reciprocal space has even parity with a parity eigenvalue of 1 whereas the Bloch field distributions at three other high-symmetry points have odd parity with parity eigenvalues of -1. With $(-1)^\nu = -1$ used to determine the Z_2 invariant, the Z_2 invariant can be readily derived: $\nu = 1$, showing a nontrivial topological state supported in our system (*SI Appendix*, Part D).

Conclusions

In summary, we study the topological property of a Tellegen photonic crystal, which has broken bosonic TRS due to Tellegen magnetoelectric coupling. Such a coupling leads to the interchange of electric field and magnetic field of an eigen state and thus gives rise to an artificial fermionic pseudo TRS T_p symmetry. This symmetry is found to be responsible for the formation of a Kramers doublet in our photonic system. The robustness of the one-way pseudospin-dependent transportations, in this case, is protected by this artificial fermionic pseudo TRS T_p instead of by the bosonic TRS T_b. Our concept of T_p symmetry-protected photonic topological phase is by no means limited only to the particular case we considered here. It may be easily applied to identify more PTIs by following the steps outlined below. For step 1, construct two degenerate photonic states that satisfy Kramers degeneracy under an artificial fermionic T_p operator; in step 2, retrieve a proper material constitutive relation by substituting a Kramers doublet and PTI's Hamiltonian for Maxwell equations; and, in step 3, find appropriate materials or design metamaterials to fulfill the required constitutive relation. We believe that our findings may expand the scope of PTIs and pave a new and viable way to classify PTIs, which can overcome the limitation based on the traditional bosonic TRS. Our results may even inspire researchers to exploit photonic topological states in unconventional systems, for instance, non-Hermitian synthetic parity-time symmetric meta-materials with balanced gain/loss modulation[40] or a purely dissipation-driven system[41]. Analogously, the concept of this T_p symmetry-protected PTI may also be applied to other classic waves, such as sound[42] and mechanic modes[43,44].

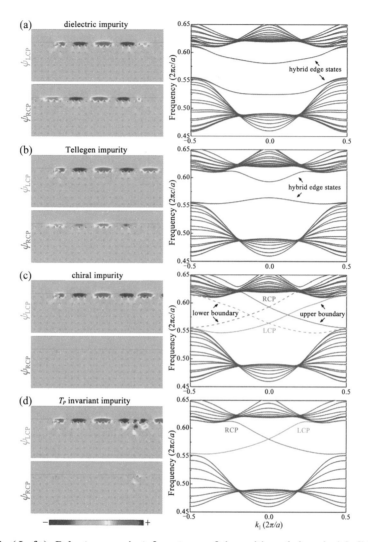

FIG.5. (*Left*) Robustness against four types of impurities: (a) uniaxial dielectric impurity ($\varepsilon_{zz}=4$), (b) Tellegen impurity ($\xi_{zz}=\zeta_{zz}=1$), (c) chiral impurity ($\xi_{zz}=-\zeta_{zz}=i$), and (D) chiral impurity with T_p invariance ($\xi_{xy}=-\xi_{yx}=\zeta_{xy}=-\zeta_{yx}=0.3i$; other parameters are the same as superlattice) remaining invariant under T_p operation. LCP point excitation source (green stars) is at a frequency of 0.6 ($2\pi c/a$). A cylindrical impurity with radius $r=0.11a$ is placed near the boundary. (*Right*) The corresponding projected bandstructures. Colorbar represents the field strength.

References and Notes

[1] Klitzing KV, Dorda G, Pepper M (1980) New method for high-accuracy determination of the fine-structure constant based on quantized Hall resistance. *Phys Rev Lett* **45**(6): 494–497.

[2] Thouless DJ, *et al*. (1982) Quantized Hall conductance in a two-dimensional periodic potential. *Phys Rev Lett* **49**(6):405–408.

[3] Tsui DC, Stormer HL, Gossard AC (1982) Two-dimensional magnetotransport in the extreme quantum limit. *Phys Rev Lett* **48**(22):1559–1562.

[4] Hatsugai Y (1993) Chern number and edge states in the integer quantum Hall effect. *Phys Rev Lett* **71**(22):3697-3700.

[5] Kane CL, Mele EJ (2005) Z_2 topological order and the quantum spin Hall effect. *Phys Rev Lett* **95**(14):146802.

[6] Bernevig BA, Hughes TL, Zhang S-C (2006) Quantum spin Hall effect and topological phase transition in HgTe quantum wells. *Science* **314**(5806):1757-1761.

[7] König M, et al. (2007) Quantum spin hall insulator state in HgTe quantum wells. *Science* **318**(5851):766-770.

[8] Hsieh D, et al. (2008) A topological Dirac insulator in a quantum spin Hall phase. *Nature* **452**(7190):970-974.

[9] Hasan MZ, Kane CL (2010) Colloquium: Topological insulators. *Rev Mod Phys* **82**(4):3045-3067.

[10] Qi X-L, Zhang S-C (2011) Topological insulators and superconductors. *Rev Mod Phys* **83**(4):1057-1110.

[11] Xia Y, et al. (2009) Observation of a large-gap topological-insulator class with a single Dirac cone on the surface. *Nat Phys* **5**(6):398-402.

[12] Zhang H, et al. (2009) Topological insulators in Bi_2Se_3, Bi_2Te_3 and Sb_2Te_3 with a single Dirac cone on the surface. *Nat Phys* **5**(6):438-442.

[13] Haldane FDM, Raghu S (2008) Possible realization of directional optical waveguides in photonic crystals with broken time-reversal symmetry. *Phys Rev Lett* **100**(1):013904.

[14] Raghu S, Haldane FDM (2008) Analogs of quantum-Hall-effect edge states in photonic crystals. *Phys Rev A* **78**(3):033834.

[15] Wang Z, Chong YD, Joannopoulos JD, Soljacic M (2008) Reflection-free one-way edge modes in a gyromagnetic photonic crystal. *Phys Rev Lett* **100**(1):013905.

[16] Wang Z, Chong Y, Joannopoulos JD, Soljacic M (2009) Observation of unidirec-tional backscattering-immune topological electromagnetic states. *Nature* **461**(7265):772-775.

[17] Kapit E, Hafezi M, Simon SH (2014) Induced self-stabilization in fractional quantum Hall states of light. *Phys Rev X* **4**(3):031039.

[18] Rechtsman MC, et al. (2013) Photonic Floquet topological insulators. *Nature* **496**(7444):196-200.

[19] Hafezi M, et al. (2011) Robust optical delay lines with topological protection. *Nat Phys* **7**(11):907-912.

[20] Hafezi M, et al. (2013) Imaging topological edge states in silicon photonics. *Nat Photon* **7**(12):1001-1005.

[21] Khanikaev AB, et al. (2013) Photonic topological insulators. *Nat Mater* **12**(3):233-239.

[22] Ma T, Khanikaev AB, Mousavi SH, Shvets G (2015) Guiding electromagnetic waves around sharp corners: Topologically protected photonic transport in meta-waveguides. *Phys Rev Lett* **114**(12):127401.

[23] Liang GQ, Chong YD (2013) Optical resonator analog of a two-dimensional topo-logical insulator. *Phys Rev Lett* **110**(20):203904.

[24] Chen W-J, et al. (2014) Experimental realization of photonic topological insulator in a uniaxial metacrystal waveguide. *Nat Commun* **5**:5782.

[25] Bliokh KY, Smirnova D, Nori F (2015) OPTICS. Quantum spin Hall effect of light. *Science* **348**(6242):1448-1451.

[26] Wu L-H, Hu X (2015) Scheme for achieving a topological photonic crystal by using dielectric material.

Phys Rev Lett **114**(22):223901.

[27] Lu L, Joannopoulos JD, Soljacic M (2014) Topological photonics. *Nat Photon* **8**(11): 821–829.

[28] Süsstrunk R, Huber SD (2015) Observation of phononic helical edge states in a me-chanical topological insulator. *Science* **349**(6243):47–50.

[29] Sihvola AH, Lindell IV (1995) Material effects in bi-anisotropic electromagnetics. *IEICE Trans Electron* **78**(10):1383–1390.

[30] Huang JH, Kuo W-S (1997) The analysis of piezoelectric/piezomagnetic composite materials containing ellipsoidal inclusions. *J Appl Phys* **81**(3):1378–1386.

[31] Pan E (2002) Three-dimensional Green's functions in anisotropic magneto-electro-elastic bimaterials. *Z Angew Math Phys* **53**(5):815–838.

[32] Kong JA (1975) *Theory of Electromagnetic Waves*(Wiley, New York).

[33] Lakhtakia A (1994) *Beltrami Fields in Chiral Media*(World Sci, Singapore).

[34] Smith DR, Pendry JB, Wiltshire MCK (2004) Metamaterials and negative refractive index. *Science* **305**(5685):788–792.

[35] Shalaev VM (2007) Optical negative-index metamaterials. *Nat Photon* **1**(1):41–48.

[36] He C, et al. (2014) Topological photonic states. *Int J Mod Phys B* **28**(02):1441001.

[37] Wang KF, Liu JM, Ren ZF (2009) Multiferroicity: The coupling between magnetic and polarization orders. *Adv Phys* **58**(4):321–448.

[38] Yang Y, et al. (2011) Time-reversal-symmetry-broken quantum spin Hall effect. *Phys Rev Lett* **107**(6):066602.

[39] Fu L, Kane CL (2007) Topological insulators with inversion symmetry. *Phys Rev B* **76**(4): 045302.

[40] Feng L, et al. (2013) Experimental demonstration of a unidirectional reflectionless parity-time metamaterial at optical frequencies. *Nat Mater* **12**(2):108–113.

[41] Bardyn CE, et al. (2012) Majorana modes in driven-dissipative atomic superfluids with a zero Chern number. *Phys Rev Lett* **109**(13):130402.

[42] Fleury R, Sounas DL, Sieck CF, Haberman MR, Alù A (2014) Sound isolation and giant linear nonreciprocity in a compact acoustic circulator. *Science* **343**(6170):516–519.

[43] Kane CL, Lubensky TC (2014) Topological boundary modes in isostatic lattices.*Nat Phys* 10(1):39–45.

[44] Chen BG, Upadhyaya N, Vitelli V (2014) Nonlinear conduction via solitons in a topo-logical mechanical insulator. *Proc Natl Acad Sci USA* **111**(36):13004–13009.

[45] Theoretical Model. A full description of the mathematic derivation and parameters of PE and PM superlattice to obtain the effect constitutive relation can be found in *SI Appendix*, *Part A*. It should be noticed that a key factor to realizing the pseudospin degenerate states for LCP and RCP is the point groups of PE and PM materials, which are chosen to be 622 and 6 mm in this work to realize off-diagonal magnetoelectric coupling. The loss of such a superlattice can be safely omitted because the operating frequency is far away from the resonance frequency. Furthermore, we also derived the theoretical model of SLPC via the tight-binding approximation approach as shown in *SI Appendix*, *Part D*.

[46] Numerical Method. Numerical investigations in this work are conducted using a hybrid RF mode of commercial FEM software (COMSOL Multiphysics). The parameters used in numerical investigations are based on the effective constitutive relation of PE and PM superlattice.

[47] We thank Dr. P. Nayar for helpful discussion. The work was jointly supported by the National Basic

Research Program of China (Grants 2012CB921503 and 2013CB632702) and the National Nature Science Foundation of China (Grants 11134006, 11474158, and 11404164). We also acknowledge support from Academic Program Development of Jiangsu Higher Education. Y. C. acknowledges support from a DARPA MESO project (187 N66001-11-1-4105). L.F. was funded by Department of Energy (DE-SC0014485) for analyzing the results of magnetoelectric coupling and photonic topological insulator.

[48] Author contributions: C.H. and M.-H.L. designed research; C.H. performed research; C.H., X.-C.S., and M.-H.L. contributed new reagents/analytic tools; C.H., X.-C.S., X.-P.L., M.-H.L., Y.C., L.F., and Y.-F.C. analyzed data; and C.H., X.-C.S., X.-P.L., M.-H.L., L.F., and Y.-F.C. wrote the paper.

[49] This article contains supporting information online at www.pnas.org/lookup/suppl/doi:10.1073/pnas.1525502113/-/DCSupplemental.

Acoustic Topological Insulator and Robust One-way Sound Transport[*]

Cheng He[1,2], Xu Ni[1], Hao Ge[1], Xiao-Chen Sun[1], Yan-Bin Chen[1],
Ming-Hui Lu[1,2], Xiao-Ping Liu[1,2] and Yan-Feng Chen[1,2]

[1] *National Laboratory of Solid State Microstructures & Department of Materials Science and Engineering,
Nanjing University, Nanjing, Jiangsu 210093, China*
[2] *Collaborative Innovation Center of Advanced Microstructures, Nanjing University,
Nanjing, Jiangsu 210093, China*

Topological design of materials enables topological symmetries and facilitates unique backscattering-immune wave transport[1-26]. In airborne acoustics, however, the intrinsic longitudinal nature of sound polarization makes the use of the conventional spin-orbital interaction mechanism impossible for achieving band inversion. The topological gauge flux is then typically introduced with a moving background in theoretical models[19-22]. Its practical implementation is a serious challenge, though, due to inherent dynamic instabilities and noise. Here we realize the inversion of acoustic energy bands at a double Dirac cone[15,27,28] and provide an experimental demonstration of an acoustic topological insulator. By manipulating the hopping interaction of neighbouring 'atoms' in this new topological material, we successfully demonstrate the acoustic quantum spin Hall effect, characterized by robust pseudospin-dependent one-way edge sound transport. Our results are promising for the exploration of new routes for experimentally studying topological phenomena and related applications, for example, sound-noise reduction.

The topological insulator, characterized by the quantum spin Hall effect (QSHE), originates from condensed matter. A prerequisite condition for such an effect, Kramers doublet, can be fulfilled thanks to the intrinsic spin-1/2 fermionic character of electrons. Thus, the key physics behind realizing a bosonic (for example, photonic or phononic) analogue of the electronic topological insulator is to increase the degrees of freedom so as to create a double Dirac cone, where Kramers doublet exists in the form of two two-fold states (pseudospin-up and pseudospin-down). For example, a photonic topological insulator often takes advantage of two polarizations of a spin-1 photon, which are utilized to construct the required Kramers doublet, for example, degenerate TM+TE/TM−TE states[11] (TM and TE represent transverse magnetic and electric polarizations, respectively), TM/TE states[29], or left/right circularly polarized states[30]. An acoustic

---

[*] Nat. Phys., 2016, 3867

system especially for longitudinal airborne sound, however, possesses intrinsic spin-0, giving rise to no such polarization-based Kramers doublet and thus hindering acoustic QSHE. To overcome this issue, Kramers doublet in the form of degenerate artificial acoustic spin-1/2 states may be constructed, for example, using clockwise and anticlockwise circulating acoustic waves[23]. Spin – orbital interaction may then be introduced in a time-dependent fashion[24]. Most recently, ref. [15] proposed a novel concept for realizing a photonic topological insulator by utilizing two pairs of degenerate Bloch modes of TM polarization(instead of two polarizations) as a result of the zone folding mechanism in a composite lattice structure[15]. Here, we show, other than the zone folding mechanism[15], a double cone can also be accidentally formed by deliberately manipulating the filling factor of a honeycomb lattice. This straightforward yet elegant approach takes advantage of acoustic systems' large index and impedance contrast, which unfortunately is usually absent in photonic systems (see Supplementary Information). On the basis of our novel concept, we experimentally demonstrate a completely passive or static two-dimensional acoustic topological insulator, utilizing phononic 'graphene' consisting of stainless-steel rods in air.

In our phononic 'graphene', an artificial spin-1/2 is emulated through mode hybridization near the dispersion degeneracy of different energy bands. Specifically, two degenerate modes M_1 and M_2 at a given frequency can be hybridized to construct two new superimposed degenerate spin-1/2 states: spin $\pm \equiv M_1 + iM_2/M_1 - iM_2$. Since in this scenario each spin-1/2 state corresponds to two degenerate modes, the two-fold degeneracy of the Dirac cone in a spin-1/2 electronic system, where the QSHE occurs, is replaced with a four-fold degeneracy or a double Dirac cone. By engineering the nearest neighbour coupling in a phononic 'graphene' with C_{6v} symmetry, an accidental double Dirac cone can be realized[27,28] (see Supplementary Information), because the C_{6v} symmetry, according to group theory, possesses two two-dimensional irreducible representations, which can be leveraged to construct a double Dirac cone by forming a four-fold accidental degeneracy[31]. Note that similar double Dirac cones have also been reported previously in other classic wave systems, such as electromagnetic waves[15,31,32] and elastic waves[18]. Different from all of these works[27,28], we show that, in addition to the double Dirac cone, the phononic 'graphene' studied here undergoes a symmetry inversion in reciprocal space, causing energy band inversion, which ultimately leads to a distinct topological phase transition[15].

The topological phase transition can be leveraged to form topologically protected gapless edge states inside the bulk frequency bandgap at the interface between two phononic crystals with topologically dissimilar phases, which can then be exploited to construct a topologically protected waveguide as shown in Fig. 1(a). Due to topological protection, the acoustic backscattering can be largely suppressed and thus robust against various kinds of defect (including cavities, disorders and even sharp bends). To identify these topological phases, the evolution of energy band structures and their acoustic states

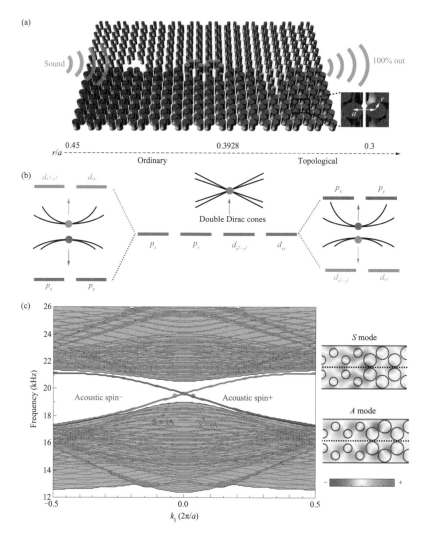

FIG. 1. Schematic of the acoustic insulator and band inversion process. a, A schematic of our proposed acoustic topological insulator constructed with two types of phononic 'graphene' with the same lattice constant a but different 'atom' (stainless-steel rod) radii r. This novel design enables one-way robust spin-dependent transportation against defects such as a cavity, a lattice disorder and a bend as indicated in the figure. The inset shows a zoom-in view of our phononic 'graphene'. b, Illustration of a band inversion process with a decreasing filling ratio. At a large filling ratio, for example, $r/a = 0.45$, the phononic crystal has two two-fold degenerate acoustic states denoted by p_x/p_y and $d_{x^2-y^2}/d_{xy}$, which are separated by a bulk bandgap; at a decreased filling ratio, for example, $r/a = 0.3928$ in our case, the bulk bandgap disappears and an accidental double Dirac cone with a four-fold degeneracy is formed; at an even smaller filling ratio, for example, $r/a = 0.3$, the bulk bandgap reappears along with two inverted two-fold degenerate acoustic states, corresponding to $d_{x^2-y^2}/d_{xy}$ and p_x/p_y. In this process, a topological phase transition occurs near the double Dirac cone. Note sketches of energy bands are shown as black curves. c, Calculated projected energy bands of a supercell consisting of a topological phononic crystal ($r = 0.3$ cm, $a = 1$ cm) stacked with an ordinary phononic crystal ($r = 0.45$ cm, $a = 1$ cm). The material parameters used for calculation are the density and longitudinal sound speed of $7,800$ kg·m^{-3} and $6,010$ m·s^{-1} for stainless steel and 1.25 kg·m^{-3} and 343 m·s^{-1}. The red and blue lines represent an acoustic spin+ and spin− edge state that is hybridized with a symmetric edge mode (S) and anti-symmetric edge mode (A) based on $S+iA$ and $S-iA$, respectively. The right panel shows a representative example of the pressure field distribution at $k_\parallel = 0.05$ for the S and A modes. The shadow regions represent the bulk energy bands.

with decreasing filling ratio are illustrated in Fig. 1(b). At a high filling ratio, for example, $r/a = 0.45$ [left panel of Fig. 1(b)], separated by a bandgap, two two-fold degeneracy, one for the lower bands p_x/p_y and the other for the upper bands $d_{x^2-y^2}/d_{xy}$, appears at the Brillouin zone centre. Spin-1/2 for the bulk states can thus be achieved through hybridizing these p/d states as $p_\pm = (p_x \pm ip_y)/\sqrt{2}$ and $d_\pm = (d_{x^2-y^2} \pm id_{xy})/\sqrt{2}$ (ref. [15]). Similar to p and d orbitals of electrons, here p_x obeys symmetry $\sigma_x/\sigma_y = -1/+1$; p_y obeys $\sigma_x/\sigma_y = +1/-1$; $d_{x^2-y^2}$ obeys $\sigma_x/\sigma_y = +1/+1$; and d_{xy} obeys $\sigma_x/\sigma_y = -1/-1$, where $\sigma_{x,y} = +1, -1$ means the even or odd symmetry along the x or y axis of the unit cell, respectively(see Supplementary Information). With a deceased filling ratio of $r/a = 0.3928$ [middle panel of Fig. 1(b)], the energy bandgap is eliminated, resulting in an accidental double Dirac cone with the desired four-fold degeneracy at the Dirac point. Further decreasing the filling ratio, for example, $r/a = 0.3$, destroys the four-fold degeneracy, and reopens the bandgap [right panel of Fig. 1(b)]. In this process, the symmetry in reciprocal space is inverted, causing the energy band inversion, which is confirmed by examining the Bloch modes before the bandgap disappears and after it reappears (see Supplementary Information). The acoustic band inversion effect here is related to the fact that the velocity of a longitudinal acoustic wave in an air background is much slower than that in stainless-steel rods. The created symmetry inversion in phononic 'graphene' further leads to a topological transition from an ordinary phononic crystal (OPC) to a topological phononic crystal (TPC) (see Supplementary Figs 1 and 2 for details). Note that the acoustic topological insulator we proposed here is based on an accidental double Dirac cone resulting from C_{6v} symmetry. In principle, other configurations, as long as they comply with the required C_{6v} symmetry, can have the same phenomenon, for example, the ring-shaped components in triangle lattices[28,32].

The projected band structure of a TPC-OPC supercell is shown in Fig. 1(c). Inside the overlapped bulk energy bandgap of the two phononic crystals, it is evident that there exists a pair of edge states(red and blue lines). These edge states are localized at the interface between the TPC and the OPC and support edge propagation of sound. Acoustic spin $\pm$ of the edge states are obtained via hybridizing a symmetric mode(S) and an anti-symmetric mode(A) shown in the right panel of Fig. 1(c) as $S+iA/S-iA$, respectively. They can also be correlated to the spins of the bulk states in the TPC and the OPC located on either side of the interface as: $S + iA/S - iA \equiv p_+ + d_+/p_- + d_-$ (see Supplementary Information). Hence, the symmetric component S can be represented as $S = (p_x + d_{x^2-y^2})/\sqrt{2}$, while the anti-symmetric component A as $A = (p_y + d_{xy})/\sqrt{2}$. The two acoustic spin edge states have the same sound velocity amplitude but in opposite directions, implying the existence of spin-orbital coupling and thus one-way spin-dependent propagation.

Generally, it is very difficult to selectively excite a particular pseudospin in experiment even with multiple excitation sources. Here, we utilize a cross-waveguide splitter[33], which allows us to study pseudospin-dependent transport with a very high fidelity even in the case of unknown pseudospin excitation state. As shown in Fig. 2(a), the splitter is divided into four sections with the TPC residing in the upper and lower sections and the OPC residing in the left and right sections. There are four input/output ports in this splitter (labelled as 1, 2, 3 and 4). For an acoustic bar state, that is, the sound transmitting straight through from port 1 to port 3 (or from port 2 to port 4), with respect to its transmission direction, the TPC is located on the left side while the OPC is located on the right side before the acoustic wave propagates across the central region of the splitter. However, such kind of structural spatial symmetry is inverted immediately after the acoustic wave propagates across the centre (OPC on the left side and TPC on the right side). Since either pseudospin state is preserved across the centre region, the edge state is not allowed to propagate in a straight through fashion, implying that the acoustic bar state in this splitter is completely prohibited for all pseudospin edge states. Similar analysis can show that for an acoustic cross state, for example, from port 1 to port 2 or to port 4, the structural spatial symmetry is always preserved and the propagation of edge states is always allowed. In addition, the acoustic cross state is spin-dependent, which is determined by the spatial symmetry. For instance, in Fig. 2(a), the acoustic cross state with input sound at port 1 or at port 3 supports only the spin+ edge state while the cross state for the other two ports supports the spin− edge state. This unique effect can be explained by checking the slope of the dispersion band, that is, the group velocity, for each individual acoustic spin edge state in Fig. 1(c). Clearly, because of their opposite slope across the whole Brillouin zone, the acoustic spin+ edge state and spin− edge state can travel only in a one-way fashion but with opposite directions, clockwise (red circular arrows) versus anticlockwise (blue circular arrows) as illustrated in Fig. 2(a). Such a transport behaviour characterizes an acoustic counterpart of the QSHE.

Figure 2(b), (c) shows the simulated pressure field distribution with inputs at port 1 and at port 2 respectively for an acoustic frequency in the bulk bandgap, which resemble clearly two cross states with almost no transmission into the through port in both cases. Such an observation is consistent with our theoretical analysis above. These configurations are then experimentally studied with the resulting transmission spectra shown in Fig. 2 (d), (e). It is found that the acoustic bar state is heavily suppressed as evidently observed from its low transmission, <-10 dB for the through port. The spin-dependent cross state for the two acoustic spins, however, remains with very high transmission for the two cross output ports. These experimental observations for the cross-waveguide splitter model confirmed clearly the unique one-way spin-dependent transportation in our acoustic topological insulator.

The demonstrated topological edge states are also symmetry protected with topological immunity against all non-spin-mixing defects. Different defects are intentionally introduced

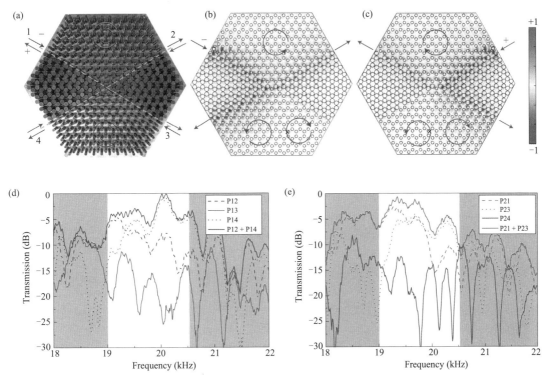

FIG. 2. Acoustic one-way spin-dependent transport. a, A photo of the cross-waveguide splitter sample used in our experiment. Topological phononic crystals reside in the upper and lower sections, while ordinary phononic crystals reside in the left and right sections. Four input/output ports labelled as 1, 2, 3 and 4 are located at the interface between these sections. In this test, only clockwise (anticlockwise) edge circulating propagation along the interfaces is allowed for acoustic spin+ (spin−) as indicated by the red (blue) circular arrow. b, c, Simulated acoustic pressure field distribution at a frequency of 19.8 kHz (within the bulk bandgap) for the acoustic cross state with an input from port 1 and port 2, respectively. d, e, Experimental transmission spectra for acoustic spin-incidence from port 1 and for acoustic spin+ incidence from port 2, respectively. Pij represents the acoustic transmission spectrum from port i to port j ($i=1,2$ and $j=1,2,3,4$) including the in and out coupling losses. The shadow regions indicate the bulk energy bands.

into our topological waveguide near the TPC/OPC interface to study the propagation robustness of our acoustic topological edge states. In comparison, a bandgap-guiding phononic crystal waveguide is also constructed with two OPCs and similar defects are introduced for a control experiment. A picture of the test sample is shown in Fig. 3(a), where the location of the topological waveguide and the phononic crystal waveguide can be seen. Three different defects, including a cavity, a lattice disorder, and a bend, are implemented in the two waveguides. Note that all these three types of defect are none-spin-mixing defects by their nature, meaning that no effective 'magnetic' impurities are presented intentionally to break the topological characteristics[15,16,18,30,34]. Simulation results show that with a spin-acoustic wave incidence from the left the edge state in the topological waveguide can detour around all these defects and maintain a very high transmission [three upper panels of Fig. 3(b)]. In contrast, the results for the bandgap-guiding phononic crystal waveguide are drastically different. The cavity in this case causes acoustic resonances, while the disorder and bend severely inhibit the acoustic forward propagation, leading to a decreased transmission or even a total reflection [three lower panels of Fig. 3(b)]. The experimentally measured transmission spectra for these two waveguides are shown in Fig. 3(d). The transmission through the topological waveguide is nearly unaffected by any of these defects especially for the spectral region near the centre of the bulk bandgap. However, their presence causes significant disturbance on the sound transportation in the bandgap-guiding phononic crystal waveguide, which could include largely reduced transmission especially on the high-frequency side of the bandgap and multiple resonance dips corresponding to the cavity resonances. Hence, the topological waveguide clearly distinguishes itself from the bandgap-guiding phononic crystal waveguide with the unparalleled advantage of robust sound transportation.

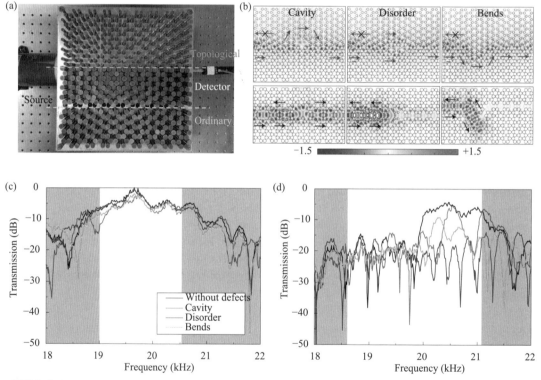

FIG.3. Robust one-way sound transport. (a), A photo of our experimental set-up. The green and yellow dashed lines indicate the location of an acoustic topological waveguide and an ordinary bandgap-guiding phononic crystal waveguide, respectively. (b), Simulated acoustic pressure field distribution for acoustic spin-incident at 20.1 kHz (within the bulk bandgap) in three different configurations, corresponding to three different defects, that is, a cavity, a lattice disorder and a bend. Blue, red and black arrows denote the propagation direction of acoustic spin−, spin+ and ordinary sound, respectively. (c),(d), Experimental transmission spectra for the topological waveguide and ordinary bandgap-guiding phononic crystal waveguide, respectively. The black curve corresponds to the case without any defects, while the red, blue and green curves indicate the transmission spectrum for the case with a cavity, a lattice disorder and a bend, respectively. The shadow regions correspond to the bulk energy bands.

In our acoustic system, the topological property of the TPC is characterized by a non-zero spin Chern number of bulk bands (see Supplementary Information), implying that the edge states are preserved under continuous deformations. Note that a tiny gap exists between two dispersions of the edge states at the $k_{\parallel} = 0$ point (see Supplementary Information), which is originated from an imperfect cladding layer rather than the TPC itself. For instance, in our case we use the topologically trivial phononic crystal as a cladding layer but its eigen states are slightly different from those of the TPC, which gives rise to a theoretically calculated absolute bandgap of 0.095 kHz or a relative bandgap less than 0.5%. Such a bandgap can hardly be probed in our swept frequency tests [Figs 2(d), (e) and 3(c),(d)] even with a frequency increment of 0.01 kHz because of the usual finite size issue for fabricated samples and the propagation loss for airborne sound. If desired, this relative bandgap can be further reduced to less than 0.02% in theory by adjusting the boundary condition[35], that is, changing the filling factor of a row of cylinders to be $r/a =$ 0.1 (see Supplementary Information). In principle, this tiny bandgap can even be closed with a deliberately designed cladding layer, if it can support the same eigen states as those of the TPC. In reality, any fabrication imperfection could lead to a tiny gap. In this regard, the backscattering is not completely forbidden in experiment. However, the real issue is whether the experimental measurement can detect such a tiny bandgap. What we have found in our work is that the transmission through our topological waveguide is nearly unaffected [Figs 2(d),(e) and 3(c),(d)], suggesting that the bandgap is hardly noticeable and the backscattering is largely suppressed.

The acoustic topological states obtained here pave a new way for studying topological properties in classic wave systems without any noises resulting from dynamical modulation. This unique topological material can in principle be leveraged to explore quantum behaviour of classical waves at a high fidelity. In addition, our design principle and measurement technique demonstrated here can be easily extended to cover a large acoustic spectrum from audible sound to ultrasonic frequencies, and they can even be extended to other classic waves such as Lamb waves. Moreover, our demonstrated topological phenomenon of sound may generate considerable impact in applications in the foreseeable future. For instance, when combined with a low propagation loss substrate/background environment, the backscattering-immune sound transportation may give rise to an ultrahigh-Q acoustic resonance, which, to date, remains a challenge in the acoustic industry.

References and Notes

[1] Klitzing, K. V., Dorda, G. & Pepper, M. New method for high-accuracy determination of the fine-structure constant based on quantized Hall resistance. *Phys. Rev. Lett.* **45**, 494–497 (1980).

[2] Thouless, D. J., Kohmoto, M., Nightingale, M. P. & den Nijs, M. Quantized Hall conductance in a two-dimensional periodic potential. *Phys. Rev. Lett.* **49**, 405–408 (1982).

[3] Hasan, M. Z. & Kane, C. L. Colloquium: topological insulators. *Rev. Mod. Phys.* **82**, 3045–3067

(2010).

[4] Qi, X.-L. & Zhang, S.-C. Topological insulators and superconductors. *Rev. Mod. Phys.* **83**, 1057–1110 (2011).

[5] Haldane, F. D. M. & Raghu, S. Possible realization of directional optical waveguides in photonic crystals with broken time-reversal symmetry. *Phys. Rev. Lett.* **100**, 013904 (2008).

[6] Wang, Z., Chong, Y., Joannopoulos, J. D. & Soljacic, M. Observation of unidirectional backscattering-immune topological electromagnetic states. *Nature* **461**, 772–775 (2009).

[7] Lu, L., Joannopoulos, J. D. & Soljacic, M. Topological photonics. *Nat. Photon.* **8**, 821–829 (2014).

[8] Hafezi, M., Demler, E. A., Lukin, M. D. & Taylor, J. M. Robust optical delay lines with topological protection. *Nat. Phys.* **7**, 907–912 (2011).

[9] Hafezi, M., Mittal, S., Fan, J., Migdall, A. & Taylor, J. M. Imaging topological edge states in silicon photonics. *Nat. Photon.* **7**, 1001–1005 (2013).

[10] Rechtsman, M. C. et al. Photonic floquet topological insulators. *Nature* **496**, 196–200 (2013).

[11] Khanikaev, A. B. et al. Photonic topological insulators. *Nat. Mater.* **12**, 233–239 (2013).

[12] Chen, W.-J. et al. Experimental realization of photonic topological insulator in a uniaxial metacrystal waveguide. *Nat. Commun.* **5**, 5782 (2014).

[13] Lu, L. et al. Experimental observation of Weyl points. *Science* **349**, 622–624 (2015).

[14] Bliokh, K. Y., Smirnova, D. & Nori, F. Quantum spin Hall effect of light. *Science* **348**, 1448–1451 (2015).

[15] Wu, L.-H. & Hu, X. Scheme for achieving a topological photonic crystal by using dielectric material. *Phys. Rev. Lett.* **114**, 223901 (2015).

[16] Süsstrunk, R. & Huber, S. D. Observation of phononic helical edge states in a mechanical topological insulator. *Science* **349**, 47–50 (2015).

[17] Wang, P., Lu, L. & Bertoldi, K. Topological phononic crystals with one-way elastic edge waves. *Phys. Rev. Lett.* **115**, 104302 (2015).

[18] Mousavi, S. H., Khanikaev, A. B. & Wang, Z. Topologically protected elastic waves in phononic metamaterials. *Nat. Commun.* **6**, 8682 (2015).

[19] Yang, Z. et al. Topological acoustics. *Phys. Rev. Lett.* **114**, 114301 (2015).

[20] Ni, X. et al. Topologically protected one-way edge mode in networks of acoustic resonators with circulating air flow. *New J. Phys.* **17**, 053016 (2015).

[21] Khanikaev, A. B., Fleury, R., Mousavi, S. H. & Alu, A. Topologically robust sound propagation in an angular-momentum-biased graphene-like resonator lattice. *Nat. Commun.* **6**, 8260 (2015).

[22] Fleury, R., Sounas, D. L., Sieck, C. F., Haberman, M. R. & Alù, A. Sound isolation and giant linear nonreciprocity in a compact acoustic circulator. *Science* **343**, 516–519 (2014).

[23] Zhu, X.-F. et al. Topologically protected acoustic helical edge states and interface states in strongly coupled metamaterial ring lattices. Preprint at http://arXiv.org/abs/1508.06243 (2015).

[24] Fleury, R., Khanikaev, A. & Alu, A. Floquet topological insulators for sound. *Nat. Commun.* **7**, 11744 (2016).

[25] Xiao, M. et al. Geometric phase and band inversion in periodic acoustic systems. *Nat. Phys.* **11**, 240–244 (2015).

[26] Xiao, M., Chen, W.-J., He, W.-Y. & Chan, C. T. Synthetic gauge flux and Weyl points in acoustic systems. *Nat. Phys.* **11**, 920–924 (2015).

[27] Chen, Z.-G. et al. Accidental degeneracy of double Dirac cones in a phononic crystal. *Sci. Rep.* **4**, 4613

[28] Li, Y., Wu, Y. & Mei, J. Double Dirac cones in phononic crystals. *Appl. Phys. Lett.* **105**, 014107 (2014).

[29] Ma, T., Khanikaev, A. B., Mousavi, S. H. & Shvets, G. Guiding electromagnetic waves around sharp corners: topologically protected photonic transport in metawaveguides. *Phys. Rev. Lett.* **114**, 127401 (2015).

[30] He, C. *et al.* Photonic topological insulator with broken time-reversal symmetry. *Proc. Natl Acad. Sci. USA* **113**, 4924–4928 (2016).

[31] Sakoda, K. Double Dirac cones in triangular-lattice metamaterials. *Opt. Express* **20**, 9925–9939 (2012).

[32] Li, Y. & Mei, J. Double Dirac cones in two-dimensional dielectric photonic crystals. *Opt. Express* **23**, 12089–12099 (2015).

[33] He, C. *et al.* Tunable one-way cross-waveguide splitter based on gyromagnetic photonic crystal. *Appl. Phys. Lett.* **96**, 111111 (2010).

[34] Fu, L. Topological crystalline insulators. *Phys. Rev. Lett.* **106**, 106802 (2011).

[35] He, C., Lu, M.-H., Wan, W.-W., Li, X.-F. & Chen, Y.-F. Influence of boundary conditions on the one-way edge modes in two-dimensional magneto-optical photonic crystals. *Solid State Commun.* **150**, 1976–1979 (2010).

[36] The work was jointly supported by the National Basic Research Program of China (Grant No. 2012CB921503, 2013CB632904 and 2013CB632702) and the National Nature Science Foundation of China (Grant No. 11134006, No. 11474158, and No. 11404164). M.-H.L. also acknowledges the support of the Natural Science Foundation of Jiangsu Province (BK20140019) and the support from the Academic Program Development of Jiangsu Higher Education (PAPD).

[37] C.H., M.-H.L. and X.-P.L. conceived the idea. C.H. performed the numerical simulation and fabricated the samples. C. H., X. N. and H. G. carried out the experimental measurements. All the authors contributed to discussion of the results and manuscript preparation. M.-H.L., X.-P.L. and Y.-F.C. supervised all aspects of this work and managed this project.

[38] Supplementary information is available in the online version of the paper.

[39] Methods: Methods, including statements of data availability and any associated accession codes and references, are available in the online version of this paper. Experiments. The phononic 'graphene' consists of commercial 304 stainless-steel rods with radii $r=0.3$ cm and $r=0.45$ cm, arranged in air in graphene-like lattices. The radii tolerance of these steel rods is ± 0.01 cm. The height of our phononic 'graphene' is chosen to be 20 cm, roughly 10 times larger than the acoustic wavelength in air at 20 kHz to make sure that two-dimensional approximation is applicable. Experiments are conducted with a large-area acoustic transducer. The transducer excites a roughly planar acoustic field, which then will be scattered at the input facet into the appropriate spin edge state according to the symmetry of the interface. A B & K – 4939 – 2670 microphone acts as a detector, which is placed 2 cm away from the output port with its response acquired and analysed in B & K – 3560 – C. Acoustic input frequencies are swept from 18 kHz to 22 kHz with an increment of 0.01 kHz. The experimentally measured transmission spectra plotted in Figs 2d, e and 3c, d are normalized to the acoustic wave transmission through the same distance in air. In theory, the transmission should be 100% (0 dB) for the topological edge states with frequencies from 19.00 kHz to 20.53 kHz. The deviation recorded in experiments (between 0 dB and -5 dB) is primarily due to the inefficient coupling into and out of the

topological waveguides.

Simulations. Numerical investigations to calculate Figs 1c, 2b, c and 3b in this letter are conducted using the acoustic mode of commercial FEM software (COMSOL MULTIPHYSICS).

Data availability. The data that support the plots within this paper and other findings of this study are available from the corresponding author on request.

A 14×14 μm² Footprint Polarization-encoded Quantum Controlled-NOT Gate Based on Hybrid Waveguide[*]

S.M. Wang[1,2], Q.Q. Cheng[1,2], Y.X. Gong[3], P. Xu[1,2], C. Sun[4], L.Li[1,2], T.Li[1,2] & S.N. Zhu[1,2]

[1] *National Laboratory of Solid State Microstructures, School of Physics, College of Engineering and Applied Sciences, Nanjing University, Nanjing 210093, China*

[2] *Collaborative Innovation Center of Advanced Microstructures, Nanjing 210093, China*

[3] *Department of Physics, Southeast University, Nanjing 211189, China*

[4] *Department of Mechanical Engineering, Northwestern University, Evanston, Illinois 60208-3111, USA*

Photonic quantum information processing system has been widely used in communication, metrology and lithography. The recent emphasis on the miniaturized photonic platform is thus motivated by the urgent need for realizing large-scale information processing and computing. Although the integrated quantum logic gates and quantum algorithms based on path encoding have been successfully demonstrated, the technology for handling another commonly used polarization-encoded qubits has yet to be fully developed. Here, we show the implementation of a polarization-dependent beam-splitter in the hybrid waveguide system. With precisely design, the polarization-encoded controlled-NOT gate can be implemented using only single such polarization-dependent beam-splitter with the significant size reduction of the overall device footprint to 14×14 μm². The experimental demonstration of the highly integrated controlled-NOT gate sets the stage to develop large-scale quantum information processing system. Our hybrid design also establishes the new capabilities in controlling the polarization modes in integrated photonic circuits.

The quantum controlled-NOT (CNOT) gate is one of the fundamental building block for quantum information system that flips the target qubit state conditional on the control qubit being in the state $|1\rangle$. It enables the construction of any quantum computing circuits when combined with single-qubit gates[1]. A widely used linear optical CNOT gate[2,3] manipulating path qubits by path interference with multiple beam-splitters has been first realized by using free-space optical components[4] and subsequently using integrated waveguides[5]. By employing the polarization-encoded scheme, the gate can be significantly simplified by requiring only three polarization-dependent beam-splitters(PDBSs) and thus, results in further size reduction and improved stability by eliminating the phase-sensitive interference[6-8]. Recently, such simplified CNOT gate has been realized using the femtosecond-laser-written directional couplers with the precise control on the splitting ratio

---

[*] Nat.Commun.,2016,7:11490

for the orthogonally polarized modes. The use of directional couplers requires quite long interaction length and thus, the overall dimension of the gate remains in a millimetre scale in order to gain independent control of the orthogonally polarized modes[9]. Hence, a more compact CNOT gate is yet to be explored for addressing the urgent needs in developing large-scale quantum information processing system.

In the following, we report the implementation of a hybrid PDBS by strategically combining the dielectric and plasmonic waveguide that each is dedicated to handle the TE (transverse electric) and TM (transverse magnetic) polarized mode all within the same component. With precisely design of the output slits, the polarization-encoded CNOT gate can be implemented using only single such PDBS with the significant size reduction of the overall device footprint to 14×14 μm^2. The gate demonstrates the good quantum CNOT functionality with a high fidelity.

Results

Classical characterization of the HW-based PDBS. The polarization-encoded CNOT gate[6–8] is schematically shown in Fig. 1a. The core part of the gate is a PDBS ($PDBS_0$) that allows 100% transmission of the TE polarized light and $\xi/3$ ($2\xi/3$) transmission (reflection) of the TM polarized light, where ξ is the total coefficient of the system. Auxiliary PDBSs (PDBSa), with transmittances (T) obeying $T_{TE}/T_{TM} = 1 : 2$, are employed to balance the contributions from the two polarizations. We have employed the design of a hybrid waveguide (HW), namely dielectric-loaded surface plasmon polariton (SPP) waveguide, which supports both TM (SPP) and TE (photon) modes[10]. The SPPs have been proven to be a valid carrier of quantum information[11,12]. Recently, experiments have further verified the bosonic nature of SPPs via an on-chip non-classical interference[13–17].

Our HW-based PDBS is presented in Fig. 1(b). The HW system is comprised of a 300-nm silver film covered with a 250-nm silica film, which is chosen to ensure that only the lowest TM mode (SPP) and the lowest TE mode are supported in the system (see Supplementary Fig. 1 and Supplementary Note 1 for details). The input couplers, gratings milled on the silver film by focused ion beam etching (Helios nanolab 600i, FEI), are utilized to couple the incident light into the waveguide modes. The p-polarized light with the electric field perpendicular to the gratings is coupled to the SPPs supported TM mode in the HW system, whereas the s-polarized light with the electric field along the gratings is coupled to the TE mode. The excited SPPs, with excellent directionality and beam quality[18], propagate along the metal surface. A 45° rotated grating milled on the silver surface 4.5 μm away from the input coupler works as an SPP beam-splitter whose transmittance and reflectance can be conveniently tailored by controlling its period, width and depth. For the convenience of extraction and measurement of the quantum information carried by the polarization-encoded waveguide modes, output couplers, five slits, located

4.5 μm away from the SPP beam-splitter are used to convert SPPs back into p-polarized light on the back side of the device. A scanning electron microscopy image of the sample configured on the surface of silver film is presented in Fig. 1(c). The output light is collected by a leaky mode microscopy system[10,18], in which an oil-immersed objective lens is used. The optical images of the back side of the device, with a p-polarized 785 nm laser illuminating each of the two input couplers, are shown in Fig. 1(d), (e), respectively. After carefully tuning parameters of the SPP beam-splitter, the PDBS can be fabricated with the measured transmittance/reffectance (T/R) ratios being 1 : 1.9 and 1 : 2.15, respectively, for different input gratings, commensurate with the ideal ratio 1 : 2 required (see Supplementary Fig. 2 and Supplementary Note 2 for details). The deviation between these T/R ratios may be due to the fabrication imperfections during focused ion beam etching.

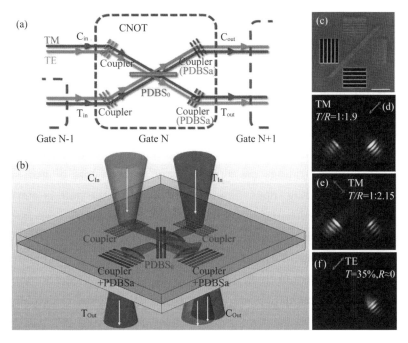

FIG.1. Classical characterization of the HW-based CNOT gate. (a) A schematic of the simplified CNOT gate composed of three PDBSs as proposed in refs [7] and [8]. The photons enter the gate from the input couplers C_{in} (control) and T_{in} (target) and get out from the output couplers C_{out} and T_{out}. (b) The sketch of the CNOT gate realized by the HW-based PDBS. (c) The scanning electron microscopy image of the PDBS on a metal film. The input coupler consists of two kinds of gratings: the working gratings having five grooves with depth $d=30$ nm, length $l=5$ μm, width $w=200$ nm and period $P=540$ nm, and the reflecting gratings with different specifics, $w=100$ nm and $P=270$ nm for high conversion efficiency. This approach aims to obtaining the high conversion efficiency from p-polarized light to SPP, in a direction-dependent way. The SPP reflecting grating comprises of three grooves with $d=50$ nm, $w=100$ nm and $P=380$ nm. The slits of the output coupler have $l=5$ μm, $w=600$ nm and $P=830$ nm. The scale bar denotes 4 μm. (d,e) The CCD images of the output setup with p-polarized light input from the left coupler (d) p-Polarized light input from the right coupler (e) and s-polarized light input from the left coupler (f) where the blue arrows mark the positions of input lights and their polarizations.

On the other hand, the HW system also supports the lowest TE mode that can be excited by s-polarized light by sharing the same input couplers for SPPs. Owing to the obvious difference in wavelengths and field distributions between the TE mode and SPP, the SPP beam-splitter has little influence on the TE mode. This leads to a nearly unity transmittance of TE mode at the SPP beam-splitter, which satisfies the perfect transmission required in the CNOT gate. In this way, one can simultaneously manipulate the two polarized waveguide modes using the HW system. Moreover, the five slits of the output couplers not only can convert the TE mode back to s-polarized light for collection but also can be used to accurately tune the intensity of TE mode to $1/3\xi$ by precisely adjusting their parameters for balancing the contributions of the TE mode and SPP in the gate. Therefore, the two auxiliary PDBSs for TE mode attenuation used in the previously scheme can be removed and the architecture of the polarization-encoded CNOT gate based on HW system is further simplified to a single PDBS. As a result, in this work, the footprint of the polarization-encoded CNOT gate is reduced to only 14×14 μm^2, which is two orders smaller than the implementations in dielectric waveguide system[5,9] and is much promising for future quantum photonic integration. The optical image of the back side of the device with s-polarized light illuminating on one input coupler is shown in Fig. 1 (f). The transmittances are $37\%\xi$ and $35\%\xi$ for input from the right and left gratings, respectively. Nearly zero reflection of the TE mode can be observed in both cases. As the results for the two input gratings are similar, here without loss of generality we only present the case of left grating input. The 45° alignment of the gate is employed for the convenience of separately collecting the photons at the two outputs.

The quantum characterization of the HW-based CNOT gate. The experimental setup is sketched in Fig. 2. A pulsed 785 nm femtosecond laser from a Ti: Sapphire oscillator is frequency-doubled using a 2-mm-thick β-barium borate (BBO) crystal to obtain 392.5 nm pulses. These pulses then pump a 2-mm-thick type-II BBO crystal for generating collinear photon pairs with orthogonal polarizations. The quality of the photon-pair source can be characterized by using the two-photon Hong-Ou-Mandel (HOM) interference[19]. The coincidence of the HOM interference of the photon-pair source is shown in Fig. 3(a). The HOM dip with high visibility $V = (C_{max} - C_{min})/C_{max} = 95.2 \pm 0.7\%$, manifests its good quality for further quantum interference utilization[13-17]. We then use this photon-pair source to characterize the PDBS we fabricated in the HW system. The orthogonally polarized photons, separated by a polarizing beam splitter, are collected with single-mode fibres and injected into the HW-based PDBS to excite SPPs with their polarizations both rotated to the p-polarization using the polarization controller 1. The excited SPPs interfere at the PDBS, with the best interference effect obtained by carefully controlling the temporal overlap in the delay line. The output SPPs are then coupled to the propagating wave by the output couplers and subsequently collected using an oil-immersed objective lens. The HOM interference of the SPPs on this HW-based PDBS is depicted in Fig. 3(b).

The visibility of 72.4±3.1% has been observed, which is close to the theoretical value of 80% for an ideal 1∶2 beam-splitter. The result also proves the bosonic nature of SPPs and demonstrates the good quality of SPPs as a quantum information carrier[13—17].

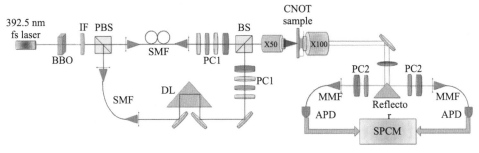

FIG.2. The experimental setup. The photon pairs at wavelength $\lambda = 785$ nm are generated via type-II spontaneous parametric down conversion (SPDC) in a 2-mm BBO crystal, pumped by 392.5 nm pulses frequency-doubled by a 785-nm femtosecond laser. A 10-nm interference filter (IF) with the central wavelength of 785 nm is used. Orthogonally polarized photon pairs are separated by a polarizing beam-splitter (PBS) and coupled to single mode fibres (SMF). A delay line (DL) is inserted to control the temporal superposition of the photons. The polarization controller (PC1, quarter wave plate (QWP)+half wave plate (HWP)+QWP+ Glan prism) is used to control the polarization of photons output from the fibres. A beam-splitter (BS) is employed for alignment of the two polarized beams. An input objective (×50, NA=0.4) and an oil objective (×100, NA=1.32) are used to excite the waveguide modes in the HW system and to collect the light from the sample, respectively. A triangular reflector is used to split the photons from the two outputs. After selected by polarization controller (PC2, HWP+Glan prism), the output photons are transmitted to the silicon avalanche photodiodes (APD) through multimode fibres (MMF) and coincidence measurements are made at the single photon counting modules (SPCM).

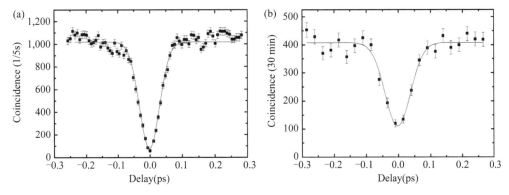

FIG.3. The HOM interference results. (a) The HOM interference pattern of the photon pairs of the source. (b) The HOM interference pattern of the SPP at the HW-based PDBS. Black dots are data and the red lines correspond to fitting curves. Error bars are drawn to represent one standard deviation from the Poisson distribution.

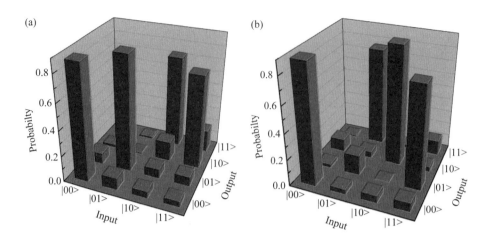

FIG. 4. Characterization of the integrated gate. Measured operation probabilities of the CNOT gate on the computational **ZZ** (a) and the **XX** (b) basis, respectively.

We then carry out the measurement of the operation of the CNOT gate. We first consider the **ZZ** basis defined as $|0_{zz}\rangle_c \equiv |s\rangle_c, |1_{zz}\rangle_c \equiv |p\rangle_c$ for the control qubit, and $|0_{zz}\rangle_t \equiv |D\rangle_t = (|s\rangle_t + |p\rangle_t)/\sqrt{2}, |1_{zz}\rangle_t \equiv |A\rangle_t = (|s\rangle_t - |p\rangle_t)/\sqrt{2}$ for the target qubit. Note that we define the control and target qubits in different basis as the gate constructed in this work is, at its core, a controlled-Z gate for s-and p-polarization, which requires two Hadamard operations on the input and output of the target qubit to become to a CNOT gate. Here, these Hadamard operations are achieved by a basis redefinition of the target qubit. Under post-selection, that is, a successful twofold coincidence measurement on the two output ports, the CNOT gate succeeds with a probability of 1/9 (refs [2],[3]). To characterize the operation of this gate, we measured the output of the gate for each of the four possible input states: $|00_{zz}\rangle_{ct}, |01_{zz}\rangle_{ct}, |10_{zz}\rangle_{ct}$ and $|11_{zz}\rangle_{ct}$ prepared by the polarization controller 1. The measured results for input-output operation probabilities, normalized by the sum of all coincidence counts obtained for each of the respective input states, is presented in Fig. 4(a). The four correct output states corresponding to four input states are evident with high probabilities. Here, the probabilities of 88.3±8.3% and 88.7±8.4% for the input states $|00_{zz}\rangle_{ct}$ and $|01_{zz}\rangle_{ct}$, respectively, are higher than those of 72.1±7.5% and 71.9±7.3% for $|10_{zz}\rangle_{ct}$ and $|11_{zz}\rangle_{ct}$, respectively. The reason lies in that the non-classical interference relying on the overlap on the PDBS is required for the two latter input states, whereas it is not required for the two former input states. Some other non-zero probabilities can be attributable to the crosstalk between two polarizations from the scattering of the TM mode (SPP) into the TE mode in addition to the inaccuracy from the transmittance and reflectance of the PDBS. The average operation fidelity of the gate can then be obtained as $F_{zz} = 80.3 \pm 7.9\%$. These results show that this HW-based gate can well present the quantum CNOT function.

We then consider the complementary diagonal **XX** basis given by $|0_{xx}\rangle_c \equiv |D\rangle_c = (|s\rangle_c + |p\rangle_c)/\sqrt{2}$, $|1_{xx}\rangle_c \equiv |A\rangle_c = (|s\rangle_c - |p\rangle_c)/\sqrt{2}$ for the control qubit, and $|0_{xx}\rangle_t \equiv |s\rangle_t$, $|1_{xx}\rangle_t \equiv |p\rangle_t$ for the target qubit. As shown in Fig. 4(b), the operation of the gate presents the correct output states $|00_{xx}\rangle_{ct}$, $|11_{xx}\rangle_{ct}$, $|10_{xx}\rangle_{ct}$, $|01_{xx}\rangle_{ct}$ corresponding to the input states $|00_{xx}\rangle_{ct}$, $|01_{xx}\rangle_{ct}$, $|10_{xx}\rangle_{ct}$, $|11_{xx}\rangle_{ct}$, respectively, with high probabilities given by 89.9±9.1%, 90.6±9.2%, 75.7±6.1%, 77.6±6.5%, respectively. The average operation fidelity can be computed as $F_{xx} = 83.5 \pm 7.7\%$. Then, the upper and lower bounds for the quantum process fidelity F_{process} of the gate can be obtained via $F_{xx} + F_{zz} - 1 \leqslant F_{\text{process}} \leqslant \min[F_{xx}, F_{zz}]$[7]. Thus, the process fidelity of the HW-based quantum CNOT gate is $63.8 \pm 7.8\% \leqslant F_{\text{process}} \leqslant 80.3 \pm 7.9\%$.

The CNOT gate can be used as an entangling gate, which produces a two-qubit entangled output state from a separable input state. The lower bound of the process fidelity also defines a lower bound of the entanglement capability of the gate, as the fidelity of entanglement generation is at least equal to the process fidelity[7]. In terms of the concurrence C that the gate can generate from the product state inputs, the minimal entanglement capability is given by $C \geqslant 2F_{\text{process}} - 1$ (ref. [7]). As our experimental results show that the minimal process fidelity of the gate is 0.638, the lower bound of the entanglement capability can be $C \geqslant 0.276$ consequently. We also produced a post-selected polarization entanglement from a separable input state (see Supplementary Note 3 and Supplementary Fig. 3 for details). The high visibility also presents the good entangling function of the gate.

Discussion

We finally discuss the integratabiltiy of this HW-based quantum logic gate. It seems that integratability of our gate suffers from the low photon counting rates due to high losses of the plasmonic component. The high loss also reduces the signal/noise of the quantum measurement results, which will affect the performance of the HW-based quantum logic gate. However, in the matter of fact, the loss is mainly brought from the conversion processes between the spatial light and waveguide modes (SPP and TE mode) at the input and output couplers of the PDBS. Such coupling loss could be reduced by improving the measurement setup and coupling approach after this proof-of-principle work, such as using tapered nano-fibre coupling[20] or single-mode fibre coupling[21]. Furthermore, in the future integrated quantum processor containing on-chip sources and detectors, such conversion processing is not necessary and the losses of conversion processes can thus be further reduced. At this time, the attenuation function of the output couplers for TE mode can easily be fulfilled by some alternative small structures, such as gratings milled on the loaded dielectric layer. In addition, considering the wavelength of 785 nm in our work, the theoretically predicted propagating length in dielectric-loaded SPP waveguide is more than 100 μm. It is sufficient for accommodating multiple logical elements in creating sophisticated quantum information processing system. For the telecom wavelength(around 1,550 nm), the loss can be further reduced, leading to the millimetre propagation length and a larger scale chip[22]. Therefore, the signal/noise ratio of the quantum measurement results will also increase. Besides, the phase-damping decoherence resulting from the index difference of the SPP and TE mode may also affect the integratability. As we discuss in the Supplementary Note 4 and Supplementary Fig. 4, the coherence length can be further increased using a narrow-band photon-pair source, which meets the scalability requirement of the quantum chip. Consequently, we believe that, with the advanced nano-fabrication techniques[23], the functional elements based on HW system can be conveniently integrated onto a chip to realize diverse logical functions or algorithms, all done within an ultra-compact footprint.

References and Notes

[1] Nielsen, M. A. & Chuang, I. L. *Quantum Computation and Quantum Information*(Cambridge Univ., 2000).

[2] Ralph, T. C., Langford, N. K., Bell, T. B. & White, A. G. Linear optical controlled-NOT gate in the coincidence basis. *Phys. Rev.* A **65**, 062324 (2002).

[3] Hofmann, H. F. & Takeuchi, S. Quantum phase gate for photonic qubits using only beam splitters and postselection. *Phys. Rev.* A **66**, 024308 (2002).

[4] O'Brien, J. L. *et al.* Demonstration of an all-optical quantum controlled-NOT gate. *Nature* **426**, 264 – 267 (2003).

[5] Politi, A., Cryan, M. J., Rarity, J. G., Yu, S. & O'Brien, J. L. Silica-on-silicon waveguide quantum circuits. *Science* **320**, 646 – 649 (2008).

[6] Langford, N. K. *et al.* Demonstration of a simple entangling optical gate and its use in bell-state analysis. *Phys. Rev. Lett.* **95**, 210504 (2005).

[7] Kiesel, N., Schmid, C., Weber, U., Ursin, R. & Weinfurter, H. Linear optics controlled-phase gate made simple. *Phys. Rev. Lett.* **95**, 210505 (2005).

[8] Okamoto, R. *et al.* Demonstration of an optical quantum controlled-NOT gate without path interference. *Phys. Rev. Lett.* **95**, 210506 (2005).

[9] Crespi, A. *et al.* Integrated photonics quantum gates for polarization qubits. Nat. *Commun* **2**, 566 (2011).

[10] Reinhardt, C. *et al.* Mode-selective excitation of laser-written dielectric-loaded surface plasmon polariton waveguides. *J. Opt. Soc. Am.* B **26**, B55 – B60 (2009).

[11] Altewischer, E., van Exter, M. P. & Woerdman, J. P. Plasmon-assisted transmission of entangled photons. *Nature* **418**, 304 – 306 (2002).

[12] Fasel, S. *et al.* Energy-time entanglement preservation in plasmon-assisted light transmission. *Phys. Rev. Lett.* **94**, 110501 (2005).

[13] Heeres, R. W., Kouwenhoven, L. P. & Zwiller, V. Quantum interference in plasmonic circuits. *Nat. Nanotech* **8**, 719 – 722 (2013).

[14] Fakonas, J. S., Lee, H., Kelaita, Y. A. & Atwater, H. A. Two-plasmon quantum interference. *Nat. Photon* **8**, 317 – 320 (2014).

[15] Di Martino, G. *et al.* Observation of quantum interference in the plasmonic Hong-Ou-Mandel effect. *Phys. Rev. Appl.* **1**, 034004 (2014).

[16] Cai, Y. J. *et al.* High-visibility on-chip quantum interference of single surface plasmons. *Phys. Rev. Appl.* **2**, 014004 (2014).

[17] Fujii, G., Fukuda, D. & Inou, S. Direct observation of bosonic quantum interference of surface plasmon polaritons using photon-number-resolving detectors. *Phys. Rev. B* **90**, 085430 (2014).

[18] Li, L. *et al.* Plasmonic Airy beam generated by in-plane diffraction. *Phys. Rev. Lett.* **107**, 126804 (2011).

[19] Hong, C. K., Ou, Z. Y. & Mandel, L. Measurement of subpicosecond time intervals between two photons by interference. *Phys. Rev. Lett.* **59**, 2044–2046 (1987).

[20] Dong, C. *et al.* In-line high efficient fiber polarizer based on surface plasmon. *Appl. Phys. Lett.* **100**, 041104 (2012).

[21] Roelkens, G. *et al.* High efficiency diffractive grating couplers for interfacing a single mode optical fiber with a nanophotonic silicon-on-insulator waveguide circuit. *Appl. Phys. Lett.* **92**, 131101 (2008).

[22] Maier, S. *Plasmonics Fundamentals and Applications* (Springer, 2007).

[23] Fu, Y. *et al.* All-optical logic gates based on nanoscale plasmonic slot waveguides. *Nano Lett.* **12**, 5784–5790 (2012).

[24] This work was supported by the National Key Projects of Basic Researches in China (No. 2012CB921501, No. 2012CB921802 and No. 2012CB933501), the National Natural Science Foundation of China (Nos. 11322439, 91321312, 11321063, 11422438 and 11474050). T. L. acknowledges the support of the Dengfeng Project B of Nanjing University and the PAPD project of Jiangsu Higher Education Institutions, C. S. acknowledges the support of State Administration of Foreign Experts Affairs "High-end foreign experts project" (No. GDW20153200149). We acknowledge Professor Xiaosong Ma and Professor Edna Cheung for their helpful discussion.

[25] S. M. W., Q. Q. C., Y. X. G., these authors contributed equally to this work. S. M. W. and Y. X. G. conceived the idea and designed the experiments. Q. Q. C. and L. L. carried out the device fabrication. S. M. W. and Q. Q. C. did characterizations. S. M. W., Y. X. G., T. L. and C. S. analysed the results. S. M. W., T. L. and S. N. Z. supervised the work. All authors discussed the results and commented on the manuscript.

[26] Supplementary Information accompanies this paper at http://www.nature.com/naturecommunications

[27] Methods: Polarization-encoded CNOT gate. The CNOT gate based on polarization-encoding is schematically shown in Fig. 1(a). The core part of the gate is a PDBS denoted as $PDBS_0$ that perfectly transmits the TE polarized light and allows for 1/3 (2/3) transmission (reflection) of the TM polarized light. In practice, when the PDBS is leaky, the CNOT gate still works if the transmittance (T) and reflectance (R) for the TE and TM polarized lights satisfy $T_{TE} = \xi$, $R_{TE} = 0$ and $T_{TM} = \xi/3$, $R_{TM} = 2\xi/3$, respectively. ξ is related to the total loss including the conversion and propagation losses and so on. Auxiliary PDBSs (PDBSa), with $T_{TE}/T_{TM} = 1:2$, are employed to balance the contributions from the two polarizations. For the input and the extraction of on-chip signals, the construction of a single CNOT gate calls for the input and output couplers (Fig. 1(a)) as used in this work. Here, ξ was estimated to be ~1% from the experimental data in Fig. 1(d)-(f).

Experimental setup. The experimental setup can be divided into three parts. The first part is the source: A 2.5 W pulsed 785 nm femtosecond laser from a Ti: Sapphire oscillator (Tsunami, Spectra-Physics Lasers) is frequency-doubled using a 2-mm-thick BBO crystal to obtain 392.5 nm pulses. The photon pairs with wavelength of 785 nm are then generated via spontaneous parametric down conversion in a 2-mm type-II BBO crystal pumped by these pulses. The photon pairs are filtered by a 10-nm interference filter with the central wavelength of 785 nm. The orthogonally polarized photon pairs are separated by a polarizing beam splitter and coupled to single-mode fibres. A delay line is inserted to control the temporal superposition of the photons. The polarization controller 1 (quarter wave plate + half wave plate + quarter wave plate + Glan prism) is used to compensate polarization transformations in the single-mode fibres and perform polarization unitary operations. The second part is the microscopy system, composed of the CNOT gate sample, an input objective ($\times 50$, numerical aperture (NA) = 0.4) used to excite different waveguide modes in the hybrid system, and an oil objective ($\times 100$, NA = 1.32) employed to collect the scattering light from the output slits of the gate. The final part comprises the collection and analysis apparatus, where photons scattering from the two outputs are separated by a triangular reflector placed at the image plane behind the oil objective. After selected by polarization controller 2 (half wave plate + Glan prism), the output photons are then delivered to the silicon avalanche photodiodes(Perkin Elmer) through multimode fibres and coincidence measurements are made at the single photon counting modules (SPCM, Becker & Hickl DPC - 230).

Wavefront Shaping Through Emulated Curved Space in Waveguide Settings[*]

Chong Sheng[1], Rivka Bekenstein[2], Hui Liu[1], Shining Zhu[1] & Mordechai Segev[2]

[1] National Laboratory of Solid State Microstructures & School of Physics, Collaborative Innovation Center of Advanced Microstructures, Nanjing University, Nanjing, Jiangsu 210093, China

[2] Department of Physics and Solid State Institute, Technion, Haifa 32000, Israel

The past decade has witnessed remarkable progress in wavefront shaping, including shaping of beams in free space, of plasmonic wavepackets and of electronic wavefunctions. In all of these, the wavefront shaping was achieved by external means such as masks, gratings and reflection from metasurfaces. Here, we propose wavefront shaping by exploiting general relativity (GR) effects in waveguide settings. We demonstrate beam shaping within dielectric slab samples with predesigned refractive index varying so as to create curved space environment for light. We use this technique to construct very narrow non-diffracting beams and shape-invariant beams accelerating on arbitrary trajectories. Importantly, the beam transformations occur within a mere distance of 40 wavelengths, suggesting that GR can inspire any wavefront shaping in highly tight waveguide settings. In such settings, we demonstrate Einstein's Rings: a phenomenon dating back to 1936.

General electromagnetic (EM) beams propagating through linear homogenous media experience diffraction broadening. However, many applications would greatly benefit from having beams that remain very narrow or shape-invariant for large distances. The past two decades have witnessed remarkable progress in wavefront shaping specifically for the purpose of generating non-diffracting beams, such as shape-preserving Bessel beams[1] and accelerating beams in free space[2—5], in plasmonics[6—9] and even in nonlinear materials[10—15]. The concept of shape-invariant wavepackets was extended beyond EM waves, for example to shaping wavefunctions of electrons[16—19] and generating shape-invariant acoustic beams[20,21], and even accelerating surface water gravity waves[22]. All of these shape-invariant wavepackets are not square integrable (they carry infinite power), hence physically they must be truncated, which implies that they stay non-diffracting only for a finite distance[2]. In a similar vein, there are other kind of beams which are a priori designed to stay shape-invariant only for a finite distance, for example, the cosine-Gauss beams[23] and a class of beams that propagate on arbitrary curved trajectories[5,24,25].

---

[*] Nat.Commun.,2016,7:10747

Naturally, all of these beams require wavefront shaping: the launch beam must be shaped in a specific structure (amplitude and phase), to stay non-diffracting for the specified distance.

Wavefront shaping for generating non-diffracting optical beams can be achieved by various methods, ranging from annular slits[1], axicon lenses[26], computer generated holograms[24], spatial light modulators[3,28], gratings[7,23,29], metasurfaces[30—32] and diffraction from nanoparticles[4,33]. Importantly, non-diffracting beams can also be generated in inhomogeneous media such as photonic crystal slabs[34—38], photonic crystals[39,40] and photonic lattices[41]. All these too require wavefront shaping, that is typically done externally, outside the medium within which the beam is propagating. However, wavefront shaping can also be done by shaping the EM environment in which the wave is propagating[42,43]. The fact that the propagation of EM waves in static curved space is analogous to that in inhomogeneous media[42—44] is the underlying principle of emulating general relativity (GR) phenomena in transformation optics[42,43,45—52]. In transformation optics, the permittivities and permeabilities are structured to vary according to the curvature of space[53—59], giving rise to unique trajectories[55—57,60] and controlling the diffraction of light[61,62].

Here, we show that using ideas inspired by GR yields efficient beam shaping in waveguide settings. The concept is general, applicable to many cases where wavefront beam shaping in a waveguide platform is required. First, we fabricate the micro-structured optical waveguide with the specific refractive index emulating the curved space environment generated by a massive gravitational object. This dielectric structure yields a very narrow beam that remains non-diffracting for many Rayleigh lengths. Second, with the same experimental system, we demonstrate the Einstein's rings phenomenon, matching Einstein's 80 years old formula. Finally, we present a general formalism to transform Gaussian beams to considerably narrower shape-invariant beams accelerating (bending) along arbitrary trajectories.

Results

Generating non-diffracting beams through gravitational collimation. The first goal is to create a narrow beam that would propagate in a non-diffracting fashion for a considerable distance in a homogeneous medium. We do that by passing a Gaussian beam through a specific refractive index structure, inspired by the gravitational lensing phenomenon occurring around massive a stars. We design a specific curvature where the emulated gravitational lensing of the light on the micro-scale can create a very narrow non-diffracting beam. The basic principles of diffraction imply that non-diffracting beams can be constructed when their plane-waves constituents accumulate phase at the same rate. The non-diffracting property of beams depends on the dimensions of the wavepackets, that is, a non-diffracting beam can be a shape-invariant solution to the wave equation in three dimensions (3D) or in two dimensions (2D). In 3D homogeneous media, beams that are structured in both their transverse dimensions exhibit shape-invariant propagation on a straight line in the third dimension include the family of Bessel beams[1]. In 2D, on the other hand, when the beams are structured in a single transverse dimension (for example, when the beam is propagating in a planar waveguide), an ideal non-diffracting beam has a unique shape: two plane waves propagating at opposite symmetric angles with respect to the propagation axis. However, whereas the Bessel beams are localized, that is, they have a main lobe carrying most of the power, the planar case is just an interference grating—which is periodic and cannot be used for applications that require a beam with a single main lobe. Interestingly, providing proper spatial bandwidth to each of the opposite waves in the one-dimensional (1D) case does lead to a localized beam displaying non-diffracting features for some finite distance. More specifically, superimposing two beams whose spectrum in k-space is small compared with the wavenumber, at opposite angles with respect to the propagation axis, gives rise to non-diffracting propagation up to a finite distance, due to the similar rate of phase accumulation of the different modal (plane waves) constituents. Here, we construct such a very narrow non-diffracting beam by drawing on intuition from GR, where it is known that light waves are deflected by the space curvature generated by a massive star[63,64]. We exploit this gravitational lensing effect to construct a field that is a superposition of two beams of a finite spatial bandwidth, propagating at opposite angles with respect to the propagation axis. Such a beam remains non-broadening for a finite distance that can be much larger than the Rayleigh length of its main lobe. An example for such a 1D non-diffracting beam and its spectrum is displayed in Fig. 1(a), (b), respectively. Figure 1(c) shows zoom-in on the spectrum, while Fig. 1(d) presents its simulated propagation—where it is clear that the main lobe remains narrow for a large distance, in spite of the fact that its width is only four wavelength. The two main peaks in the spectrum[Fig. 1(c)] represent a superposition of cosine/sine distributions, along with a central peak. The width of the spectral peaks is two orders of magnitude smaller than the wavenumber, enabling a non-diffracting property to a finite distance. This structured beam, whose full-width-half-maximum (FWHM) is 2 μm, is approximately shape-preserving for ~200 μm, which corresponds to six Rayleigh lengths [Fig. 1(d)].

To transform a broad Gaussian beam (FWHM~30 μm) into this non-diffracting beam in a planar waveguide setting, we fabricate a specific refractive index structure inspired by the concepts of curved space known from GR. Namely, curved space generated by a massive gravitational body leads to gravitational lensing, that can in principle overcome diffraction broadening and cause beam collimation. The planar waveguide has a unique width profile, causing a change in the propagation constant and effectively modifying the refractive index. The structure is shown in Fig. 2(a). During the fabrication process, a silver film is deposited on a silica (SiO_2) substrate with a thickness of 80 nm, followed by polymethyl methacrylate (PMMA) microsphere powder scattered on the substrate. The

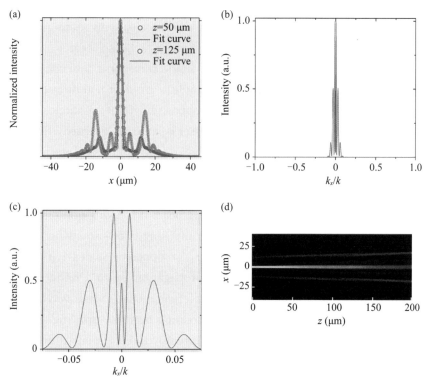

FIG. 1. Calculated propagation of gravitational collimation resulting in a non-diffracting beam. (a) The calculated non-diffracting beam fitted to the beam arising from the simulation of the experimental setting. (b) Spatial spectrum of the beam displaying two main peaks, as can be seen in (c) showing zoom-in on the central section of the spectrum. The two pronounced peaks correspond to a superposition of non-diffracting cosine and sine distributions, resulting in the narrow non-diffracting beam. (d) Simulated propagation of the non-diffracting beam of (a), for a distance of 200 μm inside a homogenous medium, revealing the non-diffracting property.

microspheres are distributed on the substrate, with a small density and large separation distance between microspheres. The sample processing includes a stage where the sample is put on the heating table (300℃) for 30 s. As the melting temperature of PMMA polymer is ~250℃, the heating process deforms the PMMA microspheres into domes, just as shown in Fig. 2(b),(c). In this process, the size of resultant PMMA domes is not uniform, and their diameters can vary greatly, from 1 to 100 μm. For the experiment presented here, we work with one of domes that has an appropriate size, as shown in its optical microscope image in Fig. 2(b). The structure is shaped as a dome protruding from the plane of the waveguide [Fig. 2(a)]. This is further confirmed by mapping the surface structure with atomic force microcopy (Asylum Research, MFP-3D-SA, USA), as shown in Fig. 2(c). Next, a set of gratings with the period 310 nm are drilled on the sliver film around the microdroplet with focused ion beam (FEI Strata FIB 201, 30 keV, 150 pA). These gratings enable to couple the light into the slab waveguide. Next, we spin-coat the sample with a PMMA photoresist mixed with rare earth (Eu^{3+}) to a thickness of ~1 μm, and

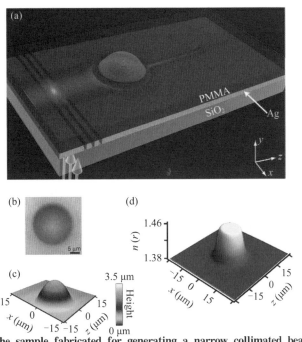

FIG. 2. The sample fabricated for generating a narrow collimated beam. (a) Schematic view of the fabricated waveguide: the inhomogeneous planar waveguide with the specifically designed refractive index structure. The structure is fabricated by depositing a thin silver film on a silica (SiO$_2$) substrate with a thickness of 80 nm, followed by PMMA microsphere powder scattered on the substrate. The blue arrows at the bottom represent the incident 457 nm blue laser light, and the bright spot marks the illumination spot where the light is incident on the grating. (b) Top-view optical microscopy image of the microdroplet. (c) The surface structure of the microdroplet, as mapped by AFM measurements. (d) The effective refractive index structure calculated from (c), based on waveguide theory.

subsequently dry the sample in the oven at 70 ℃ for 2 h. The Eu^{3+} rare earth ions are added to the sample to facilitate fluorescence imaging that will reveal the propagation dynamics of the beam. These Eu^{3+} ions absorb the beam propagating in the slab waveguide, whose wavelength (457 μm) is specifically chosen to excite the rare earth ions, that in turn emit fluorescent light at 615 nm wavelength. We note that, although the 1-μm-thick PMMA layer is not single-mode waveguide for the 457 nm beam, the designed grating allows only one mode to be excited inside the waveguide. Here, only the TM3 mode is excited in our experiment (The grating is designed that only one waveguide mode is excited. Hence, plasmonic modes are not excited in the experiment). The resultant 2D structure of the refractive index is displayed in Fig. 2(d), together with a 3D illustration of the entire sample [Fig. 2(a)]. Figure 2(c) shows the width of the PMMA waveguide as mapped by AFM measurements. From this width, we calculate the refractive index structure displayed in Fig. 2(d), which is fitted with the function $n(x,z) = n_0 + a/(1+(\sqrt{x^2+z^2}/r_c)^8)$, with $n=1.37$, $a=9.22\times10^{-2}$, $r_c=9.69$. Recall that the refractive index of bulk PMMA polymer is 1.49, hence our fabrication process reduces the refractive index according to our design. Specifically, in the region of the dome, the thickness is increased to 3.5 μm, and therefore the effective index of the TM3 waveguide mode is increased from 1.37 to 1.49.

In the experiment, we launch a Gaussian beam of 457 nm wavelength and 11.3 μm FWHM to propagate inside the PMMA layer that acts as a waveguide. The loss in this waveguide is quite small, in spite of the proximity of the thin Ag layer, enabling propagation distances of hundreds of micrometres. The specifically designed refractive index structure focuses the wide beam to a very narrow (2 μm) beam that is subsequently propagating without diffraction for ~200 μm, as expected from the theory. We emphasize that, after passing the 'star', the very narrow beam is propagating in a completely homogeneous medium, hence its non-diffracting property arises solely from the beam structure generated by passing the 'star'. Moreover, whereas most shape-preserving beams are very broad, this beam presents a narrow profile, only 2 μm wide. For comparison, we study the dynamic of a Gaussian beam passing through the same medium numerically and compare it with the experimental results (Fig. 3). We do this by numerically simulating the beam propagation, with the beam propagation method in a medium with the specific refractive index structure conforming to that of the sample used in the experiment [Fig. 2(d)]. In both the experiments and the simulations the transformation of the wide Gaussian beam to a narrow collimated beam is achieve within a very short propagation distance(~20 μm), allowing the use of this scheme in integrated photonics circuits. In Fig. 3, the diameter of the dome is roughly 25 μm. In the experiment, we can fabricate domes with different diameters, always with circular shape. Naturally, domes of different sizes yield collimation for different propagation distances and with different beam widths.

Experiments emulating the Einstein rings phenomenon. Interestingly, we find that besides producing collimated beams, the same planar 'central potential' index structure can also be used to emulate the phenomenon of Einstein's Rings, which is a famous phenomenon predicted by GR and observed in astronomy[65,66]. The Einstein Ring phenomena occurs when light from a point source is deformed by a mass distribution through gravitation lensing that causes the appearance of a ring around the mass distribution. For this case, the beam approaching the 'star' should emulate the radiation originating from a point source, namely, the wave reaching the 'star' should be a spherical wave. To emulate a point source, we fabricate (with focused ion beam) an arc-shaped grating (period of 310 nm) inside the metal film. This is shown in Fig. 4(b), where the radius of the arc is 30 μm. When a plane wave is incident (from below) on the arc grating, the grating transforms it into a spherical wave propagating inside the waveguide layer. The region of incidence on the grating acts as a point source, emitting a spherical wave diverging both to the left and to the right of that point (negative and positive z,

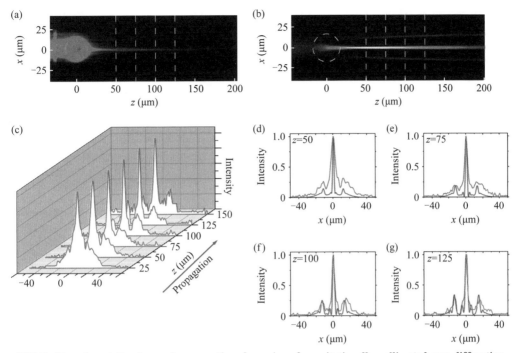

FIG.3. Experimentally observed propagation dynamics of gravitationally collimated non-diffracting beam. (a) Top-view photograph of the experimentally observed results obtained through .orescence. A broad Gaussian beam with FWWH 11.3 μm passes through the region of the dome, giving rise to the refractive index profile described in Fig. 2(c). The wide Gaussian beam focuses to a narrow collimated beam that is non-diffracting for ~ 200 μm. The entire beam transformation process occurs within 20 μm. (b) Simulated results of the same beam showing a similar effect as the experiment. The white dashed circle corresponds to the dome region. (c) Normalized intensity profile of the beam for several propagation distances, after passing though the dome region. ((d) - (g)) Measured (red) and simulated (blue) 1D intensity profiles for $z = 50$ μm, $z = 75$ μm, $z = 100$ μm, $z = 125$ μm, respectively, which correspond to the planes marked by the yellow dashed lines in (a)-(b).

respectively). In such a setting, the spherical wavefront produced by the arc-grating emulates the wave radiated outwards from a point source located at the centre of grating arc. When this 1D spherical wavefront is passing by the star—it is focused and the beam width changes as a function of the propagation distance, as extracted from the experimental data. It is important to emphasize that our optical setting represents Einstein's rings formed by a time-harmonic EM waves, hence the entire dynamics is in space (not in time). Typical results for two different 'stars' (microdroplets with two different radii) are displayed in Fig. 4. As the Radius of the 'star' is larger the convergence of the beam is more extreme, but the final beam is wider (Fig. 4). At this point it is intriguing to compare our emulation results with Einstein's prediction. The Einstein Formula for the angular diameter of the virtual ring[64] is given by $\beta = \sqrt{\alpha_0 R_0 / z}$, that depends on the convergence angle α_0, the radius of the mass distribution R_0 and the distance between the centre of the mass distribution to the observation point. We calculate the angular diameter of the Einstein Ring from the measured convergence angle of the

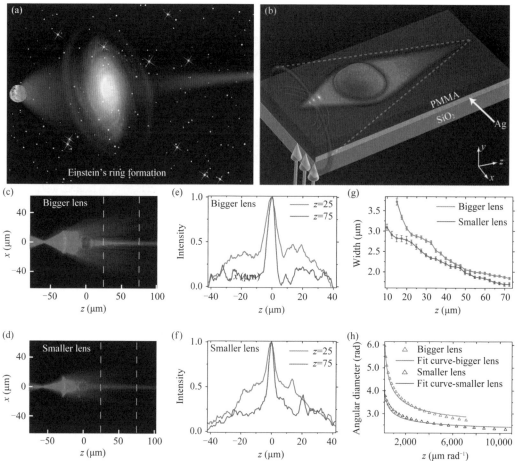

FIG. 4. Experimental emulation of the formation of Einstein's ring. (a) Einstein's vision: light from a point source is focused by a gravitational lens, and is subsequently observed as a virtual ring around the mass distribution. (b) Schematic view of the fabricated inhomogeneous waveguide. (c), (d) Experimental results (obtained through florescence) showing a spherical wave passing though the dome region, for two domes of different radii. The inhomogeneous area acts as a gravitational lens on the light. (e), (f) Measured beam profiles at $z=25$ (red line), $z=75$ (blue line), respectively, which correspond to the locations marked by the yellow dashed lines in (c), (d). (g) Measured beam width as a function of the propagation distance in the homogenous medium after the dome region. (h) Fit to Einstein's formula for the angular diameter of the Einstein rings. The calculated angular diameter from the experimental measurement is in a very good agreement with the theoretical formula.

beam, for several different observation points (propagation distances). For a given observation point, the focusing angle of the beam after passing the 'star' gives the slope, from which we calculate the angular diameter of the virtual ring that an observer located at this specific distance (from the 'star') will see. To conform with the Einstein formula, we calculate the relative angular radius between the two mass distributions (two samples). Namely, instead of calculating the absolute angular radius as a function of z, we calculate the relative angular radius between the results of each sample. We then fit the curve $\beta\left(\dfrac{z}{\alpha_0}\right)$

$=\sqrt{c/\left(\dfrac{z}{\alpha_0}\right)}$ with c as a free parameter and compare the relative constant extracted from the experiment with the constant expected from Einstein's formula. In comparing the ratio and not the absolute number, we avoid the factor 2 between the relativistic Einstein formula and our experiment that represents Newtonian dynamics. As Fig. 4(h) shows, the experiments agree well with theory, although at large z, the experimental values are slightly lower than the model. This minute discrepancy arises from the difference between the fabricated optical potential (refractive index structure) and the $1/r$ gravitational potential of a point source. Consequently, for large values of z (distances), the focusing angle of the light deviates from Einstein's formula, hence the measured focusing angle is somewhat smaller than the theoretical curve.

Shaping beams accelerating on arbitrary trajectories. Finally, we present a general formalism for transforming broad Gaussian beams to accelerating beams that bend along arbitrary (convex) trajectories in a planar waveguide setting. As above, we do that by passing an incident broad Gaussian beam (11.3 μm FWHM) through a miniature refractive index structure that is designed specifically for this task. Accelerating beams are beams with a well-defined peak intensity that propagates along some non-straight trajectory, depending on the phase of the initial beam[4,5,29]. From the point of view of GR, the peak intensity of the beam does not follow geodesics paths[67], which are the shortest paths that light propagated along (by the Fermat principle). This important property of accelerating beams had been exploited for various applications, such as curved plasma channels[65], manipulating microparticles[68,69] and micromachining[70]. We design accelerating beams by utilizing the formalism suggested in ref. [5], for finding the specific 1D phase $\phi(x)$ required for shaping the wavefront of an accelerating beam that will propagate along a specific trajectory. This 1D phase can be achieved by a 2D refractive index structure that the beam passes through, and obeys the relation

$$\phi(x) = k_0 \int_{z_i}^{z_f} n(x,z) \mathrm{d}z, \tag{1}$$

under the assumption that the propagation of the beam is in the paraxial regime. Using this method, there is no unique solution for $n(x, z)$. We therefore suggest a simple method that solves equation (1) for one specific refractive index profile to a specified phase, by assuming $n(x, z)$ is constructed from a function that is separable in x, z, namely $n(x, z) = f(x)g(z)$. For simplicity, we take $g(z) = \exp(-z^2/\sigma^2)$, and assume the Gaussian width (in z) is small compared with the propagation distance ($\sigma \ll z_f - z_i$). This allows setting the boundaries of the integral to infinity which after integrating over z yields:

$$f(x) = \dfrac{\phi(x)}{k_0 \sqrt{\pi \sigma}}. \tag{2}$$

It is important to emphasize that the approximation we used for solving the integral of the phase only, causes additional effects. Due to the 2D refractive index distribution the beam is shifted to some different direction of propagation— $z' = ze^{i\theta}$ while propagating through the inhomogeneous area. Consequently, $n(x,z) = \dfrac{\phi(x)}{k_0\sqrt{\pi}\sigma}\exp(-z^2/\sigma^2)$. To present an example for this method, we find the refractive index profile required to create the phase for an accelerating beam along the trajectory $f(z) = az'^3$. In this specific case, the propagation of the resulting beam can be solved analytically using the method presented in ref.[5]. In more complicated cases, a numerical solution for the ordinary differential equation (ODE) is required. We then use equation (1) to calculate the 2D refractive index structure that will provide the beam with the appropriate phase. By simulating the dynamic of a broad Gaussian beam passing through the designed refractive index structure, we find that the main lobe indeed accelerates along the expected trajectory, for a distance of 20 μm as displayed in Fig. 5. In this regime, it is possible to design a beam that will accelerate beam on an arbitrary trajectory. As any accelerating beam, the structure of such a beam involves a main lobe accompanied by oscillations on one side, and exponential decay on the other side. An example is shown in Fig. 5(c), where the beam cross-sections at several

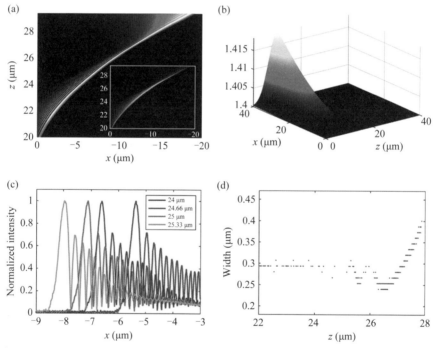

FIG.5. Accelerating beams propagating along arbitrary trajectories produced by designing the refractive index structure within the initial 10 μm propagation distances in the waveguide layer. (a) Simulated evolution of the accelerating beam, where the peak intensity is propagating along the curve $f(z) = az^3$. Inset: the evolution displayed with a non-normalized intensity (b) The calculated refractive index structure which transforms a broad Gaussian beam into the narrow non-diffracting accelerating beam of (a). (c) Structure of the accelerating beam for different propagation distance. (d) Width of the main lobe as a function of the propagation distance.

propagation distances is displayed. This technique for beam shaping inside a slab waveguide is general, and can be used to shape the wavefront of non-diffracting beams accelerating on any convex trajectory, by designing the refractive index structure using equations (1 and 2), which relates the initial phase front (assumed here to be of a broad Gaussian beam) and the desired phase front $\phi(x)$ to the refractive index structure required for such wavefront shaping.

Discussion

To conclude, we have presented a method for shaping optical wavefronts in waveguide settings. Our technique is inspired by GR and it provides a platform for emulating the spatial dynamics of EM waves in curved space. This method can be achieved in thin film waveguides and can be implemented in integrated photonics settings. Specifically, we have demonstrated experimentally the construction of a narrow non-diffracting beam, the formation of Einstein's rings, and presented a general method to construct accelerating beams propagating along arbitrary trajectories. This method can be used for shaping any general beam, thereby suggesting a new way of using transformation optics media for beam shaping in waveguide settings with a single dielectric material. In this work, we presented beam shaping in the spatial domain; consequently, our experiments employed only continuous laser beams as our input waves. However, in principle this technique can also be used to shape ultrashort laser pulses with the traditional grating pairs, the lenses and the spatial modulation at the focal plane, all implemented in a slab waveguide geometry with proper design of the planar refractive index structure. This idea will be pursued in our future research.

References and Notes

[1] Durnin, J., Miceli, J. J. & Eberly, J. H. Diffraction-free beams. *Phys. Rev. Lett.* **58**, 1499 – 1501 (1987).

[2] Siviloglou, G. A. & Christodoulides, D. N. Accelerating finite energy Airy beams. *Opt. Lett.* **32**, 979 – 981 (2007).

[3] Siviloglou, G. A., Broky, J., Dogariu, A. & Christodoulides, D. N. Observation of Accelerating Airy Beams. *Phys. Rev. Lett.* **99**, 213901 (2007).

[4] Kaminer, I., Bekenstein, R., Nemirovsky, J. & Segev, M. Nondiffracting accelerating wave packets of Maxwell's equations. *Phys. Rev. Lett.* **108**, 163901 (2012).

[5] Greenfield, E., Segev, M., Walasik, W. & Raz, O. Accelerating light beams along arbitrary convex trajectories. *Phys. Rev. Lett.* **106**, 213902 (2011).

[6] Salandrino, A. & Christodoulides, D. N. Airy plasmon: a nondiffracting surface wave. *Opt. Lett.* **35**, 2082 – 2084 (2010).

[7] Zhang, P. *et al.* Plasmonic Airy beams with dynamically controlled trajectories. *Opt. Lett.* **36**, 3191 – 3193 (2011).

[8] Minovich, A. *et al.* Generation and near-field imaging of airy surface plasmons. *Phys. Rev. Lett.* **107**,

116802 (2011).

[9] Epstein, I. & Arie, A. Arbitrary bending plasmonic light waves. *Phys. Rev. Lett.* **112**, 023903 (2014).

[10] Wulle, T. & Herminghaus, S. Nonlinear optics of Bessel beams. *Phys. Rev. Lett.* **70**, 1401–1404 (1993).

[11] Kaminer, I., Segev, M. & Christodoulides, D. N. Self-accelerating self-trapped optical beams. *Phys. Rev. Lett.* **106**, 213903 (2011).

[12] Lotti, A. et al. Stationary nonlinear Airy beams. *Phys. Rev. A* **84**, 021807 (2011).

[13] Bekenstein, R. & Segev, M. Self-accelerating optical beams in highly nonlocal nonlinear media. *Opt. Express* **19**, 23706–23715 (2011).

[14] Dolev, I., Kaminer, I., Shapira, A., Segev, M. & Arie, A. Experimental observation of self-accelerating beams in quadratic nonlinear media. *Phys. Rev. Lett.* **108**, 113903 (2012).

[15] Bekenstein, R., Schley, R., Mutza, M., Rotschild, C. & Segev, M. Optical simulations of gravitational effects in the Newton-Schrodinger system. *Nat. Phys.* **11**, 872–878 (2015).

[16] Uchida, M. & Tonomura, A. Generation of electron beams carrying orbital angular momentum. *Nature* **464**, 737–739 (2010).

[17] Voloch-Bloch, N., Lereah, Y., Lilach, Y., Gover, A. & Arie, A. Generation of electron Airy beams. *Nature* **494**, 331–335 (2013).

[18] Grillo, V. et al. Generation of nondiffracting electron bessel beams. *Phys. Rev. X* **4**, 011013 (2014).

[19] Kaminer, I., Nemirovsky, J., Rechtsman, M., Bekenstein, R. & Segev, M. Self-accelerating Dirac particles and prolonging the lifetime of relativistic fermions. *Nat. Phys.* **11**, 261–267 (2015).

[20] Zhang, P. et al. Generation of acoustic self-bending and bottle beams by phase engineering. *Nat. Commun.* **5**, 4316 (2014).

[21] Bar-Ziv, U., Postan, A. & Segev, M. Observation of shape-preserving accelerating underwater acoustic beams. *Phys. Rev. B* **92**, 100301 (2015).

[22] Fu, S., Tsur, Y., Zhou, J., Shemer, L. & Arie, A. Propagation dynamics of airy water-wave pulses. *Phys. Rev. Lett.* **115**, 034501 (2015).

[23] Lin, J. et al. Cosine-gauss plasmon beam: a localized long-range nondiffracting surface wave. *Phys. Rev. Lett.* **109**, 093904 (2012).

[24] Rosen, J. & Yariv, A. Snake beam: a paraxial arbitrary focal line. *Opt. Lett.* **20**, 2042–2044 (1995).

[25] Froehly, L. et al. Arbitrary accelerating micron-scale caustic beams in two and three dimensions. *Optics Express* **19**, 16455 (2011).

[26] Scott, G. & McArdle, N. Efficient generation of nearly diffraction-free beams using an axicon. *Opt. Eng.* **31**, 2640–2643 (1992).

[27] Rosen, J. & Yariv, A. Synthesis of an arbitrary axial field profile by computer-generated holograms. *Opt. Lett.* **19**, 843–845 (1994).

[28] Zhang, P. et al. Nonparaxial mathieu and weber accelerating beams. *Phys. Rev. Lett.* **109**, 193901 (2012).

[29] Li, L., Li, T., Wang, S. M. & Zhu, S. N. Collimated plasmon beam: nondiffracting versus linearly focused. *Phys. Rev. Lett.* **110**, 046807 (2013).

[30] Bomzon, Z., Kleiner, V. & Hasman, E. Formation of radially and azimuthally polarized light using space-variant subwavelength metal stripe gratings. *Appl. Phys. Lett.* **79**, 1587–1589 (2001).

[31] Yu, N. et al. Light propagation with phase discontinuities: generalized laws of reffection and refraction. *Science* **334**, 333–337 (2011).

[32] Kildishev, A. V., Boltasseva, A. & Shalaev, V. M. Planar photonics with metasurfaces. *Science* **339**, 1232009 (2013).

[33] Chen, Z., Taflove, A. & Backman, V. Photonic nanojet enhancement of backscattering of light by nanoparticles: a potential novel visible-light ultramicroscopy technique. *Opt. Express* **12**, 1214–1220 (2004).

[34] Yu, X. & Fan, S. Bends and splitters for self-collimated beams in photonic crystals. *Appl. Phys. Lett.* **83**, 3251–3253 (2003).

[35] Rakich, P. T. *et al.* Achieving centimetre-scale supercollimation in a large-area two-dimensional photonic crystal. *Nat. Mater.* **5**, 93–96 (2006).

[36] Shih, T.-M. *et al.* Supercollimation in photonic crystals composed of silicon rods. *Appl. Phys. Lett.* **93**, 131111 (2008).

[37] Hamam, R. E., Ibanescu, M., Johnson, S. G., Joannopoulos, J. D. & Soljacic, M. Broadband super-collimation in a hybrid photonic crystal structure. *Opt. Express* **17**, 8109–8118 (2009).

[38] Mocella, V. *et al.* Self-collimation of light over millimeter-scale distance in a quasi-zero-average-index metamaterial. *Phys. Rev. Lett.* **102**, 133902 (2009).

[39] Longhi, S. & Janner, D. X-shaped waves in photonic crystals. *Phys. Rev. B* **70**, 235123 (2004).

[40] Conti, C. & Trillo, S. Nonspreading wave packets in three dimensions formed by an ultracold bose gas in an optical lattice. *Phys. Rev. Lett.* **92**, 120404 (2004).

[41] Manela, O., Segev, M. & Christodoulides, D. N. Nondiffracting beams in periodic media. *Opt. Lett.* **30**, 2611–2613 (2005).

[42] Leonhardt, U. Optical conformal mapping. *Science* **312**, 1777–1780 (2006).

[43] Pendry, J. B., Schurig, D. & Smith, D. R. Controlling electromagnetic fields. *Science* **312**, 1780–1782 (2006).

[44] Laundau, L.D. & Lifshitz, E. M. *The Classical Theory Of Fields* (Butterworth-Heinemann, 1975).

[45] Li, J. & Pendry, J. B. Hiding under the carpet: a new strategy for cloaking. *Phys. Rev. Lett.* **101**, 203901 (2008).

[46] Alù, A. & Engheta, N. Multifrequency optical invisibility cloak with layered plasmonic shells. *Phys. Rev. Lett.* **100**, 113901 (2008).

[47] Smolyaninov, I. I., Smolyaninova, V. N., Kildishev, A. V. & Shalaev, V. M. Anisotropic metamaterials emulated by tapered waveguides: application to optical cloaking. *Phys. Rev. Lett.* **102**, 213901 (2009).

[48] Valentine, J., Li, J., Zentgraf, T., Bartal, G. & Zhang, X. An optical cloak made of dielectrics. *Nat. Mater.* **8**, 568–571 (2009).

[49] Gabrielli, L. H., Cardenas, J., Poitras, C. B. & Lipson, M. Silicon nanostructure cloak operating at optical frequencies. *Nat. Photon.* **3**, 461–463 (2009).

[50] Smolyaninova, V. N., Smolyaninov, I. I., Kildishev, A. V. & Shalaev, V. M. Experimental observation of the trapped rainbow. *Appl. Phys. Lett.* **96**, 211121 (2010).

[51] Chen, H., Chan, C. T. & Sheng, P. Transformation optics and metamaterials. *Nat. Mater.* **9**, 387–396 (2010).

[52] Zentgraf, T., Liu, Y., Mikkelsen, M. H., Valentine, J. & Zhang, X. Plasmonic luneburg and eaton lenses. *Nat. Nanotechnol.* **6**, 151–155 (2011).

[53] Smolyaninov, I. I. Surface plasmon toy model of a rotating black hole. *New J. Phys.* **5**, 147–147 (2003).

[54] Leonhardt, U. & Philbin, T. G. General relativity in electrical engineering. *New J. Phys.* **8**, 247 (2006).

[55] Genov, D. A., Zhang, S. & Zhang, X. Mimicking celestial mechanics in metamaterials. *Nat. Phys.* **5**, 687–692 (2009).

[56] Narimanov, E. E. & Kildishev, A. V. Optical black hole: broadband omnidirectional light absorber. *Appl. Phys. Lett.* **95**, 041106–041106–3 (2009).

[57] Cheng, Q., Cui, T. J., Jiang, W. X. & Cai, B. G. An omnidirectional electromagnetic absorber made of metamaterials. *New J. Phys.* **12**, 063006 (2010).

[58] Smolyaninov, I. I. & Narimanov, E. E. Metric signature transitions in optical metamaterials. *Phys. Rev. Lett.* **105**, 067402 (2010).

[59] Genov, D. A. General relativity: optical black-hole analogues. *Nat. Photon.* **5**, 76–78 (2011).

[60] Sheng, C., Liu, H., Wang, Y., Zhu, S. N. & Genov, D. A. Trapping light by mimicking gravitational lensing. *Nat. Photon.* **7**, 902–906 (2013).

[61] Batz, S. & Peschel, U. Linear and nonlinear optics in curved space. *Phys. Rev. A* **78**, 043821 (2008).

[62] Bekenstein, R., Nemirovsky, J., Kaminer, I. & Segev, M. Shape-preserving accelerating electromagnetic wave packets in curved space. *Phys. Rev. X* **4**, 011038 (2014).

[63] Einstein, A. Die Grundlage der allgemeinen relativitätstheorie. *Ann. Phys.* **354**, 769–822 (1916).

[64] Einstein, A. Lens-like action of a star by the deviation of light in the gravitational .eld. *Science* **84**, 506–507 (1936).

[65] Hewitt, J. N. *et al.* Unusual radio source MG1131+0456: a possible Einstein ring. *Nature* **333**, 537–540 (1988).

[66] King, L. J. *et al.* A complete infrared Einstein ring in the gravitational lens system B1938+666. *MNRAS* **295**, L41–L44 (1998).

[67] Polynkin, P., Kolesik, M., Moloney, J. V., Siviloglou, G. A. & Christodoulides, D. N. Curved plasma channel generation using ultraintense airy beams. *Science* **324**, 229–232 (2009).

[68] Baumgartl, J., Mazilu, M. & Dholakia, K. Optically mediated particle clearing using Airy wavepackets. *Nat. Photon.* **2**, 675–678 (2008).

[69] Schley, R. *et al.* Loss-proof self-accelerating beams and their use in non-paraxial manipulation of particles' trajectories. *Nat. Commun.* **5**, 5189 (2014).

[70] Mathis, A. *et al.* Micromachining along a curve: Femtosecond laser micromachining of curved profiles in diamond and silicon using accelerating beams. *Appl. Phys. Lett.* **101**, 071110–071113 (2012).

[71] R.B. gratefully acknowledges the support of the Adams Fellowship Programme of the Israel Academy of Sciences and Humanities. This research was also supported by the ICore Excellence center 'Circle of Light' and a grant from the US Air Force Office for Scientific Research (AFOSR). H.L. gratefully acknowledges the support of the National Natural Science Foundation of China (No's 11321063, 61425018 and 11374151), the National Key Projects for Basic Researches of China (No. 2012CB933501 and 2012CB921500), the Doctoral Programme of Higher Education (20120091140005) and Dengfeng Project B of Nanjing University. C.S. gratefully acknowledge the support of the programme A for Outstanding PhD candidate of Nanjing University.

[72] C.S. and R.B. contributed equally to this work. All authors contributed to all aspects of this work.

图书在版编目(CIP)数据

介电体超晶格. 下 / 朱永元等编著. —南京：南京大学出版社，2017.3
ISBN 978-7-305-17916-7

Ⅰ. ①介⋯ Ⅱ. ①朱⋯ Ⅲ. ①超晶格半导体-研究 Ⅳ. ①TN304.9

中国版本图书馆 CIP 数据核字(2016)第 281440 号

出版发行	南京大学出版社		
社　　址	南京市汉口路 22 号	邮　编	210093
出 版 人	金鑫荣		

书　　名	介电体超晶格·下		
编　著	朱永元　王振林　陈延峰　陆延青　祝世宁		
策划编辑	吴　汀		
责任编辑	王南雁	编辑热线	025-83593962

照　　排	南京紫藤制版印务中心
印　　刷	南京爱德印刷有限公司
开　　本	787×1092　1/16　印张 32.75　字数 956 千
版　　次	2017年3月第1版　2017年3月第1次印刷
ISBN	978-7-305-17916-7
定　　价	198.00 元

网　　址	http://www.njupco.com
官方微博	http://weibo.com/njupco
官方微信	njupress
销售热线	025-83594756

* 版权所有，侵权必究
* 凡购买南大版图书，如有印装质量问题，请与所购图书销售部门联系调换

ISBN 978-7-305-17916-7

198.00元